PRENTICE-HALL INTERNATIONAL SERIES
IN THE PHYSICAL AND CHEMICAL ENGINEERING SCIENCES

NEAL R. AMUNDSON, EDITOR, *University of Houston*

ADVISORY EDITORS

ANDREAS ACRIVOS, *Stanford University*
JOHN DAHLER, *University of Minnesota*
THOMAS J. HANRATTY, *University of Illinois*
JOHN M. PRAUSNITZ, *University of California*
L. E. SCRIVEN, *University of Minnesota*

AMUNDSON *Mathematical Methods in Chemical Engineering: Matrices and Their Application*
BALZHISER, SAMUELS, AND ELIASSEN *Chemical Engineering Thermodynamics*
BRIAN *Staged Cascades in Chemical Processing*
BUTT *Reaction Kinetics and Reactor Design*
DENN *Process Fluid Mechanics*
FOGLER *The Elements of Chemical Kinetics and Reactor Calculations: A Self-Paced Approach*
HIMMELBLAU *Basic Principles and Calculations in Chemical Engineering, 4th edition*
HOLLAND *Fundamentals and Modeling of Separation Processes: Absorption, Distillation, Evaporation, and Extraction*
HOLLAND AND ANTHONY *Fundamentals of Chemical Reaction Engineering*
LEVICH *Physicochemical Hydrodynamics*
MODELL AND REID *Thermodynamics and Its Applications*
MYERS AND SEIDER *Introduction to Chemical Engineering and Computer Calculations*
NEWMAN *Electrochemical Systems*
PRAUSNITZ ET AL. *Molecular Thermodynamics of Fluid-Phase Equilibria, 2nd edition*
PRAUSNITZ, ET AL. *Computer Calculations for Multicomponent Vapor-Liquid and Liquid-Liquid Equilibria*
RAMKRISHNA AND AMUNDSON *Linear Operator Methods in Chemical Engineering with Applications to Transport and Chemical Reaction Systems*
RUDD ET AL. *Process Synthesis*
SCHULTZ *Diffraction for Materials Scientists*
SCHULTZ *Polymer Materials Science*
VILLADSEN AND MICHELSEN *Solution of Differential Equation Models by Polynomial Approximation*
WILLIAMS *Polymer Science and Engineering*

# *Linear Operator Methods in Chemical Engineering*

*with Applications to Transport and Chemical Reaction Systems*

DORAISWAMI RAMKRISHNA

*Purdue University*

NEAL R. AMUNDSON

*University of Houston*

PRENTICE-HALL, INC.

Englewood Cliffs, New Jersey 07632

**Library of Congress Cataloging in Publication Data**

Ramkrishna, Doraiswami (date)
Linear operator methods in chemical engineering with applications to transport and chemical reaction systems.

Bibliography: p.
Includes index.
1. Chemical engineering—Mathematics. 2. Linear operators. I. Amundson, Neal R. II. Title.
TP149.R26 1985 660'.01'51 84–13452
ISBN 0–13–537341–7



Printed in the United States of America

Editorial/production: Nicholas C. Romanelli
Manufacturing buyer: Anthony Caruso

10 9 8 7 6 5 4 3 2 1

ISBN 0-13-537341-7

Prentice-Hall International, *London*
Prentice-Hall of Australia Pty. Limited, *Sydney*
Editora Prentice-Hall do Brasil, Ltda., *Rio de Janeiro*
Prentice-Hall Canada, Inc., *Toronto*
Prentice-Hall of India Private Limited, *New Delhi*
Prentice-Hall of Japan, Inc., *Tokyo*
Prentice-Hall of Southeast Asia Pte. Ltd., *Singapore*
Whitehall Books Limited, *Wellington, New Zealand*

*To Ponnu and M. R. Doraiswami*
*and*
*To Shirley D*

# *Contents*

# *Preface*

In recent years there has been growing consciousness of the role of functional analysis in the application of mathematics to the solution of scientific and engineering problems. It has manifested in the appearance of several expositions, differing in degree of rigor, to an audience whose affinity for abstraction is tempered by concern for its functional attributes. While this atmosphere should prove conducive for another text on the subject, one is left with a sense of apprehension that treatments in the area have been content with dwelling more on the expository features of functional analysis than on demonstrations of its special capabilities. Thus many applications have generally been confined to traditional problems albeit in the elegant garb of functional analysis. In writing this book our motive has been to demonstrate that the deductions of a general theory play their most important role in expanding the class of solvable problems and that elegance is merely a way of life.

Our interest in the area was spurred by a course in the middle sixties taught in the mathematics department of Minnesota by Professor G. R. Sell, who subsequently coauthored a book with A. W. Naylor, titled *Linear Operator Theory in Engineering and Science* and published by Holt, Rinehart and Winston in 1971. We would also like to record our special appreciation to the faculty of the Mathematics Department at the University of Minnesota, whose extraordinary interest in the applications of mathematics and interaction with colleagues in engineering fostered a unique mathematical culture among the engineering students and faculty there. An early book that has influenced us is B. Friedman's *Principles and Techniques of Applied Mathematics*, published by John Wiley & Sons in 1956.

In that abstraction is apparently antithetical to the art of practicality, it is but natural that the engineering community would view askance any lengthy exposition of the abstract elements of a subject such as linear operator theory. On the other hand, to establish that the cause of practicality can in fact be served most profitably by generalizations perceived through abstraction, it calls for some indulgence in the abstract theory. A happy medium seemed to us not possible given the heterogeneity of our audience, ranging from the (non-empty) set of the mathematically sophisticated to those that have little more than cursory contact with conventional mathematical techniques in engineering curricula. After teaching several versions of a graduate course in different chemical engineering departments such as the Indian Institute of Technology at Kanpur, the University of Minnesota, Purdue University, and the University of Houston, we have converged on a treatment with a proposal for different modes of coverage depending on the background of our audience. Although the problems governing applications are generally in the analysis of transport and chemical reaction systems and are of primary interest to chemical and mechanical engineers, we expect that the book may be of interest not only to other engineers but to scientists and applied mathematicians as well. Broadly, we have viewed our audience in three classes, with at least the background of an elementary course on calculus, matrix algebra, and differential equations.

Class I: Seeking detailed understanding of the elements of linear operator theory, and innovative skills.

Class II: Aspiring for a non-leisurely introduction to the benefits of the operator-theoretic method.

Class III: Seeking a quick appraisal of the special attributes of the book or solutions of certain types of problems.

It is not suggested that the foregoing sets be disjoint. It is quite conceivable that one in class I may choose to be in either II or III for a first reading. A course addressed to class I would last a year while that for class II could be given in a (hectic) semester. Class III is viewed as a heterogeneous group consisting of those that are either well-informed about operator-theoretic methods and/or those that have a particular interest in solving various types of linear boundary value problems.

A diligent coverage of the book in sequence is meant for class I. Chapter 0 covers rather elementary aspects of set theory. Particular attention is called to the Index of Symbols at the end of the book. The purpose of Chapter 1 is to provide a non-constructive development of real numbers with focus on certain properties of the real number system that carry over to the so-called Banach and Hilbert spaces on which the linear operators of interest are defined. Chapter 2 deals with the algebraic features of linear vector spaces. Chapter 3 introduces topological aspects in the setting of a metric space. Chapter 4 discusses the elements of Lebesgue integration. The treatment of linear spaces in which algebraic features are combined with topological properties is covered in Chapters 5 and 6. Spectral theory of self-adjoint operators has been treated separately in Chapter 7. The applications of linear operators have been covered in Chapters 8–11; Chapter 8 deals with applications in infinite dimensional space while Chapter 9, the longest chapter in the book, discusses differential operators and applications to a diverse variety of boundary value problems. In Chapter 10, partial differential operators and their applications are treated. Chapter 11 is an introduction to the theory of non-self-adjoint operators. We ask for repeated readings of the Foreword at the beginning of each chapter and the Concluding Remarks at the end to maintain perspective throughout the book. Special emphasis is placed on the exercises, which frequently introduce extensions and new applications.

For class II, we recommend beginning with Chapter 5 and working through Chapter 11 with ad hoc referrals to Chapters 0–4 as and when it becomes necessary.

For class III, Chapter 8, Sections 9.3–9.5 in Chapter 9, Chapter 10, with special emphasis in Sections 10.2–10.5, and Chapter 11 are recommended.

In classifying our readership and making alternative recommendations for coverage of the text material our motivation has been to prevent frustration in an enquiring reader more enthusiastic about applications than about the methods of mathematics. But we are unable to contain our pleasure when Bertrand Russell's dictum that "the best of mathematics is not merely to be learned as a task but to be assimilated as a part of daily thought," evokes a sparkling response of approval from a student.

An effort such as this is based liberally on the support of several faculty colleagues and students. Without their persistent encouragement, and occasionally embarassing enquiries about the status of our manuscript, we could well have been perpetually in the stage of rewriting, modifying, and cleaning up. Particular mention must be made of Professor James M. Caruthers, whose observations were truly most stimulating and whose continued encouragement to complete the manuscript provided the most impetus to its conclusion. The direct or indirect contributions of many of our students, Terry Papoutsakis (now at Rice), Brian Turner, and Shankar Nataraj, to mention only a few, are gratefully acknowledged. Mrs. Christa Van Etten deserves special commendation for her excellent typing of the manuscript. We gratefully acknowledge a grant from the Education Development Center at Kanpur for typing a very early version of this manuscript, the Purdue School of Chemical Engineering for free typing facilities, and the University of Houston for travel grants to enable us to meet for frequent discussions on the manuscript.

Finally, the forbearance of our wives, Geetha (who also helped compile the author index) and Shirley, made a subtle contribution too important not to receive special mention.

D.R., N.R.A.

*Linear Operator Methods in Chemical Engineering*

*with Applications to Transport and Chemical Reaction Systems*

# Sets, Mappings, and Other Preliminaries

# 0

## 0.0 Foreword

The present chapter is concerned with certain preliminaries by way of mathematical language used to lay out the material in succeeding chapters. In this sense it contains a glossary of terminology that will be used throughout the book. It must be noted that set theory represents, in fact, the very fabric of modern mathematics. In addition, set-theoretic concepts and the algebra of sets, known as Boolean algebra, have a number of useful applications. For example, electrical network synthesis draws richly from Boolean algebra. It stands to reason that the usefulness of set-theoretic concepts must be felt in situations where "structure" or "configuration" is an important consideration. Our motivation for this chapter, however, is not to develop its contents to any degree of completeness required for its applications per se but rather because of its implications to subsequent material.

## 0.1 Introduction

In this chapter we present the rudiments of mathematical abstraction. Fundamentally, mathematics deals with abstract entities, called *objects* or *elements*, which invariably arise from a collection, class, or family of objects called a *set*. The objects are therefore referred to as members or elements of the set from which they arise. The concept of an object has a high degree of abstraction, since "object" does not necessarily have to be perceived as a stone or a tree; indeed, it may be anything, such as a particular sound or a smell or an algebraic symbol. Perceptions of any kind relating to objects are only necessary for us to be able to talk about them. Also, in some circumstances *an* object may well refer to several entities or a collection of entities. Thus in one context, we may refer to sets whose elements are themselves sets from a different context.

The reader is forewarned that the term "object" will be used frequently in this book. It is not relevant to ask what these objects are, for, as we had observed earlier, such nonspecificity is a part of mathematical abstraction. The set to which the objects belong will be identified by *assertions* of *properties* of its members. There is no more (or no less) to the identity of these objects than their properties, which define them. In order to talk about them, we will represent them by algebraic symbols, and assign them names, such as *numbers*, *scalars*, *vectors*, and so on.

## 0.2 Sets, Elements, and Operation in Sets

We have, in an intuitive way, defined a set and its constituent elements. We have deliberately refrained from using specific examples of sets and elements in order to promote a bit of abstract thinking on the part of the student. It is essential to make a symbolic representation of the statement "*this* object belongs to *that* set," in other words, to provide a notational framework in which mathematical statements such as the one in quotes may be expressed conveniently and concisely. Thus *that* set may be denoted by the letter $A$ and *this* object by the letter $a$ and the foregoing statement may be expressed by the concise representation "$a \in A$," which is read "$a$ is an element of $A$." The negation of this statement, "$a$ is not an element of $A$," is represented by "$a \notin A$."

Often the qualification is made that a set is nonempty; that is, it contains at least one element. This may sound irksome to the engineering student, who might quip: "Why would anyone be interested in talking about a set that contains nothing?" It should be realized, however, that a set is frequently defined by one or more mathematical propositions involving the elements of the set and that the *empty* or *null* set is a convenient instrument for describing situations under which no elements exist satisfying the propositions stated. Capital letters such as $S$, $A$, $B$, . . . will be used to denote sets. The definition of a set by propositions will be represented as follows:

$$A = \{a\text{: propositions involving } a\}$$

The letter $a$ in braces above is a typical element of the set and the propositions involving $a$ are one or more mathematical statements concerning $a$. Sometimes, however, a set may be most conveniently defined by simply listing its members. We consider below some examples.

1. $A = \{1, 2, 3\}$.
2. $B = \{\text{integers less than } 10\}$.
3. $C = \{a: 0 \leq a \leq 1\}$.
4. $S = \{x: 0 < x < 1; (x^2 + x - 2) > 0\}$.

In representing the fact that 2 is an element of $A$, we may write $2 \in A$. It should also be evident by inspecting the sets above that

$$2 \in B, \qquad 2 \notin C, \qquad 1 \in C$$

and so on. The set $S$ can be seen to be *empty* because $x^2 + x - 2 = (x - 1)(x + 2)$, which *cannot* be positive for $x$ between 0 and 1. Thus $S$ is the *null* set. The symbol 0 is reserved to represent the null or empty set.

If every element of a set $A$ is an element of another set $B$ (note that this is so in the examples cited above), then $A$ is said to be "contained in $B$," symbolically represented as $A \subset B$. Equivalently, $B$ contains $A$ or, symbolically, $B \supset A$. If every element of $A$ is an element of $B$ and every element of $B$ is an element of $A$, then the sets $A$ and $B$ consist of the *same* elements. In this case, the sets $A$ and $B$ are said to be *equal*, in symbols, $A = B$. On the other hand, if $A \subset B$ and not all elements of $B$ are in $A$, then $A$ is said to be a *proper subset* of $B$. Since sets are abstract collections, before one can talk about "relationships between sets" (such as one set being contained in the other), it is essential to make certain that the different sets contain the same "kind" of elements. (Certainly, sets of real numbers can have nothing in common with sets of vectors or complex numbers.) A *universal set* $I$ is defined, which consists of all elements of interest. Other sets $A, B, C, \ldots$ will then be subsets of $I$. In the examples considered earlier, the universal set $I$ may have been considered as the set of all real members.

Given any two subsets $A$ and $B$ of a universal set $I$, it is not necessary that one be contained in the other. Indeed, they may have no common elements at all, as is the case between $A$ and $S$ in the examples. If $A$ and $B$ have no common elements, they are said to be *disjoint*. We are already now into a discussion of what is known as *set theory*. Each subset $A$ of a universal set $I$ has a *complementary set* denoted $A'$ and usually called the *complement of A*, which consists of all elements of $I$ not contained in $A$. Clearly, $A$ and $A'$ are disjoint sets. Again, given two sets $A$ and $B$, we may define their *union* as the set containing all elements of *A and B*. Symbolically, the union of $A$ and $B$ is denoted by $A \cup B$ or $A + B$. Furthermore, we may also define the *intersection* of two sets $A$ and $B$ as the set of the *common* elements of $A$ and $B$, and denote it by $A \cap B$ or simply $AB$. These definitions are readily understood from a

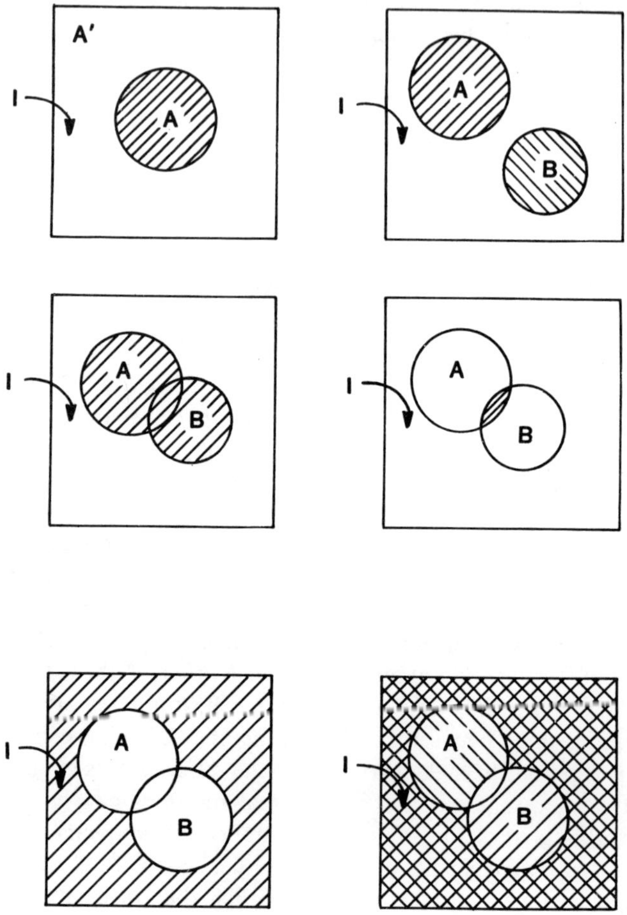


**Figure 0.2.1** Diagrammatic representation of sets and relations.

geometric visualization of sets. Figure 0.2.1 represents the universal set $I$ by an all-inclusive square and its subsets by regions enclosed by closed curves. Reference to this figure reveals certain "self-evident truths." For example, Figure 0.2.1(a) shows that

$$A \cup A' = I \quad \text{or} \quad A + A' = I$$

The fact that $A$ and $A'$ are disjoint leads to the equation

$$A \cap A' = 0 \quad \text{or} \quad AA' = 0$$

Similarly, it is not difficult to see that

$$A \cup A = A \quad \text{or} \quad A + A = A$$

$$A \cap A = A \quad \text{or} \quad AA = A$$

$$A \cup I = I \quad \text{or} \quad A + I = I$$

$$A \cap I = A \quad \text{or} \quad AI = A$$

and so on. Such equations and operations involving sets form the basis of an *algebra of sets*, known as *Boolean* algebra after its originator, G. Boole. Boole condensed *all* algebraic results involving sets to a small number of basic "self-evident truths" called *axioms*, from which could be derived all other results involving sets. We do not deal with Boolean algebra in this book.† It is of interest to note, however, that Boolean algebra provides a means of dealing with *logic* in a mathematical way. Each of the equations above may be looked upon as a logical statement if the set is replaced by a *proposition*. For example, the set $A$ may be a mathematical statement such as "$a$ is equal to $\frac{1}{2}$" or something entirely nonmathematical such as "it is raining now." The complement $A'$ will then translate in the first case to "$a$ is not equal to $\frac{1}{2}$" and in the second to "it is not raining now." The union (or sum) of two propositions $A$ and $B$ may then be regarded as "either $A$ is true or $B$ is true or both $A$ and $B$ are true" or, more briefly, "either $A$ or $B$ or both." The empty set 0 implies no truth and the universal set $I$ is the complete truth. Thus $A \cap A' = 0$ may be understood as "it cannot be true that a statement $A$ and its contradiction $A'$ are both correct." $A \cup A' = I$ may be construed as "a statement $A$ is either correct or its contradiction is correct" or alternatively, "a given statement is either true or untrue." We realize that such statements are somewhat obvious. Consider, however, the result from Figure 0.2.1(e) and (f), which gives

$$(A + B)' = A'B'$$

The foregoing equation means that "if it is not true that either $A$ or $B$ or both is true, then neither $A$ nor $B$ is true, and vice versa," which takes a little stretching of the mind.

The statement $A \subset B$ may be construed as "the truth of $A$ implies the truth of $B$."‡ "$A = B$" means that "the truth of $A$ follows from that of $B$, and vice versa" or alternatively, "$A$ is true if and only if $B$ is true," a statement that occurs frequently in mathematical theorems. Another variation of the foregoing statement is that "for $A$ to be true, it is *necessary* and *sufficient* that $B$ be true." The proof of a theorem which seeks, for example, to establish that $A \rightarrow B$ ($A$ implies $B$) proceeds by identifying intermediate propositions $C, D, \ldots$ such that $A \rightarrow C$, $C \rightarrow D, \ldots$ and finally implying $B$. Indeed, each of the intermediate results results either directly from the axioms or

†Boolean algebra is extremely useful in its application to probability theory and other theories concerning sets.

‡When dealing with propositions (rather than sets) we will use the symbol $\rightarrow$, meaning "implies," instead of $\subset$. Thus $A \rightarrow B$ is the same as $A \subset B$. Equivalence will be denoted by $\leftrightarrow$.

from prior theorems. It is well to remember that in establishing $A \rightarrow B$, it might sometimes be more convenient to prove that $B' \rightarrow A'$. That these are equivalent statements can be readily seen from a pictorial visualization such as that in Figure 0.2.1.

**EXERCISES**

With the help of diagrams, *show* that

**0.2.1** $(AB)' = A' + B'$.

**0.2.2** $A \subset B$ is equivalent to $B' \subset A'$.

**0.2.3** $A + BC = (A + B)(A + C)$.

**0.2.4** $A(B + C) = AB + AC$.

Since sets are collections of abstract objects, it is possible that elements of a given set may themselves be sets. For example, from a set $A$ it is possible to take *pairs* of elements and form another set. The new set so formed has elements that are pairs of elements from $A$. It is more usual to take *ordered pairs* of elements of $A$, in which case the new set is denoted $A \times A$ and elements from it are denoted $(a, b)$, where $a \in A$ and $b \in A$. The fact that $A \times A$ contains ordered pairs of elements from $A$ implies that the element $(a, b)$ is different from the element $(b, a)$. Similarly, one may form *ordered n-tuples* of elements of $A$ into a set $A \times A \times \ldots \times A$.

## 0.3 Mappings

So far, we have talked about relationships among sets of the *same* kind, via subsets of a universal set $I$. It is also possible to talk about a different kind of relationship between sets that are not of the same kind. For example, let $A$ and $B$ be two such sets that contain quite unlike elements. For each element $a \in A$, it is possible to associate in some manner a unique element $b \in B$. The *rule* for such association may vary from the extreme situation of a specification of the unique element from $B$ for *each* element from $A$ to the simplest situation of a simple formula universally valid for all corresponding elements. Regardless of how the rule is arrived at, it is referred to as a *mapping* which *maps* elements of $A$ into elements of $B$. To show how mappings may occur between sets of very dissimilar kinds, let us consider $A$ to be the set of all towns on earth that have a population of more than 10,000 and $B$ to be the set of ordered pairs of real numbers. A mapping may then be defined from $A$ to $B$ which measures the latitude and longitude of the town concerned. Figure 0.3.1(a) shows the representation of a mapping that maps elements of a set $A$ into elements of another set, $B$. The mapping may be represented in symbols as

$$f: A \longrightarrow B$$

where $f$ is itself referred to as the mapping. The set of points in $B$ obtained by transforming all the elements of $A$ is called the *range* of $f$ and is denoted $f(A)$. If $f(A)$ is a proper subset of $B$, then $f$ is referred to as a mapping of $A$ *into* $B$ [see Figure 0.3.1(a)]. If, however, $f(A) = B$, then $f$ is called a mapping of $A$ *onto* $B$.

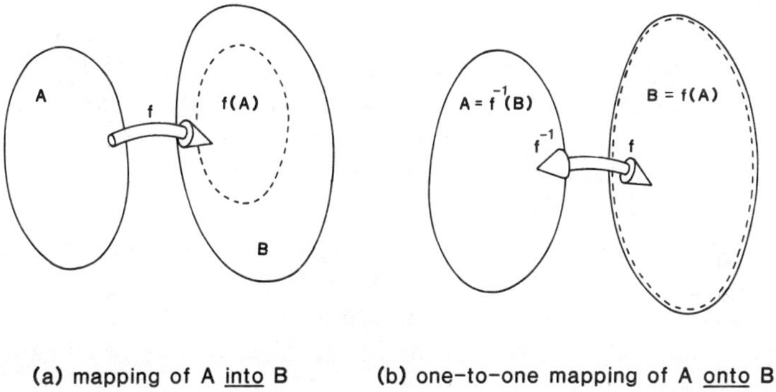


**Figure 0.3.1** Mappings.

It is natural to raise the question of whether, given an element $b \in f(A)$, it is possible to recover a *unique* element $a \in A$, which is transformed by $f$ to $b$. If this is possible for every element in $f(A)$, then it is said that the *inverse mapping* of $f$, denoted $f^{-1}$, exists, which maps elements of $f(A)$ onto $A$. Alternatively, we write

$$f^{-1}: f(A) \longrightarrow A$$

When the inverse mapping $f^{-1}$ exists, $f$ is said to be a *one-to-one mapping* of $A$ *into* $B$. Figure 0.3.1(b) shows the case where $f$ is a one-to-one mapping of $A$ *onto* $B$. In this case, $A$ and $B$ are said to be *equivalent* sets.

In the example considered earlier, the mapping of towns into their latitudes and longitudes is a one-to-one mapping of towns into the set of ordered pairs of real numbers. If, however, only the latitude (or longitude) had been considered, it should be clear that the mapping would not have been one-to-one.

## EXERCISES

**0.3.1** Investigate each of the following mappings of $A$ into $B$, identifying whether it is one-to-one, into, or onto. The mapping is defined by a relationship between $a$ and $b$.

(a) $A = \{0 \leq a < \infty\}$, $B = \{0 \leq b < \pi/2\}$, $b = \tan^{-1} a$.

(b) $A = \{-1 < a < 1\}$, $B = \{0 \leq b \leq 1\}$, $b = a^2$.

(c) $A = \{0 \leq a < \infty\}$, $B = \{1, 2, 3, \ldots\}$, $b =$ smallest integer larger than $a$.

**0.3.2** Let $B$ be a set comprising a finite number of objects and $A$ be a proper subset of $B$. Is it possible to find a one-to-one mapping of $A$ onto $B$?

**0.3.3** Let $A = \{2, 4, 6, \ldots\}$ and $B = \{1, 2, 3, \ldots\}$. Is $A$ a proper subset of $B$? Can you find a one-to-one mapping of $A$ onto $B$?

## 0.4 Concluding Remarks

We have dealt with some mathematical preliminaries concerning sets and relationships among sets. We did not go into the details of set theory or Boolean algebra in a systematic manner. Most examples of sets were those involving real numbers. Boolean algebra is particularly useful in application to probability theory, which will not be of interest to this book. However, as noted in the foreword to this chapter, the material in this chapter is not merely an abstract preliminary to subsequent mathematical developments, but is itself a useful framework in certain applications. The prerequisite is that the engineer formulate the proper setting for such applications to be possible. It cannot be argued that where the framework is useful, it is *indispensable*; rather, its merit is to provide an efficient way of thinking about things.

## FURTHER READING

The axiomatic method is the *modus operandi* of mathematics. An interesting account of the axiomatic method may be found in

*The Anatomy of Mathematics*, by R. B. Kershner and L. R. Wilcox, Ronald Press, New York, 1950.

A particularly readable account of the methods of mathematics is available in

*What Is Mathematics?* by R. Courant and H. Robbins, Oxford University Press, New York, 1941.

A number of mathematicians were involved in the development of the axiomatic method. Chief among these were Russell, Hilbert, Frege, Cantor, Peano, and Boole. That the axiomatic method was not without difficulties was established by Gödel's epoch-making contributions to mathematical logic in 1931. An elementary exposition of this very complex subject is presented in

*Gödel's Proof*, by E. Nagel and J. R. Newman, New York University Press, New York, 1958.

# *The Number System. Real Numbers* 1

## 1.0 Foreword

Numbers are the key elements of quantification of all human activity. Engineering students will probably feel the least need to be educated on numbers, whose manipulation is a main source of pride in the "quantitative outlook" of such students. It is also true that much of the material in this chapter will be familiar to most students, but its purpose is to focus on certain aspects of numbers that may hitherto have appeared inconsequential. The value of this presentation lies in exposing a structure possessed by familiar entities (numbers) which extend to the more complicated function spaces to be discussed subsequently.

## 1.1 Introduction

We begin with a brief discussion of the number system, in particular, the real numbers. In the ordinary course of events, a treatment of the real numbers would start with the natural numbers (i.e., the positive integers $1, 2, \ldots$), leading to the rational numbers (which are expressible as ratios of integers), and eventually introducing the irrational numbers. Such a procedure would essentially be a constructive development of the real number system in a logical framework. But the procedure we adopt here is one of presenting a set of assertions or *axioms* that an abstract number system must satisfy. In so doing we will have provided a *description* of the structure of the real number system without reference to the systematic *evolution* of this structure.

An abstract number system is referred to as a *field.* In the following section we present the axiomatic structure of a field, followed by that of an *ordered field,* and conclude with the very important axiom of continuity for ordered fields.

## 1.2 Fields

A *field* $\mathfrak{F}$ is a collection of objects called numbers such that *two* binary operations, *addition* and *multiplication,* are defined with the following properties. If $a \in \mathfrak{F}$, $b \in \mathfrak{F}$, then

$$(a + b) \in \mathfrak{F} \qquad \text{(addition)}$$

$$ab \in \mathfrak{F} \qquad \text{(multiplication)}$$

Further, the following axiomatic laws of addition and multiplication must hold.

A1 Commutativity: $a + b = b + a$, $ab = ba$.
A2 Associativity: $(a + b) + c = a + (b + c)$, $(ab)c = a(bc)$.
A3 Distributivity: $a(b + c) = ab + ac$.
A4 There are two distinct elements, denoted 0 and 1, respectively, with the special properties

$$a + 0 = a \quad \text{and} \quad a \cdot 1 = a \qquad \forall \; a \in \mathfrak{F}$$

A5 The *additive inverse* of any number $a$ is defined by a number $x$ such that

$$a + x = 0$$

$x$ is denoted $-a$.
A6 The *multiplicative inverse* of $a (a \neq 0)$ is defined by a number $x$ such that

$$a \cdot x = 1$$

$x$ is denoted $a^{-1}$.

The operations of *subtraction* and *division* are merely extensions of addition and multiplication, respectively. Thus the *subtraction* of $a$ *from* $b$ is defined by $b + (-a)$ and is denoted $b - a$. The *division* of $a$ by $b (b \neq 0)$ is defined by $a \cdot b^{-1}$ and is denoted $a/b$.

These rules and definitions provide the basis for *all* algebraic manipulations involving equalities. We consider some examples below.

*Example 1.* *Show* that the zero element is unique.

A uniqueness proof frequently proceeds as follows. If the zero element is *not* unique, let $0_1$ and $0_2$ be *two* zeros. Since $0_1$ is a zero element and $0_2 \in \mathfrak{F}$ from axiom A4, we have

$$0_2 + 0_1 = 0_2$$

Similarly, because $0_2$ is a zero element and $0_1 \in \mathfrak{F}$, axiom A4 gives

$$0_1 + 0_2 = 0_1$$

The commutativity axiom A1 gives $0_1 = 0_2$, so that the zero element is unique.

*Example 2.* *Show* that $a \cdot 0 = 0$.

Axiom A4 applied to the element 1 gives

$$1 + 0 = 1$$

Multiplying by $a$ and using axioms A3 and A4, one has

$$a + a \cdot 0 = a$$

Adding the additive inverse of $a$ on both sides, we have

$$a \cdot 0 = 0$$

The concept of one number being greater or less than another number is not encompassed by the aforementioned axioms, and comes within the purview of ordered fields, which we consider in the next section.

## EXERCISES

Given $a, b \in \mathfrak{F}$, *show* that

**1.2.1** The additive inverse of 0 is 0 itself.

**1.2.2** The multiplicative inverse of 1 is 1 itself.

**1.2.3** $(-1)(a) = -a$; $(-1)(-1) = 1$.

**1.2.4** $-(a^{-1}) = (-a)^{-1}$ $(a \neq 0)$.

**1.2.5** $(ab)^{-1} = a^{-1}b^{-1}$ $(a \neq 0, b \neq 0)$.

**1.2.6** *Show* that the set of complex numbers forms a field. [A complex number $(x, y)$ is *defined* as an ordered pair of real numbers $x$ and $y$.] Define addition, multiplication, the zero element, the unity element, and additive and multiplicative inverses appropriately for complex numbers so that the field axioms are satisfied.

## 1.3 Ordered Fields

An *ordered field* is a pair of objects $(\mathfrak{F}, A)$, where $\mathfrak{F}$ is a field and $A$ is a subset of $\mathfrak{F}(A \subset \mathfrak{F})$, implying that $A$ is a partial collection of numbers belonging to $\mathfrak{F}$ with the following properties. If $a, b \in A$,

A7 $(a + b) \in A$.
A8 $ab \in A$.
A9 Given a number $a \in \mathfrak{F}$, *one* and *only one* of the following statements is true.
 (i) $a = 0$.
 (ii) $a \in A$.
 (iii) $-a \in A$.

The elements of $A$ are called *positive numbers*. Note that axiom A9 excludes 0 from the set $A$. The following definitions refer to numbers from an ordered field.

***Definition 1.3.1*** A number $a$ is said to be *less than* another number $b$, denoted $a < b$ (or alternatively, $b$ is said to be *greater than* $a$, denoted $b > a$), if $(b - a) \in A$.

It should be obvious from the definition above that if $a$ is a positive number, $0 < a$ or, equivalently, $a > 0$.

***Definition 1.3.2*** The *absolute value* of a number $a$, denoted $|a|$, is defined by

$$|a| = \begin{cases} a & \text{if } a > 0 \\ 0 & \text{if } a = 0 \\ -a & \text{if } a < 0 \end{cases}$$

The preceding rules and definitions suffice to establish all inequalities involving numbers and their absolute values. We prove below the important *triangular inequality*:

$$||a| - |b|| \leq |a + b| \leq |a| + |b|, \qquad a, b \in \mathfrak{F}$$

We first prove that $|a + b| \leq |a| + |b|$. Since the result is clearly true for $a > 0, b > 0$, we need only consider the case $a > 0, b < 0$ (why?). Thus

$$|a| + |b| = a - b$$

Now

$$b < 0 \rightarrow b < -b \rightarrow a + b < a - b = |a| + |b|$$

$$a > 0 \rightarrow -a < a \rightarrow -a - b < a - b = |a| + |b|$$

which together yield

$$|a+b| \leq |a| + |b|$$

To prove the second part, we let $x = -a$, $y = a + b$ and apply the proven inequality to $x$ and $y$ so that

$$|b| \leq |a| + |a+b|$$

so that

$$|b| - |a| \leq |a+b|$$

Similarly,

$$|a| - |b| \leq |a+b| \qquad \text{(how?)}$$

Thus

$$||a| - |b|| \leq |a+b|$$

Some simple inequalities are left for the exercises.

**EXERCISES**

*Show* that

**1.3.1** $0 < a^2$ if $a \neq 0$.

**1.3.2** If $a < b$, $b < c$, then $a < c$.

**1.3.3** If $a < b$, then for any $c$, $a + c < b + c$.

**1.3.4** If $a < b$, $0 < c$, then $ac < bc$.

**1.3.5** $|ab| = |a||b|$.

## 1.4 The Axiom of Continuity

The axiom of continuity, which is due to Dedekind, is an abstract version of the assertion in regard to the existence of the irrational numbers in the real number system and is stated as follows:

A10 Let $(\mathfrak{F}, A)$ be an ordered field. Divide all numbers into two classes $L$ and $R$ such that
- (i) Each class contains at least one number (i.e., neither $L$ nor $R$ is *empty*).
- (ii) Every number is *either* in $L$ *or* in $R$.
- (iii) If $a \in L$, $b \in R$, $a < b$.

*Then* there *exists* a *unique cut number* $c$ (which is called a *Dedekind cut*) in either $L$ or $R$ such that all numbers *less* than $c$ are in $L$ and all numbers *greater* than $c$ are in $R$.

As observed earlier, the axiom of continuity is fundamental to the existence of irrational numbers in the real number system. If, however, the irrational numbers are defined to exist in some alternative manner (e.g., see

Section 1.5), then the existence of a Dedekind cut must be *proved.* That the cut is unique is easily seen, for if it is not unique, let there be two cut numbers $c_1$ and $c_2$ such that $c_1 < c_2$. Let $a = \frac{1}{2}(c_1 + c_2)$. Since $a < c_2$, $a \in L$; but $a > c_1 \rightarrow a \in R$, which is impossible since $a$ is *either* in $L$ *or* $R$. Thus the cut number is unique.

## 1.5 Real Numbers

As observed earlier, we do not undertake a constructive development of the real number system. We will, however, spotlight briefly some of the salient developmental features. Basic to the real number system are, of course, the *natural numbers* 1, 2, 3, . . . , also known as the positive integers. Addition and multiplication of the natural numbers are readily defined and their extensions to the (positive) fractions (which are ratios of natural numbers) are also relatively straightforward. A *rational number* may be defined as the set of all fractions equivalent to some fixed fraction. Thus the set of fractions $\{\frac{1}{9}, \frac{2}{18}, \frac{3}{27}, \ldots\}$, which is presumed to include all possible such fractions equivalent to $\frac{1}{9}$, represents the rational number, say $\frac{1}{9}$, or expressed in decimals as 0.1111. . . . Addition and multiplication of rationals are no different from those of fractions, so that it is readily shown that the set of rationals forms an ordered field.† However, this set will not satisfy the axiom of continuity. For example, if we divide the rationals into two classes $L$ and $R$ such that

$$L = \{x : x^2 < 2\}, \qquad R = x : \{x^2 > 2\}$$

then $L$ and $R$ include *all* the rational numbers since $\sqrt{2}$ can be shown not to be rational, as follows. If $\sqrt{2}$ is rational, there exist integers $p$ and $q$ such that $\sqrt{2} = p/q$, where $p$ and $q$ have no common factor. Squaring, $p^2 = 2q^2$, which implies that $p^2$ and hence $p$ are even; $q$ should therefore be odd. But since $p$ is even, $p = 2r$, where $r$ is an integer, and $p^2 = 4r^2$, so that $q^2 = 2r^2$, which makes $q$ even, producing a contradiction resulting from the assumption that $\sqrt{2}$ is rational. Thus $\sqrt{2}$ is not a rational number. It is possible to find rational numbers whose squares are as close to 2 (either greater or less than 2) as we would like but never exactly 2. Thus no rational number qualifies for a cut, which shows that the axiom of continuity is invalid for rational numbers.

From a logical standpoint, the introduction of irrational numbers represents the most difficult problem. Several approaches that are equivalent to each other are available. Thus Dedekind's axiom of continuity defines the irrational number. In the example of the preceding paragraph we may assert

†The additive and multiplicative inverses of rationals may be defined by axioms A5 and A6 in Section 1.2.

the existence of some number denoted $\sqrt{2}$ in order to validate the axiom of continuity. We may do no better than write this number symbolically as $\sqrt{2}$, for there is no other *precise* representation, such as by the decimal system. The existence of *all* irrationals may thus be asserted by the validity of the axiom of continuity.

An alternative way to introduce the irrational numbers is via the decimal representation of numbers. The rational number may possess either a finite or an infinite number of digits following the decimal point. The rational number $\frac{1}{8}$ is 0.125, which represents a finite decimal, whereas the number $\frac{11}{13} = 0.846153846153\ldots$ is an infinite decimal number. Notice that the second number has a periodic recurrence of identical digits. It is left as an exercise to show that the decimal representation of a number is *periodic if* and *only if* the number is rational; the finite decimals may be regarded as periodic with zeros occurring repeatedly. The irrational numbers may then be regarded as *nonperiodic infinite decimals.* Thus the real numbers may be regarded as made up of rational numbers which are periodic decimals and irrational numbers which are nonperiodic decimals. In view of its infinite nonperiodic nature, it is not possible to *write down* precisely the decimal representation of an irrational number. What can be done is to write the *rational approximations* of the irrational number as closely as we please. Thus 1.4, 1.41, 1.414, ... represent increasingly better rational approximations of the irrational number $\sqrt{2}$. Similarly, 3.14, 3.142, 3.14159, ... represent rational approximations of the irrational number $\pi$.

Another definition, due to Cantor, regards the real numbers as *limits* of converging sequences of rational numbers. A *sequence* is merely an infinite array $a_1, a_2, \ldots$. By *convergence* of this sequence is meant that $|a_m - a_n|$ can be made as small as we please by increasing the integers $m$ and $n$ sufficiently and arbitrarily. Thus the numbers in this sequence eventually become almost (but never exactly) equal to one another. The "limit" of this sequence, which we shall define more precisely later, represents a *real number.* Of course, rational sequences may also have rational limit numbers. But sequences such as those which represent rational approximations of irrationals will obviously not converge to a rational number. The property of the rationals to approximate irrational numbers to any arbitrary degree of accuracy is referred to as "denseness" of the rational numbers in the real number field.

Before concluding this section, we observe that addition and multiplication of irrational numbers can be defined as follows. Let $\alpha$ and $\beta$ be two irrational numbers with rational sequences $\{a_n\}$ and $\{b_n\}$ converging to $\alpha$ and $\beta$, respectively. Then $(\alpha + \beta)$ and $\alpha\beta$ may be regarded as the respective limits of the sequences $\{a_n + b_n\}$ and $\{a_n b_n\}$. Thus the real numbers can now be shown to be an ordered field satisfying the axiom of continuity.

In speaking of the real number system we shall often use the term "real line." Furthermore, we will denote the set of real numbers by $\mathcal{R}$.

**EXERCISES**

*Show* that

**1.5.1** $\sqrt{3}$ is irrational.

**1.5.2** 0.99999 . . . is the same as 1.

**1.5.3** A number is a periodic decimal *if and only if* it is rational.

## 1.6 Mathematical Induction

The principle of mathematical induction is a useful implement for the proof of a sequence of mathematical propositions $A_1, A_2, \ldots$ and is built on an axiom associated with the natural numbers known as the *axiom of induction.* This axiom is stated as follows.

Let $M$ be a collection (set) of natural numbers such that 1 belongs to $M$ and if any natural number $n$ belongs to $M$, the successor of $n$, given by $(n + 1)$, also belongs to $M$. *Then* $M$ contains all the natural numbers.

We shall now formulate the principle of mathematical induction for establishing the validity of each of an infinite set of mathematical propositions $A_1, A_2, \ldots$. Let

$$M = \{n\colon\ A_n \text{ is true}\}$$

where $n$ is, of course, a positive integer. $M$ is thus a collection of natural numbers $n$ such that the proposition $A_n$ is true. If none of the propositions $A_1, A_2, \ldots$ is true, then $M$ is clearly an *empty set*. If *all* the propositions are to be true, then $M$ has to be shown to consist of *all* the natural numbers. The following principle of mathematical induction is essentially a restatement of the axiom of induction.

Let $A_1, A_2, \ldots$ be an infinite sequence of mathematical propositions such that
(i) $A_1$ is true.
(ii) If $A_n$ is true for any $n$, then $A_{n+1}$ is true.
*Then* $A_1, A_2, A_3, \ldots$ are *all* true.

We illustrate the use of this principle by an example.

*Example*

Consider the inequality

$$(1 + p)^r \geq 1 + rp$$

where $p$ is any real number greater than $-1$ and $r$ is a positive integer. It is obviously true that

$$(1 + p)^1 \geq 1 + p$$

which is the proposition when $r = 1$. Now suppose that it is true for $r = n$, that is,

$$(1 + p)^n \geq (1 + np)$$

Since $(1 + p) > 0$ we have

$$(1 + p)^{n+1} \geq (1 + p)(1 + np) = 1 + np + p + np^2$$

We may drop the positive term $np^2$ from the right-hand side of the expression above to write

$$(1 + p)^{n+1} \geq 1 + (n + 1)p$$

which shows that *if* the equality is true for $n$, then it is true for $n + 1$. Since we have shown that the result is true for $n = 1$, the inequality is established for all positive integers $n$.

### EXERCISES

Prove by mathematical induction.

**1.6.1** $\left| \sum_{j=1}^{n} a_j \right| \leq \sum_{j=1}^{n} |a_j|$, where $a_1, a_2, \ldots, a_n$ are real numbers.

**1.6.2** $1^2 + 2^2 + \ldots + n^2 = \dfrac{n(n + 1)(2n + 1)}{6}$.

## 1.7 Denumerability

The term *denumerability* applies to *sets*. We recall from the preceding chapter the definition of two *equivalent* sets as those for which there exists a one-to-one mapping of one onto the other. We also recall that equivalence was not possible between finite sets containing different numbers of objects, but that it is possible between infinite sets even when one of them may have "more" elements than the other. Thus the set of *even* numbers is equivalent to the set of all integers.† By arranging the rational numbers in a specific manner, they may be shown to be equivalent to the set of integers (see Further Reading at the end of this chapter).

***Definition 1.7.1*** A set is said to be *denumerable* if it is equivalent to the set of integers.

It is clear that denumerability is associated essentially with infinite sets. We have already stated the result, which may at first seem surprising, that the rationals form a denumerable set. We leave it to the reader to show that the

†The set $\{2, 4, 6, \ldots\}$ is equivalent to $\{1, 2, 3, \ldots\}$ because the one-to-one mapping of the first onto the second is obtained by dividing elements of the former by 2 or multiplying those of the latter by 2.

union of two denumerable sets is also a denumerable set. We now state and prove the following theorem.

***Theorem 1.7.1*** The set of real numbers is nondenumerable.

*Proof*: The real numbers comprise the denumerable set of rational numbers and the set of irrational numbers so that it would suffice to show that the irrationals form a nondenumerable set.

Assume that the set of irrationals is denumerable. Viewing the irrationals as nonperiodic infinite decimals, we list the correspondence between the integers and irrationals as follows:

| *Integer* | *Irrational Number* |
|---|---|
| 1 | $n_1 \cdot a_1 a_2 a_3 \ldots$ |
| 2 | $n_2 \cdot b_1 b_2 b_3 \ldots$ |
| 3 | $n_3 \cdot c_1 c_2 c_3 \ldots$ |
| . | . |
| . | . |
| . | . |

where $n$ is the integral part and the letters $a, b, c, \ldots$ denote the digits following the decimal place. Suppose that we now form an irrational number

$$0 \cdot abcde \ldots$$

where $a$ is different from $a_1$, $b$ is different from $b_2$, and so on; then there is no way that this number, by the very nature of its construction, may be equal to any of the listed numbers above. Since the original list was presumed to contain *all* the irrational numbers, we conclude that such listing is impossible and that the irrationals and hence the real numbers are nondenumerable. ∎

The property of rational numbers to approximate any irrational number as closely as we please is an interesting one and is referred to as denseness. Thus the denumerable set of rational numbers is dense in the nondenumerable set of real numbers. This property of the real number system may be referred to as *separability*. These properties will be encountered again in the discussion of what are called abstract metric spaces.

## EXERCISES

**1.7.1** Show that any bounded interval $[a, b]$ on the real line is equivalent to the interval $[0, 1]$.

**1.7.2** Show that the real line is equivalent to $(-\pi/2, \pi/2)$ and hence to any bounded interval. (*Hint:* Find a correspondence rule through an appropriate mathematical relationship between elements of the two sets.)

**1.7.3** Let $\{A_n\}$ be a denumerable family of sets each of which is denumerable. Show that the set $\bigcup_{n=1}^{\infty} A_n$ is denumerable.

## 1.8 Bounded Sets

We will concern ourselves with sets of real numbers.

***Definition 1.8.1*** A set of real numbers $S$ is said to be *bounded from above* if $\exists$ a number $m \ni s < m \ \forall \ s \in S$. $m$ is called an *upper bound* of $S$.

*Example 1*

An example of a set bounded from above is $S_1 = (-\infty, 1)$ and 1 is an upper bound of $S_1$. Clearly, $S_1$ has many upper bounds since all numbers greater than 1 are upper bounds.

*Example 2*

Another example of a set bounded from above is

$$S_2 = \{y = \log x \colon \ 0 < x \leq 1\}$$

Note that any positive number is an upper bound of this set.

We now define the least upper bound or the supremum of a set that is bounded from above.

***Definition 1.8.2*** Let $S$ be a set bounded from above and $m$ be a number such that $s \leq m \ \forall \ s \in S$. If $\forall \ m' < m \ \exists \ s \in S \ni s > m'$, then $m$ is called the *least upper bound* or *supremum* of $S$. It is written as $m = \sup S$.

It is readily shown that the set $S_1$ mentioned earlier has $\sup S_1 = 1$. In the second example it should be obvious that $\sup S_2 = 0$.

The following theorem is a direct consequence of the axiom of continuity.

***Theorem 1.8.1*** If a set is bounded from above, then it has a supremum.

*Proof*: Let $S$ be the set bounded from above. Divide the real numbers into two classes $L$ and $R$ as follows:

$$L = \{x \colon \ x < s \text{ for some } s \in S\}$$

$$R = \{y \colon \ y \geq s \ \forall \ s \in S\}$$

Clearly, every real number must be in either $L$ or $R$. Neither $L$ nor $R$ is empty since $s \in S \rightarrow (s - 1) \in L$, and since $S$ is bounded from above, all upper bounds of $S$ belong to $R$. Moreover, $x \in L, y \in R \rightarrow x < y$, so that $L$ and $R$ form a Dedekind cut.

Let $c$ be the unique cut number guaranteed by the axiom of continuity. We show that $c$ is an upper bound of $S$.

Since $c$ is the cut number, $x < c \rightarrow x \in L$. Now if $c$ is not an upper bound of $S$, then $\exists\ s \in S \ni s > c$. Let $z$ be a number such that $s > z > c$. Then

$$z < s \rightarrow z \in L$$

$$z > c \rightarrow z \in R$$

which are contradictory statements. Thus $c$ is an upper bound of $S$. Now we show that $c$ is the least upper bound or supremum of $S$. If $c$ is not sup $S$, let $b$ be an upper bound such that $b < c$.

Now $b < c \rightarrow b \in L$. But $b > s \ \forall\ s \in S \rightarrow b \in R$, which is impossible. Hence $b \geq c \rightarrow c = \sup S$. ▌

In some treatments Theorem 1.8.1 is given an axiomatic status and the axiom of continuity is then proved as a theorem.

We now define convergence of a sequence of real numbers. A sequence is merely a denumerable set of numbers.

***Definition 1.8.3*** A sequence of numbers $\{x_n\}$ is said to *converge* to a number $a$ if $\forall\ \epsilon > 0\ \exists$ an integer $n_\epsilon$ (which depends on $\epsilon$) $\ni n > n_\epsilon \rightarrow |x_n - a| < \epsilon$, where $\epsilon$ in a small positive number. $a$ is called the limit of the sequence.

This definition implies that a sequence $\{x_n\}$ will converge to $a$ if the members of this sequence will *eventually* be as close to $a$ as desired.

The sequence $\{1/n\}$ converges to zero since $|1/n - 0| = 1/n$ can be made as small as desired by taking $n_\epsilon = [(1/\epsilon) + 1]$,† for $n \geq n_\epsilon$,

$$\frac{1}{n} \leq \frac{1}{n_\epsilon} < \frac{1}{(1/\epsilon) + 1 - 1} = \epsilon$$

It is often not possible or convenient to use this criterion for testing the convergence of any given sequence. Standard tests of convergence, which we will not go into here, are discussed in texts on advanced calculus.‡

We now prove the following theorem.

***Theorem 1.8.2*** Let $\{x_n\}$ be a sequence of real numbers such that $x_1 \leq x_2 \leq \ldots \leq x_n \leq x_{n+1} \leq \ldots$ bounded from above; that is, $x_n \leq m \ \forall\ n$. Then the sequence converges to the supremum of the sequence (set).

*Proof*: Since the sequence is bounded from above, there exists a supremum $a$ of the set from Theorem 1.8.1. Because $a$ is the least upper bound, $\forall\ \epsilon > 0$ for some $n$, say $= n_\epsilon$, $a - \epsilon < x_{n_\epsilon}$. But

$$x_{n_\epsilon} \leq x_n \qquad \text{when } n \geq n_\epsilon$$

---

†The square bracket is used to define the following function: $[x]$ is the largest integer that does not exceed the real number $x$. Thus $[x] \leq x$ and $[x + 1] > x$.

‡See, for example, *Advanced Calculus*, by A. E. Taylor, Ginn, Boston, 1955.

so that

$$a - \epsilon < x_n \leq a$$

Thus

$$0 \leq a - x_n < \epsilon \rightarrow \{x_n\} \text{ converges to } a \quad \blacksquare$$

The following corollary to Theorem 1.8.2 is useful.

***Corollary*** Let $S$ be a set bounded from above and $a$ be its supremum. Then one can find a sequence entirely contained in $S$ which converges to $a$.

*Proof*: If $a$ belongs to $S$, the required sequence is immediately obtained by repeating $a$ infinitely. If $a$ does not belong to $S$, then $S$ is necessarily infinite, denumerable or nondenumerable. If $S$ is denumerable, a sequence is obtained by enumerating $S$ in a strictly ascending order and applying Theorem 1.8.2.

If $S$ is nondenumerable, the rationals of $S$ may be arranged in an ascending order, so that Theorem 1.8.2 again becomes applicable, yielding a sequence converging to $a$. ▮

So far we have dealt with sets that have been bounded from above. We leave it as an exercise to the student to define sets bounded from below, the *greatest lower bound* or *infimum* of such sets, and to show by using the axiom of continuity that every set bounded from below has an infimum. Obviously, Theorem 1.8.2 and the accompanying corollary also have their counterparts with regard to sets bounded from below. We conclude this section with the following definition.

***Definition 1.8.4*** A set that is both bounded from above and below is called a *bounded set.*

## EXERCISES

**1.8.1** Define a set bounded from below and its greatest lower bound or infimum.

**1.8.2** Show that every set bounded from below has an infimum.

**1.8.3** State and prove the counterparts for Theorem 1.8.2 and its corollary for sets bounded from below.

**1.8.4** Find the supremum and infimum of each of the following sets, establishing existence first.

(a) $S = \{1/x: x \in (0, 1)\}$.

(b) $S = \{y: y = x^2 - 2x + 4, 0 < x < 2\}$.

(c) $S = \{x: x^3 - 9x^2 + 26x - 24 < 0\}$.

**1.8.5** Let $\{a_n\}$ and $\{b_n\}$ be two convergent sequences such that $a_n \leq b_n$. Show that $\lim a_n \leq \lim b_n$. [*Hint:* Let $\lim a_n = a$, $\lim b_n = b$. If $b < a$, let $\epsilon = \frac{1}{2}(a - b)$. Now show that for sufficiently large $n$, $a_n > b_n$, which is impossible.]

**1.8.6** Let $S$ and $S'$ be two sets of real numbers bounded from above such that $\forall a \in S$, $\exists$ some $a' \in S' \ni a' \geq a$. Show that $\sup S' \geq \sup S$.

**1.8.7** Let $S$ be a set bounded from above and $S'$ be another set bounded from below such that every element of $S$ is a lower bound of $S'$, and every element of $S'$ is an upper bound of $S$. If, further, $\forall \epsilon > 0$, $\exists\ a \in S, a' \in S', \ni a' - a < \epsilon$, show that $\sup S = \inf S'$.

**1.8.8** Let $\{I_n\}$, $I_n = \{x\colon a_n \leq x \leq b_n\}$, be a sequence of *nested intervals*, defined by $I_{n+1} \subset I_n$, such that $\lim_{n\to\infty} (b_n - a_n) = 0$. Show that there exists *one* and *only one* point in the set $\bigcap_{n=1}^{\infty} I_n$. What is that point? (*Hint:* Use Theorem 1.8.2.)

**1.8.9** Let $\{a_n\}$ be a sequence bounded above and below. Form the sequences

$$\begin{aligned} \bar{a}_n &= \sup \{a_{n+1}, a_{n+2}, \ldots\} \\ \underline{a}_n &= \inf \{a_{n+1}, a_{n+2}, \ldots\} \end{aligned} \qquad n = 1, 2, \ldots$$

Show that $\{\bar{a}_n\}$ and $\{\underline{a}_n\}$ are convergent sequences. Note that it is customary to denote their respective limits as

$$\overline{\lim_{n\to\infty}}\, a_n = \lim_{n\to\infty} \bar{a}_n, \qquad \underline{\lim_{n\to\infty}}\, a_n = \lim_{n\to\infty} \underline{a}_n$$

Alternatively, $\overline{\lim}$ is called lim sup and $\underline{\lim}$ is called lim inf.

## 1.9 Concluding Remarks

This chapter has paid special attention to the real numbers as a particular example of an abstract field. The properties of real numbers were assumed to be given. For example, nothing was said about how the positive integers are constructed or how even and odd numbers are defined. Such features were presumed to be known facts. The essence of the chapter is an exposition of real numbers with known properties as being in conformity with the concept of a field. Special attention must be paid to the result of how every real number (rational or irrational) may be closely approximated by the denumerable (or countable) rationals. Also *every* Cauchy sequence of real (especially rational) numbers (i.e., sequences in which numbers become *arbitrarily* close to each other eventually) will converge to some number. This property, known as *completeness* of the real numbers, is a particularly important concept that has extensions with profound consequences.

In the hierarchy of mathematical sets, there are "more primitive" collections than a field. By "more primitive" is meant that the axiomatic stipulations, for example, of a *ring* cover only a *part* of those of a field. Similarly, *ideals* are subsets of rings that are also more primitive than fields in the foregoing sense. Rings and ideals are important components of modern algebra. That no attention has been paid to them implies a limitation of the scope of this book rather than of their importance to the general subject of linear operators.

## FURTHER READING

That the rationals are denumerable or countably infinite is established by demonstrating a method of listing them. This is not difficult to do and is covered in many texts. A book of special interest, however, is

*What Is Mathematics?* by R. Courant and H. Robbins, Oxford University Press, New York, 1941.

Many textbooks are available on advanced calculus that discuss convergences of sequences of numbers, standard tests for convergence (since the definition of convergence is not always convenient to test for convergence), and so on. We especially recommend

*Advanced Calculus*, by A. E. Taylor, Ginn, Boston, 1955.

A very simple discussion of rings, ideals, and other concepts of modern algebra may be found in

"Groups and Other Algebraic Systems," by A. J. Malćev, Chapter XX of *Mathematics: Its Contents, Methods, and Meaning*, Vol. III, edited by A. D. Alexandrov, A. N. Kolmogorov, and M. A. Lavrentév, translated by K. Hirsh, MIT Press, Cambridge, Mass., 1969.

# Linear Spaces 2

## 2.0 Foreword

It is in this chapter that we provide the first exposure to quantities that are of direct interest to us: *vectors*. Many engineering systems naturally submit to a vectorial description; that is, the *state* variables usually form an "$n$-dimensional vector." The system *behavior* is then modeled by suitable *equations* in the vectors, usually derived on physical grounds. Such model equations must be solved for the unknown vector that represents the model's viewpoint of the system behavior. Insofar as the solution vector originates from a *set* of vectors, its identification requires familiarity with the *territory* to which it belongs. It is the exploration of this territory or set of vectors with which we will be concerned in this chapter. The exploration, however, is quite preliminary here in that only certain simple features connected with *algebraic relationships* between vectors will be of interest.

The engineering student's early acquaintance with vectors is normally through their role in describing physical quantities that have *magnitude* and *direction*. The student trying to make a connection between the foregoing concept of a vector and those in this chapter is likely to run into some confusion. First, we do not recognize the existence of a "magnitude" for the vectors to be encountered here. This is not to say that they *cannot* have magnitudes but simply that we are not concerned about them for the present. Second, the concept of direction is tied up with that of angle between lines, which cannot be deduced from the properties of vectors that are considered in this chapter. The only properties that are of interest here are those relating to sums of vectors, scalar multiplication, and linear combinations of vectors. The main motivation for this limited scope is the simplicity with which certain properties may be studied in isolation from others.

Another important feature that is covered in this chapter is the subject of *linear transformations*. Indeed, this is the very essence of our subject matter. Linear engineering models lead to equations involving vectors and linear transformations. Generally, the solution of the model equation involves identifying a vector that has been transformed by a linear transformation to some specified vector. It is reasonable, then, that a systematic approach to the solution of such equations should be to explore vectors in their "natural habitat" and the effect of linear transformations on them.

## 2.1 Introduction

We consider here the abstract generalization of three-dimensional vector space. As observed in the foreword, we will be concerned only with the *algebraic structure* of the space, such as addition and subtraction of vectors, scalar multiplication, and so on. Such algebraic structure is to be distinguished from geometric or *topological* structure, which has to do with "magnitude" of a vector, distance between points, and so on. The topological aspects will be dealt with in a subsequent chapter. Prior to defining a linear space we observe that a linear space is always associated with a field $\mathfrak{F}$.

A *linear space* $\mathfrak{L}$ is a collection of elements called vectors denoted by boldface letters $\mathbf{x}, \mathbf{y}, \mathbf{z}, \ldots$ on which two fundamental operations, called sum and scalar multiplication, are defined. For every pair of elements $\mathbf{x}, \mathbf{y} \in \mathfrak{L}$, there exists a third element, $\mathbf{z} \in \mathfrak{L}$, which represents the *sum* of $\mathbf{x}$ and $\mathbf{y}$, denoted $\mathbf{x} + \mathbf{y}$.† For every $\mathbf{x} \in \mathfrak{L}$ and $\alpha \in \mathfrak{F}$ there exists an element $\mathbf{z} \in \mathfrak{L}$, called the *scalar multiplication* of $\mathbf{x}$ by $\alpha$, denoted $\alpha\mathbf{x}$.‡ In the following additional stipulations regarding the sum and scalar multiplication, $\mathbf{x}$, $\mathbf{y}$, and $\mathbf{z}$ are arbitrary elements of $\mathfrak{L}$ and $\alpha$ and $\beta$ are arbitrary scalars in the associated field $\mathfrak{F}$.

A1 Commutativity: $\mathbf{x} + \mathbf{y} = \mathbf{y} + \mathbf{x}$.
A2 Associativity: $(\mathbf{x} + \mathbf{y}) + \mathbf{z} = \mathbf{x} + (\mathbf{y} + \mathbf{z})$

$$\alpha(\beta\mathbf{x}) = (\alpha\beta)\mathbf{x}.$$

†Alternatively, *addition* of vectors may be regarded as a mapping $S\colon \mathfrak{L} \times \mathfrak{L} \longrightarrow \mathfrak{L}$, where $S$ is the mapping representing the sum.

‡Scalar multiplication may also be regarded as S.M.: $\mathfrak{F} \times \mathfrak{L} \longrightarrow \mathfrak{L}$, which maps the *pair* of objects comprising a number (scalar) and a vector into another vector.

A3 Distributivity: $(\alpha + \beta)\mathbf{x} = \alpha\mathbf{x} + \beta\mathbf{x}$

$$\alpha(\mathbf{x} + \mathbf{y}) = \alpha\mathbf{x} + \alpha\mathbf{y}.$$

A4 There exists a *zero element* $\mathbf{0} \in \mathfrak{L}$ such that

$$\mathbf{x} + \mathbf{0} = \mathbf{x}$$

A5 For every element $\mathbf{x} \in \mathfrak{L}$ there exists an *additive inverse* $\mathbf{y} \in \mathfrak{L}$ such that

$$\mathbf{x} + \mathbf{y} = \mathbf{0}$$

$\mathbf{y}$ is denoted $-\mathbf{x}$. *Subtraction* of a vector $\mathbf{y}$ from $\mathbf{x}$, denoted $\mathbf{x} - \mathbf{y}$, is defined as

$$\mathbf{x} - \mathbf{y} = \mathbf{x} + (-\mathbf{y})$$

A6 Scalar multiplication is such that

$$1\mathbf{x} = \mathbf{x} \qquad \forall \quad \mathbf{x} \in \mathfrak{L}$$

where "1" is the unity element in $\mathfrak{F}$.

Axioms A1 to A6 provide the basis for all algebraic manipulations involving vectors. The following simple theorems demonstrate the use of these axioms to obtain results that are not so obvious as to be taken for granted.

***Theorem 2.1.1*** The *zero* element of a linear space is unique.

*Proof*: Assume that there are two zero elements $\mathbf{0}_1$ and $\mathbf{0}_2$.

$$\text{For } \mathbf{0}_1, \qquad \mathbf{0}_2 + \mathbf{0}_1 = \mathbf{0}_2$$

$$\text{For } \mathbf{0}_2, \qquad \mathbf{0}_1 + \mathbf{0}_2 = \mathbf{0}_1$$

By axiom A1 we have $\mathbf{0}_1 = \mathbf{0}_2$. ∎

***Theorem 2.1.2*** In any linear space $\mathfrak{L}$, $0\mathbf{x} = \mathbf{0}$ for every $\mathbf{x} \in \mathfrak{L}$, where 0 is the zero element of the associated field $\mathfrak{F}$.

*Proof*: From axiom A3 we have for every $\mathbf{x}$,

$$(0 + 1)\mathbf{x} = 0\mathbf{x} + 1\mathbf{x} = 0\mathbf{x} + \mathbf{x}$$

It is also true that

$$(0 + 1)\mathbf{x} = 1\mathbf{x} = \mathbf{x}$$

from which

$$\mathbf{x} = 0\mathbf{x} + \mathbf{x}$$

Adding the additive inverse of $\mathbf{x}$ to both sides

$$\mathbf{0} = (0\mathbf{x} + \mathbf{x}) + (-\mathbf{x}) = 0\mathbf{x} + (\mathbf{x} + (-\mathbf{x})) = 0\mathbf{x} + \mathbf{0} = 0\mathbf{x}$$

so that

$$0\mathbf{x} = \mathbf{0} \quad ∎$$

**EXERCISES**

**2.1.1** Prove that in any linear space the additive inverse of each element is unique.

**2.1.2** Show that in a linear space, $-\mathbf{x} = (-1)\mathbf{x}$.

*Examples of Linear Spaces*

$\mathcal{R}_3$: The three-dimensional vector space in which each element $\mathbf{x}$ is an ordered set of three real numbers denoted as

$$\mathbf{x} \equiv \begin{bmatrix} x_1 \\ x_2 \\ x_3 \end{bmatrix}$$

The associated field is that of the real numbers. It is easy to see that the elements of $\mathcal{R}_3$ form a linear space with the zero element

$$\mathbf{0} \equiv \begin{bmatrix} 0 \\ 0 \\ 0 \end{bmatrix}$$

by establishing axioms A1 to A6, using the following definitions of vector addition and scalar multiplication:†

$$\mathbf{x} + \mathbf{y} \equiv \begin{bmatrix} x_1 + y_1 \\ x_2 + y_2 \\ x_3 + y_3 \end{bmatrix}, \qquad \alpha\mathbf{x} \equiv \begin{bmatrix} \alpha x_1 \\ \alpha x_2 \\ \alpha x_3 \end{bmatrix}$$

and the property of real numbers.

The extension of this space to $n$-dimensional vector spaces is natural. Thus $\mathcal{R}_n$ represents a linear space in which the individual element is an ordered set of $n$ real numbers, that is,

$$\mathbf{x} \equiv \begin{bmatrix} x_1 \\ x_2 \\ \cdot \\ \cdot \\ \cdot \\ x_n \end{bmatrix}$$

Again, the associated field is the set of real numbers. When the real numbers in $\mathbf{x}$ are replaced by complex numbers, the resulting linear space is denoted $\mathcal{C}_n$.

---

†See also *Mathematical Methods in Chemical Engineering: Matrices and Their Applications*, Vol. 1, by N. R. Amundson, Prentice-Hall, Englewood Cliffs, N.J., 1964.

The linear spaces above may also be extended to infinite arrays of either real numbers, called $\mathcal{R}_\infty$, or complex numbers, denoted by $\mathcal{C}_\infty$.

$\mathcal{C}[a, b]$: This is an important example since it establishes the notion of a function as a *vector*. Consider the set of all functions $f(t)$ defined on the *closed* interval $[a, b]$ (which implies that the end points $a$ and $b$ are included) and *continuous* in $[a, b]$.

We recall from elementary calculus the definition of a continuous function. A function $f(t)$ is continuous at $t = t_0$ if $\forall\, \varepsilon > 0,\ \exists\ \delta_\varepsilon > 0 \ni |t - t_0| < \delta_\varepsilon \rightarrow |f(t) - f(t_0)| < \varepsilon$. A function is said to be continuous in $[a, b]$ if it is continuous at every point in $[a, b]$. At $a$, $f(t)$ is continuous from the right and at $b$ it is continuous from the left. Figure 2.1.1(a) shows a continuous function, while 2.1.1(b) shows a function with a finite number of discontinuities.

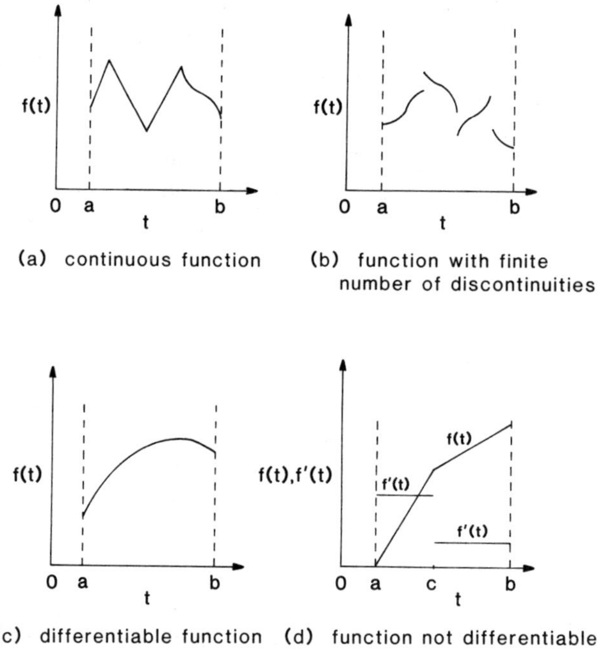


**Figure 2.1.1** Differentiable, continuous, and discontinuous functions.

If the functions are real-valued, the associated field of the space must also be real. If, however, the functions are complex-valued, then the associated field may be real or complex, although the latter is generally the case.

We introduce here an important notation concerning functions. Since vectors will be denoted by boldface symbols, we will let **f**, **g**, . . . represent the elements of $\mathcal{C}[a, b]$. Thus

$$\mathbf{f} \equiv \{f(t)\colon\ a \leq t \leq b\} \quad \text{or more simply} \quad \{f(t)\}$$

The foregoing notation brings to focus how the specification of vector **f** requires the values of $f(t)$ for *all values* of $t$ in the interval $[a, b]$ in the same way that a vector **x** in $\mathcal{R}_n$ requires the stipulation of all components $x_1, x_2, \ldots, x_n$. It also shows the distinction between $f(t)$ and **f** in that the former is concerned only with the function evaluated at a fixed value $t$, whereas the latter includes all the values of the function in the interval $[a, b]$. In a sense $f(t)$ may be regarded as a component of **f**. The addition of two elements **f** and **g** is defined by

$$\mathbf{f} + \mathbf{g} \equiv \{f(t) + g(t)\colon\ a \leq t \leq b\}$$

which implies that the components of $\mathbf{f} + \mathbf{g}$ are obtained from $f(t) + g(t)$. The scalar multiplication is defined by

$$\alpha\mathbf{f} \equiv \{\alpha f(t)\colon\ a \leq t \leq b\}$$

The zero element is defined by

$$\mathbf{0} \equiv \{0\colon\ a \leq t \leq b\}$$

In order to establish that $\mathcal{C}[a, b]$ is a linear space, it is only necessary to show that $[f(t) + g(t)]$ and $\alpha f(t)$ are continuous functions in $[a, b]$. The proof is straightforward and may be found in any standard treatment of calculus.†

In a similar way, we may define linear spaces of functions that are *differentiable*. A function $f(t)$ is said to be differentiable at $t$ if the limits

$$\lim_{h \to 0} \frac{f(t+h) - f(t)}{h}, \qquad \lim_{h \to 0} \frac{f(t) - f(t-h)}{h}$$

exist and are *equal*. At the end point $a$, however, we consider the limit only from the right end since $f(t)$ is not defined for $t < a$. Similarly, at the point $b$, only the left limit is considered. Figure 2.1.1(c) shows a function differentiable in $[a, b]$. At an interior point $t$ in the interval $[a, b]$, notice that the two limits may exist without being equal for a function continuous at $t$ but with a "sharp corner" [see Figure 2.1.1(d)].

There are functions smooth enough to be differentiable such that their derivatives are continuous functions. Such functions are also referred to as continuously differentiable functions. The set of all continuously differentiable functions is denoted $\mathcal{C}^{(1)}[a, b]$ and forms a linear space using the same definitions of addition and scalar multiplication as those in $\mathcal{C}[a, b]$. Similarly, we denoted the linear space of functions with $n$ derivatives (i.e., $n$ times differentiable), with the $n$th derivative continuous, by $\mathcal{C}^{(n)}[a, b]$. More concisely, this space is known as the class of all functions with $n$ continuous derivatives.

---

†See, for example, *Advanced Calculus*, by A. E. Taylor, Ginn, Boston, 1955.

One may also have infinitely differentiable functions which form a linear space, denoted $\mathcal{C}^{\infty}[a, b]$. There are also other linear spaces of functions; for example, we may consider functions that have a *finite* number of discontinuities [see Figure 2.1.1(b)] and form a linear space.

**EXERCISES**

**2.1.3** Explain why $\mathcal{R}_n$ cannot be a linear space when the associated field is complex. Can $\mathcal{C}_n$ be a linear space with the real numbers as the associated field?

**2.1.4** Enumerate the differences between $\mathcal{R}_2$ and the field of complex numbers.

**2.1.5** Show that $\mathcal{C}_n$ with the field of complex numbers is a linear space.

**2.1.6** Show that $\mathcal{C}^{(1)}[a, b]$ with $\mathcal{R}$ as the associated field is a linear space.

**2.1.7** Consider the function $f(t)$ defined in [0, 1] by

$$f(t) = \begin{cases} t^2, & 0 \leq t \leq \frac{1}{2} \\ \frac{1}{2} - (1 - t)^2, & \frac{1}{2} < t \leq 1 \end{cases}$$

Let $\mathbf{f} \equiv \{f(t): 0 \leq t \leq 1\}$. Is $\mathbf{f} \in \mathcal{C}[0, 1]$? Is $\mathbf{f} \in \mathcal{C}^{(1)}[0, 1]$? Is $\mathbf{f} \in \mathcal{C}^{(2)}[0, 1]$?

**2.1.8** Show that functions with a finite number of discontinuities in $[a, b]$ form a linear space.

## 2.2 Linear Dependence

Consider a linear space $\mathcal{L}$ associated with a field $\mathcal{F}$. The associativity axiom makes it meaningful to define a vector $\mathbf{y}$ by

$$\mathbf{y} = \alpha_1 \mathbf{x}_1 + \alpha_2 \mathbf{x}_2 + \ldots + \alpha_k \mathbf{x}_k$$

where $\alpha_j \in \mathcal{F}$, $\mathbf{x}_j \in \mathcal{L}$, $j = 1, 2, \ldots, k$. The vector $\mathbf{y}$ is then said to be a *linear combination* of vectors $\mathbf{x}_1, \mathbf{x}_2, \ldots, \mathbf{x}_k$. One also refers to such a *linear combination* as a linear expansion of $\mathbf{y}$ in terms of $\mathbf{x}_1, \mathbf{x}_2, \ldots, \mathbf{x}_k$ with *coefficients* $\alpha_1, \alpha_2, \ldots, \alpha_k$. Observe that if $\alpha_j = 0, j = 1, 2, \ldots, k$, $\mathbf{y}$ is the zero vector itself. We now define linear dependence of a set of vectors.

***Definition 2.2.1*** A set of vectors $(\mathbf{x}_1, \mathbf{x}_2, \ldots, \mathbf{x}_k)$ of a linear space is said to be *linearly dependent* if there exist $\alpha_1, \alpha_2, \ldots, \alpha_k \in \mathcal{F}$ not all equal to zero such that

$$\mathbf{0} = \alpha_1 \mathbf{x}_1 + \alpha_2 \mathbf{x}_2 + \ldots + \alpha_k \mathbf{x}_k \tag{2.2.1}$$

If a set of vectors is not linearly dependent, then it is *linearly independent.* Thus for a linearly independent set of vectors $\mathbf{x}_1, \mathbf{x}_2, \ldots, \mathbf{x}_k$, expansion (2.2.1) is possible only when $\alpha_1 = \alpha_2 = \ldots = \alpha_k = 0$. It is obvious that if any subset of a set of vectors is linearly dependent, then the set is itself linearly dependent.

We consider below some examples of linear spaces in which we examine conditions under which a given set of vectors may be considered linearly dependent or independent.

*Example 1*

Let

$$\mathbf{x}_j = \begin{bmatrix} x_{1,j} \\ x_{2,j} \\ \cdot \\ \cdot \\ \cdot \\ x_{n,j} \end{bmatrix}, \qquad j = 1, 2, \ldots, k$$

be a set of vectors in $\mathcal{R}_n$ so that $x_{1,j}, x_{2,j}, \ldots, x_{n,j}$ are real numbers. When is the set $(\mathbf{x}_1, \mathbf{x}_2, \ldots, \mathbf{x}_k)$ linearly dependent or independent? To answer this question we examine the existence of numbers $\alpha_1, \alpha_2, \ldots, \alpha_k$ such that

$$\alpha_1\mathbf{x}_1 + \alpha_2\mathbf{x}_2 + \ldots + \alpha_k\mathbf{x}_k = \mathbf{0}$$

which yields the set of homogeneous equations in $\alpha_1, \alpha_2, \ldots, \alpha_k$:

$$\begin{aligned} x_{1,1}\alpha_1 + x_{1,2}\alpha_2 + \ldots + x_{1,k}\alpha_k &= 0 \\ x_{2,1}\alpha_1 + x_{2,2}\alpha_2 + \ldots + x_{2,k}\alpha_k &= 0 \\ &\cdot \\ &\cdot \\ &\cdot \\ x_{n,1}\alpha_1 + x_{n,2}\alpha_2 + \ldots + x_{n,k}\alpha_k &= 0 \end{aligned}$$

which always possesses a solution.† The solution may be trivial ($\alpha_1 = \alpha_2 = \ldots = \alpha_k = 0$) or nontrivial, depending on the rank of the coefficient matrix

$$\begin{bmatrix} x_{1,1} & x_{1,2} & \ldots & x_{1,k} \\ x_{2,1} & x_{2,2} & \ldots & x_{2,k} \\ \cdot \\ \cdot \\ \cdot \\ x_{n,1} & x_{n,2} & \ldots & x_{n,k} \end{bmatrix}$$

†Recall that a set of linear simultaneous, algebraic equations has a solution if and only if the ranks of the coefficient matrix and the augmented matrix are identical (see, for example, *Mathematical Methods in Chemical Engineering*: *Matrices and Their Applications*, Vol. 1, by N. R. Amundson, Prentice-Hall, Englewood Cliffs, N.J., 1964). Thus a set of linear homogeneous simultaneous algebraic equations must always have a solution.

Denoting the rank by $r$, we note that $r \leq \min\{n, k\}$. Let $k < n$. If $r = k$, then $\alpha_1 = \alpha_2 = \ldots = \alpha_k = 0$, so that the set $\{\mathbf{x}_j\}_1^k$ is linearly independent. If $r < k$, the set is linearly dependent, since $(k - r)$ of the $\alpha_j$'s can be nonzero.

If $n < k$, then $r \leq n$, for which some of the $\alpha_j$'s are necessarily nonzero, so that $\{\mathbf{x}_j\}_1^k$ is linearly dependent.

For $n = k$, $\{\mathbf{x}_j\}_1^k$ is linearly dependent if $r < n$ and independent when $r = n$.

*Example 2*

For the second example, we consider the linear space $\mathcal{C}^{(k-1)}[a, b]$ and the set of vectors $\{\mathbf{f}_j\}_1^k$, $\mathbf{f}_j \equiv \{f_j(t)\}$ $(j = 1, 2, \ldots, k)$, such that each of the $k$ functions $f_1(t), f_2(t), \ldots, f_k(t)$ can be differentiated $(k - 1)$ times. We derive a *sufficient* condition for the set of functions to be linearly independent. Consider the equation

$$\alpha_1 f_1(t) + \alpha_2 f_2(t) + \ldots + \alpha_k f_k(t) = 0$$

One may differentiate the equation above, once, twice, and so on, up to $(k - 1)$ times to obtain the equations

$$\begin{aligned} &\alpha_1 f'_1(t) + \alpha_2 f'_2(t) + \ldots + \alpha_k f'_k(t) = 0 \\ &\quad\vdots \\ &\alpha_1 f_1^{(k-1)}(t) + \alpha_2 f_2^{(k-1)} + \ldots + \alpha_k f_k^{(k-1)}(t) = 0 \end{aligned}$$

Thus we obtain $k$ equations in $k$ unknowns $\alpha_1, \alpha_2, \ldots, \alpha_k$. If the *determinant* of the matrix†

$$\begin{bmatrix} f_1(t) & f_2(t) & \ldots & f_k(t) \\ f'_1(t) & f'_2(t) & \ldots & f'_k(t) \\ \vdots & & & \\ f_1^{(k-1)}(t) & f_2^{(k-1)}(t) & \ldots & f_k^{(k-1)}(t) \end{bmatrix}$$

is *nonzero* for any $t \in [a, b]$, then clearly $\alpha_1 = \alpha_2 = \ldots = \alpha_k = 0$, which implies that $\{\mathbf{f}_j\}_1^k$ is a linearly independent set. Note that the condition above is only one of sufficiency; that is, the vanishing of the determinant above does not necessarily imply linear dependence of $\{\mathbf{f}_j\}_1^k$.

---

†This matrix is normally called the *Wronskian* matrix of the set of functions $\{f_j(t)\}_{j=1}^k$. Its determinant is called the *Wronskian*.

**EXERCISES**

**2.2.1** Show that a set of vectors $\{\mathbf{x}_j\}_1^k$ is linearly dependent if and only if one of the vectors can be expressed as a linear combination of the remaining vectors.

**2.2.2** Show that every vector in a linearly independent set is a nonzero vector.

**2.2.3** Show that the set $\{t^j\}_0^k$ of the linear space $\mathcal{C}[a, b]$ is linearly independent for arbitrary integer $k$.

**2.2.4** Is the set of functions $(\sin^2 t, \cos^2 t, 1)$ in $\mathcal{C}[0, \pi]$ linearly dependent or independent?

**2.2.5** Let $\{\mathbf{x}_j\}_1^k$ be a linearly independent set in a linear space $\mathcal{L}$. Form a new set of vectors $\{\mathbf{y}_j\}_1^k$ such that

$$\mathbf{y}_i = \sum_{j=1}^{k} \alpha_{i,j}\mathbf{x}_j, \qquad i = 1, 2, \ldots, k$$

where $\alpha_{i,j} \in \mathfrak{F}$, $i, j = 1, 2, \ldots, k$. Derive a condition for $\{\mathbf{y}_j\}_1^k$ to be linearly independent. If $\alpha_{i,j} = 0$ for $i < j$, establish that $\{\mathbf{y}_j\}_1^k$ is linearly independent.

## 2.3 Bases, Components, and Dimension

The concepts herein arise as a direct generalization of expressing a physical vector as the vector sum of components along the $x$, $y$, and $z$ coordinate axes.

***Definition 2.3.1*** Let $\mathcal{L}$ be a linear space associated with a field $\mathfrak{F}$. A linearly independent set of vectors $\{\mathbf{x}_j\}_1^k$ is a *basis* in $\mathcal{L}$ if $\forall\, \mathbf{x} \in \mathcal{L}$, $\exists\ \alpha_1, \alpha_2, \ldots, \alpha_k$ $\ni$

$$\mathbf{x} = \alpha_1\mathbf{x}_1 + \alpha_2\mathbf{x}_2 + \ldots + \alpha_k\mathbf{x}_k$$

The numbers $\alpha_1, \alpha_2, \ldots, \alpha_k$ are called *components* of $\mathbf{x}$ with respect to the basis $\{\mathbf{x}_j\}_1^k$.

Note that the definition above involves a *finite* number of vectors in the basis. Linear spaces in which a finite basis does not exist are called infinite-dimensional spaces.† Infinite-dimensional spaces may possess bases, but we are not in a position to discuss this at the present time, since infinite sums are defined in terms of limits that are dependent on the topological structure of the space.

A given linear space has any number of bases and the following theorem holds.

---

†The existence of a basis for every linear space depends on Zorn's axiom, which is beyond the scope of this book (see Further Reading).

***Theorem 2.3.1*** Let $A_1 = \{\mathbf{x}_j\}_1^{k_1}$ and $A_2 = \{\mathbf{y}_j\}_1^{k_2}$ be two different bases in a linear space $\mathfrak{L}$. The number of vectors in each basis must be identical, that is, $k_1 = k_2$.†

*Proof*: If $k_1 \neq k_2$, let $k_2 > k_1$. Since $A_1$ is a basis, we have

$$\mathbf{y}_i = \sum_{j=1}^{k_1} \alpha_{i,j}\mathbf{x}_j, \qquad i = 1, 2, \ldots, k_2$$

The coefficients $\alpha_{i,j}$ $(i = 1, 2, \ldots, k_2; j = 1, 2, \ldots, k_1)$ form a $k_2 \times k_1$ matrix

$$\begin{bmatrix} \alpha_{1,1} & \alpha_{1,2} & \cdots & \alpha_{1,k_1} \\ \alpha_{2,1} & \alpha_{2,2} & \cdots & \alpha_{2,k_1} \\ \vdots & & & \\ \alpha_{k_1,1} & \alpha_{k_1,2} & \cdots & \alpha_{k_1,k_1} \\ \vdots & & & \\ \alpha_{k_1+r,1} & \alpha_{k_1+r,2} & \cdots & \alpha_{k_1+r,k_1} \\ \vdots & & & \\ \alpha_{k_2,1} & \alpha_{k_2,2} & \cdots & \alpha_{k_2,k_1} \end{bmatrix}$$

whose rank is $k_1$. From matrix theory we may write any of the rows below the $k_1$th row, say the $(k_1 + r)$th row as a linear combination of the top $k_1$ rows.‡ Thus

$$\alpha_{k_1+r,s} = \sum_{j=1}^{k_1} \beta_{r,j}\alpha_{j,s}, \qquad s = 1, 2, \ldots, k_1$$

Now

$$\mathbf{y}_{k_1+r} = \sum_{j=1}^{k_1} \alpha_{k_1+r,s}\mathbf{x}_s$$

On combining the two aforementioned equations, we get

$$\mathbf{y}_{k_1+r} = \sum_{j=1}^{k_1} \beta_{r,j}\mathbf{y}_j$$

which is a contradiction. Thus we have $k_1 \geq k_2$. Similarly, applying the same arguments to the sets $A_1$ and $A_2$ with their roles reversed, we obtain $k_1 \leq k_2$. Thus $k_1 = k_2$. ▌

†$k_1$ and $k_2$ are called the cardinal numbers of sets $A_1$ and $A_2$, respectively.

‡See, for example, *Mathematical Methods in Chemical Engineering: Matrices and Their Applications*, by N. R. Amundson, Prentice-Hall, Englewood Cliffs, N.J., 1964.

The components of any vector with respect to a given basis $\{\mathbf{x}_j\}_1^k$ in $\mathcal{L}$ are readily shown to be unique, for if they are not, let

$$\mathbf{x} = \sum_{j=1}^{k} \alpha_j \mathbf{x}_j \quad \text{and} \quad \mathbf{x} = \sum_{j=1}^{k} \beta_j \mathbf{x}_j$$

On subtraction of one from the other, it is readily seen that

$$\mathbf{0} = \sum_{j=1}^{k} (\alpha_j - \beta_j)\mathbf{x}_j$$

The linear independence of $\{\mathbf{x}_j\}_1^k$ implies that $\alpha_j = \beta_j, j = 1, 2, \ldots, k$.

## EXERCISES

**2.3.1** Show that

$$\mathbf{e}_1 = \begin{bmatrix} 1 \\ 0 \\ \cdot \\ \cdot \\ \cdot \\ 0 \end{bmatrix}, \quad \mathbf{e}_2 = \begin{bmatrix} 0 \\ 1 \\ \cdot \\ \cdot \\ \cdot \\ 0 \end{bmatrix}, \quad \ldots, \quad \mathbf{e}_n = \begin{bmatrix} 0 \\ 0 \\ \cdot \\ \cdot \\ \cdot \\ 1 \end{bmatrix}$$

form a basis in $\mathcal{R}_n$.

**2.3.2** Show that any set of $n$ linearly independent vectors is a basis in $\mathcal{R}_n$.

We have so far loosely used the word "dimension" in referring to "$n$-dimensional spaces" such as $\mathcal{R}_n$ and $\mathcal{C}_n$ and "infinite-dimensional spaces" such as $\mathcal{C}[a, b]$. In this section we provide a definition of the dimension of a linear space. There are several ways of definiing dimensions which are equivalent.

Theorem 2.3.1 suggests the following definition.

***Definition 2.3.2*** The *dimension* of a linear space is the number of vectors in any basis. Thus linear spaces which have no finite basis are referred to as infinite-dimensional spaces.

We shall now prove the following theorem.

***Theorem 2.3.2*** Let $\mathcal{L}$ be a linear space of dimension $n$. Every set of $(n + 1)$ vectors is linearly dependent.

*Proof*: Let $(\mathbf{e}_1, \mathbf{e}_2, \ldots, \mathbf{e}_n)$ be a basis in $\mathcal{L}$ (since the dimension is $n$). Consider the set of vectors $\{\mathbf{y}_j\}_1^{n+1}$, each of which may be expressed in terms of the given basis set.

$$\mathbf{y}_j = \alpha_{j,1}\mathbf{e}_1 + \alpha_{j,2}\mathbf{e}_2 + \ldots + \alpha_{j,n}\mathbf{e}_n, \qquad j = 1, 2, \ldots, n + 1$$

The proof is now completely analogous to that of Theorem 2.3.1. Thus at least one of the rows of the matrix

$$\begin{bmatrix} \alpha_{1,1} & \alpha_{1,2} & \cdots & \alpha_{1,n} \\ \alpha_{2,1} & \alpha_{2,2} & \cdots & \alpha_{2,n} \\ \cdot & & & \\ \cdot & & & \\ \cdot & & & \\ \alpha_{n+1,1} & \alpha_{n+1,2} & \cdots & \alpha_{n+1,n} \end{bmatrix}$$

is expressible as a linear combination of the remaining $n$ rows. Thus $\{\mathbf{y}_j\}_1^{n+1}$ is linearly dependent. ▮

A corollary of Theorem 2.3.2 is that every linearly independent set of $n$ vectors in an $n$-dimensional linear space is a basis, the proof of which is immediately evident, for let $\{\mathbf{x}_j\}_1^n$ be a linearly independent set and further let $\mathbf{x}$ be any arbitrary element of the linear space. From Theorem 2.3.2 the set $(\mathbf{x}, \mathbf{x}_1, \mathbf{x}_2, \ldots, \mathbf{x}_n)$ is linearly dependent so that $\mathbf{x}$ is expressible as a linear combination of $\mathbf{x}_1, \mathbf{x}_2, \ldots, \mathbf{x}_n$ (see Exercise 2.2.1). Thus $\{\mathbf{x}_j\}_1^n$ is a basis.

**EXERCISES**

**2.3.3** Show that the class of polynomials of degree up to $n$ forms a linear space of dimension $n + 1$. Find a suitable basis for this space.

## 2.4 Linear Subspaces

The subspace is an important concept of mathematical analysis, and is defined as follows.

***Definition 2.4.1*** A *linear subspace* (or simply *subspace*) $L$ of a linear space $\mathcal{L}$ is a subset of $\mathcal{L}$ such that $\mathbf{x} \in L, \mathbf{y} \in L \rightarrow (\mathbf{x} + \mathbf{y}) \in L$ and $\alpha\mathbf{x} \in L, \forall \alpha \in \mathfrak{F}$.

Clearly, $\mathbf{0}$ is an element of every subspace. Thus $L$ is itself a linear space. Also, the set consisting only of $\mathbf{0}$, denoted $\{\mathbf{0}\}$, is itself a subspace; in fact, it is the smallest subspace of $\mathcal{L}$. The largest subspace of $\mathcal{L}$ is $\mathcal{L}$ itself. A subspace that is a proper subset of the linear space is called a proper subspace. Clearly, $\{\mathbf{0}\}$ is a proper subspace of $\mathcal{L}$, but $\mathcal{L}$ itself is not.

*Examples of Subspaces*

$\mathcal{R}_n$: The sets $L_i \equiv \{\mathbf{x} \equiv (x_1, x_2, \ldots, x_n) \in \mathcal{R}_n : x_i = 0\}, i = 1, 2, \ldots, n$, are all subspaces of $\mathcal{R}_n$. This is easily established from the definition. (Note that in ordinary three-dimensional space the subspaces above are the three mutually perpendicular planes whose normals are the coordinate axes. Of course, a similar interpretation also holds for $\mathcal{C}_n$). In general, any set defined by

$$L \equiv \{\mathbf{x} \in \mathcal{R}_n: \alpha_1 x_1 + \alpha_2 \mathbf{x}_2 + \ldots + \alpha_n \mathbf{x}_n = 0\}$$

is a subspace of $\mathcal{R}_n$.

$\mathcal{C}[a, b]$: Since all differentiable functions must necessarily be continuous, $\mathcal{C}^{(1)}[a, b]$ is a subspace of $\mathcal{C}[a, b]$. Furthermore, since continuity does not necessarily imply differentiability, $\mathcal{C}^{(1)}[a, b]$ is a proper subspace of $\mathcal{C}[a, b]$.

One may also define subspaces in $\mathcal{C}[a, b]$ such as

$$L = \{\mathbf{f} \in \mathcal{C}[a, b]: \alpha f(a) + \beta f(b) = 0\}$$

where $\alpha$ and $\beta$ are elements of the field associated with $\mathcal{C}[a, b]$.

The set of all polynomials of degree at most $n$ is a finite-dimensional subspace of $\mathcal{C}[a, b]$.

It is obvious that the dimension of a subspace $L$ of a linear space $\mathcal{L}$ is at most equal to the dimension of $\mathcal{L}$. The proper subspaces of an infinite-dimensional linear space may be finite- or infinite-dimensional. Thus the examples of subspaces of $\mathcal{C}[a, b]$ in the preceding discussion covered both finite- and infinite-dimensional subspaces. The following theorem is of interest.

***Theorem 2.4.1*** Let $L$ be an $r$-dimensional subspace of an $n$-dimensional linear space $\mathcal{L}$ and let $\{\mathbf{x}_j\}_1^r$ be a basis in $L$. Then there exist vectors $\mathbf{x}_{r+1}, \mathbf{x}_{r+2}, \ldots, \mathbf{x}_n \in \mathcal{L} \ni \{\mathbf{x}_j\}_1^n$ forms a basis in $\mathcal{L}$.

*Proof*: Evidently, $r < n$. Hence $\{\mathbf{x}_j\}_1^r$ is not a basis of $\mathcal{L}$, so that there exists a vector $\mathbf{x}_{r+1} \in \mathcal{L}$ which is *not* expressible as a linear combination of $(\mathbf{x}_1, \mathbf{x}_2, \ldots, \mathbf{x}_r)$. (Obviously, $\mathbf{x}_{r+1}$ is not in $L$.) Thus $\{\mathbf{x}_j\}_1^{r+1}$ is a linearly independent set (why?). This is *not* a basis in $\mathcal{L}$ unless $r + 1 = n$. Using the same argument progressively, one finds vectors $\mathbf{x}_{r+2}, \mathbf{x}_{r+3}, \ldots, \mathbf{x}_n$ so that ultimately the linearly independent set $\{\mathbf{x}_j\}_1^n$ forms a basis in $\mathcal{L}$. ▮

We now define the *algebraic sum* of two subspaces. Let $L_1$ and $L_2$ be subspaces of linear space $\mathcal{L}$.

***Definition 2.4.2*** Let $L_1$ and $L_2$ be subspaces of linear space $\mathcal{L}$. The *algebraic sum* of $L_1$ and $L_2$, denoted $L_1 + L_2$, is the set of all elements of $\mathcal{L}$ obtained by summing elements of $L_1$ and $L_2$. Notationally,

$$L_1 + L_2 = \{\mathbf{z} = \mathbf{x}_1 + \mathbf{x}_2: \mathbf{x}_1 \in L_1, \mathbf{x}_2 \in L_2\}$$

It is a trivial exercise to show that $L_1 + L_2$ is a linear space, that is, a subspace of $\mathcal{L}$.

Given two linear subspaces $L_1$ and $L_2$, the intersection $L_1 \cap L_2$ consists of elements each of which is both in $L_1$ and $L_2$, and can be shown to be a linear subspace of $\mathfrak{L}$; for let $\mathbf{x}, \mathbf{y} \in L_1 \cap L_2$. Then $\mathbf{x} \in L_1, \mathbf{y} \in L_1$, so that $\mathbf{x} + \mathbf{y} \in L_1$.

Similarly, $\mathbf{x} \in L_2, \mathbf{y} \in L_2$, so that $\mathbf{x} + \mathbf{y} \in L_2$. Thus $\mathbf{x} + \mathbf{y} \in L_1 \cap L_2$. Also, $\forall\, \alpha \in \mathfrak{F}$, $\alpha\mathbf{x} \in L_1$ (since $\mathbf{x} \in L_1$) and $\alpha\mathbf{x} \in L_2$ (since $\mathbf{x} \in L_2$). Hence $\alpha\mathbf{x} \in L_1 \cap L_2$. Thus $L_1 \cap L_2$ is a linear subspace of $\mathfrak{L}$.

It is important to note that the union $L_1 \cup L_2$ of two subspaces $L_1$ and $L_2$ is *not* a linear subspace, whereas the algebraic sum $L_1 + L_2$ is a linear subspace. The equivalence of notation used in Chapter 0 between the symbols "$\cup$" and "$+$" does *not* extend to this chapter.

*Examples*

$\mathfrak{R}_3$: The algebraic sum of two subspaces is exemplified by the set of vectors obtained by summing vectors along, say, the $x$-axis and the $y$-axis. The vectors resulting from this sum confine to the $x - y$ plane, which is the algebraic sum of the subspaces represented by the $x$ and the $y$ axes.

$\mathfrak{C}[a, b]$: The algebraic sum of the subspaces of $r$th or lower-degree polynomials and $s$th or lower-degree polynomials will yield a subspace of polynomials of degree $\leq \max(r, s)$. The proof of this is self-evident. Finally, we define the complement of a subspace.

***Definition 2.4.3*** Let $L$ be a subspace of a linear space $\mathfrak{L}$. The *complement* of $L$, denoted by $L'$, is defined as the set of elements in $\mathfrak{L}$ that are not in $L$. Notationally,

$$L' = \{\mathbf{x} \in \mathfrak{L}: \mathbf{x} \notin L\}$$

Since $\mathbf{0} \in L, \mathbf{0} \notin L'$, so that $L'$ is not a linear space.

**EXERCISES**

**2.4.1** Show that if $L_1$ is an $r_1$-dimensional subspace of $\mathfrak{L}$ and $L_2$ is an $r_2$-dimensional subspace of $\mathfrak{L}$, then $L_1 + L_2$ is a subspace of dimension $(r_1 + r_2 - r)$, where $r$ is the dimension of $L_1 \cap L_2$. (*Hint:* Theorem 2.4.1 may be of use.)

**2.4.2** Let $\mathbf{A}$ be an $m \times n$ matrix of rank $r < m < n$. Show that the solution of the homogeneous equation $\mathbf{Ax} = \mathbf{0}$ forms a subspace of $\mathfrak{R}_n$. What is the dimension of this subspace?

## 2.5 Linear Manifolds. Hyperplanes

A linear manifold is a subspace of a linear space generated by *all* linear combinations of a given set of vectors in $\mathfrak{L}$. Thus if $(\mathbf{x}_1, \mathbf{x}_2, \ldots, \mathbf{x}_n)$ represents a set of vectors in $\mathfrak{L}$, the *linear manifold spanned* by these vectors is denoted

$L(\mathbf{x}_1, \mathbf{x}_2, \ldots, \mathbf{x}_n)$. Clearly, the linear manifold is a subspace of $\mathcal{L}$. In fact, it is the *smallest* subspace containing the generating vectors $\mathbf{x}_1, \mathbf{x}_2, \ldots, \mathbf{x}_n$. Note that the generating set is not necessarily linearly independent, so that the manifold may be of a dimension smaller than $n$. The reader should be satisfied that if the generating set is linearly independent, then it is a basis for the generated manifold.

*Examples*

$\mathcal{R}_3$: Consider any two vectors in $\mathcal{R}_3$. The linear manifold spanned by these two vectors is the plane containing the two vectors.

$\mathcal{C}[a, b]$: The linear manifold spanned by $\{\mathbf{x}_j\}_0^k$, where $x_j(t) = t^j$, represents the class of all polynomials of degree $\leq k$.

We now define a hyperplane. In $\mathcal{R}_3$, a plane containing the origin is a subspace. If $\mathbf{x} = (x_1, x_2, x_3)$ is a point on this plane, the coordinates will satisfy an equation of the form

$$\alpha_1 x_1 + \alpha_2 x_2 + \alpha_3 x_3 = 0$$

A plane parallel to this plane but not passing through the origin consists of vectors $\mathbf{y} = (y_1, y_2, y_3)$ such that

$$\alpha_1 y_1 + \alpha_2 y_2 + \alpha_3 y_3 = \gamma \tag{2.5.1}$$

Let $\mathbf{y}_0 = (y_{1,0}, y_{2,0}, y_{3,0})$ be a particular vector belonging to the plane not passing through the origin. Thus

$$\alpha_1 y_{1,0} + \alpha_2 y_{2,0} + \alpha_3 y_{3,0} = \gamma \tag{2.5.2}$$

Subtracting (2.5.2) from (2.5.1), we obtain

$$\alpha_1(y_1 - y_{1,0}) + \alpha_2(y_2 - y_{2,0}) + \alpha_3(y_3 - y_{3,0}) = 0$$

which implies that the vector $(\mathbf{y} - \mathbf{y}_0)$ belongs to the plane passing through the origin. Thus all points on the plane away from the origin are described by vectors $\mathbf{y}$ such that

$$\mathbf{y} = \mathbf{y}_0 + \mathbf{x}$$

where $\mathbf{x}$ lies on the plane through the origin. This provides the motivation for the definition of a hyperplane in an abstract linear space.

***Definition 2.5.1*** Let $L$ be a subspace of a linear space $\mathcal{L}$ and let $\mathbf{y}_0 \in \mathcal{L}$ not necessarily in $L$. Then a *hyperplane* $H$ is associated with the subspace $L$ and $\mathbf{y}_0$ defined by

$$H = \{\mathbf{y}\colon\ \mathbf{y} = \mathbf{x} + \mathbf{y}_0;\ \mathbf{x} \in L\}$$

Clearly, $H$ is not a subspace in general.

**EXERCISES**

**2.5.1** Show that $L(x_1, x_2, \ldots, x_n)$ is the *smallest* subspace containing the vectors $\mathbf{x}_1, \mathbf{x}_2, \ldots, \mathbf{x}_n$.

**2.5.2** In an $n$-component reaction mixture, the composition vector $\mathbf{y} = (y_1, y_2, \ldots, y_n)$, where $y_j =$ mass fraction of $j$th species, describes the state of the system at any time. Show that $\mathbf{y}$ lies on a hyperplane associated with the initial composition vector $\mathbf{y}_0$ and the subspace represented by the $(n-1)$-dimensional linear manifold $L(\mathbf{z}_1, \mathbf{z}_2, \ldots, \mathbf{z}_{n-1})$, where

$$\mathbf{z}_1 = (1, -1, 0, \ldots, 0), \quad \mathbf{z}_2 = (1, 0, -1, 0, \ldots, 0), \quad \ldots,$$
$$\mathbf{z}_{n-1} = (1, 0, \ldots, 0, -1)$$

## 2.6 Linear and Bilinear Functionals. Quadratic Forms.

A functional maps elements of a linear space $\mathcal{L}$ into its associated field $\mathcal{F}$. A *linear functional* $l: \mathcal{L} \longrightarrow \mathcal{F}$ has the properties

$$l(\mathbf{x} + \mathbf{y}) = l(\mathbf{x}) + l(\mathbf{y}) \qquad \forall \ \mathbf{x}, \mathbf{y} \in \mathcal{L}$$
$$l(\alpha\mathbf{x}) = \alpha l(\mathbf{x}) \qquad \forall \ \mathbf{x} \in \mathcal{L}, \quad \alpha \in \mathcal{F}$$

The second of the relations above, when applied to $\alpha = 0$, yields $l(\mathbf{0}) = 0$. However, $l(\mathbf{x}) = 0$ does not necessarily imply that $\mathbf{x} = 0$. In fact, the set

$$S = \{\mathbf{x} \in \mathcal{L}: l(\mathbf{x}) = 0\}$$

forms a linear subspace of $\mathcal{L}$, for let $\mathbf{x}, \mathbf{y} \in S$. $l(\mathbf{x} + \mathbf{y}) = l(\mathbf{x}) + l(\mathbf{y}) = 0 \rightarrow \mathbf{x} + \mathbf{y} \in S$. Similarly, $l(\alpha\mathbf{x}) = \alpha l(\mathbf{x}) = 0 \rightarrow \alpha\mathbf{x} \in S$, so that $S$ is a linear subspace of $\mathcal{L}$. It is usual to denote $S$ by $N(l)$. We now consider some examples of linear functionals.

*Examples*

$\mathcal{R}_n$: Define for every $\mathbf{x} = (x_1, x_2, \ldots, x_n) \in \mathcal{R}_n$.

$$l(\mathbf{x}) = \sum_{j=1}^{n} \alpha_j x_j, \qquad \alpha_j \in \mathcal{R}, \quad j = 1, 2, \ldots, n$$

Clearly, $l$ is a linear functional mapping $\mathcal{R}_n$ into $\mathcal{R}$.

$\mathcal{C}[a, b]$: For every $\mathbf{f} \equiv \{f(t)\} \in \mathcal{C}[a, b]$, define

$$l(\mathbf{f}) \equiv f(t_0)$$

where $t_0$ is a fixed element of $[a, b]$. It is obvious that $l$ is a linear functional.

A second functional on $\mathcal{C}[a, b]$ is given by

$$l(\mathbf{f}) \equiv \int_a^b g(t) f(t)\, dt$$

where $\{g(t)\}$ is a fixed element of $\mathcal{C}[a, b]$.

$\mathcal{C}^{(1)}[a, b]$: An example of a linear functional on this space is given by

$$l(\mathbf{f}) \equiv \alpha_1 f(a) + \alpha_2 f'(a) + \alpha_3 f(b) + \alpha_4 f'(b)$$

where $\alpha_1$, $\alpha_2$, $\alpha_3$, and $\alpha_4$ are elements of the associated field of $\mathcal{C}^{(1)}[a, b]$, and the prime denotes differentiation of $f(t)$ with respect to $t$.

We now define a *bilinear functional* as a mapping of ordered pairs of elements of $\mathcal{L}$ into the associated field $\mathcal{F}$, generally the set of complex numbers; that is,

$$b\colon (\mathcal{L} \times \mathcal{L}) \to \mathcal{C}$$

such that $b(\mathbf{x}, \mathbf{y})$ is a linear functional of $\mathbf{x}$ for every fixed $\mathbf{y} \in \mathcal{L}$; that is,

$$b(\mathbf{x}_1 + \mathbf{x}_2, \mathbf{y}) = b(\mathbf{x}_1, \mathbf{y}) + b(\mathbf{x}_2, \mathbf{y});\ b(\alpha\mathbf{x}, \mathbf{y}) = \alpha b(\mathbf{x}, \mathbf{y})$$

and for every fixed $\mathbf{x} \in \mathcal{L}$, $b(\mathbf{x}, \mathbf{y})$ satisfies

$$b(\mathbf{x}, \mathbf{y}_1 + \mathbf{y}_2) = b(\mathbf{x}, \mathbf{y}_1) + b(\mathbf{x}, \mathbf{y}_2); \qquad b(\mathbf{x}, \alpha\mathbf{y}) = \alpha^* b(\mathbf{x}, \mathbf{y})$$

If, further, $b(\mathbf{x}, \mathbf{y}) = b^*(\mathbf{y}, \mathbf{x})$ the bilinear functional is said to be *symmetric*; then the quantity $b(\mathbf{x}, \mathbf{x})$ is clearly real-valued and is called a *quadratic form* in $\mathbf{x}$, denoted $q(\mathbf{x})$. Indeed $q$ is a functional but clearly not a linear functional. For $\alpha \in \mathcal{C}$, we have

$$q(\alpha\mathbf{x}) = b(\alpha\mathbf{x}, \alpha\mathbf{x}) = |\alpha|^2 q(\mathbf{x})$$

Evidently $q(\mathbf{0}) = 0$. A quadratic form $q(\mathbf{x})$ is said to be nonnegative† (*nonpositive*) if $\forall\, \mathbf{x} \in \mathcal{L}$, $q(\mathbf{x}) \geq 0$ $[q(\mathbf{x}) \leq 0]$. It is said to be *positive* (*negative*) if for $\mathbf{x} \neq \mathbf{0}$, $q(\mathbf{x}) > 0$ $[q(\mathbf{x}) < 0]$.

We now consider some examples of bilinear functionals and the quadratic forms generated by them. The associated field is real in each case.

*Examples*

$\mathcal{R}_n$: Define for every $\mathbf{x} \equiv (x_1, x_2, \ldots, x_n) \in \mathcal{R}_n$ the bilinear functional

$$b(\mathbf{x}, \mathbf{y}) \equiv \sum_{i=1}^{n} \sum_{j=1}^{n} a_{ij} x_i y_j$$

where $\{a_{ij}\}$ are real numbers. It is easily verified that $b(\mathbf{x}, \mathbf{y})$ is a bilinear functional; further, the quadratic form generated by it is

$$q(\mathbf{x}) = \sum_{i=1}^{n} \sum_{j=1}^{n} a_{ij} x_i x_j$$

†Frequently, the term "positive definite" is used to describe a nonnegative functional and "strictly positive definite" is used for a positive functional. When $q$ can assume both positive and negative values, it is said to be indefinite.

$\mathcal{C}[0, 1]$: For each pair $\mathbf{f}$, $\mathbf{g} \in \mathcal{C}[0, 1]$, where $\mathbf{f} \equiv \{f(t)\}$ and $\mathbf{g} \equiv \{g(t)\}$, define

$$b(\mathbf{f}, \mathbf{g}) = \int_0^1 \int_0^1 K(t, s) f(t) g(s)\, dt\, ds$$

where $K(t, s)$ is a continuous function of $s$ and $t$ in $[0, 1] \times [0, 1]$. All functions are assumed to be real-valued. It is left to the reader to verify that $b$ is indeed a bilinear functional.

**EXERCISES**

**2.6.1** Let $l\colon \mathcal{L} \longrightarrow \mathfrak{F}$ be a linear functional and $\mathbf{x}_0 \in \mathcal{L}$ be a fixed vector. Show that the set $S$ defined by

$$S = \{\mathbf{x} \in \mathcal{L}\colon l(\mathbf{x}) = l(\mathbf{x}_0)\}$$

is a hyperplane. What is the subspace associated with this hyperplane?

**2.6.2** Boundary conditions for second-order differential equations that involve functions $\{y(t), a \leq t \leq b\}$ are often expressed as

$$\alpha_{11} y(a) + \alpha_{12} y'(a) + \alpha_{13} y(b) + \alpha_{14} y'(b) = \gamma_1$$
$$\alpha_{21} y(a) + \alpha_{22} y'(a) + \alpha_{23} y(b) + \alpha_{24} y'(b) = \gamma_2$$

where the matrix of coefficients has rank 2. Show that the boundary conditions above define a set of vectors $S \subset \mathcal{C}^{(1)}[a, b]$, which is the intersection of two hyperplanes in $\mathcal{C}^{(1)}[a, b]$, by identifying the hyperplanes.

**2.6.3** Let $l\colon \mathcal{L} \longrightarrow \mathfrak{F}$ be a linear functional. Which of the following are bilinear functionals?

(a) $l(\mathbf{x}) + l(\mathbf{y})$.

(b) $l(\mathbf{x} + \mathbf{y})$.

(c) $l(\mathbf{x}) l(\mathbf{y})$.

**2.6.4** Show that

$$b(\mathbf{u}, \mathbf{v}) \equiv \int_0^1 u''(x) v(x)\, dx$$

is a bilinear functional on $\mathcal{C}^{(2)}[0, 1]$. Show that the set

$$S = \{\mathbf{f} \in \mathcal{C}^{(2)}[0, 1]\colon f'(0) = \alpha f(0), f'(1) + \beta f(1) = 0\}$$

where $\alpha$ and $\beta$ are positive real numbers, is a subspace of $\mathcal{C}^{(2)}[0, 1]$, and further establish that the foregoing bilinear functional (a) is *symmetric* in $S$, that is, $b(\mathbf{u}, \mathbf{v}) = b(\mathbf{v}, \mathbf{u})$, and (b) generates a negative quadratic form in $S$.

## 2.7 Linear Transformations or Operators

The concept of a linear transformation or linear operator is most vital to this book. A linear transformation maps elements of a linear space into elements of the same space. In symbols we write

$$\mathbf{T}\colon \mathcal{L} \rightarrow \mathcal{L}$$

$\mathbf{T}$ is called a *linear transformation* or a *linear operator* if it further satisfies the conditions

$$\mathbf{T}(\mathbf{x}+\mathbf{y}) = \mathbf{T}(\mathbf{x}) + \mathbf{T}(\mathbf{y}) \qquad \forall \quad \mathbf{x}, \mathbf{y} \in \mathfrak{L}$$
$$\mathbf{T}(\alpha\mathbf{x}) = \alpha\mathbf{T}(\mathbf{x}) \qquad \forall \quad \alpha \in \mathfrak{F}, \mathbf{x} \in \mathfrak{L}$$

The definitions above imply that $\mathbf{T}(\mathbf{0}) = \mathbf{0}$. We consider below some examples of linear transformations.†

*Example 1*

The *zero operator* transforms every element of $\mathfrak{L}$ into the zero element $\mathbf{0}$ of $\mathfrak{L}$; that is,

$$\mathbf{0x} = \mathbf{0} \qquad \forall \quad \mathbf{x} \in \mathfrak{L}$$

*Example 2*

The identity operator transforms every element of $\mathfrak{L}$ into itself. Thus

$$\mathbf{Ix} = \mathbf{x} \qquad \forall \quad \mathbf{x} \in \mathfrak{L}$$

One may distinguish a *similarity operator* from the identity operator by defining

$$\mathbf{Sx} = \sigma\mathbf{x} \qquad \forall \quad \mathbf{x} \in \mathfrak{L}$$

where $\sigma$ is a fixed element of the field $\mathfrak{F}$.

*Example 3*

Consider the space $\mathfrak{C}[a, b]$ and the operator $\mathbf{K}$ such that

$$g(t) = \int_a^b K(t, s)f(s)\,ds, \qquad a \leq t \leq b$$

which in operator notation may be written as

$$\mathbf{g} = \mathbf{Kf}$$

The function $K(t, s)$ is called a *kernel* and is assumed to be continuous in $[a, b] \times [a, b]$ so that the function $g(t)$ is in $\mathfrak{C}[a, b]$. Thus $\mathbf{K}$ transforms elements of $\mathfrak{C}[a, b]$ into elements of $\mathfrak{C}[a, b]$.

*Example 4*

Often the transformation may be such that only selected elements of the linear space may be operated on.‡ This is especially true of operators on

†In representing the action of a linear transformation $\mathbf{T}$ on $\mathbf{x}$, we write $\mathbf{Tx}$ instead of the more correct $\mathbf{T}(\mathbf{x})$. The parentheses will be used only when vectorial sums such as $\mathbf{x}+\mathbf{y}$ or scalar multiples such as $\alpha\mathbf{x}$ are involved as arguments of $\mathbf{T}$.

‡The set to which operation of linear transformation $\mathbf{T}$ is restricted is called the *domain* of $\mathbf{T}$.

infinite-dimensional spaces. Thus consider the differential operator $\mathbf{T} \equiv d/dt$ acting on elements in $\mathcal{C}^{(1)}[a, b]$. Obviously, only those elements that are differentiable can be used for this transformation. In order that the differentiated function be in $\mathcal{C}[a, b]$, only functions with continuous derivatives may be transformed.

*Example 5*

The *projection operator* is an important generalization of the geometrical projection of physical vectors in $\mathcal{R}_3$ along any of the three coordinate axes. The projection operator plays a very important role in analysis. It is defined as a *linear* operator $\mathbf{P}$ such that

$$\mathbf{P}(\mathbf{Px}) = \mathbf{Px} \qquad \forall \quad \mathbf{x} \in \mathcal{L}$$

There are several examples of projection operators which we shall encounter in later stages; for the present, we cite the obvious generalization to $\mathcal{R}_n$, by defining

$$\mathbf{Px} = (x_1, x_2, \ldots, x_m, 0, \ldots, 0) \qquad (m < n)$$

where $\mathbf{x} = (x_1, x_2, \ldots, x_n) \in \mathcal{R}_n$. That $\mathbf{P}$ is indeed a projection follows from the fact that

$$\begin{aligned}\mathbf{P}(\mathbf{Px}) &= (x_1, x_2, \ldots, x_m, 0, \ldots, 0)\\ &= \mathbf{Px}\end{aligned}$$

**EXERCISES**

**2.7.1** Let $\mathcal{L}$ be a linear space of dimension $n$, and $\{\mathbf{e}_j\}_1^n$ be a basis in $\mathcal{L}$. Define an operator $\mathbf{T}$ such that $\forall\ \mathbf{x} \in \mathcal{L}$,

$$\mathbf{Tx} = \sum_{j=1}^{n} \lambda_j \alpha_j \mathbf{e}_j$$

where $\alpha_j \in \mathcal{F}, j = 1, 2, \ldots, n$, are the components of $\mathbf{x}$ with respect to the basis set $\{\mathbf{e}_j\}_1^n$; that is,

$$\mathbf{x} = \sum_{j=1}^{n} \alpha_j \mathbf{e}_j$$

and $\lambda_j$'s are fixed constants belonging to $\mathcal{F}$. Show that $\mathbf{T}$ is a linear transformation. (Note that $\mathbf{T}$ is called a *diagonal operator*.)

**2.7.2** Obtain a projection operator on an $n$-dimensional linear space with the basis set $\{\mathbf{e}_j\}_1^n$.

## 2.8 Null and Range Spaces of a Linear Transformation

We observed in the preceding section that if $\mathbf{T}$ is a linear operator acting on elements of a linear space $\mathcal{L}$, then $\mathbf{T0} = \mathbf{0}$. However, $\mathbf{Tx} = \mathbf{0}$ does not necessarily imply that $\mathbf{x} = \mathbf{0}$. The set of vectors $\mathbf{x}$ that satisfy the condition $\mathbf{Tx} = \mathbf{0}$

is seen to be a linear space, a subspace of $\mathfrak{L}$. To prove this we denote the set by $S$; that is,

$$S = \{\mathbf{x} \in \mathfrak{L}: \mathbf{Tx} = \mathbf{0}\}$$

Let $\mathbf{x}, \mathbf{y} \in S$. Then $\mathbf{T}(\mathbf{x} + \mathbf{y}) = \mathbf{Tx} + \mathbf{Ty} = \mathbf{0} + \mathbf{0} = \mathbf{0}$, which implies that $(\mathbf{x} + \mathbf{y}) \in S$. Similarly, let $\alpha \in \mathfrak{F}$. $\mathbf{T}(\alpha\mathbf{x}) = \alpha\mathbf{Tx} = \mathbf{0}$, so that $\alpha\mathbf{x} \in S$. Thus $S$ is a subspace of $\mathfrak{L}$ and it is called the *null space* of the operator $\mathbf{T}$, denoted $N(\mathbf{T})$.

We now consider the set of elements obtained by transforming all elements of $\mathfrak{L}$ by the linear transformation $\mathbf{T}$, and denote this set by $R(\mathbf{T})$. This set is called the range space of the operator. Clearly, $\mathbf{y} \in R(\mathbf{T}) \rightarrow \exists\ \mathbf{x} \in \mathfrak{L} \ni \mathbf{Tx} = \mathbf{y}$. That $R(\mathbf{T})$ is a linear space emerges from what follows. Let

$$\mathbf{y}_1, \mathbf{y}_2 \in R(\mathbf{T});\ \exists\ \mathbf{x}_1, \mathbf{x}_2 \in \mathfrak{L} \ni \mathbf{y}_1 = \mathbf{Tx}_1$$

and $\mathbf{y}_2 = \mathbf{Tx}_2$. Then

$$\mathbf{y}_1 + \mathbf{y}_2 = \mathbf{Tx}_1 + \mathbf{Tx}_2 = \mathbf{T}(\mathbf{x}_1 + \mathbf{x}_2)$$

Since

$$\mathbf{x}_1 + \mathbf{x}_2 \in \mathfrak{L}, \qquad \mathbf{y}_1 + \mathbf{y}_2 \in R(\mathbf{T})$$

Again let $\alpha \in \mathfrak{F}$, $\mathbf{y} \in R(\mathbf{T})$, and $\mathbf{y} = \mathbf{Tx}$. It follows that $\alpha\mathbf{y} = \mathbf{T}(\alpha\mathbf{x})$ and since $\alpha\mathbf{x} \in \mathfrak{L}$, $\alpha\mathbf{y} \in R(\mathbf{T})$. Thus $R(\mathbf{T})$ is a subspace of $\mathfrak{L}$, so that it is referred to as the *range space* of the operator $\mathbf{T}$.

It was observed earlier (see Example 4 of Section 2.7) that in infinite-dimensional spaces some operators may be constrained to act on some subspace $D(\mathbf{T})$ of $\mathfrak{L}$, called the domain space. In this case $\mathbf{T}$ maps $D(\mathbf{T})$ onto $R(\mathbf{T})$. We examine below the null and range spaces of the operators considered in the preceding section.

*Example 1*

The zero operator has the entire space $\mathfrak{L}$ as its null space. The range space consists only of the zero element. Thus the intersection of the two subspaces contains only the zero vector.

*Example 2*

The identity operator has for its null space $\{\mathbf{0}\}$, which contains only the zero element, while the range space is the entire linear space. Again the intersection of the two spaces comprises only the zero vector. These statements also hold for the similarity operator.

*Example 3*

The null space of the integral operator $\mathbf{K}$ (see Example 3 of Section 2.7) consists of the linear manifold spanned by the linearly independent solutions of the homogeneous integral equation

$$\int_a^b K(t, s)f(s)\,ds = 0, \qquad a \leq t \leq b$$

Unless the form of the kernel $K(t, s)$ is specified, further identification of the null space or the range space is not possible. The kernel $K(t, s)$ is called *degenerate* when it is of the form

$$K(t, s) = \sum_{j=1}^{n} p_j(t)q_j(s), \qquad a \leq t, \quad s \leq b$$

where $\{p_j\}_1^n$ and $\{q_j\}_1^n$ are independent sets contained in $\mathcal{C}[a, b]$. The homogeneous integral equation then becomes

$$\sum_{j=1}^{n} \alpha_j p_j(t) = 0$$

where $\alpha_j = \int_a^b q_j(s)f(s)\,ds$. The linear independence of the set $\{\mathbf{p}_j\}_1^n$ implies that

$$\int_a^b q_j(s)f(s)\,ds = 0, \qquad j = 1, 2, \ldots, n$$

each of which represents a subspace of $\mathcal{C}[a, b]$ mapped by the corresponding linear functional

$$l_j(\mathbf{f}) = \int_a^b q_j(s)f(s)\,ds$$

into the zero element of the associated field $\mathfrak{F}$. Thus the null space of $\mathbf{K}$ is the intersection of the subspaces $L_j = \{\mathbf{f}: l_j(\mathbf{f}) = 0\}, j = 1, 2, \ldots, n$.† To identify the range space of $\mathbf{K}$, we observe that $\mathbf{g} = \mathbf{Kf}$ implies that

$$g(t) = \sum_{j=1}^{n} \alpha_j p_j(t)$$

so that $g(t)$ belongs to the linear manifold spanned by $\{\mathbf{p}_j\}_1^n$. Thus $R(\mathbf{K}) \subset L(\mathbf{p}_1, \mathbf{p}_2, \ldots, \mathbf{p}_n)$. We are not in a position to improve this result any further by a more pointed identification of $R(\mathbf{K})$.

*Example 4*

The null space of the differential operator $\mathbf{T} \equiv d/dt$ acting on elements in $\mathcal{C}^{(1)}[a, b]$ is the one-dimensional linear manifold spanned by the function

$$f(t) = 1, \qquad a \leq t \leq b$$

The range space of $\mathbf{T}$ depends on the domain of $\mathbf{T}$. If the domain is $\mathcal{C}^{(1)}[a, b]$, then the range space $R(\mathbf{T})$ is $\mathcal{C}[a, b]$.

---

†Since we are dealing with only algebraic aspects of linear spaces, we are not in a position to recognize the condition $\int_a^b q_j(s)f(s)\,ds = 0$ as one of *orthogonality* $\mathbf{q}_j$ and $\mathbf{f}$. This is the subject matter of Chapter 6.

*Example 5*

Consider the projection operator **P** in Section 2.7, which has some interesting properties. Let $N(\mathbf{P})$ and $R(\mathbf{P})$ be the null and range spaces of **P**, which are subspaces of linear space $\mathcal{L}$.

First, we observe that for each $\mathbf{x} \in \mathcal{L}$, $(\mathbf{x} - \mathbf{Px}) \in N(\mathbf{P})$, by definition of the projection operator. Second, if $\mathbf{x} \in R(\mathbf{P})$, $\mathbf{Px} = \mathbf{x}$. This is readily seen since there exists a $\mathbf{y} \in \mathcal{L}$ such that $\mathbf{Py} = \mathbf{x}$, which yields $\mathbf{P}(\mathbf{Py}) = \mathbf{Px}$. But $\mathbf{P}(\mathbf{Py}) = \mathbf{Py}$, again by definition. Thus it is clear that $\mathbf{Px} = \mathbf{x} \ \forall\ \mathbf{x} \in R(\mathbf{P})$.

An interesting and important fact which emerges is that every vector $\mathbf{z} \in \mathcal{L}$ can be expressed as the sum of a vector **x** in $R(\mathbf{P})$ and a vector **y** in $N(\mathbf{P})$. Thus $\mathbf{x} = \mathbf{Pz}$ and $\mathbf{y} = \mathbf{z} - \mathbf{Pz}$. We further show that this "decomposition" of **z** into an algebraic sum of **x** and **y** is unique; that is, there is only one **x** in $R(\mathbf{P})$ and only one **y** in $N(\mathbf{P})$ for which $\mathbf{z} = \mathbf{x} + \mathbf{y}$. If the decomposition is not unique, let $\mathbf{x}_1, \mathbf{x}_2 \in R(\mathbf{P})$, $\mathbf{x}_1 \neq \mathbf{x}_2$, $\mathbf{y}_1, \mathbf{y}_2 \in N(\mathbf{P})$, $\mathbf{y}_1 \neq \mathbf{y}_2$, and

$$\mathbf{z} = \mathbf{x}_1 + \mathbf{y}_1 \tag{2.8.1}$$

$$\mathbf{z} = \mathbf{x}_2 + \mathbf{y}_2 \tag{2.8.2}$$

On subtracting (2.8.2) from (2.8.1), we obtain

$$\mathbf{0} = (\mathbf{x}_1 - \mathbf{x}_2) + (\mathbf{y}_1 - \mathbf{y}_1) \tag{2.8.3}$$

Operating on (2.8.3) by **P**, we obtain

$$\mathbf{0} = \mathbf{P}(\mathbf{x}_1 - \mathbf{x}_2) + \mathbf{P}(\mathbf{y}_1 - \mathbf{y}_2) \tag{2.8.4}$$

Since $R(\mathbf{P})$ and $N(\mathbf{P})$ are subspaces, $(\mathbf{x}_1 - \mathbf{x}_2) \in R(\mathbf{P})$ and $(\mathbf{y}_1 - \mathbf{y}_2) \in N(\mathbf{P})$ so that $\mathbf{P}(\mathbf{x}_1 - \mathbf{x}_2) = (\mathbf{x}_1 - \mathbf{x}_2)$ and $\mathbf{P}(\mathbf{y}_1 - \mathbf{y}_2) = \mathbf{0}$. Equation (2.8.4) therefore yields $\mathbf{x}_1 - \mathbf{x}_2 = \mathbf{0}$ or $\mathbf{x}_1 = \mathbf{x}_2$, which further obtains from (2.8.3) $\mathbf{y}_1 = \mathbf{y}_2$. These are contradictory to the earlier stipulation that $\mathbf{x}_1 \neq \mathbf{x}_2$, $\mathbf{y}_1 \neq \mathbf{y}_2$, so the decomposition is unique. This decomposition allows us to write

$$\mathcal{L} = R(\mathbf{P}) + N(\mathbf{P})$$

Note that this is a familiar result for physical vectors in $\mathcal{R}_3$. Thus if we consider the projection of vectors on the $x$-axis, then the range space is the entire $x$-axis, while the null space is the $y - z$ plane. All vectors on the $y - z$ plane will be projected onto the origin. It is also well known that every vector in $\mathcal{R}_3$ can be written as the algebraic sum of a vector on the $y - z$ plane and a vector along the $x$-axis.

## EXERCISES

**2.8.1** Let **T** be a linear transformation; $\mathbf{T}: \mathcal{L} \longrightarrow \mathcal{L}$. If $\{\mathbf{e}_j\}_1^n$ is a basis of $\mathcal{L}$, show that $R(\mathbf{T}) = L(\mathbf{Te}_1, \mathbf{Te}_2, \ldots, \mathbf{Te}_n)$. Show that the set $\{\mathbf{Te}_j\}_1^n$ is linearly independent *if and only if* $N(\mathbf{T}) = \{\mathbf{0}\}$. This is an important result. [*Hint:* First assume that $N(\mathbf{T}) = \{\mathbf{0}\}$. Let $\alpha_1, \alpha_2, \ldots, \alpha_n \in \mathcal{F} \ni \mathbf{0} = \sum_{j=1}^n \alpha_j \mathbf{Te}_j$. Now show that $\alpha_1$

$= \alpha_2 = \ldots = \alpha_n = 0$, which will establish the "if" condition. For the "only if" part, assume that $\{\mathbf{Te}_j\}_1^n$ are independent. Then establish that $\mathbf{Tx} = \mathbf{0} \rightarrow \mathbf{x} = \mathbf{0}$.]

**2.8.2** Consider the linear space $\mathcal{R}_n$ with the basis $\{\mathbf{i}_j\}_1^n$, where $\mathbf{i}_j = (0, \ldots, 0, 1, 0, \ldots, 0)$, the unity element occurring in the $j$th position, and the projection operator $\mathbf{P}$ defined by $\mathbf{Px} = \mathbf{P}(x_1, x_2, \ldots, x_n) = (x_1, x_2, \ldots, x_m, 0, \ldots, 0)$. Identify the null and range spaces of $\mathbf{P}$ as appropriate linear manifolds of selected vectors from the set $\{\mathbf{i}_j\}_1^n$.

**2.8.3** Let $\mathbf{P}$ be a projection operator defined on a linear space $\mathcal{L}$. Show that $N(\mathbf{P}) = \{\mathbf{0}\}$ *if* and *only if* $\mathbf{P}$ is the identity operator $\mathbf{I}$. (*Hint:* Use results from Example 5 of this section.)

**2.8.4** Show that an $n \times n$ matrix is a linear operator on $\mathcal{R}_n$ using the rules of matrix multiplication and addition. Show that the dimension of the range space of the operator is the rank of the matrix $r$, while the null space has dimension $(n - r)$. How does one identify a basis in the range space?

**2.8.5** In the reaction mixture of Exercise 2.5.2 assume that first-order reactions occur of the type

$$A_i \underset{k_{i,j}}{\overset{k_{j,i}}{\rightleftharpoons}} A_j$$

where $A$ represents the species and $k$ is the rate constant. The composition vector $\mathbf{y}$ depends on time and the rate of change of $\mathbf{y}$, denoted $\dot{\mathbf{y}}$, is given by $\dot{\mathbf{y}} = \mathbf{Ky}$, where $\mathbf{K}$ is the matrix of rate constants:†

$$\mathbf{K} = \begin{bmatrix} k_{1,1} & k_{1,2} & \cdots & k_{1,n} \\ k_{2,1} & k_{2,2} & \cdots & k_{2,n} \\ \vdots & & & \\ k_{n,1} & k_{n,2} & \cdots & k_{n,n} \end{bmatrix}$$

We define $k_{i,i} = -\sum_{j=1,i}^{n} k_{j,i}$; the comma under the summation is used to exclude $j = i$. Show that the steady states of the system lie in the set $S = H \cap N(\mathbf{K})$, where $H$ is the hyperplane to be determined in Exercise 2.5.2. When is the steady state unique?

## 2.9 Operations on Linear Transformations. Linear Space of Operators

Consider a linear space $\mathcal{L}$ and the set of all linear transformations on it. First we observe that two operators $\mathbf{T}$ and $\mathbf{T}'$ are equal if $\mathbf{Tx} = \mathbf{T}'\mathbf{x} \ \forall \ \mathbf{x} \in \mathcal{L}$.

†See, for example, *Mathematical Methods in Chemical Engineering: Matrices and Their Applications*, by N. R. Amundson, Prentice-Hall, Englewood Cliffs, N.J., 1964, Chap. 8.

We may define the sum of two operators as follows:

***Definition 2.9.1*** Let $\mathbf{T}_1$ and $\mathbf{T}_2$ be linear operators on $\mathfrak{L}$. The *sum* of $\mathbf{T}_1$ and $\mathbf{T}_2$ is an operator $\mathbf{T}$, designated $\mathbf{T}_1 + \mathbf{T}_2$, defined by

$$\mathbf{Tx} \equiv (\mathbf{T}_1 + \mathbf{T}_2)\mathbf{x} = \mathbf{T}_1\mathbf{x} + \mathbf{T}_2\mathbf{x} \qquad \forall \quad \mathbf{x} \in \mathfrak{L}$$

The sum is readily seen to be a linear operator, for

$$\begin{aligned}(\mathbf{T}_1 + \mathbf{T}_2)(\mathbf{x} + \mathbf{y}) &= \mathbf{T}_1(\mathbf{x} + \mathbf{y}) + \mathbf{T}_2(\mathbf{x} + \mathbf{y}) \\ &= \mathbf{T}_1\mathbf{x} + \mathbf{T}_1\mathbf{y} + \mathbf{T}_2\mathbf{x} + \mathbf{T}_2\mathbf{y} \\ &= \mathbf{T}_1\mathbf{x} + \mathbf{T}_2\mathbf{x} + \mathbf{T}_1\mathbf{y} + \mathbf{T}_2\mathbf{y} \\ &= (\mathbf{T}_1 + \mathbf{T}_2)\mathbf{x} + (\mathbf{T}_1 + \mathbf{T}_2)\mathbf{y}\end{aligned}$$

for every $\mathbf{x}, \mathbf{y} \in \mathfrak{L}$, and

$$\begin{aligned}(\mathbf{T}_1 + \mathbf{T}_2)(\alpha\mathbf{x}) &= \mathbf{T}_1(\alpha\mathbf{x}) + \mathbf{T}_2(\alpha\mathbf{x}) \\ &= \alpha\mathbf{T}_1\mathbf{x} + \alpha\mathbf{T}_2\mathbf{x} \\ &= \alpha(\mathbf{T}_1\mathbf{x} + \mathbf{T}_2\mathbf{x}) \\ &= \alpha(\mathbf{T}_1 + \mathbf{T}_2)\mathbf{x}\end{aligned}$$

for every $\alpha \in \mathfrak{F}$ and $\mathbf{x} \in \mathfrak{L}$. Thus $(\mathbf{T}_1 + \mathbf{T}_2)$ is indeed a linear transformation. That the sum has the property of commutativity obviously follows from the corresponding property of vectors in $\mathfrak{L}$.

We now define the scalar multiplication of an operator.

***Definition 2.9.2*** Let $T$ be a linear operator on $\mathfrak{L}$, and $\alpha \in \mathfrak{F}$. Then the operator $\alpha\mathbf{T}$, called the *scalar multiple* of $\mathbf{T}$, is defined by

$$(\alpha\mathbf{T})\mathbf{x} = \alpha\mathbf{Tx} \qquad \forall \quad \mathbf{x} \in \mathfrak{L}$$

The operator $\alpha\mathbf{T}$ is a linear operator because

$$\begin{aligned}(\alpha\mathbf{T})(\mathbf{x} + \mathbf{y}) &= \alpha\mathbf{T}(\mathbf{x} + \mathbf{y}) = \alpha\mathbf{Tx} + \alpha\mathbf{Ty} = (\alpha\mathbf{T})\mathbf{x} + (\alpha\mathbf{T})\mathbf{y} \\ (\alpha\mathbf{T})(\lambda\mathbf{x}) &= \alpha\mathbf{T}(\lambda\mathbf{x}) = \alpha\lambda\mathbf{Tx} = \lambda\alpha\mathbf{Tx} = \lambda(\alpha\mathbf{T})\mathbf{x}\end{aligned}$$

$$\forall \quad \mathbf{x}, \mathbf{y} \in \mathfrak{L}, \quad \lambda \in \mathfrak{F}$$

Since the zero operator has been defined in Section 2.7, it would seem that the set of linear operators on a given linear space $\mathfrak{L}$ has all the attributes of a linear space. Thus we have

***Theorem 2.9.1*** Let $\mathfrak{L}$ be a linear space over the field $\mathfrak{F}$. Then the set of all linear transformation forms a linear space over the same field $\mathfrak{F}$, and is designated $\mathfrak{I}(\mathfrak{L})$ or simply $\mathfrak{I}$.†

---

†Similarly, the linear space of all operators $\mathbf{T}: \mathfrak{L} \longrightarrow \mathfrak{L}'$ is denoted $\mathfrak{I}(\mathfrak{L} \longrightarrow \mathfrak{L}')$.

The proof of this theorem, which essentially depends on the properties of vectors in $\mathfrak{L}$, is left as an exercise. Thus a linear transformation is a "vector" in the linear space $\mathfrak{J}$. However, linear transformations may also be multiplied together, a property not possessed by ordinary vectors. We define the product of two operators as follows:

***Definition 2.9.3*** Let $\mathbf{T}_1, \mathbf{T}_2, \in \mathfrak{J}$; that is, $\mathbf{T}_1$ and $\mathbf{T}_2$ are linear transformations on $\mathfrak{L}$. The product of $\mathbf{T}_1$ and $\mathbf{T}_2$ is an operator $\mathbf{T}$ defined by

$$\mathbf{Tx} \equiv (\mathbf{T}_1\mathbf{T}_2)\mathbf{x} = \mathbf{T}_1(\mathbf{T}_2\mathbf{x}) \qquad \forall \quad \mathbf{x} \in \mathfrak{L}$$

It is easy to see that $\mathbf{T}_1\mathbf{T}_2$ is a linear operator, since

$$\begin{aligned}\mathbf{T}_1\mathbf{T}_2(\mathbf{x}+\mathbf{y}) &= \mathbf{T}_1(\mathbf{T}_2(\mathbf{x}+\mathbf{y})) = \mathbf{T}_1(\mathbf{T}_2\mathbf{x}+\mathbf{T}_2\mathbf{y}) \\ &= \mathbf{T}_1(\mathbf{T}_2\mathbf{x}) + \mathbf{T}_1(\mathbf{T}_2\mathbf{y}) \\ &= \mathbf{T}_1\mathbf{T}_2(\mathbf{x}) + \mathbf{T}_1\mathbf{T}_2(\mathbf{y})\end{aligned}$$

and

$$\begin{aligned}\mathbf{T}_1\mathbf{T}_2(\alpha\mathbf{x}) &= \mathbf{T}_1(\mathbf{T}_2(\alpha\mathbf{x})) = \mathbf{T}_1(\alpha\mathbf{T}_2\mathbf{x}) \\ &= \alpha\mathbf{T}_1(\mathbf{T}_2\mathbf{x}) = \alpha\mathbf{T}_1\mathbf{T}_2(\mathbf{x})\end{aligned}$$

for every $\mathbf{x}, \mathbf{y} \in \mathfrak{L}$, $\alpha \in \mathfrak{J}$. Hence $\mathbf{T}_1\mathbf{T}_2 \in \mathfrak{J}$. It is evident that the product of two operators is not commutative; that is, in general, $\mathbf{T}_1\mathbf{T}_2 \neq \mathbf{T}_2\mathbf{T}_1$.

The extension to multiple products obeys the rules:

$$\left.\begin{aligned}&1.\ \alpha(\mathbf{T}_1\mathbf{T}_2) = (\alpha\mathbf{T}_1)\mathbf{T}_2 \\ &2.\ \mathbf{T}_1(\mathbf{T}_2\mathbf{T}_3) = (\mathbf{T}_1\mathbf{T}_2)\mathbf{T}_3\end{aligned}\right\} \text{associativity}$$

$$\left.\begin{aligned}&3.\ \mathbf{T}_1(\mathbf{T}_2+\mathbf{T}_3) = \mathbf{T}_1\mathbf{T}_2 + \mathbf{T}_1\mathbf{T}_3 \\ &4.\ (\mathbf{T}_1+\mathbf{T}_2)\mathbf{T}_3 = \mathbf{T}_1\mathbf{T}_3 + \mathbf{T}_2\mathbf{T}_3\end{aligned}\right\} \text{distributivity}$$

We shall prove the second result as a demonstrative example. Let $\mathbf{x} \in \mathfrak{L}$.

$$\mathbf{T}_1(\mathbf{T}_2\mathbf{T}_3)\mathbf{x} = \mathbf{T}_1(\mathbf{T}_2\mathbf{T}_3\mathbf{x}) = \mathbf{T}_1(\mathbf{T}_2(\mathbf{T}_3\mathbf{x}))$$

Also,

$$(\mathbf{T}_1\mathbf{T}_2)\mathbf{T}_3\mathbf{x} = \mathbf{T}_1\mathbf{T}_2(\mathbf{T}_3\mathbf{x}) = \mathbf{T}_1(\mathbf{T}_2(\mathbf{T}_3\mathbf{x}))$$

Hence

$$\mathbf{T}_1(\mathbf{T}_2\mathbf{T}_3)\mathbf{x} = (\mathbf{T}_1\mathbf{T}_2)\mathbf{T}_3\mathbf{x} \qquad \forall \quad \mathbf{x} \in \mathfrak{L}$$

so that

$$\mathbf{T}_1(\mathbf{T}_2\mathbf{T}_3) = (\mathbf{T}_1\mathbf{T}_2)\mathbf{T}_3$$

which is the required result.

This property of associativity under multiplication allows us to define *powers* of operators. Thus

$$\mathbf{T}^2 = \mathbf{TT}, \qquad \mathbf{T}^n = \mathbf{TT}\ldots\mathbf{T} \quad (n \text{ times})$$

Clearly, the laws of indices must hold

$$\mathbf{T}^m\mathbf{T}^n = \mathbf{T}^{m+n}$$

and

$$(\mathbf{T}^m)^n = \mathbf{T}^{mn}$$

where $m$ and $n$ are integers. One may also define the zeroth power of any operator as the identity operator, that is,

$$\mathbf{T}^0 \equiv \mathbf{I}$$

It is a trivial exercise to show that $\mathbf{TI} = \mathbf{IT} = \mathbf{T}$.

### EXERCISES

**2.9.1** Prove Theorem 2.9.1.

**2.9.2** Show that

$$\alpha(\mathbf{T}_1\mathbf{T}_2) = (\alpha\mathbf{T}_1)\mathbf{T}_2 = \mathbf{T}_1(\alpha\mathbf{T}_2)$$

$$\mathbf{T}_1(\mathbf{T}_2 + \mathbf{T}_3) = \mathbf{T}_1\mathbf{T}_2 + \mathbf{T}_1\mathbf{T}_3$$

$$(\mathbf{T}_1 + \mathbf{T}_2)\mathbf{T}_3 = \mathbf{T}_1\mathbf{T}_3 + \mathbf{T}_2\mathbf{T}_3$$

$$(\alpha + \beta)\mathbf{T} = \alpha\mathbf{T} + \beta\mathbf{T}$$

where $\mathbf{T}, \mathbf{T}_1, \mathbf{T}_2, \mathbf{T}_3 \in \mathfrak{J}(\mathfrak{L})$ and $\alpha, \beta \in \mathfrak{F}$.

**2.9.3** Obtain expressions for $(\mathbf{T}_1 + \mathbf{T}_2)^2$ and $(\mathbf{T}_1 + \mathbf{T}_2)^3$.

**2.9.4** Let $\mathbf{T}$ be a linear operator on $\mathfrak{L}$ with range space $R(\mathbf{T})$. If $\mathbf{P}$ is a projection operator whose range space $R(\mathbf{P})$ is identical with $R(\mathbf{T})$ [or alternatively, $\mathbf{P}$ is a projection operator mapping $\mathfrak{L}$ onto $R(\mathbf{T})$], show that $\mathbf{PT} = \mathbf{T}$.

**2.9.5** Show that $\mathbf{T}^2 = \mathbf{0}$ if and only if $R(\mathbf{T}) \subset N(\mathbf{T})$.

**2.9.6** Prove that $N(\mathbf{T}_2) \subset N(\mathbf{T}_1\mathbf{T}_2)$ and $R(\mathbf{T}_1\mathbf{T}_2) \subset R(\mathbf{T}_1)$. Further show that $N(\mathbf{T}_2) = N(\mathbf{T}_1\mathbf{T}_2)$ if $N(\mathbf{T}_1) = \{\mathbf{0}\}$.

**2.9.7** If $N(\mathbf{T}_1) = \{\mathbf{0}\}$ and $N(\mathbf{T}_1\mathbf{T}_2) = \mathfrak{L}$ (i.e., $\mathbf{T}_1\mathbf{T}_2 = \mathbf{0}$), then $N(\mathbf{T}_2) = \mathfrak{L}$ (or $\mathbf{T}_2 = \mathbf{0}$).

## 2.10 The Inverse Operator

The concept of the inverse of a linear operator is fundamental to the solution of linear mathematical equations of physical phenomena. Much of the theory of linear spaces is, in fact, concerned with the construction and representation of the inverse of a linear operator. We define the inverse of a linear operator $\mathbf{T}$ as follows:

***Definition 2.10.1*** Let $\mathbf{T} \in \mathfrak{J}(\mathfrak{L})$. The *inverse* of $\mathbf{T}$ is an operator $\mathbf{T}'$ such that $\mathbf{T}'\mathbf{T} = \mathbf{TT}' = \mathbf{I}$. The inverse operator is denoted $\mathbf{T}^{-1}$.

It is readily shown that when the inverse exists it is a linear operator. Because $\forall\, x, y \in \mathfrak{L}$, $\mathbf{x} = \mathbf{TT}^{-1}\mathbf{x}$, $\mathbf{y} = \mathbf{TT}^{-1}\mathbf{y}$, so that $\mathbf{x} + \mathbf{y} = \mathbf{TT}^{-1}\mathbf{x} + \mathbf{TT}^{-1}\mathbf{y} = \mathbf{T}(\mathbf{T}^{-1}\mathbf{x} + \mathbf{T}^{-1}\mathbf{y})$, which obtains from the linearity of $\mathbf{T}$.

Operating on both sides by $\mathbf{T}^{-1}$, we have $\mathbf{T}^{-1}(\mathbf{x} + \mathbf{y}) = \mathbf{T}^{-1}\mathbf{x} + \mathbf{T}^{-1}\mathbf{y}$ (since $\mathbf{T}^{-1}\mathbf{T} = \mathbf{I}$). Similarly, $\mathbf{T}^{-1}(\alpha\mathbf{x}) = \alpha\mathbf{T}^{-1}\mathbf{x}$ for each $\alpha \in \mathfrak{F}$. Hence $\mathbf{T}^{-1}$ is a linear operator.

Not every linear operator $\mathbf{T}$ has an inverse, for let $\mathbf{T}$ be an operator such that $\mathbf{T}\mathbf{x} = \mathbf{0}$ for some $\mathbf{x} \neq \mathbf{0}$. Then clearly for every linear operator $\mathbf{T}'$, $\mathbf{T}'\mathbf{T}\mathbf{x} = \mathbf{0}$, from which $\mathbf{T}'\mathbf{T} \neq \mathbf{I}$, so that $\mathbf{T}$ has no inverse.

If an operator $\mathbf{T}$ does have an inverse $\mathbf{T}^{-1}$, then it is readily shown that it is unique, for let $\mathbf{T}_1$ and $\mathbf{T}_2$ be two inverses of $\mathbf{T}$. Then $\mathbf{T}\mathbf{T}_1 = \mathbf{I}$. Multiplying on the left by $\mathbf{T}_2$, we obtain $\mathbf{T}_2 = \mathbf{T}_1$. From this it is immediately evident that $\mathbf{T}$ is the inverse of $\mathbf{T}^{-1}$. In some cases the operator $\mathbf{T}$ may have an operator $\mathbf{T}'$ such that only one of the equalities, say $\mathbf{T}'\mathbf{T} = \mathbf{I}$, is true. This holds only for operators on infinite-dimensional spaces. Consider, for instance, the space $\mathcal{R}_\infty$ and the operators $\mathbf{T}$ and $\mathbf{T}'$ defined as follows. For

$$\begin{aligned} \mathbf{x} &= (x_1, x_2, \ldots) \\ \mathbf{T}\mathbf{x} &= (0, x_1, x_2, \ldots) \\ \mathbf{T}'\mathbf{x} &= (x_2, x_3, \ldots) \end{aligned}$$

It should then be clear that

$$\mathbf{T}'\mathbf{T}\mathbf{x} = \mathbf{x} \qquad \forall \quad \mathbf{x} \in \mathcal{L}$$

so that $\mathbf{T}'\mathbf{T} = \mathbf{I}$. However, it is obvious that

$$\mathbf{T}\mathbf{T}'\mathbf{x} = (0, x_2, \ldots) \neq \mathbf{x}$$

which implies that $\mathbf{T}\mathbf{T}' \neq \mathbf{I}$. Thus from Definition 2.10.1, $\mathbf{T}$ does not possess an inverse. However, $\mathbf{T}$ is said to have a "left inverse."

There is another reason why an operator on an infinite-dimensional space may have a "one-sided" inverse. Thus consider the linear space $\mathcal{C}[a, b]$ and the operator $\mathbf{T}$ defined on this space as follows:

$$\mathbf{g} = \mathbf{T}\mathbf{f}; \qquad g(t) = \int_a^t f(t')\,dt'$$

The range space of the operator $\mathbf{T}$ consists of functions $g(t)$ which have continuous first derivatives and subject to the constraint that $g(a) = 0$. Clearly, $R(\mathbf{T})$ is a linear subspace of $\mathcal{C}^{(1)}[a, b]$. Now consider the operator $\mathbf{T}' \equiv d/dt$, acting on $R(\mathbf{T})$. Then

$$\mathbf{T}'\mathbf{T}\mathbf{f} \equiv \left\{\frac{d}{dt}\int_a^t f(t')\,dt'\right\} = \{f(t)\} = \mathbf{f}$$

for every function $f(t)$ in $\mathcal{C}[a, b]$.† Clearly, therefore, $\mathbf{T}'\mathbf{T} = \mathbf{I}$ and $\mathbf{T}'$ represents the *left* inverse of $\mathbf{T}$. On the other hand, since $\mathbf{T}'$ is constrained to act on $\mathcal{C}^{(1)}[a, b]$, we *do not* have $\mathbf{T}\mathbf{T}' = \mathbf{I}$ with respect to the entire space $\mathcal{C}[a, b]$.

†This follows from the fundamental theorem of the calculus. See, for example *Advanced Calculus*, by A. E. Taylor, Ginn, Boston, 1955.

However, if the domain of $\mathbf{T}'$ is restricted to $R(\mathbf{T})$, then $\mathbf{TT}'$ behaves as the identity there. From the viewpoint of the operator $\mathbf{T}'$, one notices that it has its *right* inverse in $\mathbf{T}$ but has *no left* inverse in the sense of the equality $\mathbf{TT}' = \mathbf{I}$. In fact, even if $\mathbf{T}'$ is considered on $\mathcal{C}^{(1)}[a, b]$, it has no left inverse. Thus $\mathbf{T}'$ will have a left inverse only when its domain is restricted to $R(\mathbf{T})$.

The foregoing discussion prompts an alternative definition of the left inverse of an operator $\mathbf{T}$ as an operator $\mathbf{T}'$ such that $\mathbf{T}'\mathbf{T}$ behaves as the identity operator on $D(\mathbf{T})$. The domain of $\mathbf{T}'$ is then $R(\mathbf{T})$, so that $\mathbf{TT}'$ acts as the identity on $R(\mathbf{T})$. At this stage it becomes evident that this left inverse $\mathbf{T}'$ may itself be called the inverse of $\mathbf{T}$ and denoted by $\mathbf{T}^{-1}$. Henceforth, the term "inverse" will be used to connote this inverse. The only requirement for such an inverse to exist is a one-to-one correspondence between $D(\mathbf{T})$ and $R(\mathbf{T})$, which means that $\mathbf{Tx}_1 = \mathbf{Tx}_2 \rightarrow \mathbf{x}_1 = \mathbf{x}_2$. The following theorem is almost a restatement of the preceding statement.

***Theorem 2.10.1*** Let $\mathbf{T} \in \mathfrak{J}(\mathcal{L})$. The inverse of $\mathbf{T}$ exists if and only if $N(\mathbf{T}) \cap D(\mathbf{T}) = \{\mathbf{0}\}$.

*Proof*: We first note that if $N(\mathbf{T}) \subset D(\mathbf{T})$, then $N(\mathbf{T}) \cap D(\mathbf{T}) = N(\mathbf{T})$. If $N(\mathbf{T}) \cap D(\mathbf{T})$ contains nonzero elements, we already know that no inverse operator exists so that the "only if" condition is proved.

To prove the "if" condition, assume that $N(\mathbf{T}) \cap D(\mathbf{T}) = \{\mathbf{0}\}$. Let $\mathbf{x}, \mathbf{y} \in D(\mathbf{T})$ such that $\mathbf{Tx} = \mathbf{Ty}$. Then $\mathbf{T}(\mathbf{x} - \mathbf{y}) = \mathbf{0}$, so that $(\mathbf{x} - \mathbf{y}) \in N(\mathbf{T})$. Thus $(\mathbf{x} - \mathbf{y}) \in N(\mathbf{T}) \cap D(\mathbf{T})$, which implies that $\mathbf{x} - \mathbf{y} = \mathbf{0}$ or $\mathbf{x} = \mathbf{y}$. Hence $\forall\, \mathbf{z} \in R(\mathbf{T})\ \exists$ a unique $\mathbf{x} \in D(\mathbf{T}) \ni \mathbf{Tx} = \mathbf{z}$. Now define $\mathbf{T}'\mathbf{z} = \mathbf{x}$ so that $\mathbf{T}'\mathbf{T}$ acts as the identity on $D(\mathbf{T})$, implying that $\mathbf{T}'$ is the inverse. ∎

The domain of an operator $\mathbf{T}$ may be constrained either naturally or by imposed additional conditions. Thus the operator $\mathbf{T} \equiv d/dt$ is naturally constrained to act on $\mathcal{C}^{(1)}[a, b]$, whereas additional constraints such as, say, $f(a) = 0$ (where $\mathbf{f} \in \mathcal{C}^{(1)}[a, b]$) will make the domain of $\mathbf{T}$ a proper subspace of $\mathcal{C}^{(1)}[a, b]$. Note that $\mathbf{T}$ has no inverse on $\mathcal{C}^{(1)}[a, b]$ since nonzero elements in this space such as $f(t) =$ constant, $a \leq t \leq b$, belong to the null space of $\mathbf{T}$. However, when $D(\mathbf{T})$ is restricted to functions with $f(a) = 0$, then $N(\mathbf{T}) \cap D(\mathbf{T})$ contains only the zero element, so that $\mathbf{T}$ has an inverse on this new domain. This is the essential content of Theorem 2.10.1.

In finite-dimensional spaces, the domain of the operator is always the entire space $\mathcal{L}$ and the inverse exists when (and only when) $R(\mathbf{T}) = \mathcal{L}$. This result obtains from the application of Exercise 2.8.1 and Theorem 2.10.1. Thus the existence of the inverse of a linear operator $\mathbf{T}$ defined on a finite-dimensional space $\mathcal{L}$ is equivalent to the existence of a one-to-one mapping from $\mathcal{L}$ onto $\mathcal{L}$. This is also precisely the statement of Definition 2.10.1.

We may now take stock of the situation in regard to the inverse of an operator T. Figure 2.10.1 represents a linear space by the region within a

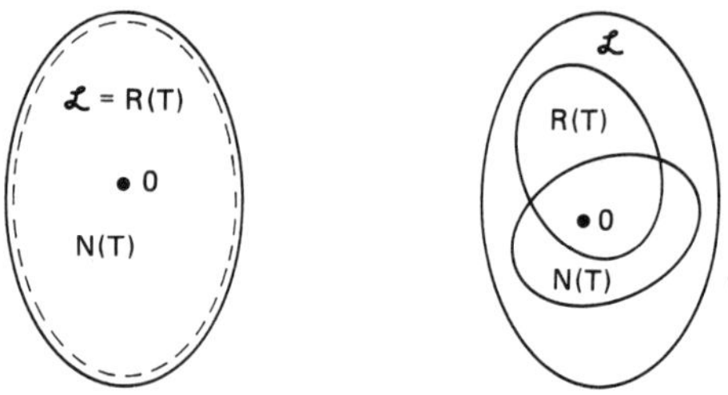


(a) Inverse exists (b) No inverse exists

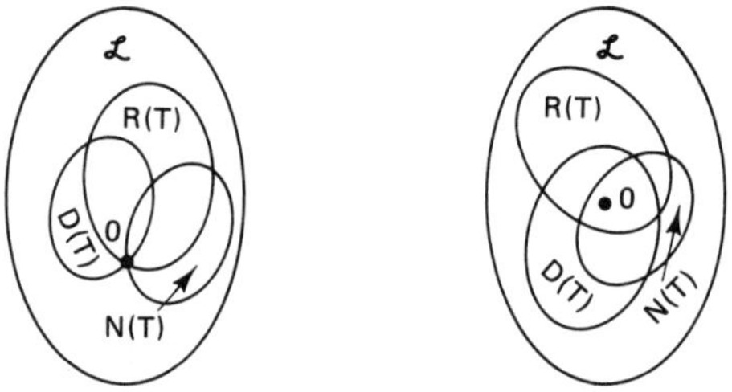


(c) Inverse exists (d) No inverse exists

**Figure 2.10.1** Linear mappings.

closed curve, the point at the center representing the zero vector. Figure 2.10.1(a) depicts the situation of a finite-dimensional space $\mathfrak{L}$ on which is defined a linear transformation **T** whose inverse exists; Figure 2.10.1(b) shows the situation when the inverse does not exist. Figures 2.10.1(c) and (d) are applicable to infinite-dimensional spaces, the first representing the situation when an inverse exists and the second displaying the case where the inverse does not exist.

The reader is warned, however, that these diagrams should not be taken too literally, since the realism with which they represent actual situations is substantially diminished by considerations of a topological nature in subsequent chapters. For the present, however, Figure 2.10.1 sums up the features concerning the inverse of a linear operator reasonably well.

Operating on the *left* of an operator by its inverse is of particular significance in the solution of mathematical equations arising in the description of physical systems. Thus a large number of problems are of the form

$$\mathbf{Tx} = \mathbf{y} \tag{2.10.1}$$

where **T** is a linear operator, **y** is a known vector belonging to $R(\mathbf{T})$, and **x** is an unknown element in $D(\mathbf{T})$. The domain space $D(\mathbf{T})$ and the range space $R(\mathbf{T})$ are, of course, subspaces of some linear space $\mathfrak{L}$ pertinent to the problem. When **T** does have an inverse $\mathbf{T}^{-1}$, the "solution" of the mathematical equation may be obtained by

$$\mathbf{T}^{-1}\mathbf{T}\mathbf{x} = \mathbf{x} = \mathbf{T}^{-1}\mathbf{y}$$

Note that the solution emerged by operating on the left of **T**. It is therefore immaterial whether or not **T** has an inverse in the sense of Definition 2.10.1 for the solution of equation (2.10.1).

More recently, the concept of a *generalized* inverse has been developed for an operator that does not have an inverse in the sense of the earlier definitions.† We will briefly deal with this concept for finite-dimensional spaces only.

When an operator **T** has an inverse **T**′, then $\mathbf{T}'\mathbf{T} = \mathbf{I}$, from which it follows that $\mathbf{T}\mathbf{T}'\mathbf{T} = \mathbf{T}$. However, the latter equality does not imply the former, which provides a basis for defining the generalized inverse of an operator **T**.

***Definition 2.10.2*** Let $\mathbf{T} \in \mathfrak{J}(\mathfrak{L})$ such that it has no inverse. The generalized inverse of **T** is a linear operator **T**′ such $\mathbf{T}\mathbf{T}'\mathbf{T} = \mathbf{T}$ and is denoted **T** .

Such a generalized inverse has the merit of solving linear equations of the type (2.10.1), through the relation $\mathbf{x} = \mathbf{T}^{-}\mathbf{y}$. When **T** does not have an inverse, the equation

$$\mathbf{T}\mathbf{x} = \mathbf{y}$$

will possess a solution only when $\mathbf{y} \in R(\mathbf{T})$. If $\mathbf{x} = \mathbf{T}^{-}\mathbf{y}$ must represent the solution, then $\mathbf{T}\mathbf{T}^{-}\mathbf{y} = \mathbf{y}$ whenever $\mathbf{y} \in R(\mathbf{T})$. We will now show that the generalized inverse satisfies this requirement. First we show that $\mathbf{T}\mathbf{T}^{-}\mathbf{T} = \mathbf{T} \rightarrow \mathbf{T}\mathbf{T}^{-}\mathbf{y} = \mathbf{y}$ when $\mathbf{y} \in R(\mathbf{T})$.

Let $\mathbf{y} \in R(\mathbf{T})$. $\exists\ \mathbf{z} \in \mathfrak{L} \ni \mathbf{T}\mathbf{z} = \mathbf{y}$ (so that **z** is a solution). Then $\mathbf{T}\mathbf{T}^{-}\mathbf{T}\mathbf{z} = \mathbf{T}\mathbf{T}^{-}\mathbf{y}$. But $\mathbf{T}\mathbf{T}^{-}\mathbf{T}\mathbf{z} = \mathbf{T}\mathbf{z} = \mathbf{y}$, which implies that $\mathbf{T}\mathbf{T}^{-}\mathbf{y} = \mathbf{y}$.

Next we prove that $\mathbf{T}\mathbf{T}^{-}\mathbf{y} = \mathbf{y}\ \forall\ \mathbf{y} \in R(\mathbf{T}) \rightarrow \mathbf{T}\mathbf{T}^{-}\mathbf{T} = \mathbf{T}$.

Let $z \in \mathfrak{L}$. $\mathbf{T}\mathbf{z} \in R(\mathbf{T})$, so that $\mathbf{T}\mathbf{T}^{-}\mathbf{T}\mathbf{z} = \mathbf{T}\mathbf{z}$. This implies that $\mathbf{T}\mathbf{T}^{-}\mathbf{T} = \mathbf{T}$.

Hence we have shown that $\mathbf{T}^{-}$ is a generalized inverse **T** *if and only if* $\mathbf{T}\mathbf{T}^{-}\mathbf{y} = \mathbf{y}\ \forall\ \mathbf{y} \in R(\mathbf{T})$. We will not pursue this much further in this book.

## EXERCISES

**2.10.1** Consider the diagonal operator in Exercise 2.7.1. Show that an inverse exists *if and only if* $\lambda_j \neq 0, j = 1, 2, \ldots, n$.

---

†See, for example, *Generalized Inverse of Matrices and Its Applications*, by C. R. Rao and S. K. Mitra, Wiley, New York, 1971.

**2.10.2** Consider the linear space $\mathcal{C}^{(2)}[0, 1]$ on which we define the operator $\mathbf{T}$ by

$$\mathbf{T}f \equiv \left\{\frac{d^2 f}{dt^2}(t)\right\}$$

(a) What is the range space $R(\mathbf{T})$?
(b) If $\mathbf{T}f = \mathbf{g}$, show that

$$f(t) = (1 - t)f(0) + tf(1) - t\int_0^1 (1 - t')g(t')\,dt' + \int_0^t (t - t')g(t')\,dt'$$

Why does the foregoing expression imply that no inverse $\mathbf{T}^{-1}$ exists mapping $R(\mathbf{T})$ onto $\mathcal{C}^{(2)}[0, 1]$?
(c) Find some $D(\mathbf{T}) \subset \mathcal{C}^{(2)}[0, 1]$ such that $\mathbf{T}^{-1}\colon R(\mathbf{T}) \longrightarrow D(\mathbf{T})$ exists. Can you find more than one such domain?

## 2.11 Isomorphisms. Isomorphic Spaces

The preceding sections dealt with linear transformations which mapped elements of a linear space into other elements of the same space. It is also possible to entertain transformations from a linear space $\mathcal{L}$ into another linear space $\mathcal{L}'$ provided that the associated fields of the two linear spaces are identical. Clearly, a linear transformation $\mathbf{T}\colon \mathcal{L} \longrightarrow \mathcal{L}'$ is one for which $\mathbf{T}(\mathbf{x} + \mathbf{y}) = \mathbf{Tx} + \mathbf{Ty}$ and $\mathbf{T}(\alpha\mathbf{x}) = \alpha\mathbf{Tx}\ \forall\ \mathbf{x}, \mathbf{y} \in \mathcal{L}$ and $\alpha \in \mathcal{F}$. Such a transformation may also possess an inverse that maps elements of $\mathcal{L}'$ into $\mathcal{L}$. This inverse exists if $\mathbf{Tx} = \mathbf{Ty} \rightarrow \mathbf{x} = \mathbf{y}$, in which case the mapping of $\mathcal{L}$ into $\mathcal{L}'$ is *one-to-one*. If, further, the range of $\mathbf{T}$ is all of $\mathcal{L}'$, the mapping is one-to-one and onto and $\mathbf{T}$ is called an *isomorphism*. Two linear spaces $\mathcal{L}$ and $\mathcal{L}'$ possessing the same associated field are said to be *isomorphic* if there exists an isomorphism $T\colon \mathcal{L} \longrightarrow \mathcal{L}'$. It is obvious that if $(\mathcal{L}, \mathcal{L}')$ and $(\mathcal{L}', \mathcal{L}'')$ are each isomorphic pairs, then $(\mathcal{L}, \mathcal{L}'')$ is an isomorphic pair. We now prove the following theorem.

***Theorem 2.11.1*** If $\mathcal{L}$ and $\mathcal{L}'$ are isomorphic linear spaces, they have the same dimension.

*Proof*: Let $\{\mathbf{x}_j\}_1^n$ be a linearly independent set in $\mathcal{L}$, and $\{\mathbf{x}'_j\}_1^n$ be the corresponding set in $\mathcal{L}'$ where $\mathbf{x}'_j = \mathbf{Tx}_j$, $\mathbf{T}$ being an isomorphism. Clearly, $\{\mathbf{x}'_j\}_1^n$ is an independent set (see Exercise 2.8.1). Thus linearly independent sets from either space map into linearly independent sets in the other. Similarly, linearly dependent sets from one space map into linearly dependent sets in the other. The preceding two statements imply that $\mathcal{L}$ and $\mathcal{L}'$ have the same dimension. ∎

An even stronger result is the content of the next theorem.

***Theorem 2.11.2*** Every pair of finite-dimensional linear spaces of identical dimension possessing a common associated field is isomorphic.

*Proof*: Let the common associated field be the complex numbers and the dimension of the spaces be $n$. We will establish that every $n$-dimensional linear space with a complex associated field is isomorphic with $\mathcal{C}_n$, which is sufficient to prove the theorem. Consider space $\mathcal{L}$. Let $\{\mathbf{e}_j\}_1^n$ be a basis in $\mathcal{L}$. Then $\forall\ \mathbf{x} \in \mathcal{L}\ \exists\ \alpha_1, \alpha_2, \ldots, \alpha_n$, which are unique complex numbers such that

$$\mathbf{x} = \alpha_1\mathbf{e}_1 + \alpha_2\mathbf{e}_2 + \ldots + \alpha_n\mathbf{e}_n$$

The vector $\boldsymbol{\alpha} = (\alpha_1, \alpha_2, \ldots, \alpha_n)$ is an element of $\mathcal{C}_n$ and we may define a transformation $\mathbf{T}\colon \mathcal{L} \longrightarrow \mathcal{C}_n$ such that $\mathbf{Tx} = \boldsymbol{\alpha}$. It is easy to see that $\mathbf{T}$ is a linear transformation. The inverse mapping, say $\mathbf{T}'\colon \mathcal{C}_n \longrightarrow \mathcal{L}$, is obtained by taking a linear combination of $\{\mathbf{e}_j\}_1^n$ using the ordered array of complex numbers in $\boldsymbol{\alpha}$. $\mathbf{T}$ is therefore an isomorphism (note that $\mathbf{T}$ is relative to the basis $\{\mathbf{e}_j\}_1^n$) so that $\mathcal{L}$ and $\mathcal{C}_n$ are isomorphic. Since $\mathcal{L}$ is an arbitrary $n$-dimensional space, every pair on $n$-dimensional spaces is isomorphic. ∎

An important class of operators on $\mathcal{C}_n$ (or $\mathcal{R}_n$) is the set of $n \times n$ matrices containing complex numbers (or real numbers). It is easy to show from the rules of matrix addition and scalar multiplication that the set of matrices forms a linear space denoted $\mathcal{A}(\mathcal{C}_n)$ or $\mathcal{A}(\mathcal{R}_n)$. The following theorem is of interest.

***Theorem 2.11.3*** Let $\mathcal{L}$ be a linear space of dimension $n$ with the complex numbers as the associated field, and let $\mathfrak{J}(\mathcal{L})$ be the space of linear transformations defined on $\mathcal{L}$ with range in $\mathcal{L}$. Then $\mathfrak{J}(\mathcal{L})$ and $\mathcal{A}(\mathcal{C}_n)$ are isomorphic.

*Proof*: Let $\{\mathbf{e}_j\}_1^n$ be a basis in $\mathcal{L}$. Let $\mathbf{T} \in \mathfrak{J}(\mathcal{L})$. Since $\mathcal{L}$ and $\mathcal{C}_n$ are isomorphic, $\forall\ x \in \mathcal{L},\ \exists$ a corresponding vector $(\alpha_1, \alpha_2, \ldots, \alpha_n) \equiv \boldsymbol{\alpha} \in \mathcal{C}_n$, such that $\mathbf{x} = \alpha_1\mathbf{e}_1 + \alpha_2\mathbf{e}_2 + \ldots + \alpha_n\mathbf{e}_n$. For $\mathbf{Tx} = \mathbf{y} \in R(\mathbf{T})\ \exists$ a corresponding vector $(\beta_1, \beta_2, \ldots, \beta_n) \equiv \boldsymbol{\beta} \in \mathcal{C}_n$ such that $\mathbf{y} = \mathbf{e}_1\beta_1 + \beta_2\mathbf{e}_2 + \ldots + \beta_n\mathbf{e}_n$. Now for $\mathbf{Te}_j$, $\exists\ (\gamma_{1j}, \gamma_{2j}, \ldots, \gamma_{nj}) \equiv \boldsymbol{\gamma}_j \in \mathcal{C}_n$ such that $\mathbf{Te}_j = \gamma_{1j}\mathbf{e}_1 + \gamma_{2j}\mathbf{e}_2 + \ldots + \gamma_{nj}\mathbf{e}_n$, $j = 1, 2, \ldots, n$. Since

$$\mathbf{y} = \alpha_1\mathbf{Te}_1 + \alpha_2\mathbf{Te}_2 + \ldots + \alpha_n\mathbf{Te}_n = \sum_{j=1}^{n} \alpha_j\mathbf{Te}_j$$

$$= \sum_{j=1}^{n} \alpha_j[\gamma_{1j}\mathbf{e}_1 + \gamma_{2j}\mathbf{e}_2 + \ldots + \gamma_{nj}\mathbf{e}_n]$$

$$= \mathbf{e}_1 \sum_{j=1}^{n} \gamma_{1j}\alpha_j + \mathbf{e}_2 \sum_{j=1}^{n} \gamma_{2j}\alpha_j + \ldots + \mathbf{e}_n \sum_{j=1}^{n} \gamma_{nj}\alpha_n$$

and the expansion of $\mathbf{y}$ is unique, we must have

$$\beta_i = \sum_{j=1}^{n} \gamma_{ij}\alpha_j$$

Representing the matrix whose *ij*th element is $\gamma_{ij}$ by the preceding equation may be represented by $\boldsymbol{\beta} = \boldsymbol{\Gamma\alpha}$. Thus as $\mathbf{x} \in \mathfrak{L}$ corresponds to $\boldsymbol{\alpha} \in \mathfrak{C}_n$, $\mathbf{Tx} \in \mathfrak{L}$ corresponds to $\boldsymbol{\Gamma\alpha}$. Hence the action of $\mathbf{T}$ on $\mathfrak{L}$ may be *represented* by $\boldsymbol{\Gamma}$ on $\mathfrak{C}_n$. Note that the *uniqueness* of $\boldsymbol{\Gamma}$ is only *relative* to the *basis* set $\{\mathbf{e}_j\}_1^n$ in $\mathfrak{C}_n$. We call $\boldsymbol{\Gamma}$ the *matrix representation* of $\mathbf{T}$ relative to the basis set $\{\mathbf{e}_j\}_1^n$. We leave the reader to establish that this correspondence between $\mathbf{T} \in \mathfrak{J}$ and $\boldsymbol{\Gamma} \in \mathfrak{A}$ is an isomorphism so that $\mathfrak{J}$ and $\mathfrak{A}$ are isomorphic spaces. ▮

**EXERCISES**

**2.11.1** Show that the matrix representation of the diagonal operator in Exercise 2.7.1 relative to the given basis is a diagonal matrix (hence the term "diagonal" to the operator).

**2.11.2** Let $\mathbf{T} \in \mathfrak{J}(\mathfrak{L})$. If its matrix representation is to be diagonal, with respect to a basis, say $\{\mathbf{e}_j\}_1^n$, find $\mathbf{Te}_j, j = 1, 2, \ldots, n$.

**2.11.3** Show that the identity operator in $\mathfrak{J}(\mathfrak{L})$ has as its representation the identity matrix relative to any basis set.

**2.11.4** If $\mathbf{T}_1, \mathbf{T}_2 \in \mathfrak{J}(\mathfrak{L})$ and $\boldsymbol{\Gamma}_1, \boldsymbol{\Gamma}_2$ are their respective matrix representations, show that the matrix representation of $\mathbf{T}_1\mathbf{T}_2$ is $\boldsymbol{\Gamma}_1\boldsymbol{\Gamma}_2$.

**2.11.5** Show that matrix representation of any operator $\mathbf{T} \in \mathfrak{J}(\mathfrak{L})$ is singular if and only if $R(\mathbf{T})$ is a proper subspace of $\mathfrak{L}$.

**2.11.6** Let $\mathbf{T} \in \mathfrak{J}(\mathfrak{L})$, which has a matrix representation $\boldsymbol{\Gamma}$ relative to a given basis in $\mathfrak{L}$. If $\mathbf{T}$ has an inverse, show that the matrix representation of $\mathbf{T}^{-1}$ is $\boldsymbol{\Gamma}^{-1}$.

## 2.12 Direct Sums and Tensor Products of Linear Spaces

In this section we show how to construct, by suitably combining two or more given linear spaces, certain other linear spaces. The concepts that arise have found useful applications. We begin with the following definition.

***Definition 2.12.1*** Let $\mathfrak{L}_1$ and $\mathfrak{L}_2$ be two linear spaces over the *same* associated field $\mathfrak{F}$. The *direct sum* of $\mathfrak{L}_1$ and $\mathfrak{L}_2$ is a linear space $\mathfrak{L}$, denoted $\mathfrak{L}_1 \oplus \mathfrak{L}_2$, whose elements are the set $\mathfrak{L}_1 \times \mathfrak{L}_2$ [i.e., $\mathbf{z} \equiv (\mathbf{x}, \mathbf{y}) \in \mathfrak{L}$, where $\mathbf{x} \in \mathfrak{L}_1$, and $\mathbf{y} \in \mathfrak{L}_2$], on which summation and scalar multiplication are defined as follows:

$$\text{For } \mathbf{z}_1 \equiv (\mathbf{x}_1, \mathbf{y}_1), \quad \mathbf{z}_2 \equiv (\mathbf{x}_2, \mathbf{y}_2), \quad \mathbf{z}_1 + \mathbf{z}_2 = (\mathbf{x}_1 + \mathbf{x}_2, \mathbf{y}_1 + \mathbf{y}_2)$$

$$\text{For } \mathbf{z} \equiv (\mathbf{x}, \mathbf{y}), \quad \alpha \in \mathfrak{F}, \quad \alpha\mathbf{z} = (\alpha\mathbf{x}, \alpha\mathbf{y})$$

The zero element $\mathbf{0}$ of $\mathfrak{L}$ is the element $(\mathbf{0}_1, \mathbf{0}_2)$, where $\mathbf{0}_1$ and $\mathbf{0}_2$ are the zero vectors of $\mathfrak{L}_1$ and $\mathfrak{L}_2$, respectively.

It is left to the reader to prove that the preceding definition does satisfy all the axioms of a linear space.

One may also have direct sums of subspaces, say, $L_1$ and $L_2$ of the same linear space $\mathfrak{L}$ in which case a distinction must be drawn between the *algebraic sum* $L_1 + L_2$ and the direct sum $L_1 \oplus L_2$. While $L_1 \oplus L_2$ consists of *pairs* of elements, the first element from each pair coming from $L_1$ and the other from $L_2$, $L_1 + L_2$ consists of *single* elements each of which can be expressed as the sum of an element from $L_1$ and an element from $L_2$. Thus $L_1 + L_2$ is a subspace of $\mathfrak{L}$, but $L_1 \oplus L_2$, although a linear space in itself, is not a linear subspace of $\mathfrak{L}$. The direct sum of subspaces is a useful concept, for example, when it is necessary to identify for a given element $\mathbf{z}$ of the algebraic sum $L_1 + L_2$, the vectors $\mathbf{x}_1 \in \mathbf{L}_1$ and $\mathbf{x}_2 \in \mathbf{L}_2$ which yield $\mathbf{z} = \mathbf{x}_1 + \mathbf{x}_2$. Such identification may be regarded as a mapping of $(L_1 + L_2)$ into $L_1 \oplus L_2$. Thus consider the example of the projection operator $\mathbf{P}$ in Section 2.8 defined on $\mathfrak{L}$ and the subsequent unique decomposition of each element of $\mathfrak{L}$ into the algebraic sum of an element from $R(\mathbf{P})$ and another from $N(\mathbf{P})$. This *decomposition* may be regarded as a mapping of $\mathfrak{L}$ onto $R(\mathbf{P}) + N(\mathbf{P})$.

We prove the following theorem, which refers to the dimension of a direct sum space.

***Theorem 2.12.1*** Let $\mathfrak{L} = \mathfrak{L}_1 \oplus \mathfrak{L}_2$, where $\mathfrak{L}_1$ and $\mathfrak{L}_2$ are linear spaces whose dimensions are $n_1$ and $n_2$, respectively. Then the dimension of $\mathfrak{L}$ is $n_1 + n_2$.

*Proof*: Let $\{\mathbf{e}_j\}_1^{n_1}$ be a basis in $\mathfrak{L}_1$ and $\{\mathbf{f}_i\}_1^{n_2}$ be a basis in $\mathfrak{L}_2$. Form the set $\{\mathbf{z}_k\}_1^{n_1+n_2}$, where

$$\mathbf{z}_j = (\mathbf{e}_j, \mathbf{0}_2), \qquad j = 1, 2, \ldots, n_1$$

$$\mathbf{z}_{n_1+i} = (\mathbf{0}_1, \mathbf{f}_i), \qquad i = 1, 2, \ldots, n_2$$

First we show that $\{\mathbf{z}_k\}_1^{n_1+n_2}$ is linearly independent. Consider the linear expansion of $\mathbf{0} = (\mathbf{0}_1, \mathbf{0}_2)$ in terms of $\mathbf{z}_1, \mathbf{z}_2, \ldots, \mathbf{z}_{n_1+n_2}$:

$$\alpha_1\mathbf{z}_1 + \alpha_2\mathbf{z}_2 + \ldots + \alpha_{n_1+n_2}\mathbf{z}_{n_1+n_2} = \mathbf{0}$$

which implies that

$$\alpha_1\mathbf{e}_1 + \alpha_2\mathbf{e}_2 + \ldots + \alpha_{n_1}\mathbf{e}_{n_1} = \mathbf{0}_1$$

and

$$\alpha_{n_1+1}\mathbf{f}_1 + \alpha_{n_2+2}\mathbf{f}_2 + \ldots + \alpha_{n_1+n_2}\mathbf{f}_{n_2} = \mathbf{0}_2$$

Since $\{\mathbf{e}_j\}_1^{n_1}$ is independent, the first of the two preceding equations implies that $\alpha_1 = \alpha_2 = \ldots = \alpha_{n_1} = 0$, while the second implies that $\alpha_{n_1+1} = \alpha_{n_2+2} = \ldots = \alpha_{n_1+n_2} = 0$ because of the independence of $\{\mathbf{f}_i\}_1^{n_2}$. Thus $\{\mathbf{z}_k\}_1^{n_1+n_2}$ is linearly independent.

We now show that $\{\mathbf{z}_j\}_1^{n_1+n_2}$ is a basis in $\mathfrak{L}$. Let $\mathbf{z} = (\mathbf{x}, \mathbf{y}) \in \mathfrak{L}$, where $\mathbf{x} \in \mathfrak{L}_1$ and $\mathbf{y} \in \mathfrak{L}_2$. Since $\{\mathbf{e}_j\}_1^{n_1}$ is a basis in $\mathfrak{L}_1$, $\exists\ \alpha_1, \alpha_2, \ldots, \alpha_{n_1} \in \mathfrak{F}$,

$$\mathbf{x} = \alpha_1\mathbf{e}_1 + \alpha_2\mathbf{e}_2 + \ldots + \alpha_{n_1}\mathbf{e}_{n_1}$$

Also $\exists$ $n_2$ numbers $\alpha_{n_1+1}, \alpha_{n_1+2}, \ldots, \alpha_{n_1+n_2}$

$$\mathbf{y} = \alpha_{n_1+1}\mathbf{f}_1 + \alpha_{n_1+2}\mathbf{f}_2 + \ldots + \alpha_{n_1+n_2}\mathbf{f}_{n_2}$$

Thus we have

$$\mathbf{z} = \alpha_1\mathbf{z}_1 + \alpha_2\mathbf{z}_2 + \ldots + \alpha_{n_1+n_2}\mathbf{z}_{n_1+n_2}$$

so that $\{\mathbf{z}_k\}_1^{n_1+n_2}$ is a basis in $\mathcal{L}$. Hence the dimension of $\mathcal{L}_1 \oplus \mathcal{L}_2$ is $(n_1 + n_2)$. ∎

We consider below some examples of direct sums.

*Examples*

$\mathcal{R}_n \oplus \mathcal{R}_m$: The direct sum of $\mathcal{R}_n$ and $\mathcal{R}_m$ consists of elements $(\mathbf{x}, \mathbf{y})$, where $\mathbf{x} = (x_1, x_2, \ldots, x_n) \in \mathcal{R}_n$ and $\mathbf{y} = (y_1, y_2, \ldots, y_m) \in \mathcal{R}_m$, which is essentially the same space as $\mathcal{R}_{n+m}$.

$\mathcal{C}[a, b] \oplus \mathcal{R}_n$: The elements of this direct sum may be represented by $[\mathbf{f}, \mathbf{x}]$, where $\mathbf{f} \equiv \{f(t)\}$ and $\mathbf{x} \equiv (x_1, x_2, \ldots, x_n)$. This space is infinite-dimensional.

It should be evident that direct sums may be extended to several linear spaces, say $\mathcal{R}_1, \mathcal{R}_2, \ldots, \mathcal{R}_n$, defined over a common associated field $\mathcal{F}$. Thus we may define a linear space $\mathcal{L} = \mathcal{L}_1 \oplus \mathcal{L}_2 \oplus \ldots \oplus \mathcal{L}_n$, in which an element $\mathbf{z}$ would be given by

$$\mathbf{z} \equiv (\mathbf{x}_1, \mathbf{x}_2, \ldots, \mathbf{x}_n), \qquad \mathbf{x}_i \in \mathcal{L}_i, \quad i = 1, 2, \ldots, n$$

Obviously, the associated field of $\mathcal{L}$ is $\mathcal{F}$. The definitions of vector addition, scalar multiplication, the zero element, and so on, should be self-evident.

We next turn to the *tensor product* of two linear spaces defined over the same associated field. This concept is somewhat more complicated than that of the direct sum. Before presenting a formal definition of the tensor product some discussion would be in order. The student familiar with mechanics will recognize the unit vectors $\mathbf{i}, \mathbf{j}, \mathbf{k}$ in three-dimensional physical space, in terms of which cartesian vectors are usually expressed. Similarly, one also has *dyads* of the type $\mathbf{ii}, \mathbf{ij}, \ldots$ in terms of which *second*-order tensors or *dyadics* may be expressed. For example, vectors $\mathbf{a} \equiv (a_1\mathbf{i} + a_2\mathbf{j} + a_3\mathbf{k})$ and $\mathbf{b} \equiv (b_1\mathbf{i} + b_2\mathbf{j} + b_3\mathbf{k})$ lead to the "product" $\mathbf{ab}$, which may be written as

$$\mathbf{ii}a_1b_1 + \mathbf{ij}a_1b_2 + \ldots + \mathbf{kk}a_3b_3$$

so that $\mathbf{ab}$ may be expressed as a linear combination of the *nine* dyads. Thus dyadics may be regarded as elements of a *nine*-dimensional space. Notice that this method of forming the product of two vectors satisfies the following relationships:

$$\mathbf{a}(\mathbf{b}+\mathbf{c}) = \mathbf{ab}+\mathbf{ac}$$
$$(\mathbf{a}+\mathbf{b})\mathbf{c} = \mathbf{ac}+\mathbf{bc}$$
$$(\alpha\mathbf{a})\mathbf{b} = \alpha(\mathbf{ab}) = \mathbf{a}(\alpha\mathbf{b})$$

where by a tensor equality, we have implied equality of the nine *corresponding* components. It is obvious that **ab** is not necessarily **ba**. The foregoing equalities imply that for fixed **b**, the product **ab** is linear in **a** and for fixed **a**, **ab** is linear in **b**; in other words, **ab** is a *bilinear* mapping of ordered pairs of vectors into the space of dyadics. We may then consider the collection of *all* linear combinations of the nine dyads and regard them as dyadics. It is of interest to note that not all dyadics may be *factorizable* into a product of the type **ab**. However, each dyadic may be expressible as a linear combination of *factorizable* dyadics.

We have now all the implements to define the tensor product of two linear spaces $\mathfrak{L}_1$ and $\mathfrak{L}_2$ which have the same associated field. First, there is a matter of notation. It is common to insert the symbol $\otimes$ between $\mathfrak{L}_1$ and $\mathfrak{L}_2$ to denote the tensor product $\mathfrak{L}_1 \otimes \mathfrak{L}_2$, and between vectors $\mathbf{x} \in \mathfrak{L}_1$ and $\mathbf{y} \in \mathfrak{L}_2$, to represent the product $\mathbf{x} \otimes \mathbf{y}$.

***Definition 2.12.2***† Let $\mathfrak{L}_1$ and $\mathfrak{L}_2$ be linear spaces over the same associated field $\mathfrak{F}$. The *tensor product* between $\mathfrak{L}_1$ and $\mathfrak{L}_2$, denoted $\mathfrak{L}_1 \otimes \mathfrak{L}_2$, consists of *pairs* of elements denoted $\mathbf{x} \otimes \mathbf{y}$, where $\mathbf{x} \in \mathfrak{L}_1$, $\mathbf{y} \in \mathfrak{L}_2$ and other elements that may be formally represented as linear combinations $\sum_i \alpha_i(\mathbf{x}_i \otimes \mathbf{y}_i)$, $\alpha_i \in \mathfrak{F}$, $\mathbf{x}_i \in \mathfrak{L}_1$, $\mathbf{y}_i \in \mathfrak{L}_2$, $i = 1, 2, \ldots$. The summation and scalar multiplication implied in the foregoing linear combination must satisfy the laws of commutativity, associativity, and distributivity *and* must satisfy the following rules of bilinearity:

(i) $(\mathbf{x}_1 + \mathbf{x}_2) \otimes \mathbf{y} = \mathbf{x}_1 \otimes \mathbf{y} + \mathbf{x}_2 \otimes \mathbf{y}$.
(ii) $\mathbf{x} \otimes (\mathbf{y}_1 + \mathbf{y}_2) = \mathbf{x} \otimes \mathbf{y}_1 + \mathbf{x} \otimes \mathbf{y}_2$.
(iii) $\alpha(\mathbf{x} \otimes \mathbf{y}) = (\alpha\mathbf{x}) \otimes \mathbf{y} = \mathbf{x} \otimes (\alpha\mathbf{y})$.

The definitions above lead to the fact that $\mathfrak{L}_1 \otimes \mathfrak{L}_2$ is a linear space with the zero element $\mathbf{0}_1 \otimes \mathbf{y}$ ($\mathbf{0}_1$ is the zero element of $\mathfrak{L}_1$ and $\mathbf{y}$ is an arbitrary element of $\mathfrak{L}_2$) or $\mathbf{x} \otimes \mathbf{0}_2$ ($\mathbf{0}_2$ is the zero element of $\mathfrak{L}_2$ and $\mathbf{x}$ is an arbitrary element of $\mathfrak{L}_1$). Note that they are not distinct choices because

$$\mathbf{0}_1 \otimes \mathbf{y} = 0\mathbf{x} \otimes \mathbf{y} = \mathbf{x} \otimes (0\mathbf{y}) = \mathbf{x} \otimes \mathbf{0}_2$$

which follows from property (iii) in Definition 2.12.2. That either element is

†There are other somewhat more abstract definitions of the tensor product also. See, for example, *Finite Dimensional Vector Spaces*, by P. R. Halmos, D. Van Nostrand, Princeton, N.J., 1958, p. 40.

the zero element of $\mathcal{L}_1 \otimes \mathcal{L}_2$, denoted $\mathbf{0}$, is left to be proved as an exercise. We consider below some examples of tensor product spaces.

*Example 1*

$\mathcal{R}_3$: The dyadics, discussed as a prelude to the definition of a tensor product, is an obvious example of it given by $\mathcal{R}_3 \otimes \mathcal{R}_3$. The dyads must now be represented by $\mathbf{i} \otimes \mathbf{j}, \mathbf{j} \otimes \mathbf{k}$, (instead of $\mathbf{ij}, \mathbf{jk}$), . . . .

*Example 2*

$\mathcal{R}_n \otimes \mathcal{R}_m$ is composed of elements $\mathbf{x} \otimes \mathbf{y}$, where $\mathbf{x} \equiv (x_1, x_2, \ldots, x_n) \in \mathcal{R}_n$ and $\mathbf{y} \equiv (y_1, y_2, \ldots, y_m) \in \mathcal{R}_m$. The vector $\mathbf{x} \otimes \mathbf{y}$ may also be represented by $(x_1y_1, x_1y_2, \ldots, x_1y_m, x_2y_1, x_2y_2, \ldots, x_2y_m, \ldots, x_ny_1, x_ny_2, \ldots, x_ny_m)$. This vector consists of $nm$ components.

*Example 3*

Let $I_1 \equiv \{x \in \mathcal{R}\colon a_1 \leq x \leq b_1\}$ and $I_2 \equiv \{y \in \mathcal{R}\colon a_2 \leq y \leq b_2\}$. The tensor product space $\mathcal{C}(I_1) \otimes \mathcal{C}(I_2)$ consists of elements of the type

$$\mathbf{f} \otimes \mathbf{g} = \{f(x)g(y)\colon x \in I_1, y \in I_2\}$$

and linear combinations of such elements. Thus a typical vector in $\mathcal{C}(I_1) \otimes \mathcal{C}(I_2)$, $\mathbf{u}$ will be given by

$$\mathbf{u} = \{u(x, y) = \sum_i f_i(x)g_i(y)\colon x \in I_1, y \in I_2\}$$

We now return to our discussion of the tensor product $\mathcal{L}_1 \otimes \mathcal{L}_2$ of abstract linear spaces $\mathcal{L}_1$ and $\mathcal{L}_2$.

It is readily shown from the Definition 2.12.2 that

$$(\sum_i \alpha_i \mathbf{x}_i) \otimes (\sum_j \beta_j \mathbf{y}_j) = \sum_i \sum_j \alpha_i \beta_j (\mathbf{x}_i \otimes \mathbf{y}_j)$$

If $\{\mathbf{x}_i\}_1^k$ is linearly independent in $\mathcal{L}_1$, then for each $y\ (\neq \mathbf{0}_2) \in \mathcal{L}_2$, the set $\{\mathbf{x}_i \otimes \mathbf{y}\}_1^k$ is readily shown to be linearly independent in $\mathcal{L}_1 \otimes \mathcal{L}_2$. A similar result holds using linearly independent sets, say $\{\mathbf{y}_j\}_1^l$ from $\mathcal{L}_2$. It is also easily proved that the set $\{\mathbf{x}_i \otimes \mathbf{y}_j\}$; $i = 1, 2, \ldots, k$; $j = 1, 2, \ldots, l\}$ comprising $kl$ vectors is linearly independent in $\mathcal{L}_1 \otimes \mathcal{L}_2$, for let $\alpha_{ij} \in \mathcal{F}$, $i = 1, 2, \ldots, k$; $j = 1, 2, \ldots, l$, such that

$$\sum_{i=1}^{k} \sum_{j=1}^{l} \alpha_{ij}(\mathbf{x}_i \otimes \mathbf{y}_j) = \mathbf{0}$$

Clearly,

$$\sum_{i=1}^{k} (\mathbf{x}_i \otimes \sum_{j=1}^{m} \alpha_{ij}\mathbf{y}_j) = \mathbf{0}$$

which implies that $\{\mathbf{x}_i \otimes \sum_{j=1}^{l} \alpha_{ij}\mathbf{y}_j\}_1^k$ is linearly dependent, which is impossible unless (see Exercise 2.12.3)

$$\sum_{j=1}^{l} \alpha_{ij}\mathbf{y}_j = \mathbf{0}_2, \qquad i = 1, 2, \ldots, k$$

The linear independence of $\{\mathbf{y}_j\}_1^l$ leads to the result that $\alpha_{ij} = 0, i = 1, 2, \ldots, k; j = 1, 2, \ldots, l$, thus implying that the set $\{\mathbf{x}_i \otimes \mathbf{y}_j; i = 1, 2, \ldots, k; j = 1, 2, \ldots, l\}$ is linearly independent. The following theorem utilizes the results above to establish the dimension of a tensor product space.

***Theorem 2.12.2*** Let $\mathcal{L}_1$ and $\mathcal{L}_2$ be linear spaces of dimensions $n_1$ and $n_2$, respectively. The dimension of $\mathcal{L}_1 \otimes \mathcal{L}_2$ is $n_1n_2$.

*Proof*: Let $\{\mathbf{x}_i\}_1^{n_1}$ and $\{\mathbf{y}_j\}_1^{n_2}$ be the respective bases in $\mathcal{L}_1$ and $\mathcal{L}_2$. We have shown that the set $\{\mathbf{x}_i \otimes \mathbf{y}_j; i = 1, 2, \ldots, n_1; j = 1, 2, \ldots, n_2\}$ is linearly independent in $\mathcal{L}_1 \otimes \mathcal{L}_2$. We will now show that this is a basis in $\mathcal{L}_1 \otimes \mathcal{L}_2$. We need to consider only elements of the type $\mathbf{x} \otimes \mathbf{y}$ for expansion in terms of the foregoing basis since the more general linear combinations of such elements will be automatically accounted for in the process.

Since $\mathbf{x} \in \mathcal{L}_1$, $\exists\ \alpha_1, \alpha_2, \ldots, \alpha_{n_1} \in \mathfrak{F} \ni$

$$\mathbf{x} = \sum_{i=1}^{n_1} \alpha_i\mathbf{x}_i$$

Since $\mathbf{y} \in \mathcal{L}_2$, $\exists\ \beta_1, \beta_2, \ldots, \beta_{n_2} \in \mathfrak{F} \ni$

$$\mathbf{y} = \sum_{j=1}^{n_2} \beta_j\mathbf{y}_j$$

Now

$$\mathbf{x} \otimes \mathbf{y} = \sum_{i=1}^{n_1}\sum_{j=1}^{n_2} \alpha_i\beta_j(\mathbf{x}_i \otimes \mathbf{y}_j)$$

which shows that $\{\mathbf{x}_i \otimes \mathbf{y}_j; i = 1, 2, \ldots, n_1; j = 1, 2, \ldots, n_2\}$ is a basis in $\mathcal{L}_1 \otimes \mathcal{L}_2$. Since there are $n_1n_2$ vectors, the dimension of $\mathcal{L}_1 \otimes \mathcal{L}_2$ is $n_1n_2$. ▮

Clearly, the space of dyadics discussed earlier, $\mathcal{R}_3 \otimes \mathcal{R}_3$, has the basis set $\{\mathbf{i} \otimes \mathbf{i}, \mathbf{i} \otimes \mathbf{j}, \mathbf{i} \otimes \mathbf{k}, \ldots, \mathbf{k} \otimes \mathbf{j}, \mathbf{k} \otimes \mathbf{k}\}$. Similarly, the basis set in $\mathcal{R}_n \otimes \mathcal{R}_m$ may be derived from the bases in $\mathcal{R}_n$ and $\mathcal{R}_m$. We are not in a position to discuss bases of infinite-dimensional spaces at the present stage, so that the discussion of their tensor products in regard to bases and so on must be deferred to a later stage.

Before concluding the chapter, we make some important observations of the operator spaces $\mathfrak{I}(\mathcal{L}_1 \oplus \mathcal{L}_2)$ and $\mathfrak{I}(\mathcal{L}_1 \otimes \mathcal{L}_2)$, that is, lincar transformations, which may be defined on the direct sum and tensor product spaces. We consider first the direct sum space $\mathcal{L}_1 \oplus \mathcal{L}_2$.

To identify an element of $\mathfrak{I}(\mathcal{L}_1 \oplus \mathcal{L}_2)$, we denote it by $\mathbf{T}$ and write

$$\mathbf{T}(\mathbf{x}_1, \mathbf{x}_2) = (\mathbf{y}_1, \mathbf{y}_2)$$

where $\mathbf{x}_1, \mathbf{y}_1 \in \mathfrak{L}_1$; $\mathbf{x}_2, \mathbf{y}_2 \in \mathfrak{L}_2$, $(\mathbf{x}_i, \mathbf{y}_i) \in \mathfrak{L}_1 \oplus \mathfrak{L}_2$, $i = 1, 2$. In general, we expect that $\mathbf{y}_1$ and $\mathbf{y}_2$ may depend on both $\mathbf{x}_1$ and $\mathbf{x}_2$. Thus elements from the operator spaces $\mathfrak{J}(\mathfrak{L}_1 \longrightarrow \mathfrak{L}_2)$ (i.e., operators mapping $\mathfrak{L}_1$ into $\mathfrak{L}_2$) and $\mathfrak{J}(\mathfrak{L}_2 \longrightarrow \mathfrak{L}_1)$ may also be required to specify $\mathbf{T}$ in addition to those from $\mathfrak{J}(\mathfrak{L}_1)$ and $\mathfrak{J}(\mathfrak{L}_2)$. A convenient way to represent $\mathbf{T}$ is to use a "matrix" form of the type

$$\mathbf{T} \equiv \begin{bmatrix} \mathbf{T}_{1,1} & \mathbf{T}_{1,2} \\ \mathbf{T}_{2,1} & \mathbf{T}_{2,2} \end{bmatrix}$$

where $\mathbf{T}_{1,1} \in \mathfrak{J}(\mathfrak{L}_1)$, $\mathbf{T}_{1,2} \in \mathfrak{J}(\mathfrak{L}_2 \longrightarrow \mathfrak{L}_1)$, $\mathbf{T}_{2,1} \in \mathfrak{J}(\mathfrak{L}_1 \longrightarrow \mathfrak{L}_2)$, and $\mathbf{T}_{2,2} \in \mathfrak{J}(\mathfrak{L}_2)$. The action of $\mathbf{T}$ on $\mathfrak{L}_1 \oplus \mathfrak{L}_2$ is represented by writing the vector $(\mathbf{x}_1, \mathbf{x}_2)$ as a column. Thus

$$\begin{bmatrix} \mathbf{T}_{1,1} & \mathbf{T}_{1,2} \\ \mathbf{T}_{2,1} & \mathbf{T}_{2,2} \end{bmatrix} \begin{bmatrix} \mathbf{x}_1 \\ \mathbf{x}_2 \end{bmatrix} = \begin{bmatrix} \mathbf{y}_1 \\ \mathbf{y}_2 \end{bmatrix}$$

from which one obtains $\mathbf{y}_i = \mathbf{T}_{i,1}\mathbf{x}_1 + \mathbf{T}_{i,2}\mathbf{x}_2$; $i = 1, 2$. Similarly, the operator in $\mathfrak{J}(\mathfrak{L}_1 \oplus \mathfrak{L}_2 \oplus \ldots \oplus \mathfrak{L}_n)$ may be represented by the matrix form†

$$\begin{bmatrix} \mathbf{T}_{1,1} & \mathbf{T}_{1,2} & \ldots & \mathbf{T}_{1,n} \\ \mathbf{T}_{2,1} & \mathbf{T}_{2,2} & \ldots & \mathbf{T}_{2,n} \\ \cdot & & & \\ \cdot & & & \\ \cdot & & & \\ \mathbf{T}_{n,1} & \mathbf{T}_{n,2} & \ldots & \mathbf{T}_{n,n} \end{bmatrix}$$

where $\mathbf{T}_{ij} \in \mathfrak{J}(\mathfrak{L}_j \longrightarrow \mathfrak{L}_i)$.

Since $\mathfrak{J}_j \equiv \mathfrak{J}(\mathfrak{L}_j)$ are themselves linear spaces, one can inquire into the direct sum space $\mathfrak{J}_1 \oplus \mathfrak{J}_2 \oplus \ldots \oplus \mathfrak{J}_n$. An element $\mathbf{T}$ of this space may be represented by the $n$-tuple $(\mathbf{T}_1, \mathbf{T}_2, \ldots, \mathbf{T}_n)$, which is defined by

$$\mathbf{T}(\mathbf{x}_1, \mathbf{x}_2, \ldots, \mathbf{x}_n) \equiv (\mathbf{T}_1\mathbf{x}_1, \mathbf{T}_2\mathbf{x}_2, \ldots, \mathbf{T}_n\mathbf{x}_n)$$

where $(\mathbf{x}_1, \mathbf{x}_2, \ldots, \mathbf{x}_n) \in \mathfrak{L}_1 \oplus \mathfrak{L}_2 \oplus \ldots \oplus \mathfrak{L}_n$. It is easy to see that when cast in a matrix form, the elements of $\mathfrak{J}_1 \oplus \mathfrak{J}_2 \oplus \ldots \oplus \mathfrak{J}_n$ are only the *diagonal* forms of those encountered in $\mathfrak{J}(\mathfrak{L}_1 \oplus \mathfrak{L}_2 \oplus \ldots \oplus \mathfrak{L}_n)$, which means that all the off-diagonal operators are zero operators.

We now consider the operator space $\mathfrak{J}(\mathfrak{L}_1 \otimes \mathfrak{L}_2)$. In order to identify a general linear transformation $\mathbf{T}$ on $\mathfrak{L}_1 \otimes \mathfrak{L}_2$, we need only define its action on an element of the type $(\mathbf{x}_1 \otimes \mathbf{x}_2) \in \mathfrak{L}_1 \otimes \mathfrak{L}_2$. The transformed element may be a linear combination of vector products. As in the case of the direct sum, we expect operators from $\mathfrak{J}(\mathfrak{L}_i \longrightarrow \mathfrak{L}_j)$, $i \neq j$, to play a role in the general element of $\mathfrak{J}(\mathfrak{L}_1 \oplus \mathfrak{L}_2)$. Thus we may write

$$\mathbf{T}(\mathbf{x}_1 \otimes \mathbf{x}_2) \equiv (\mathbf{T}_{1,1}\mathbf{x}_1 + \mathbf{T}_{1,2}\mathbf{x}_2) \otimes (\mathbf{T}_{2,1}\mathbf{x}_1 + \mathbf{T}_{2,2}\mathbf{x}_2)$$

†Note that this concept is similar to that of partitioned matrices in which the elements are themselves matrices.

where we have used operators $\mathbf{T}_{1,1} \in \mathfrak{I}_1$, $\mathbf{T}_{1,2} \in \mathfrak{I}(\mathfrak{L}_2 \longrightarrow \mathfrak{L}_1)$, $\mathbf{T}_{2,1} \in \mathfrak{I}(\mathfrak{L}_1 \longrightarrow \mathfrak{L}_2)$, and $\mathbf{T}_{2,2} \in \mathfrak{I}_2$ in defining $\mathbf{T}$. On the other hand, the space of operators $\mathfrak{I}_1 \otimes \mathfrak{I}_2$ consists of elements of the type $\mathbf{T}_1 \otimes \mathbf{T}_2$, $\mathbf{T}_i \in \mathfrak{I}_i$, $i = 1, 2, \ldots,$ defined by

$$(\mathbf{T}_1 \otimes \mathbf{T}_2)(\mathbf{x}_1 \otimes \mathbf{x}_2) \equiv \mathbf{T}_1\mathbf{x}_1 \otimes \mathbf{T}_2\mathbf{x}_2$$

which is characteristically free from the operators $\mathbf{T}_{1,2}$ and $\mathbf{T}_{2,1}$.

Clearly, $n$-fold tensor products of linear spaces of the type $\mathfrak{L}_1 \otimes \mathfrak{L}_2 \otimes \ldots \otimes \mathfrak{L}_n$ are readily defined and linear operators may be defined on them. The tensor product of operator spaces $\mathfrak{I}_1 \otimes \mathfrak{I}_2 \otimes \ldots \otimes \mathfrak{I}_n$ are also defined in a manner analogous to the binary case.

We are now in a position to consider some practical examples of operators on direct sum and tensor product spaces.

*Example 1. Temperature Distribution in a Peripherally Cooled Sphere*

We consider an internally heated sphere of unit radius surrounded by a *well-stirred* fluid which loses heat to the surrounding air at the rate $h_0A_0(t_f - t_a)$, where $h_0$, $A_0$, and $t_a$ are known constants representing the heat transfer coefficient to the surrounding air, the heat transfer area, and the air temperature, respectively. The fluid temperature $t_f$ is unknown and must be obtained by solving an equation to be stated presently. The steady-state temperature distribution in the sphere represented by $t(r)$ satisfies the heat conduction equation

$$\frac{k}{r^2}\frac{d}{dr}\left(r^2\frac{dt}{dr}\right) + s = 0, \qquad 0 < r < 1 \tag{2.12.1}$$

where $k$ is the conductivity of the sphere and $s$ is the volumetric heat generation rate which may be considered to be a function of $r$. The heat transfer to the fluid is described by the equation

$$-4\pi k t'(1) = h_0A_0(t_f - t_a) \tag{2.12.2}$$

The unknowns are the temperature distribution $t(r)$ *and* the fluid temperature $t_f$. Clearly, $\{t(r)\colon 0 \leq r \leq 1\} \equiv \mathbf{t}$ is a vector in a *function* space (i.e., an infinite-dimensional linear space) $\mathfrak{L}$. $\mathfrak{L}$ may be roughly identified as a subspace of vectors $\mathbf{f}$ in $\mathfrak{C}^{(1)}[0, 1]$ which allows $r^2f'(r)$ to be differentiated once more. Actually, the details here are considerably more involved, but it suffices our purpose to stipulate $\mathbf{t}$ to be from a linear space $\mathfrak{L}$ without further identification. The unknowns $\{t(r)\}$ and $t_f$ together constitute a *vector* in the direct sum space $\mathfrak{L} \oplus \mathfrak{R}$, where $\mathfrak{R}$ can be regarded as a linear space of unit dimension. Denoting a typical element of $\mathfrak{L} \oplus \mathfrak{R}$ by $\mathbf{u}$, we may write $[\{u_1(r)\}, u_2]$ for $\mathbf{u}$ or

$$\mathbf{u} \equiv \begin{bmatrix} \{u_1(r)\} \\ u_2 \end{bmatrix}$$

where $\{u_1(r)\} \in \mathfrak{L}$ and $u_2 \in \mathfrak{R}$. The element of special interest in the space above is $[\{t(r)\}, t_f]$. Next we show that (2.12.1) and (2.12.2) define a linear operator on $\mathfrak{L} \oplus \mathfrak{R}$. Rewriting (1) and (2) as†

$$\begin{bmatrix} -\frac{1}{r^2}\frac{d}{dr}\left(r^2\frac{d}{dr}\right) & 0 \\ \lim\limits_{r\to 1-}\frac{d}{dr} & \beta \end{bmatrix}\begin{bmatrix} t(r) \\ t_f \end{bmatrix} = \begin{bmatrix} \sigma(r) \\ \beta t_a \end{bmatrix}$$

where we have set $\sigma(r) \equiv s(r)/k$ and $\beta \equiv h_0 A_0/4\pi k$, we recognize an operator $\mathbf{T}$ on $\mathfrak{L} \oplus \mathfrak{R}$ defined by

$$\mathbf{Tu} \equiv \begin{bmatrix} \left\{-\frac{1}{r^2}\frac{d}{dr}\left(r^2\frac{du_1}{dr}\right)\right\} \\ u_1'(1) + \beta u_2 \end{bmatrix}$$

It is left for the reader to verify that $\mathbf{T}$ is indeed a linear transformation. The heat transfer problem may now be represented by the operator equation

$$\mathbf{T}\begin{bmatrix} \{t(r)\} \\ t_f \end{bmatrix} = \begin{bmatrix} \{\sigma(r)\} \\ \beta t_a \end{bmatrix} \tag{2.12.3}$$

The right-hand side of (2.12.3) is a vector in $R(\mathbf{T})$, on which we do not elaborate further since the purpose of the example is to demonstrate how applications give rise to operator equations on direct sum spaces. We turn to our next example, which shows the application of tensor products of spaces.

*Example 2. Two-dimensional Steady-state Conduction in a Rectangular Plate*

The rectangular domain is represented by $[0 \le x \le a; 0 \le y \le b]$. The steady-state temperature distribution $t(x, y)$ satisfies the heat conduction equation

$$\frac{\partial^2 t}{\partial x^2} + \frac{\partial^2 t}{\partial y^2} = 0, \qquad 0<x<a, \quad 0 < y < b$$

where no heat generation is assumed. Now $\{\partial^2/\partial x^2\}$ may be considered as an operator on $\mathfrak{C}^{(2)}[0, a]$ and $\{\partial^2/\partial y^2\}$ may be regarded as an operator on $\mathfrak{C}^{(2)}[0, b]$. If $I_x$ and $I_y$ denote the identity operators on $\mathfrak{C}^{(2)}[0, a]$ and $\mathfrak{C}^{(2)}[0, b]$, respectively, then the operator $\{\partial^2/\partial x^2 + \partial^2/\partial y^2\}$ may be looked upon as

$$\mathbf{T}_x \otimes \mathbf{I}_y + \mathbf{I}_x \otimes \mathbf{T}_y \equiv \mathbf{T}$$

where $\mathbf{T}_x \equiv \{\partial^2/\partial x^2\}$ and $\mathbf{T}_y \equiv \{\partial^2/\partial y^2\}$. The operator $\mathbf{T}$ operates on the tensor product space $\mathfrak{C}^{(2)}[0, a] \otimes \mathfrak{C}^{(2)}[0, b]$.

---

†The symbol $\lim_{r\to 1-}$ is used to indicate that the limit is approached from $r < 1$.

There are other ways of looking at this operator in the original equation but that the foregoing tensor product representation has distinct advantages will be established later in the book.

**EXERCISES**

**2.12.1** Recognize the following direct sum spaces by identifying typical elements.
(a) $\mathcal{C}[a, b] \oplus \mathcal{C}[b, c] \oplus \mathcal{R}$.
(b) $\mathcal{R}_n \oplus \mathcal{R}_n \oplus \ldots \oplus \mathcal{R}_n$ ($m$ times).

**2.12.2** Let $M$ and $N$ be subspaces of a linear space $\mathcal{L}$. Show that the mapping $M \oplus N \longrightarrow M + N$ is invertible if and only if $M \cap N = \{\mathbf{0}\}$.

**2.12.3** Let $\mathcal{L}_1 \otimes \mathcal{L}_2$ be the tensor product of linear spaces $\mathcal{L}_1$ and $\mathcal{L}_2$ whose zero elements are $\mathbf{0}_1$ and $\mathbf{0}_2$, respectively.
(a) Show that $\mathbf{x} \otimes \mathbf{y} = \mathbf{0} \longrightarrow \mathbf{x} = \mathbf{0}_1$ or $\mathbf{y} = \mathbf{0}_2$.
(b) Let $\{\mathbf{x}_j\}_{j=1}^k$ be a linearly independent set in $\mathcal{L}_1$. Show that the set $\{\mathbf{x}_j \otimes \mathbf{y}\}_{j=1}^k$ is linearly independent in $\mathcal{L}_1 \otimes \mathcal{L}_2$ if and only if $\mathbf{y} \neq \mathbf{0}_2$.
(c) Show that $(\sum_j \alpha_j \mathbf{x}_j) \otimes (\sum_i \beta_i \mathbf{y}_i) = \sum_j \sum_i \alpha_j \beta_i (\mathbf{x}_j \otimes \mathbf{y}_i)$.
(d) Let $\{\mathbf{u}_j\}_{j=1}^{n_1}$ be a basis in $\mathcal{L}_1$ and $\{\mathbf{v}_k\}_{k=1}^{n_2}$ be a basis in $\mathcal{L}_2$. Consider the element

$$\mathbf{z} \equiv \sum_{r=1}^{m} \alpha_r (\mathbf{x}_r \otimes \mathbf{y}_r) \in \mathcal{L}_1 \otimes \mathcal{L}_2$$

Show that $\mathbf{z}$ is the zero element in $\mathcal{L}_1 \otimes \mathcal{L}_2$ if and only if

$$\sum_{r=1}^{m} \alpha_r \beta_{rj} \gamma_{rk} = 0, \qquad j = 1, 2, \ldots, n_1; \quad k = 1, 2, \ldots, n_2$$

where $\{\beta_{rj}\}_{j=1}^{n_1}$ and $\{\gamma_{rk}\}_{k=1}^{n_2}$ are given by

$$\mathbf{x}_r = \sum_{j=1}^{n_1} \beta_{rj} \mathbf{u}_j \quad \text{and} \quad \mathbf{y}_r = \sum_{k=1}^{n_2} \gamma_{rk} \mathbf{v}_k$$

**2.12.4** Show that the following spaces are isomorphic.
(a) $\mathcal{R}_{n+m}$ and $\mathcal{R}_n \oplus \mathcal{R}_m$.
(b) $\mathcal{R}_{nm}$ and $\mathcal{R}_n \otimes \mathcal{R}_m$.

**2.12.5** Let $\mathcal{L}_1$ be the linear space of $n$th-degree polynomials and $\mathcal{L}_2$ be that of $m$th-degree polynomials. What does $\mathcal{L}_1 \otimes \mathcal{L}_2$ consist of?

## 2.13 Concluding Remarks

The vectors which we studied here came from both finite- and infinite-dimensional spaces. Insofar as the status of a vector is attained by an entity and its associates by collective observance of a stipulated set of axioms characterizing a linear space, we have shown that *functions*, too, are vectors as much as the more familiar finite-dimensional vectors. Although infinite-dimensional spaces were clearly defined to be those in which an arbitrary number of vectors could be linearly independent, the concept of a basis set could be defined only for a finite-dimensional space in this chapter. This is an inherent constraint from being limited to considerations of purely algebraic relationships between vectors, which does not allow one to interpret the meaning of an *infinite sum* of vectors.

Examples of function spaces were limited to no "worse" than continuous functions. Discontinuous functions too can form vector spaces. For example, the class of functions integrable in a given interval is an important linear space for our purposes. They have not been introduced in this chapter because they will arise more naturally in subsequent development.

The subject of linear transformations represents an important aspect of this chapter. It was pursued to a stage where algebraic relationships between *transformed* vectors helped to determine whether an inverse exists. For example, an operator with an inverse maps linearly independent sets into linearly independent sets. On the other hand, when an operator maps a linear space into a proper subspace of it, no inverse exists. The concept of an inverse is extremely important to the solution of linear equations featuring the operator.

It is important to recapitulate that the set of linear operators on a linear space is itself a linear space. When topological properties (such as distance between points) are added to this space, a useful framework is made available for approximate methods in the solution of mathematical equations. Indeed, there occurred no discussions on methods for determining the inverse of an operator. Partly this is because the requisite apparatus for the same is yet to be proposed. On finite-dimensional spaces, however, we recall the facts that each $n$-dimensional space $\mathcal{L}$ over the real field is isomorphic with $\mathcal{R}_n$ and the space of linear operators on $\mathcal{L}$ is isomorphic with the class of $n \times n$ matrices. This leads to the interesting consequence that the matrix representation of the inverse of an operator on $\mathcal{L}$ is the inverse of its matrix representation. Since the methods for determining the inverse of matrices are presumed to be known here, we have in effect provided a method for the inverse of an operator on finite-dimensional space. More generally, however, there remains considerably more to be said of operators on both finite- and infinite-dimensional spaces.

The result in Section 2.8 pertaining to the *decomposition* of a linear space $\mathcal{L}$ into the algebraic sum of the null and range spaces of a given projection operator is an important one. This idea has deep implications for subsequent stages of development of our subject.

## FURTHER READING

The engineering student will find comfortable reading in

*Finite-Dimensional Vector Spaces*, by P. R. Halmos, 2nd ed., D. Van Nostrand, Princeton, N.J., 1958.

This book is particularly useful since most formal results are preceded by lucid discussions.

The book

*Linear Operator Theory in Engineering and Science*, by A. W. Naylor and G. R. Sell, Holt, Rinehart and Winston, New York, 1971,

is somewhat more advanced on the subject but still provides a readable account. For example, the foregoing book sheds light on Zorn's lemma in regard to the existence of a basis in finite-dimensional space (in Section 2.3). The reader will find strong similarities between Zorn's lemma, which is equivalent to an "axiom of choice," and Theorem 1.8.1 or its equivalent, the axiom of continuity in Chapter 1 of this book.

An elementary treatment of the theory of linear spaces may be found in

*An Introduction to the Theory of Linear Spaces*, by G. E. Shilov, Prentice-Hall, Englewood Cliffs, N.J. 1961.

# Metric Spaces 3

## 3.0 Foreword

The linear operators encountered in Chapter 2 were somewhat general. Since only algebraic properties of linear spaces were considered, a large class of operators which have very interesting properties (related to the topological structure of linear spaces) that are useful in the solution of equations involving them were not recognized. These operators are extremely important in engineering applications.

The present chapter takes the first step in consideration of topological structure by treating it in isolation from algebraic structure.

## 3.1 Introduction

In Chapter 2 we were concerned with purely algebraic aspects of linear spaces and did not recognize the existence of such features as the "distance" between points or "magnitude" of a vector. In dealing with sums of vectors, we were naturally constrained to finite sums, since no meaning could be assigned to infinite sums of vectors within that framework. Also, one could not talk about such properties as continuity of functions defined on linear spaces. These properties, which are referred to as *topological properties* of a space, form the subject matter of the present chapter. Since the algebraic structure is independent of the topological structure of a space, we need not recognize the existence of the former in developing the subject of the latter. Such spaces, which are assigned only topological structure but not algebraic structure, are called *metric* spaces. Most of our examples of metric spaces are, however, derived from linear spaces, in which the definition of the distance between any two points (vectors) in the space may exploit algebraic properties such as sums of vectors. A formal axiomatic definition of a metric space follows.

***Definition 3.1.1*** A *metric space* is a pair of objects $(X, d)$, where $X$ is a non-empty set and $d$ is a real-valued function that maps pairs of elements from $X$ into the real numbers, that is, $d: X \times X \longrightarrow \mathcal{R}$. Further, the function $d$, called the *metric*, satisfies the following axioms.

A1 Positivity: $d(x, y) \geq 0, d(x, y) = 0 \leftrightarrow x = y$,
A2 Symmetry: $d(x, y) = d(y, x)$,
A3 Triangular inequality: $d(x, z) \leq d(x, y) + d(y, z)$, where $x, y$, and $z$ are arbitrary elements of $X$.

*Examples of Metric Spaces*

1. For our first example, we consider the set $X$ as the real line and the metric $d$ given by $d(x, y) = |x - y|$. It is easy to see that all the properties of a metric are satisfied by $d$.
2. Let $X = \mathcal{R}_n$ and define the metric $d$ by†

$$d(\mathbf{x}, \mathbf{y}) = \left\{\sum_{i=1}^{n} |x_i - y_i|^p\right\}^{1/p}, \qquad p \geq 1$$

That the function above satisfies axioms A1 and A2 is self-evident. The triangular inequality (axiom A3) is a restatement of the *Minkowski inequality* given by

$$\left\{\sum_{i=1}^{n} |x_i - y_i|^p\right\}^{1/p} \leq \left\{\sum_{i=1}^{n} |x_i|^p\right\}^{1/p} + \left\{\sum_{i=1}^{n} |y_i|^p\right\}^{1/p}$$

For a proof of the inequality above, we refer the reader to Taylor.‡ Note that $p = 2$ represents the familiar Euclidean distance.

---

†We use boldface letters for elements of $X$ when it is a linear space.

‡*Advanced Calculus*, by A. E. Taylor, Ginn, Boston, 1955. The proof of the Minkowski inequality depends on a more general inequality called the Hölder inequality, which is given by

$$\sum_{i=1}^{n} a_i b_i \leq \left[\sum_{i=1}^{n} a_i^p\right]^{1/p}\left[\sum_{i=1}^{n} b_i^q\right]^{1/q}, \qquad p > 0, \quad q > 0$$

and $1/p + 1/q = 1$; $a_i$ and $b_i$ are nonnegative real numbers.

3. Consider $X = \mathcal{C}[a, b]$ and the metric $d$ defined by

$$d(\mathbf{f}, \mathbf{g}) = \sup_{t \in [a,b]} |f(t) - g(t)|$$

Again axioms A1 and A2 are obviously satisfied, while axiom A3 is readily shown to hold. Let $\mathbf{f}, \mathbf{g}, \mathbf{h} \in \mathcal{C}[a, b]$. Now $\forall\ t \in [a, b]$ we have from the triangular inequality for real numbers

$$|f(t) - g(t)| \leq |f(t) - h(t)| + |h(t) - g(t)|$$

But

$$|f(t) - h(t)| \leq d(\mathbf{f}, \mathbf{h}) \quad \text{and} \quad |h(t) - g(t)| \leq d(\mathbf{h}, \mathbf{g})$$

so that

$$|f(t) - g(t)| \leq d(\mathbf{f}, \mathbf{h}) + d(\mathbf{h}, \mathbf{g})$$

Since $d(\mathbf{f}, \mathbf{g})$ is the lowest upper bound of $|f(t) - g(t)|$, $t \in [a, b]$, it follows that

$$d(\mathbf{f}, \mathbf{g}) \leq d(\mathbf{f}, \mathbf{h}) + d(\mathbf{h}, \mathbf{g})$$

## EXERCISES

**3.1.1** Let $X$ be any nonempty set and define $d\colon (X \times X) \longrightarrow \mathcal{R}$ by

$$d(x, y) = \begin{cases} 0 & \text{if } x = y \\ 1 & \text{if } x \neq y \end{cases}$$

Show that $(X, d)$ is a metric space.

**3.1.2** Let $(X, d)$ be a metric space. Define another function $d\colon (X \times X) \longrightarrow \mathcal{R}$ by

$$d'(x, y) = \frac{d(x, y)}{1 + d(x, y)}$$

Show that $(X, d')$ is a metric space.

**3.1.3** If $X = \mathcal{R}_n$ and $d(\mathbf{x}, \mathbf{y}) = \max |x_i - y_i|$, $i = 1, 2, \ldots, n$, prove that $(X, d)$ is a metric space.

**3.1.4** Let $X = \mathcal{C}_n$ and $d(\mathbf{x}, \mathbf{y}) = \{\sum_{i=1}^{n} (x_i - y_i)(x_i - y_i)^*\}^{1/2}$, where $(x_i - y_i)^*$ is the complex conjugate of $(x_i - y_i)$. Show that $(X, d)$ is a metric space.

**3.1.5** Let $X = \mathcal{R}_2$ and $d(\mathbf{x}, \mathbf{y}) = |x_1 - y_1|$. Is $(X, d)$ a metric space?

**3.1.6** Let $X = \mathcal{C}[a, b]$ and $d(\mathbf{f}, \mathbf{g}) = \int_a^b |f(t) - g(t)|\, dt$. Is $(X, d)$ a metric space?

**3.1.7** Consider $X$ as the set of all functions $f(t)$ defined for $t \in [a, b]$ such that $\int_a^b |f(t)|\, dt$ exists. Regard two functions $\mathbf{f}, \mathbf{g} \in X$ as equal if $f(t) = g(t)\ \forall\ t \in [a, b]$. Define $d$ by

$$d(\mathbf{f}, \mathbf{g}) = \int_a^b |f(t) - g(t)|\, dt$$

Is $(X, d)$ a metric space?

## 3.2 Convergence of Sequences

In Chapter 1 we defined convergence of a sequence of real numbers (Section 1.8). We are now in a position to define in much the same way convergence of a sequence derived from elements of an abstract metric space.

***Definition 3.2.1*** Let $\{x_n\}$ be a sequence contained in a metric space $(X, d)$ and $\tilde{x} \in X$. The sequence $\{x_n\}$ converges to $\tilde{x}$ (written as $x_n \longrightarrow \tilde{x}$) if $\forall\, \varepsilon > 0\ \exists$ an integer $n_\varepsilon \ni n > n_\varepsilon \rightarrow d(x_n, \tilde{x}) < \varepsilon$.

The definition means that members of the sequence *eventually approach* $\tilde{x}$ as *closely* as desired. The "close approach" is measured in terms of the metric $d$. The concept of convergence is relative to the metric; thus convergence of a sequence in a metric space $(X, d)$ relative to a given metric $d$ does not necessarily imply convergence relative to any other metric, say $d'$, that may be defined on $X \times X$. We shall demonstrate this aspect of convergence by a suitable example at a subsequent stage. A sequence is said to *diverge* if it does not converge.

It is also possible to define a sequence whose elements are eventually spaced arbitrarily close to each other. Such a sequence is called a Cauchy sequence and has the following formal definition.

***Definition 3.2.2*** A sequence $\{x_n\}$ contained in a metric space $(X, d)$ is called a *Cauchy sequence* if $\forall\, \varepsilon > 0\ \exists$ an integer $n_\varepsilon \ni m, n > n_\varepsilon \rightarrow d(x_m, x_n) < \varepsilon$.

Any convergent sequence is necessarily a Cauchy sequence since if $\{x_n\}$ is a sequence in $(X, d)$ converging to $\tilde{x} \in X$, then $\forall\, \varepsilon > 0 \exists$ an integer $n_{\varepsilon/2} \ni m$, $n > n_{\varepsilon/2} \rightarrow d(x_n, \tilde{x}) < \varepsilon/2$ and $d(x_m, \tilde{x}) < \varepsilon/2$. From the triangular inequality A3 we have $d(x_m, x_n) < \varepsilon$, so that $\{x_n\}$ is a Cauchy sequence.

The reverse question of whether or not a Cauchy sequence does converge to some element in $X$ is more subtle. Since eventually the elements of a Cauchy sequence are arbitrarily close to each other, it is suggestive of convergence, but this question is deferred to Section 3.6.

*Examples*

Some common examples of convergent and divergent sequences in the metric space $(X, d)$, where $X = \mathcal{R}$, $d(x, y) = |x - y|$, follow:

1. The series $1 - \frac{1}{2} + \frac{1}{3} - \frac{1}{4} + \ldots$ may be written as a sequence

$$x_n = \sum_{k=1}^{n} \frac{(-1)^{k-1}}{k}$$

which converges to ln 2.

2. The series $1 + \frac{1}{2} + \frac{1}{3} + \ldots$ or the sequence

$$x_n = \sum_{k=1}^{n} \frac{1}{k}$$

diverges and is called the harmonic series.

3. The series $1 + 1/2^r + 1/3^r + \ldots$ or the sequence

$$x_n = \sum_{k=1}^{n} \frac{1}{k^r}$$

converges for every $r > 1$ but diverges for $r \leq 1$. These results may be found in any text on calculus.†

**EXERCISES**

**3.2.1** Show that the point of convergence of a sequence in a metric space is unique.

**3.2.2** Consider the metric spaces $(X, d)$ and $(X, d')$ from Exercise 3.1.2, and a sequence $\{x_n\} \subset X$. If $\{x_n\}$ converges to $\tilde{x}$ relative to metric $d'$, does the sequence converge relative to metric $d$?

**3.2.3** Let $f$ be a real positive-valued function defined on the nonnegative real line such that $f(x) = 0$ if and only if $x = 0$. Further, $f$ is right continuous at $x = 0$. If $(X, d)$ is a metric space and $d' = f(d)$ also defines a metric on $X$, show that convergence of a sequence in $X$ with respect to $d$ implies convergence with respect to $d'$.

## 3.3 Continuity of Functions

A function defined on a metric space $(X, d)$ maps elements of $X$ into other elements of $X$ or into elements of another set $X'$, where $(X', d')$ is a second metric space. Symbolically, we may write $f\colon X \longrightarrow X$ or $f\colon X \longrightarrow X'$. A continuous function is defined as follows.

***Definition 3.3.1*** Let $(X, d)$ be a metric space and $f\colon X \longrightarrow X$. $f$ is said to be *continuous* at $x_0 \in X$, if $\forall\, \varepsilon > 0\ \exists\ \delta_\varepsilon > 0 \ni d(x, x_0) < \delta_\varepsilon \rightarrow d(f(x), f(x_0)) < \varepsilon$. $f$ is said to be continuous *on* a set $A \subset X$ if $f$ is continuous at every element, say, $x_0 \in A$. The $\delta_\varepsilon$ may depend on $x_0$. If, however, $\delta_\varepsilon$ is independent of $x_0$, $f$ is said to be *uniformly continuous* on $A$.

If $f$ maps elements of a metric space $(X_1, d_1)$ into another metric space $(X_2, d_2)$, the foregoing definitions of continuity are easily extended. Thus con-

†See, for example, *Advanced Calculus*, by A. E. Taylor, Ginn, Boston, 1955.

tinuity at $x_0 \in X_1$ implies that $\forall\, \varepsilon > 0\ \exists\ \delta_\varepsilon > 0 \ni d_1(x, x_0) < \delta_\varepsilon \rightarrow d_2(f(x), f(x_0)) < \varepsilon$.

We consider some examples of continuous functions.

*Example 1*

Let the metric space $(X, d)$ be given by $X = [a, b]$, $d(x, y) = |x - y|$, and $f: X \longrightarrow X$ be defined by $f = px^2 + qx + r$, where $p$, $q$, and $r$ are real numbers and $x \in X$. The continuity of $f$ at any point $x_0$ is readily established. Thus

$$\begin{aligned} d(f(x), f(x_0)) &= |f(x) - f(x_0)| = |px^2 - px_0^2 + qx - qx_0| \\ &\leq (|p||x + x_0| + |q|)|x - x_0| \qquad (3.3.1) \\ &\leq \{|p|(|x| + |x_0|) + |q|\}d(x, x_0) \end{aligned}$$

where the inequalities originate from the use of the triangular inequality for real numbers. If we impose the condition $|x - x_0| \leq 1$, then $|x| \leq 1 + |x_0|$, again from the triangular inequality. Thus inequality (3.3.1) becomes

$$d(f(x), f(x_0)) \leq \{|p|(1 + 2|x_0|) + |q|\}d(x, x_0)$$

Clearly, if

$$\delta_\varepsilon = \min\left[1, \frac{\varepsilon}{\{|p|(1 + 2|x_0|) + |q|\}}\right]$$

we have the result $d(x, x_0) < \delta_\varepsilon \rightarrow d(f(x), f(x_0)) < \varepsilon$, which implies the continuity of $f(x)$ at $x_0$.

Since $[a, b]$ is a bounded interval $\exists$ a number $m > 0 \ni |x| \leq m, \forall\, x \in [a, b]$. Thus inequality (3.3.1) may be rewritten as

$$d(f(x), f(x_0)) \leq \{2m|p| + |q|\}d(x, x_0)$$

Thus if $\delta_\varepsilon = \varepsilon/\{2m|p| + |q|\}$, $d(x, x_0) < \delta_\varepsilon \rightarrow d(f(x), f(x_0)) < \varepsilon$, which implies that $f(x)$ is uniformly continuous on $[a, b]$ since $\delta_\varepsilon$ does not depend on $x_0$. In fact, any function continuous on a closed bounded interval is uniformly continuous there.

*Example 2*

Consider the metric space $(X, d)$, where $X = \mathcal{C}[a, b]$ and $d(\mathbf{f}, \mathbf{g}) = \sup_{t \in [a,b]} |f(t) - g(t)|$, and the function $\phi$ defined by

$$\phi(f(t)) = \int_a^t f^2(t')\, dt'$$

We have $\phi: X \longrightarrow X$, since the integral is a continuous function of $t$. We will show below that $\boldsymbol{\phi}$ is continuous on $X$. Let $\mathbf{g}$ be a fixed element of $X$.

$$
\begin{aligned}
d(\boldsymbol{\phi}(\mathbf{f}), \boldsymbol{\phi}(\mathbf{g})) &= \sup_{t\in[a,b]} \left| \int_a^t [f^2(t') - g^2(t')]\, dt' \right| \\
&\leq \sup_{t\in[a,b]} \int_a^t | f^2(t') - g^2(t') |\, dt' \\
&= \int_a^b | f^2(t') - g^2(t') |\, dt' \qquad \text{(why?)} \\
&\leq \int_a^b | f(t') + g(t') || f(t') - g(t') |\, dt' \\
&\leq (b-a)[\sup_{t\in[a,b]} | f(t) | + \sup_{t\in[a,b]} | g(t) |]\, d(\mathbf{f}, \mathbf{g}) \qquad (3.3.2)
\end{aligned}
$$

Now

$$d(\mathbf{f}, \mathbf{g}) = \sup_{t\in[a,b]} | f(t) - g(t) |$$

Since $\mathbf{g} \in \mathcal{C}[a, b]$, $g_0 \equiv \sup_{t[a,b]} | g(t) |$ exists. Suppose that we restrict $\mathbf{f}$ such that $d(\mathbf{f}, \mathbf{g}) < 1$; then $\forall\ t \in [a, b]$, $| f(t) - g(t) | < 1$, so that $| f(t) | \leq 1 + | g(t) | \leq 1 + g_0$.

Proceeding with inequality (3.3.2), we have

$$d(\boldsymbol{\phi}(\mathbf{f}),\boldsymbol{\phi}(\mathbf{g})) \leq (b-a)(1+2g_0)d(\mathbf{f}, \mathbf{g})$$

If

$$\delta_\varepsilon = \min\left[1, \frac{\varepsilon}{(b-a)(1+2g_0)}\right]$$

then

$$d(\mathbf{f}, \mathbf{g}) < \delta_\varepsilon \longrightarrow d(\boldsymbol{\phi}(\mathbf{f}), \boldsymbol{\phi}(\mathbf{g})) < \varepsilon$$

This implies the continuity of $\boldsymbol{\phi}$ at "point" $\mathbf{g}$ in $\mathcal{C}[a, b]$. Since $\mathbf{g}$ was an arbitrary element of $\mathcal{C}[a, b]$, $\boldsymbol{\phi}$ is continuous everywhere in $\mathcal{C}[a, b]$.

The following theorem is of interest.

***Theorem 3.3.1*** Let $(X, d)$ be a metric space and $f\colon X \longrightarrow X$. $f$ is continuous at a point $x_0$ *if and only if* $\lim f(x_n) = f(x_0)$ for *every* sequence $\{x_n\}$ with $x_n \longrightarrow x_0$.

*Proof*: We prove the "only if" part first. Thus assume that $f(x)$ is continuous at $x_0$. Let $\{x_n\}$ be a sequence with $x_n \longrightarrow x_0$. Given $\varepsilon > 0$, $\exists\ \delta_\varepsilon > 0 \ni d(x, x_0) < \delta_\varepsilon \rightarrow d(f(x), f(x_0)) < \varepsilon$. Given $\delta_\varepsilon$, $\exists\ n_{\delta_\varepsilon} \ni n > n_{\delta_\varepsilon} \rightarrow d(x_n, x_0) < \delta_\varepsilon \rightarrow d(f(x_n), f(x_0)) < \varepsilon$. Thus $f(x_n) \longrightarrow f(x_0)$, which is the required result.

To prove the "if" part, assume that $\lim f(x_n) = f(x_0)$ for every sequence $\{x_n\}$ with $x_n \longrightarrow x_0$. We must prove that $f(x)$ is continuous at $x_0$. If $f$ is not continuous at $x_0$, then there exists a certain $\varepsilon_1 > 0 \ni$ for every $\delta > 0$ there is some $x$, satisfying $d(x, x_0) < \delta$, for which $d(f(x), f(x_0)) \geq \varepsilon_1$. Now consider a sequence $\delta_1, \delta_2, \ldots, \delta_n, \ldots$ such that $\delta_n \longrightarrow 0$ and form a sequence

$\{x_n\}$ by selecting $x_n$ such that $d(x_n, x_0) < \delta_n$ and $d(f(x_n), f(x_0)) \geq \varepsilon_1$. As $\delta_n \longrightarrow 0$, $x_n \longrightarrow x_0$ (why?). Since for every $x_n$, $d(f(x_n), f(x_0)) \geq \varepsilon_1$, $f(x_n)$ does not converge to $f(x_0)$, which is a contradication. Thus $f$ is continuous at $x_0$. ∎

**EXERCISES**

**3.3.1** Consider a sequence of functions $\{f_n(t)\}$ defined and continuous on $[a, b]$ such that for each fixed $t$ the sequence of real numbers converges to some number $f_0(t)$. The convergence is said to be *uniform* if $\forall\, \varepsilon > 0\ \exists\ n_\varepsilon$ *independent* of $t \ni n > n_\varepsilon \longrightarrow |f_n(t) - f_0(t)| < \varepsilon$. When $\{f_n(t)\}$ converges uniformly to $f_0(t)$, show that if each $f_n(t)$ is continuous at $t_0 \in [a, b]$, $f_0(t)$ is also continuous at $t_0$. (Thus if each $f_n(t)$ is continuous on $[a, b]$, $f_0(t)$ is continuous on $[a, b]$.)

Further show that the uniform convergence of continuous functions above is equivalent to convergence of $\{\mathbf{f}_n\} \subset \mathcal{C}[a, b]$ to $\mathbf{f}_0 \in \mathcal{C}[a, b]$ relative to the metric

$$d(\mathbf{f}, \mathbf{g}) = \sup_{t \in [a,b]} |f(t) - g(t)|$$

**3.3.2** Using Theorem 3.3.1, justify the equality

$$\int_a^b f(t)\, dt = \lim_{n \to \infty} \int_a^b f_n(t)\, dt$$

where $\{\mathbf{f}_n\} \subset \mathcal{C}[a, b]$ converges to $\mathbf{f}$ relative to the metric

$$d(\mathbf{f}, \mathbf{g}) = \sup_{t \in [a,b]} |f(t) - g(t)|$$

## 3.4 Pseudometric Spaces

The concept of a pseudometric space is an important element of analysis. Although the concept is itself not difficult to understand, the motivation for introducing it may not be readily appreciated at this stage; a careful study of the examples may in part provide the motivation.

***Definition 3.4.1*** A *pseudometric space* is a pair of objects $(X, \rho)$ where $X$ is a nonempty set and $\rho\colon (X \times X) \longrightarrow \mathcal{R}$ is called a *pseudometric*, which satisfies the following axioms.

- A1 Positivity: $\rho(x, y) \geq 0, \quad \rho(x, x) = 0,$
- A2 Symmetry: $\rho(x, y) = \rho(y, x),$
- A3 Triangular inequality: $\rho(x, z) \leq \rho(x, y) + \rho(y, z)$, where $x$, $y$, and $z$ are arbitrary elements of $X$.

In particular note that $\rho(x, y) = 0$ does *not* imply that $x = y$. It is here where a pseudometric differs from a metric.

*Example 1*

Let $X = \mathcal{R}_2$ and $\mathbf{x} = (x_1, x_2) \in \mathcal{R}_2$. Define $\rho: (X \times X) \to \mathcal{R}$ by

$$\rho(x, y) = |x_1 - y_1|$$

Clearly, $\rho$ satisfies all the axioms of a pseudometric space. In particular, note that $\rho(\mathbf{x}, \mathbf{y}) = 0 \Rightarrow x_1 = y_1$, which says nothing about $x_2$ and $y_2$; thus it does *not* follow that $\mathbf{x} = \mathbf{y}$. Geometrically, the pseudometric here refers to the projection of what is normally considered as distance between two points on the horizontal axis. Thus all pairs of distinct points lying along the vertical direction have zero pseudometric.

*Example 2*

For the second example (also see Exercise 3.1.7) we consider $X$ to be the space of functions defined on the closed interval $[a, b]$ that are absolutely integrable in the sense of Riemann.† We denote this space by $\mathcal{R}_1[a, b]$.‡ Define a function $\rho: \mathcal{R}_1[a, b] \times \mathcal{R}_1[a, b] \to \mathcal{R}$ by

$$\rho(\mathbf{f}, \mathbf{g}) = \int_a^b |f(t) - g(t)| \, dt$$

The positivity and symmetry properties of $\rho$ are immediately established. The triangular inequality follows from that for real numbers and the property that the integral of a positive-valued function is positive. That $\rho$ is a pseudometric is the result of Exercise 3.1.7.

We will now show that every pseudometric space can be converted into a metric space, called a *quotient space*, in the following manner.

Define an equivalence relationship between elements $x$ and $y$ of $X$, denoted $x \sim y$, by $\rho(x, y) = 0$. Thus $x \sim y \Rightarrow y \sim x$, which obtains from the symmetry of $\rho$. Also, $x \sim x$.

Further, $\forall\, x \in X$ form the set

$$[x] = \{y \in X\colon x \sim y\}$$

which is called an *equivalence class* belonging to $x$; of course, an equivalence class may be represented by any of its members. Two equivalent classes $[x]$ and $[y]$ may be defined to be *equal* if $x \in [x]$, $y \in [y] \Rightarrow x \sim y$; that is, they are identical sets.

---

†The normal integral with which the student is familiar at this stage is indeed the Riemann integral. The definition of the Riemann integral is dealt with in Chapter 4. Absolute integrability of $f(x)$ implies that $|f(x)|$ is integrable.

‡It is easy to show that $\mathcal{R}_1[a, b]$ is a linear space, which is why boldface letters have been used to denote its elements.

For any two equivalence classes $[x]$ and $[y]$, and $x_1, x_2 \in [x]$, $y_1, y_2 \in [y]$, it is easy to show that $\rho(x_1, y_1) = \rho(x_2, y_2)$ because $\rho(x_1, y_1) \leq \rho(x_1, x_2) + \rho(x_2, y_1) = \rho(x_2, y_1) \leq \rho(x_2, y_2) + \rho(y_2, y_1) = \rho(x_2, y_2)$, so that $\rho(x_1, y_1) \leq \rho(x_2, y_2)$. Analogously, starting from $\rho(x_2, y_2)$ it is easily seen that $\rho(x_2, y_2) \leq \rho(x_1, y_1)$, which proves that $\rho(x_1, y_1) = \rho(x_2, y_2)$.

Consider the set of *all* equivalence classes of $X$, and denote the set by $\hat{X}$. Note the difference between $X$ and $\hat{X}$ in the symbolic statement $[x] \subset X$; $[x] \in \hat{X}$. We now define a function $d: (\hat{X} \times \hat{X}) \longrightarrow \mathcal{R}$ by

$$d([x], [y]) = \rho(x, y), \qquad [x], [y] \in \hat{X}$$

where $x \in [x]$ and $y \in [y]$ are arbitrary elements. Obviously, $d$ inherits all the properties of $\rho$. Besides

$$\begin{aligned} d([x], [y]) = 0 &\rightarrow \rho(x, y) = 0 \\ &\rightarrow x \sim y \;\forall\; x \in [x], \quad y \in [y] \\ &\rightarrow [x] = [y] \end{aligned}$$

Thus $(\hat{X}, d)$ is indeed a metric space and is a quotient metric space of the pseudometric space $(X, \rho)$. We now investigate the quotient metric spaces of the examples presented earlier.

In the first example in which $X = \mathcal{R}_2$ and $\rho(\mathbf{x}, \mathbf{y}) = |x_1 - y_1|$, the equivalence class $[\mathbf{x}]$ for each $\mathbf{x} = (x_1, x_2)$ is defined by

$$[\mathbf{x}] = \{\mathbf{y} \in \mathcal{R}_2 \colon x_1 = y_1\}$$

Geometrically, this includes all the vectors whose projections on the $x_1$-axis are equal (see Figure 3.4.1). Thus each vertical line may be regarded as an equivalence class represented by any point on the line. The quotient metric space $\hat{X}$ consists of vertical lines as its elements and the metric defined is the perpendicular distance between the lines.

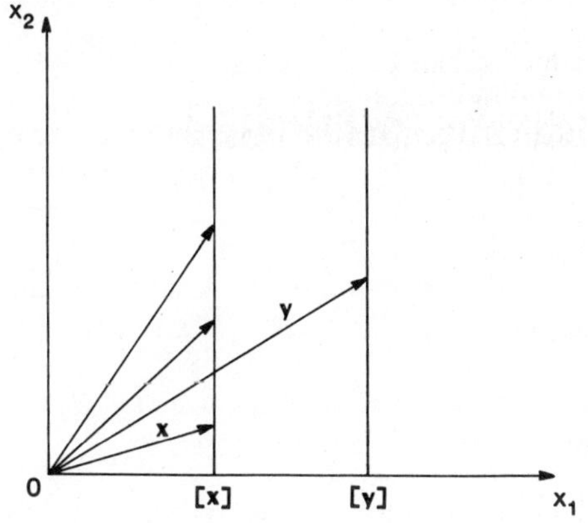


**Figure 3.4.1** Example of a pseudometric space.

The second example is of particular importance in analysis. The equivalence class belonging to $\mathbf{f} \in \mathcal{R}_1[a, b]$ consists of functions "equidistant" from $\mathbf{f}$ in the sense of the integral pseudometric. Thus two functions are to be regarded as *equal* even if they are not equal at a "selected" number of points in the interval. The basis of such "selection" is deferred to a subsequent chapter, although it should be clear at the present stage that inequality at a finite number of points may be allowed.

**EXERCISE**

**3.4.1** Let $X$ be any nonempty set and define $\rho: (X \times X) \longrightarrow \mathcal{R}$ such that $\rho = 0$ or 1 and is symmetric. If $\rho(x, x) = 0$, show that $(X, \rho)$ is a pseudometric space. Find a quotient metric space.

## 3.5 Subsets of Metric Spaces

If every element of a set $A$ is an element of metric space $(X, d)$, $A$ is said to be a *subset* of $X$. The *complement of a set* $A$, denoted $A'$, is the set of all points in $X$ that do not belong to $A$. Thus for every subset $A$ of $X$, we have $A \cup A' = X$. Naturally, the subsets of a metric space inherit the topological structure of $X$, which forms the basis of classifying different types of sets in metric spaces. We first define an *ε-neighborhood* about a point $x_0 \in X$.

***Definition 3.5.1*** Let $(X, d)$ be a metric space and $x_0 \in X$. An *ε-neighborhood* about $x_0$ is defined as the set

$$B_\varepsilon(x_0) = \{x \in X: d(x, x_0) < \varepsilon\}$$

Note that the "boundary" points given by $d(x, x_0) = \varepsilon$ are excluded from $B_\varepsilon(x_0)$. Obviously, $x_0 \in B_\varepsilon(x_0)$. If $x_0$ is excluded from $B_\varepsilon(x_0)$, it is referred to as a *deleted ε-neighborhood.*

We now define an interior point of a set.

***Definition 3.5.2*** Let $(X, d)$ be a metric space and $A \subset X$. $x_0 \in A$ is said to be an *interior point* of $A$ if $\exists\ \varepsilon > 0 \ni B_\varepsilon(x_0) \subset A$.

This definition merely asserts that an interior point of a set is one which has a neighborhood (sufficiently small) that is entirely contained in the set; the concept of an open set emerges from that of an interior point.

***Definition 3.5.3*** A set $A \subset X$ of a metric space $(X, d)$ is said to be an *open set* if *every* element $x \in A$ is an interior point.

Clearly, an ε-neighborhood $B_\varepsilon(x_0)$ is an open set, for let $x_1 \in B_\varepsilon(x_0)$. Then $d(x_1, x_0) < \varepsilon$. Let $\varepsilon_1 = \varepsilon - d(x_1, x_0) > 0$ and take $\varepsilon_0$ to be any number such that $0 < \varepsilon_0 < \varepsilon_1$. We will show that $B_{\varepsilon_0}(x_1) \subset B_\varepsilon(x_0)$, which will imply that $B_\varepsilon(x_0)$ is an open set. Let $x \in B_{\varepsilon_0}(x_1)$. Then $d(x, x_1) < \varepsilon_0$. Now $d(x, x_0) < d(x, x_1) + d(x_1, x_0) < \varepsilon_0 + d(x_1, x_0) < \varepsilon_1 + d(x_1, x_0) = \varepsilon \rightarrow x \in B_\varepsilon(x_0)$, so that $B_\varepsilon(x_0)$ is an open set.

*Examples of Open Sets*

1. The interval $(a, b)$ which excludes the end points $a$ and $b$ is an open subset of the metric space formed by the set of real numbers with the metric $d(x, y) = |x - y|$.
2. Another example is the rectangle

$$A = \begin{cases} a_1 < x < a_2 \\ b_1 < y < b_2 \end{cases}$$

which excludes its boundaries and is a subset of $\mathcal{R}_2$.

The concept of open sets allows an important characterization of convergence of sequences and continuity of functions. We first recast the definition of convergence and continuity in terms of $\varepsilon$-neigborhoods. Thus a sequence $\{x_n\} \subset X$ of a metric space $(X, d)$ converges to $x_0$ if $\forall\, \varepsilon > 0\ \exists$ an integer $n_\varepsilon \ni n > n_\varepsilon \rightarrow x_n \in B_\varepsilon(x_0)$. Similarly, a function is continuous at $x_0 \in X$ if $\forall\, \varepsilon > 0\ \exists\ \delta_\varepsilon > 0 \ni f(B_{\delta_\varepsilon}(x_0)) \subset B_\varepsilon(f(x_0))$.† The following theorems can now be established.

***Theorem 3.5.1*** A sequence $\{x_n\}$ in a metric space $(X, d)$ converges to $x_0$ *if and only if* the sequence is eventually contained in every open set containing $x_0$.

*Proof*: We prove the "if" condition first; that is, assume that the sequence is eventually contained in every open set containing $x_0$. Since $\forall\, \varepsilon > 0$ $B_\varepsilon(x_0)$ is an open set, $\exists\ n_\varepsilon \ni n > n_\varepsilon \rightarrow x_n \in B_\varepsilon(x_0)$. Thus we have $x_n \longrightarrow x_0$.

To prove the "only if" condition, assume that $x_n \longrightarrow x_0$. Let $A$ be any open set containing $x_0$. Then $\exists\ \varepsilon > 0 \ni B_\varepsilon(x_0) \subset A$. Now $\exists\ n_\varepsilon \ni n > n_\varepsilon \rightarrow x_n \in B_\varepsilon(x_0) \rightarrow x_n \in A$. Thus the sequence is eventually contained in every open set containing $x_0$. ∎

***Theorem 3.5.2*** Let $f$: $(X_1, d_1) \longrightarrow (X_2, d_2)$. Then the function $f$ is continuous on $X_1$ *if and only if* the inverse image of every open set in $X_2$ is an open set in $X_1$.

*Proof*: Denote the inverse image of a set $A \subset X_2$ by $f^{-1}(A)$. We prove the "only if" condition first; that is, assume that $f$ is continuous. Let $A \subset X_2$ be open. Consider $x_0 \in f^{-1}(A) \subset X_1$. Then $f(x_0) \in A \subset X_2$. Since $A$ is

---

†When a set $A$ is used as the argument of a function $f$, then $f(A)$ represents the set of points obtained by transforming all points of $A$.

open, $\exists\ \varepsilon > 0 \ni B_\varepsilon(f(x_0)) \subset A$. As $f$ is continuous at $x_0$, $\exists\ \delta_\varepsilon > 0 \ni f(B_{\delta_\varepsilon}(x_0)) \subset B_\varepsilon(f(x_0))$. Thus $B_{\delta_\varepsilon}(x_0) \subset f^{-1}(B_\varepsilon(f(x_0)) \subset f^{-1}(A)$. Thus the arbitrary point $x_0 \in f^{-1}(A)$ is an interior point, so that $f^{-1}(A)$ is open.

Next we consider the "if" condition. Assume that $f^{-1}(A)$ is open in $X_1$ for every open set $A$ in $X_2$. Let $x_0 \in X_1$ and $f(x_0) \in X_2$. For any given $\varepsilon > 0$, $B_\varepsilon(f(x_0))$ is an open set in $X_2$ and its inverse image $f^{-1}(B_\varepsilon(f(x_0)))$ is open in $X_1$. Since $x_0 \in f^{-1}(B_\varepsilon(f(x_0)))$, $\exists$ a $\delta_\varepsilon$-neighborhood $B_{\delta_\varepsilon}(x_0) \subset f^{-1}(B_\varepsilon(f(x_0)))$. Thus $f(B_{\delta_\varepsilon}(x_0)) \subset B_\varepsilon(f(x_0))$, so that $f$ is continuous at $x_0$. Since $x_0$ is arbitrary, $f$ is continuous on $X_1$. ▮

Theorems 3.5.1 and 3.5.2 are important in the characterization of convergence and continuity in topological spaces in which all topological properties are derived from the class of open sets without reference to a metric.†

A concept complementary to that of an open set is of a closed set, which is defined below.

***Definition 3.5.4*** A set $A \subset X$ of a metric space $(X, d)$ is said to be a *closed set* if its complement $A'$ is open.

Closed sets have properties strikingly different from those of open sets. Closedness is particularly important in regard to a property relating to the accumulation point of a set.

***Definition 3.5.5*** Let $A \subset X$ where $(X, d)$ is a metric space. $x_0 \in X$ is said to be an *accumulation point* (also called a *limit point*) if $\forall\ \varepsilon > 0$ the *deleted* $\varepsilon$-neigborhood about $x_0$ contains points of $A$.

Note that an interior point of a set is an accumulation point of the set, but an accumulation point may or may not belong to the set. It is usual to denote the set of all accumulation points of the set $A$ by $A^*$. The following theorem establishes the property that a closed set contains all its accumulation points.

***Theorem 3.5.3*** Let $A \subset X$ where $(X, d)$ is a metric space. $A$ is closed *if and only if* $A^* \subset A$.

*Proof*: To prove the "if" condition, assume that $A^* \subset A$. Consider any $x_0 \in A'$. Since $x_0 \notin A^*$, $\exists$ some $\varepsilon > 0 \ni B_\varepsilon(x_0) \subset A'$ (why?) which implies that $A'$ is open, so that $A$ is closed.

---

†See, for example, *Introduction of Topology and Modern Analysis,* by G. F. Simmons, McGraw-Hill, New York, 1963.

To prove the "only if" condition, assume that $A$ is closed. Thus $A'$ is open. Let $x_0 \in A^*$. If $x_0 \in A'$, then $\exists$ some $\varepsilon > 0 \ni B_\varepsilon(x_0) \subset A'$. Thus $x_0 \notin A^*$, which is a contradiction. Hence $x_0 \in A$, so that $A^* \subset A$. ∎

It is clear that for an arbitrary set $A$, $A \cup A^*$ is a closed set. The set $\bar{A} = A \cup A^*$ is called the *closure* of $A$. Theorem 3.5.3 provides a useful test for establishing closedness of a set (see Exercise 3.5.4).

It is interesting to ask whether accumulation points will exist for any subset $A$ of a metric space, or alternatively whether $A^*$ is nonempty. In this connection, it is useful to consider the metric space $X = \mathcal{R}$, $d(x, y) = |x - y|$ and observe the following fact through the *Bolzano–Weierstrass theorem.*

***Theorem 3.5.4*** Every bounded infinite set on the real line has at least one accumulation point.

*Proof*: We use the result of Exercise 1.8.7 in the proof. First, we observe that a bounded set of real numbers, say $A$, can always be enclosed in a closed interval $I_1 \equiv [a_1, b_1]$. Next we divide $I_1$ into two intervals $[a_1, (b_1 + a_1)/2]$ and $[(b_1 + a_1)/2, b_1]$. At least *one* of these intervals must contain infinitely many points of $A$. Denote this interval by $I_2$. Note that this interval has length $\frac{1}{2}(b_1 - a_1)$. Proceed exactly as before by dividing $I_2$ and obtain $I_3$ with length $(1/2^2)(b_1 - a_1)$ containing infinite points of $A$. The interval $I_n$ with length $(1/2^{n-1})(b_1 - a_1)$ has infinite points of $A$. Obviously, $\{I_n\}$ is a sequence of *nested intervals* (see Exercise 1.8.7) because $I_n \supset I_{n+1}$ and the length of $I_n$ contains infinitely many points of $A$. There is exactly one point, say $x_0$, that is in *every* interval in the sequence $\{I_n\}$, each of which contains points of $A$. Thus $x_0$ is an accumulation point of $A$. If $A$ is closed, then $x_0 \in A$. ∎

It is significant to note that the Bolzano–Weierstrass theorem does not hold for *arbitrary* metric spaces. In an abstract metric space a *bounded* set $A$ is defined by the existence of a number $m > 0$ such that $d(x, y) \leq m \; \forall \; x, y \in A$.

Suppose that we assign a property for a subset $A$ of a metric space that *every* infinite *subset* of $A$ has at least one accumulation point in $A$. It is important to note that the limit point is *in A*. This property, known as the *Bolzano–Weierstrass property*, is assured by Theorem 3.5.4 to exist for *closed, bounded, infinite* sets on the real line. However, the property is not assured for such sets in an abstract metric space. Instead, we must find a new class of sets for which the Bolzano–Weierstrass property would hold. While these sets, called *compact sets*, may be *defined* by the foregoing statement, we take an alternative approach to them.

**EXERCISES**

**3.5.1** Consider the metric space $(X, d)$ with $X = \mathcal{R}$, $d(x, y) = |x - y|$. Determine whether the following sets are closed, open, neither closed nor open, or both closed *and* open.
(a) $[0, 1)$.
(b) $[0, 1]$.
(c) $(0, 1)$.
(d) $(-\infty, 1]$.
(e) $(-\infty, 1)$.
(f) $(-\infty, \infty)$.

**3.5.2** Is the null set (i.e., the empty set) of a metric space $(X, d)$ closed or open?

**3.5.3** Show that the intersection and union of two open sets are open, and extend the result to any finite number of open sets.

**3.5.4** Let $A$ be a subset of $X$ in a metric space $(X, d)$. Consider any element $x_0 \in A^*$. Show that it is possible to find a sequence $\{x_n\} \subset A$ such that $x_n \longrightarrow x_0$. Thus prove that a set $A$ is closed *if and only if* every convergent sequence contained in $A$ converges in $A$; that is, the limit point lies in $A$.

**3.5.5** Consider a continuous function $f(x, y)$ defined on $\mathcal{R}_2$ and show that the set

$$A = \{(x_1, x_2) \in \mathcal{R}_2 : f(x_1, x_2) = 0\}$$

is closed in $\mathcal{R}_2$.

**3.5.6** To establish that closedness and openness of sets in a metric space is related to the metric, consider $X = \mathcal{C}[a, b]$ with the two metrics

$$d_1(\mathbf{f}, \mathbf{g}) = \sup_{t \in [a, b]} |f(t) - g(t)|$$

$$d_2(\mathbf{f}, \mathbf{g}) = \int_a^b |f(t) - g(t)|\, dt$$

Show that the set $A = \{\mathbf{f} \in \mathcal{C}[a, b]: f(a) = 0\}$ is closed with respect to $d_1$ but not closed with respect to $d_2$.

### 3.5.1 Compact Sets

The concept of compactness is a deep one and has several equivalent definitions. Some of these, which are designed for general topological spaces, are more abstract and consequently more difficult to understand than others. We give only one definition of compactness that does not require any additional concept. Before giving the definition, however, we shall provide the motivation for the same.

The closed bounded interval $[0, 1]$ in the real line has some special properties distinct from those of the corresponding intervals $(0, 1)$, $[0, 1)$, $(0, 1]$, none of which is closed. The distinguishing properties of the closed bounded set that are of interest to us are:

(i) A function that is continuous on the closed, bounded interval $[0, 1]$ is bounded above and below. Moreover, it is very important to recognize that each bound is actually attained by the function for some point (or points) inside the interval. This is not so for, say, the interval (0, 1]; for example, the function $f(x) = 1/x$, which is continuous in (0, 1], is unbounded as $x \longrightarrow 0$.

(ii) Every sequence in the closed bounded interval has a convergent subsequence. A subsequence is obtained by merely omitting points from a sequence. If one marks off points (belonging to a sequence not necessarily converging) on a bounded interval, these points will eventually "fill out" the interval and it may be possible to isolate a subsequence of points that will converge to some limit point in the set. Thus consider the sequence of points $(1 - 1/n)$, $n = 1, 2, \ldots$, in the interval [0, 1). It is clear that the only limit point of any convergent subsequence is the point 1, which does not belong to the set [0, 1). Thus there is no convergent subsequence for this sequence, a defect instantly cured by the inclusion of the point 1 in the interval.

Property (i) is a very respectable one from the point of view of applications, engineering or other. Property (ii) may seem inconsequential to applications, but it is the one in truth that assures property (i). In this sense, the two properties above are not independent.

At this stage, the student may well ask: "Why not be content with closed, bounded intervals, since it appears sufficient for property (i)?" The answer to this question is that the *inherent* property that assures property (i) is *not* true of continuous functions defined on closed, bounded subsets of metric spaces derived from linear spaces of infinite dimension. However, property (ii) (or any other equivalent property) would imply property (i) for *any* metric space $(X, d)$, which is suggestive of the following definition of compactness.

***Definition 3.5.6*** A metric space $(X, d)$ is said to be *compact* if *every* sequence has a convergent subsequence. A subset $A$ of $X$ is said to be *compact* if $(A, d)$ is a compact metric space.

It turns out that when $X = \mathcal{R}_n$, closed bounded sets are compact. Certainly, compact sets in an arbitrary metric space must be closed and bounded. The following theorem is also true.

***Theorem 3.5.5*** Let $(X, d)$ be a compact metric space and $A \subset X$ be a closed set. Then $A$ is compact.

*Proof*: Let $\{x_n\}$ be an arbitrary sequence in $A$. Since $\{x_n\} \subset X$, there is a subsequence that converges to some $x_0 \in X$. But $x_0 \in A^*$ and from Exercise 3.5.4, we have $A^* \subset A$, so that $x_0 \in A$. Thus every sequence in $A$ has a convergent subsequence, so that $A$ is compact. ∎

Property (i), which we referred to earlier, can now be shown to hold for functions continuous on a compact set. To prove this, the following theorem is a prerequisite.

***Theorem 3.5.6*** Let $(X, d)$ be a metric space and $f: X \longrightarrow X$ be continuous. If $A \subset X$ is compact, then $f(A)$ is compact.

*Proof*: Let $\{y_n\}$ be a sequence in $f(A)$. Form the sequence $\{x_n\} \subset A \ni y_n = f(x_n)$. Since $A$ is compact, $\exists$ a convergent subsequence $\{x_{k_n}\} \subset A$, which converges to some $x_0 \in A$. The sequence $y_{k_n} = f(x_{k_n})$ converges to $y_0 = f(x_0)$ since $f$ is continuous. Thus $f(A)$ is compact. ∎

Thus compactness is preserved by continuous mappings. It is easy to see how Theorem 3.5.6 is also true for continuous mappings of one metric space into another. Properties that are preserved under continuous mappings are called *topological properties.* Compactness is a topological property. We now state and prove the following important theorem.

***Theorem 3.5.7*** Let $(X, d)$ be a compact metric space and $f: X \longrightarrow \mathcal{R}$ be continuous. Then $f(x)$ is bounded; that is, $\exists$ real numbers $\bar{m}$ and $\underline{m}$ such that $\bar{m} = \sup_{x \in X} f(x)$ and $\underline{m} = \inf_{x \in X} f(x)$. Furthermore, $\exists$ $x_{\max}, x_{\min} \in X \ni$

$$\bar{m} = f(x_{\max})$$
$$\underline{m} = f(x_{\min})$$

*Proof*: Note that the result of this theorem is what was referred to earlier as property (i).

From Theorem 3.5.6, $f(X)$ is a compact interval on the real line, and is therefore bounded so that $\sup_{x \in X} f(x) = \bar{m}$ and $\inf_{x \in X} f(x) = \underline{m}$ exist (see Theorem 1.8.1).

Since $f(X)$ is closed, $\bar{m}, \underline{m} \in f(X)$ (why?). From Theorem 1.8.2 we have sequences $\{f_n^{(1)}\}$ and $\{f_n^{(2)}\}$ in $f(X)$, respectively, converging to $\bar{m}$ and $\underline{m}$. The corresponding sequences $\{x_n^{(1)}\}$ and $\{x_n^{(2)}\}$ defined by $f(x_n^{(1)}) = f_n^{(1)}$ and $f(x_n^{(2)}) = f_n^{(2)}$ must possess subsequences $\{x_{k_n}^{(1)}\}$ and $\{x_{k_n}^{(2)}\}$ converging, respectively, to say $x_{\max}$ and $x_{\min}$. Since $f$ is continuous,

$$\bar{m} = \lim f_n^{(1)} = \lim f_{k_n}^{(1)} = \lim f(x_{k_n}^{(1)}) = f(x_{\max})$$
$$\underline{m} = \lim f_n^{(2)} = \lim f_{k_n}^{(2)} = \lim f(x_{k_n}^{(2)}) = f(x_{\min}) \quad ∎$$

The theorem is easily understood for the more general situations in which $f$ is a continuous function mapping elements of one metric into the same or any other metric space.

Indeed, from our discussion of the motivation for defining compact sets, it must be evident that closed, bounded sets of a metric space derived from, say, $X = \mathcal{R}_n$, must be compact. To demonstrate that this is not true of metric spaces derived from, say, infinite-dimensional linear spaces, the following example is of interest.

Consider the metric space $X = \mathcal{C}[0, 1]$, $d(\mathbf{f}, \mathbf{g}) = \sup_{t \in [0,1]} |f(t) - g(t)|$, and the set $A \subset X$ defined by

$$A = \{\mathbf{f} \in \mathcal{C}[0, 1]: |f(t)| \leq \alpha, t \in [0, 1], \alpha > 0\}$$

Clearly, the set $A$ is bounded,† since $\forall$ $\mathbf{f}, \mathbf{g} \in X$, $d(\mathbf{f}, \mathbf{g}) \leq d(\mathbf{f}, \mathbf{0}) + d(\mathbf{0}, \mathbf{g}) \leq 2\alpha$. Also, we show that $A$ is closed as follows. Let $\mathbf{f} \in A^*$. Then $\exists$ a sequence $\{\mathbf{f}_n\} \subset A$ which converges to $\mathbf{f}$. If $\mathbf{f} \notin A$, then $d(\mathbf{f}, \mathbf{0}) > \alpha$ (why?). Let $\varepsilon = d(\mathbf{f}, \mathbf{0}) - \alpha$, for which $\exists$ $n_\varepsilon \ni n > n_\varepsilon \rightarrow d(\mathbf{f}, \mathbf{f}_n) < \varepsilon$. Since

$$d(\mathbf{f}, \mathbf{0}) \leq d(\mathbf{f}, \mathbf{f}_n) + d(\mathbf{f}_n, \mathbf{0}) < \varepsilon + \alpha = d(\mathbf{f}, \mathbf{0})$$

which is impossible. Hence we must have $\mathbf{f} \in A$ so that $A^* \subset A$; hence $A$ is closed.

Although $A$ is closed and bounded, we will show that it is *not* compact by producing a sequence such as that in Figure 3.5.1, which has no convergent subsequence. Let $\{t_n\}$ be a sequence of positive real numbers such that $t_n < 1$

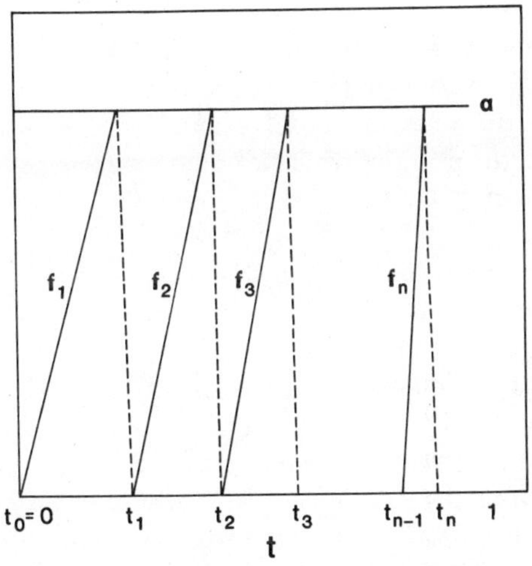


**Figure 3.5.1** Bounded sequence of continuous functions, belonging to a closed set, that has no convergent subsequence.

†To distinguish between the boundedness of a *single* function and the boundedness of a *set* of functions by a *common* bound, the term *uniform* boundedness of the set has been used (see, e.g., *Methods of Mathematical Physics*, Vol. 1, by R. Courant and D. Hilbert, Interscience, New York, 1953, p. 59). We prefer not to use this term here since mere boundedness in $\mathcal{C}[0, 1]$ with respect to the chosen metric implies uniform boundedness of the set.

and $t_n \to 1$ [e.g., $t_n = 1 - 1/(n+1)$]. Define $\mathbf{f}_n$ by

$$f_n(t) = \begin{cases} 0, & 0 \le t \le t_{n-1} \\ \alpha \dfrac{t - t_{n-1}}{t_n - t_{n-1}}, & t_{n-1} \le t \le t_n \\ \alpha, & t_n \le t \le 1 \end{cases}$$

Clearly, $\mathbf{f}_n \in A$, and $d(\mathbf{f}_m, \mathbf{f}_n) = \alpha$ for arbitrary $m$ and $n$. Thus $\{f_n\}$ has no convergent subsequence, so that $A$ is not compact. We have therefore established that closed, bounded sets of $\mathcal{C}[0, 1]$ are not necessarily compact.

Compactness in the space of continuous functions such as $\mathcal{C}[0, 1]$ can also be identified in terms of other properties by the Arzela-Ascoli theorem, which requires the following definition.

***Definition 3.5.7*** A set $A \subset \mathcal{C}[a, b]$ is said to be *equicontinuous* if $\forall\, \varepsilon > 0\ \exists\ \delta_\varepsilon > 0 \ni$ for $t, s \in [a, b]$,

$$|t - s| < \delta_\varepsilon \rightarrow |f(t) - f(s)| < \varepsilon \qquad \text{for } all\ \mathbf{f} \in A$$

The $\delta_\varepsilon$ is independent of $t$, $s$, and $\mathbf{f}$.

Note that equicontinuity refers to a *set* of functions and not to individual functions. Obviously, every member of an equicontinuous set is a continuous function. Certainly, the noncompact set $A$ considered in the example above is not equicontinuous, the proof of which is left as an exercise. We now state the *Arzela–Ascoli theorem*.

***Theorem 3.5.8*** Let $X = \mathcal{C}[a, b]$, $d(\mathbf{f}, \mathbf{g}) = \sup_{t \in [a,b]} |f(t) - g(t)|$. A *closed* set $A \subset X$ is compact *if and only if* it is *bounded* and *equicontinuous*.

The proof of this theorem is somewhat lengthy and we refer the reader to Simmons.†

Instead of giving examples of compact sets in $\mathcal{C}[a, b]$, we allow the student to discover them from the exercises. Also, Theorem 3.5.7 is important in application to optimization theory since it deals with *existence* of optimal solutions. However, note that compactness of a set is a severe condition (especially in metric spaces derived from linear spaces of infinite dimension), unnecessary for a function defined on the set to be bounded. Equicontinuity is a rather strong condition which is indicative of the natural difficulty of compressing elements of a "vast" and "spread out" infinite-dimensional space into a compact set. We will have more to say about com-

---

†*Introduction to Topology and Modern Analysis*, by G. F. Simmons, McGraw-Hill, New York, 1963, p. 126.

pactness when dealing with linear operators on linear spaces on which a metric is defined.

Compact spaces have the important property of "completeness," with which we shall deal in the next section.

**EXERCISES**

**3.5.7** Show that every compact set of a metric space is bounded.

**3.5.8** Consider the metric space $X = \mathcal{C}[a, b]$, $d(\mathbf{f}, \mathbf{g}) = \sup_{t \in [a,b]} |f(t) - g(t)|$, and the following set:

$$A = \left\{f(t) = \frac{t}{\alpha}, t \in [a, b]; \alpha \in (0, 1]\right\}$$

Show that $A \subset X$. Is $A$ a bounded set?

**3.5.9** Show that the set $A$ related to Figure 3.5.1 is not equicontinuous.

**3.5.10** In the metric space $X = \mathcal{C}[0, 1]$, $d(\mathbf{f}, \mathbf{g}) = \sup_{t \in [0,1]} |f(t) - g(t)|$, show that the set

$$A = \{f(t) = e^{-\alpha t}, \alpha \geq 0\}$$

is bounded but *not* equicontinuous.

**3.5.11** Show that the set $A$ in Exercise 3.5.10 is compact if $\alpha$ is in a compact interval on the real line. Extend this result to the set

$$A = \left\{f(t) = \sum_{j=1}^{n} \beta_j e^{-\alpha_j t}; \alpha_j \in I_1, \beta_j \in I_2, j = 1, 2, \ldots, n\right\}$$

where $I_1$ and $I_2$ are compact intervals on the real line.

## 3.6 Completeness of Metric Spaces

It was observed in Section 3.2 that every convergent sequence is a Cauchy sequence but that the reverse question has deeper implications. If we consider the metric space $X = [0, 1)$, $d(x, y) = |x - y|$, then the sequence $\{1 - 1/n\}$, which is contained entirely in $X$, converges to 1. This sequence is obviously a Cauchy sequence, but its limit point does not exist in $X$ so that the sequence cannot be said to converge in $X$. We say that a Cauchy sequence in a metric space $(X, d)$ converges if the limit point is contained in $X$, which leads to the following definition.

***Definition 3.6.1*** A metric space is said to be *complete* if *every* Cauchy sequence in the space converges to some point $x \in X$.

Completeness is a very important concept in mathematical analysis, which is fundamental to existence proofs in general. We present some examples of complete metric spaces below.

*Examples*

1. $X = \mathcal{R}$, $d(x, y) = |x - y|$, represents a complete metric space. Note that this completeness is derived from the axiom of continuity (see Section 1.4), which asserts the existence of irrational numbers.
2. $X = \mathcal{R}_n$ forms a complete metric space with any of the following metrics.

$$d_\infty(\mathbf{x}, \mathbf{y}) = \max\{|x_1 - y_1|, |x_2 - y_2|, \ldots, |x_n - y_n|\}$$

$$d_p(\mathbf{x}, \mathbf{y}) = \left\{\sum_{i=1}^{n} |x_i - y_i|^p\right\}^{1/p}, \qquad p \geq 1$$

These statements also hold for $\mathcal{C}_n$. The completeness properties of $\mathcal{R}_n$ and $\mathcal{C}_n$ are indeed derived from that of the real numbers.

3. That the completeness property is relative to a metric is established by this example. Thus $X = \mathcal{C}[a, b]$ is *complete* with respect to the metric†

$$d_\infty(\mathbf{f}, \mathbf{g}) = \sup_{t \in [a,b]} |f(t) - g(t)|$$

which follows from the well-known result that if a sequence of functions continuous on $[a, b]$ converges *uniformly*, it converges to a continuous function.‡

However, $\mathcal{C}[a, b]$ is *not* complete with any of the following metrics:

$$d_p(\mathbf{f}, \mathbf{g}) = \left\{\int_a^b |f(t) - g(t)|^p \, dt\right\}^{1/p}, \qquad p \geq 1$$

This is easily seen by constructing a Cauchy sequence of continuous functions, which does not converge to a continuous function. For simplicity, let $p = 1$. Consider the sequence of functions $\{\mathbf{f}_n\} \subset \mathcal{C}[0, 1]$, given by

$$\mathbf{f}_n(t) = \begin{cases} 0, & 0 \leq t \leq t_n \\ 1 - \dfrac{\frac{1}{2} - t}{\frac{1}{2} - t_n}, & t_n \leq t \leq \frac{1}{2} \\ 1, & \frac{1}{2} \leq t \leq 1 \end{cases}$$

where $\{\mathbf{t}_n\}$ is any strictly increasing sequence in $[0, 1]$ converging to $\frac{1}{2}$, for example, $t_n = \frac{1}{2} - 1/(n + 1)$. The sequence $\{\mathbf{f}_n\}$ is plotted in Figure 3.6.1. For arbitrary $m$ and $n$ such that $m > n$,

---

†The subscript $\infty$ is used in analogy with respect to the corresponding metric in $\mathcal{R}_n$. See Exercise 3.1.3.

‡See, for example, *Advanced Calculus*, by A. E. Taylor, Ginn, Boston, 1955, p. 598.

$$d_1(\mathbf{f}_m, \mathbf{f}_n) = \int_0^1 |f_m(t) - f_n(t)|\, dt = \int_0^1 [f_n(t) - f_m(t)]\, dt$$
$$= \tfrac{1}{2}(\tfrac{1}{2} - t_n) + \tfrac{1}{2} - \tfrac{1}{2}(\tfrac{1}{2} - t_m) - \tfrac{1}{2}$$
$$= \tfrac{1}{2}(t_m - t_n)$$

Since $\{t_n\}$ is a sequence converging to $\frac{1}{2}$, it is a Cauchy sequence so that $\{\mathbf{f}_n\}$ is also a Cauchy sequence in $\{\mathcal{C}[0, 1], d_1\}$. It is clear, however, that $\{\mathbf{f}_n\}$ converges to the function

$$f_0(t) = \begin{cases} 0, & 0 \leq t < \frac{1}{2} \\ 1, & \frac{1}{2} < t \leq 1 \end{cases}$$

which is discontinuous at $t = \frac{1}{2}$, so that $\mathbf{f}_0$ does not belong to $\mathcal{C}[0, 1]$. Thus $\{\mathcal{C}[0, 1], d_1\}$ is not a complete metric space.

4. Every compact metric space $(X, d)$ is complete, for let $\{x_n\} \subset X$ be a Cauchy sequence. If $\{x_{k_n}\}$ is a convergent subsequence converging to, say, $x_0 \in X$, we show that $\{x_n\}$ converges to $x_0$. Let $\varepsilon > 0$ be given; $\exists\ n'_{\varepsilon/2} \ni k_n > n'_{\varepsilon/2} \rightarrow d(x_{k_n}, x_0) < \varepsilon/2$. Also, $\exists\ n''_{\varepsilon/2} \ni n, m > n''_{\varepsilon/2} \rightarrow d(x_n, x_m) < \varepsilon/2$. Now let $n_\varepsilon = \max\,[n'_{\varepsilon/2}, n''_{\varepsilon/2}]$. If $n, k_n > n_\varepsilon$, then
$$d(x_n, x_0) \leq d(x_n, x_{k_n}) + d(x_{k_n}, x_0) < \varepsilon$$
Thus $\{x_n\}$ converges to $x_0$, so that $(X, d)$ is complete.
5. Let $(X, d)$ be a complete metric space and $A \subset X$ be closed. Then $(A, d)$ is a complete metric space. The proof is straightforward.

Incomplete metric spaces may always be completed by appending to the space the "missing" elements. Thus the axiom of continuity for real numbers adds the "missing irrationals" to the rational numbers to complete the real number system. There are other important completions with which we will be concerned. Before dealing with completion of abstract (incomplete) metric space, there are some additional concepts to be introduced.

We define a *dense* subset of a metric space.

***Definition 3.6.2*** In a metric space $(X, d)$ a subset $A \subset X$ is said to be *dense* in $X$ if *every* element $x_0 \in X$ can be represented as the limit of a sequence $\{x_n\} \subset A$.

It is easily shown that an equivalent definition of a dense subset $A$ is that for each $x_0 \in X$, there exists $x \in A$ such that $d(x, x_0) < \varepsilon$ for every $\varepsilon > 0$ (see Exercise 3.6.2).

*Examples of Dense Sets*

1. The set of rational numbers is dense in the real line. For every irrational number, one can find a sequence of rational numbers converging to the irrational number; alternatively, every irrational number can be approximated as closely as required by a rational number.

2. The set of polynomials, that is, the linear manifold of the set of functions $\{t^n\}$ is dense in, say, $\mathcal{C}[a, b]$ with the metric $d_\infty(\mathbf{f}, \mathbf{g}) = \sup_{t \in [a,b]} |f(t) - g(t)|$. This is the well-known Weierstrass theorem, which states that a function continuous in $[a, b]$ can be uniformly approximated by a sequence of polynomials in the interval $[a, b]$.† Indeed, this theorem is the basis of "polynomial fitting" in numerical work dealing with numbers (e.g., experimental data) which are related or presumed to be related by continuous functions. Also, another important consequence of the Weierstrass theorem is that $\mathcal{C}^{(n)}[a, b]$, the space of functions with $n$ continuous derivatives in $[a, b]$, where $n$ is arbitrarily large, is dense with respect to the metric $d_\infty$ in $\mathcal{C}[a, b]$. This is because polynomials can be differentiated any number of times.

A further concept of interest is that of *separability* of a metric space.

***Definition 3.6.3*** A metric space $(X, d)$ is said to be *separable* if there exists a *countable*, dense subset of $X$.

Clearly, the real line is separable as the rationals have been shown to be countable (see Section 1.7).

Also, $\mathcal{C}[0, 1]$ with the metric $d_\infty$ is a separable metric space if the linear manifold of $\{t^n\}$ is restricted to rational coefficients of expansion.

There are other examples, which we shall refer to at a later stage.

## EXERCISES

**3.6.1** Show that $(\mathcal{R}_n, d_p)$ is a complete metric space and that it is separable.

**3.6.2** Show that Definition 3.6.2 of a dense set and the alternative definition following it are equivalent.

**3.6.3** Let $(X, d)$ be a metric space and $A, B \subset X$. If $A$ is dense in $B$ and $B$ is dense in $X$, show that $A$ is dense in $X$.

**3.6.4** Consider $\mathcal{C}[0, 1]$ with the two metrics.

(1) $$d_\infty(\mathbf{f}, \mathbf{g}) = \sup_{t \in [0,1]} |f(t) - g(t)|$$

(2) $$d_p(\mathbf{f}, \mathbf{g}) = \left\{\int_0^1 |f(t) - g(t)|^p \, dt\right\}^{1/p}, \qquad p \geq 1$$

(a) Show that the set

$$A = \{\mathbf{f} \in \mathcal{C}[0, 1]: f(0) = 0\}$$

is *not* dense in $\{\mathcal{C}[0, 1], d_\infty\}$ but is dense in $\{\mathcal{C}[0, 1], d_p\}$.

---

†See, for example, *Methods of Mathematical Physics*, Vol. 1, by R. Courant and D. Hilbert, Interscience, New York, 1953, p. 65.

(b) Let $t_1, t_2, \ldots, t_n$ be a finite set of points in [0, 1]. Show that the set

$$A = \{\mathbf{f} \in \mathcal{C}[0, 1]: f(t_j) = 0, j = 1, 2, \ldots, n\}$$

is dense in $\{\mathcal{C}[0, 1], d_p\}$.

(c) Let $A = \{\mathbf{f} \in \mathcal{C}^{(1)}[0, 1]: f'(0) = 0\}$. Show that $A$ is dense in $\{\mathcal{C}[0, 1], d_p\}$.

(d) Let $A = \{\mathbf{f} \in \mathcal{C}^{(2)}[0, 1]: f'(0) = 0, f(1) = 0\}$. Show that $A$ is dense in $\{\mathcal{C}[0, 1], d_p\}$.

**3.6.5** Let $(X, d)$ be a complete metric space and $A \subset X$ be closed. Show that $(A, d)$ is a complete metric space.

### 3.6.1 Equivalence of Metric Spaces

We have observed that properties possessed by a metric space such as convergence of sequences and continuity of functions defined on the space are relative to the metric. For example, the sequence of continuous functions shown in Figure 3.6.1 does not converge at all with respect to the metric $d_\infty$, whereas the same sequence can be shown to converge with respect to the metric $d_p$. On the other hand, in metric spaces derived from finite-dimensional

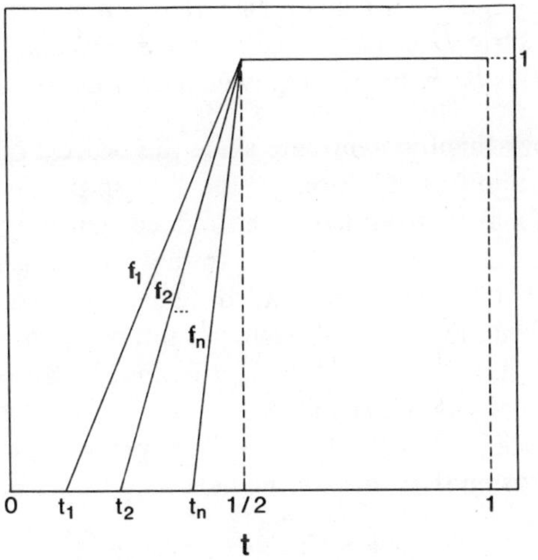


**Figure 3.6.1** Sequence of continuous functions converging to a discontinuous function with respect to the integral metric.

linear spaces such as $\mathcal{R}_n$, the metrics $d_p$ and $d_\infty$ are equivalent in the sense that convergence of a sequence with respect to one metric implies convergence with respect to the other. Similarly, continuity of a function defined on $\mathcal{R}_n$ with respect to one metric implied continuity with respect to the other. The preceding statements obviously suggest the following definition of equivalent metrics in an abstract metric space.

***Definition 3.6.4*** Two metrics $d$ and $d'$ on a (nonempty) set $X$ are said to be *equivalent* if (i) the class of converging sequences is the same for $d$ and $d'$, and (ii) the class of continuous functions is the same for $d$ and $d'$.† The metric spaces $(X, d)$ and $(X, d')$ are then said to be equivalent.

The equivalence of two metric spaces can be extended to situations in which there are two different sets $X$ and $X'$, as follows.

***Definition 3.6.5*** Two metric spaces $(X, d)$ and $(X', d')$ are said to be *equivalent* if there is a *continuous* mapping $f: X \longrightarrow X'$ which is onto $X'$ with a *continuous* inverse $f^{-1}: X' \longrightarrow X$. (That is, $f$ is a one-to-one continuous mapping of $X$ onto $X'$ with a continuous inverse.‡ The function $f$ is said to be a *homeomorphism*.) If, further, $d(x, y) = d'(f(x), f(y)) \ \forall\ x, y \in X$, then the $X$ and $X'$ are said to be *isometrically equivalent*.

There are several examples of equivalent metric spaces which could be given, but we direct the student to discover them through the exercises.

We now state an important theorem concerning the completion of an incomplete metric space.

***Theorem 3.6.1*** Let $(X, d)$ be a metric space that is not complete. Then there exists a complete metric space $(X', d')$ called the *completion* of $(X, d)$ with the property that $(X, d)$ is isometrically equivalent to a dense subset of $(X', d')$.

We do not prove this theorem here since the process of completion can indeed be looked upon as addition of the "missing" elements. Thus the irrational numbers may be regarded as the "added elements" to complete the rationals. In the next chapter it will be shown that a more generalized form of integration than that of Riemann, called Lebesgue integration, must be introduced to complete the space of Riemann-integrable functions.

The completion has a metric $d'$ distinct from $d$ because $d$ has not been defined over the set of missing elements.

We had observed that completeness is a property which played an important role in existence proofs. In the next section, we consider one such application.

## EXERCISES

**3.6.6** Show that any compact interval on the real line forms an equivalent metric space with any other compact interval, where the metric for both sets is given

---

†In view of Theorems 3.5.1 and 3.5.2, the equivalence of two metrics $d$ and $d'$ on a set $X$ can be elegantly characterized by the requirement that they generate the same class of open sets.

‡Again Theorems 3.5.1 and 3.5.2 show that there is a one-to-one correspondence between open sets in $(X, d)$ and those in $(X', d')$.

by $d(x, y) = |x - y|$. How would you define metrics on each of them so that the two metric spaces are isometrically equivalent?

**3.6.7** Let $X = (-\pi/2, \pi/2)$. Consider the two metrics

$$d_1(x, y) = |x - y|$$
$$d_2(x, y) = |\tan x - \tan y|$$

(a) Are $d_1$ and $d_2$ equivalent?

(b) Show that $(X, d_2)$ is isometrically equivalent to $(\mathcal{R}, d_1)$ and hence that every open interval with a suitably chosen metric is isometrically equivalent to $(\mathcal{R}, d_1)$.

## 3.7 A Fixed-Point Theorem

A fixed point belongs to a function $f: X \longrightarrow X$, where $(X, d)$ is a metric space. It is defined as follows.

***Definition 3.7.1*** $x_0$ is a *fixed point* of $f$ if $f(x_0) = x_0$.

Indeed, a solution of an equation of the kind $g(x) = 0$ can be looked upon as a fixed point of the function $f(x) = x + g(x)$. Fixed-point theorems are concerned with the existence of fixed points of functions which satisfy suitable sufficiency conditions; one such condition is provided by the following definition.

***Definition 3.7.2*** $f$ is said to be a *contraction mapping* or simply a contraction if $\forall\, x, y \in X$, $d(f(x) f(y)) \leq k\, d(x, y)$, where $0 \leq k < 1$.

Obviously, a contraction mapping is continuous. The following fixed-point theorem may now be enunciated.

***Theorem 3.7.1*** Let $(X, d)$ be a complete metric space and $f: X \longrightarrow X$ be a contraction. Then $f$ has a unique fixed point.

*Proof*: Let $x \in X$. Define a sequence $\{x_n\}$ by the recursive relation

$$x_1 = f(x)$$
$$x_{n+1} = f(x_n), \qquad n = 1, 2, \ldots$$

Now

$$d(x_2, x_1) = d(f(x_1), f(x)) \leq k\, d(x_1, x)$$
$$d(x_3, x_2) = d(f(x_2), f(x_1)) \leq k\, d(x_2, x_1) \leq k^2\, d(x_1, x)$$
$$\vdots$$
$$d(x_m, x_{m-1}) \leq k^{m-1}\, d(x_1, x)$$

From the triangular inequality, we have for $m > n$,

$$d(x_m, x_n) \leq [k^{m-1} + k^{m-2} + \ldots + k^n]\, d(x_1, x)$$

from which

$$d(x_m, x_n) \leq k^n \sum_{i=1}^{\infty} k^i\, d(x_1, x) = \frac{k^n}{1-k}\, d(x_1, x)$$

By making $n$ sufficiently large, it is clear that $d(x_m, x_n)$ can be made as small as possible. Thus $\{x_n\}$ is a Cauchy sequence which converges to some $x_0 \in X$, since $X$ is complete. We now show that $x_0$ is a fixed point of $f$:

$$x_0 = \lim x_{n+1} = \lim f(x_n) = f(\lim x_n) = f(x_0)$$

where the continuity of $f$ in $X$ has been exploited (see Theorem 3.3.1). That this fixed point $x_0$ is unique is established below.

If $x_0$ is not unique, let $x_0'$ denote another fixed point of $f$, that is, $f(x_0') = x_0'$. Clearly,

$$d(x_0, x_0') = d(f(x_0), f(x_0')) \leq k\, d(x_0, x_0')$$

so that

$$d(x_0, x_0')(1 - k) \leq 0$$

Since $k < 1$,

$$d(x_0, x_0') \leq 0$$

The nonnegativity of $d$ then implies that $d(x_0, x_0') = 0$, so that $x_0 = x_0'$. Thus the fixed point is unique. ∎

Successive mappings of $x \in X$ with $f$ may be defined by $f^{(p)}(x) = f(f(\ldots f(x)\ldots))$, where $f$ has been applied $p$ times. For any $p$ a fixed point $x_0$ of $f$ is a fixed point of $f^{(p)}$, for

$$f(f(x_0)) = f(x_0) = x_0$$
$$f(f(f(x_0))) = f(f(x_0)) = f(x_0) = x_0$$
$$\vdots$$
$$f^{(p)}(x_0) = x_0$$

However, a fixed point, say $y_0$ of $f^{(p)}$, is not necessarily a fixed point of $f$. Recognizing that $f(f^{(p)}(x)) = f^{(p)}(f(x))\ \forall\ x \in X$, it follows that

$$f^{(p)}(f(y_0)) = f(f^{(p)}(y_0)) = f(y_0)$$

which shows that $f(y_0)$ is also a fixed point of $f^{(p)}$.

Clearly, the procedure establishes that $f^{(2)}(y_0) \ldots f^{(m)}(y_0)$ are all fixed points of $f^{(p)}$ if $y_0$ is a fixed point of $f^{(p)}$. The following Corollary to Theorem 3.7.1 can now be proved.

***Theorem 3.7.2*** Let $(X, d)$ be a complete metric space and $f\colon X \to X$ be not necessarily continuous. If there exists an integer $p$ such that $f^{(p)}$ is a contraction, then $f$ has a unique fixed point.

*Proof*: Since $f^{(p)}$ is a contraction it has a unique fixed point, say, $x_0$. Since $f(x_0)$ is also a fixed point of $f^{(p)}$, uniqueness of $x_0$ implies that $f(x_0) = x_0$. Thus $x_0$ is a fixed point of $f$. The uniqueness follows immediately because every fixed point of $f$ is a fixed point of $f^{(p)}$. ▮

The possibility that $f^{(p)}$ may be a contraction even if $f$ is not continuous is left for the student to discover through the exercises.

It is important to note that the contraction mapping theorem (Theorem 3.7.1) provides a condition of *sufficiency* for the existence of a unique fixed point. The condition of contraction is often quite conservative, although Theorem 3.7.2 represents an improvement. There are other fixed-point theorems which provide weaker conditions for existence of a fixed point at the sacrifice of uniqueness.†

We now consider some examples of applications of the fixed-point theorem.

*Example 1. Steady States of a Stirred Tank Reactor*

We consider a stirred tank reactor, in which an exothermic, irreversible first-order reaction is carried out adiabatically. It is well known that under some operating conditions, multiple steady states may exist.‡ These are normally determined by plotting the heat generation and dissipation curves versus temperature and inspecting the intersections of the two curves. It would be desirable to obtain some a priori conditions under which there is a unique steady state for the reactor. We let

$x \equiv$ concentration of reactant in the reactor

$x_f \equiv$ reactant concentration in the feed

$y \equiv$ absolute temperature in reactor

$y_f \equiv$ absolute temperature of feed

$\Theta \equiv$ holding time, $J \equiv (-\Delta H/C_p)$, $\Delta H =$ heat of reaction

$C_p \equiv$ heat capacity of reaction mixture per unit volume

$k \equiv$ rate constant $= Ae^{-B/y}$, $A$, $B$ constants

†Thus Schauder's fixed-point theorem is a much more general theorem with respect to the existence of a fixed point. See, for example, *Introduction to Topology and Modern Analysis* by G. F. Simmons, McGraw-Hill, New York, 1963, p. 338. The theorem, however, is not applicable to abstract metric spaces.

‡See, for example, *Elementary Chemical Reactor Analysis*, by R. Aris, Prentice-Hall, Englewood Cliffs, N.J., 1969, p. 173.

The steady mass and energy balance equations are then

$$0 = x_f - x - \Theta A e^{-B/y} x \tag{3.7.1}$$

$$0 = y_f - y + \Theta A J e^{-B/y} x \tag{3.7.2}$$

Equation (3.7.1) may be solved for $x$ in terms of $y$ to obtain

$$0 = y_f - y + \Theta \frac{JAe^{-B/y} x_f}{1 + \Theta A e^{-B/y}} \tag{3.7.3}$$

A bound for the reactor temperature is obtained easily. For fixed values of $x_f$ and $y_f$, (3.7.3) shows that $y$ must have maximum value of $y^* = y_f + Jx_f$, so that $y_f \leq y \leq y^*$. The metric space $X = [y_f, y^*]$, $d(y_1, y_2) = |y_1 - y_2|$ is a closed interval in $\mathcal{R}$ and is therefore a complete metric space. Define the function $f(y)$ by

$$f(y) = y_f + \Theta \frac{JAe^{-B/y}}{1 + \Theta A e^{-B/y}} x_f \tag{3.7.4}$$

Clearly, $f$: $X \longrightarrow X$ and the steady states of the reactor are determined by $f(y) = y$ (i.e., the fixed points of $f$).

For uniqueness, we examine the conditions under which $f$ is a contraction. Thus for $u, v \in [y_f, y^*]$.

$$|f(u) - f(v)| \leq \Theta J A x_f \left| \frac{e^{-B/u}}{1 + \Theta A e^{-B/u}} - \frac{e^{-B/v}}{1 + \Theta A e^{-B/v}} \right| \tag{3.7.5}$$

Using the inequality $1 - e^{-x} \geq 0$ for $x \geq 0$, it is readily shown that inequality (3.7.5) transforms to

$$|f(u) - f(v)| < \frac{BJ\Theta k^* x_f}{y_f^2(1 + \Theta k_f)^2} |u - v| \tag{3.7.6}$$

where $k^* = Ae^{-B/y^*}$ and $k_f = Ae^{-B/y_f}$. The strictness of the inequality holds for finite $\Theta$. For Theorem 3.7.1, $f$ will have a unique fixed point; that is, the reactor will have a unique steady state if

$$\frac{BJ\Theta k^* x_f}{y_f^2(1 + \Theta k_f)^2} \leq 1 \tag{3.7.7}$$

which is a quadratic inequality in $\Theta$. If we let

$$P \equiv \frac{BJx_f k^*}{2y_f^2 k_f^2} - \frac{1}{k_f}, \qquad Q \equiv \sqrt{P^2 - \frac{1}{k_f^2}}$$

inequality (3.7.7) is satisfied if $\Theta < P - Q$ or $\Theta > P + Q$. For multiple steady states, it is necessary but not sufficient that $P - Q < \Theta < P + Q$.†

---

†Better conditions for uniqueness have been obtained by Luss. See *Chem. Eng. Sci.*, **26**, 1713, 1971.

*Example 2. Existence of Solution to a Differential Equation*

We consider the solution of the differential equation

$$\frac{dx}{dt} = f(x, t), \qquad t > 0 \tag{3.7.8}$$

subject to

$$x(0) = x_0, \qquad a < x_0 < b \tag{3.7.9}$$

given that $f(x, t)$ is continuous in the closed rectangle $\{0 \leq t \leq T;\ a \leq x \leq b\}$. We further assume that $f(x, t)$ satisfies the *Lipschitz condition*, which states that there is a $k > 0$ such that

$$|f(x, t) - f(y, t)| \leq k|x - y| \tag{3.7.10}$$

$\forall\ x, y \in [a, b], t \in [0, T]$, where $k$ is independent of $x$, $y$, and $t$.

By a solution of (3.7.8) in some interval $0 \leq t \leq t_0$, $t_0 \leq T$, subject to the initial condition (3.7.9), where $a < x_0 < b$, we mean a function $x(t)$ with a continuous derivative in $[0, t_0]$ such that $x(0) = x_0$ and $\forall\ t$ the equality $dx/dt = f(x(t), t)$ is satisfied.

Subject to the aforementioned Lipschitz condition, we will establish that there is some interval $0 \leq t \leq t_0$ in which there exists a unique solution of (3.7.8) subject to (3.7.9).

It is readily seen that a solution of the differential equation must satisfy the integral equation

$$x(t) = x_0 + \int_0^t f(x(s), s)\, ds \tag{3.7.11}$$

Conversely, a solution of (3.7.11) also satisfies the differential equation and the initial condition, which follows from the fundamental theorem of calculus.† The existence of solution to (3.7.11) would then imply existence of the solution to (3.7.8) subject to (3.7.9).

Consider the metric space $\{\mathcal{C}[0, t_0], d_\infty\}$, where $t_0 \leq T$ is yet to be identified, and the function $g: \mathcal{C}[0, t_0] \longrightarrow \mathcal{C}[0, t_0]$ defined by $\mathbf{v} = \mathbf{g}(\mathbf{u})$, where

$$v(t) = x_0 + \int_0^t f(u(s), s)\, ds \tag{3.7.12}$$

Since $f(x, t)$ is assumed to be continuous on $[a, b] \times [0, T]$, there exists an $\alpha > 0$ such that

$$|f(x, t)| \leq \alpha \qquad \forall \quad x \in [a, b], \quad t \in [0, T]$$

We further restrict the domain of $\mathbf{g}$ to the set

$$A = \{\mathbf{u} \in \mathcal{C}[0, t_0]\colon |u(t) - \gamma| \leq \beta\ \forall\ t \in [0, t_0]\}$$

---

†See, for example, *Advanced Calculus*, by A. E. Taylor, Ginn, Boston, 1955, p. 519.

where $\beta \equiv \frac{1}{2}(b - a)$ and $\gamma \equiv \frac{1}{2}(a + b)$.† Choose $t_0 = (\beta - |x_0 - \gamma|)/\alpha$. The set $A$ is a closed subset of $\mathcal{C}[0, t_0]$ (see the example of Figure 3.5.1) and forms a complete metric space with $d_\infty$. Since from (3.7.12)

$$|v(t) - \gamma| \leq |x_0 - \gamma| + \int_0^{t_0} |f(u(s), s)|\, ds \leq |x_0 - \gamma| + \alpha t_0 = \beta$$

so that $\mathbf{v} \in A$ and $\mathbf{g}$ maps $A$ into $A$. Let $\mathbf{v}_1 = \mathbf{g}(\mathbf{u}_1)$, $\mathbf{v}_2 = \mathbf{g}(\mathbf{u}_2)$, where $\mathbf{u}_1, \mathbf{u}_2 \in A$. Then

$$|v_1(t) - v_2(t)| \leq \int_0^t |f(u_1(s), s) - f(u_2(s), s)|\, ds \tag{3.7.13}$$

The Lipschitz condition transforms inequality (3.7.13) into

$$|v_1(t) - v_2(t)| \leq k \int_0^t |u_1(s) - u_2(s)|\, ds \leq kt\, d_\infty(\mathbf{u}_1, \mathbf{u}_2) \tag{3.7.14}$$

Since $\mathbf{v}_1, \mathbf{v}_2 \in A$, the Lipschitz condition may be applied again to obtain

$$|f(v_1(s), s) - f(v_2(s), s)| \leq k|v_1(s) - v_2(s)| \quad \forall\, s \in [0, t_0] \tag{3.7.15}$$

Letting $\mathbf{w}_1 = \mathbf{g}(\mathbf{v}_1) = \mathbf{g}^{(2)}(\mathbf{u}_1)$ and $\mathbf{w}_2 = \mathbf{g}(\mathbf{v}_2) = \mathbf{g}^{(2)}(\mathbf{u}_2)$, analogous to (3.7.14), we obtain

$$\begin{aligned} |w_1(t) - w_2(t)| &\leq k \int_0^t |v_1(s) - v_2(s)|\, ds \\ &\leq k^2 d_\infty(\mathbf{u}_1, \mathbf{u}_2)\frac{t^2}{2} \leq \frac{k^2 t_0^2}{2}\, d_\infty(\mathbf{u}_1, \mathbf{u}_2) \end{aligned} \tag{3.7.16}$$

Thus we have shown that

$$d_\infty(\mathbf{g}^{(2)}(\mathbf{u}_1), \mathbf{g}^{(2)}(\mathbf{u}_2)) \leq \frac{k^2 t_0^2}{2}\, d_\infty(\mathbf{u}_1, \mathbf{u}_2) \tag{3.7.17}$$

In a completely analogous manner, the result for every integer $p$ is

$$d_\infty(\mathbf{g}^{(p)}(\mathbf{u}_1), \mathbf{g}^{(p)}(\mathbf{u}_2)) \leq \frac{k^p t_0^p}{p!}\, d_\infty(\mathbf{u}_1, \mathbf{u}_2) \tag{3.7.18}$$

Clearly, it is always possible to find a sufficiently large integer $p$ such that $k^p t_0^p/p! < 1$, so that $\mathbf{g}^{(p)}$ would be a contraction. By Theorem 3.7.2, then, $\mathbf{g}$ has a unique fixed point so that $\exists\ \mathbf{x} \in A \ni$

$$x(t) = x_0 + \int_0^t f(x(s), s)\, ds$$

which is indeed the *unique* solution of the differential equation (3.7.8) subject to the initial condition (3.7.9). This solution may be obtained by successive iterations starting from an initial element $\mathbf{u} \in A$.

---

†Note that $a \leq x \leq b$ if and only if $|x - \frac{1}{2}(a + b)| \leq \frac{1}{2}(b - a)$.

This uniqueness result is readily extended to systems of first-order differential equations such as

$$\frac{dx_i}{dt} = f_i(x_1, x_2, \ldots, x_n, t), \qquad t > 0 \tag{3.7.19}$$

$$x_i(0) = x_{i,0}, \qquad i = 1, 2, \ldots, n \tag{3.7.20}$$

The solution of (3.7.19) subject to (3.7.20) in some interval $[0, t_0]$ is an element of the $n$-fold direct sum space $X = \mathcal{C}[0, t_0] \oplus \mathcal{C}[0, t_0] \oplus \ldots \oplus \mathcal{C}[0, t_0]$ with any of the following equivalent metrics. For $\mathbf{u} = (\mathbf{x}_1, \mathbf{x}_2, \ldots, \mathbf{x}_n)$, $\mathbf{y} = (\mathbf{y}_1, \mathbf{y}_2, \ldots, \mathbf{y}_n)$, where $\mathbf{x}_i, \mathbf{y}_i \in \mathcal{C}[0, t_0]$, $i = 1, 2, \ldots, n$,

$$d_p(\mathbf{u}, \mathbf{v}) = \left\{\sum_{i=1}^{n} [d_\infty(x_i, y_i)]^p\right\}^{1/p}, \qquad p \geq 1 \tag{3.7.21}$$

The Lipschitz condition is now imposed on all functions $f_i$ in some closed, bounded, and *connected* set.† Since $f_i(x_1(t), x_2(t), \ldots, x_n(t), t)$ is continuous in $\mathcal{C}[0, t_0]$, $\mathbf{f} \equiv (f_1, f_2, \ldots, f_n) \in X$. The Lipschitz conditions on $f_i$ can then be shown to imply the existence of a constant $k > 0$:

$$d_p(\mathbf{f}(\mathbf{u}), \mathbf{f}(\mathbf{v})) \leq k d_p(\mathbf{u}, \mathbf{v}) \tag{3.7.22}$$

where the arguments of $\mathbf{f}$ have been contracted into single vectors $\mathbf{u}, \mathbf{v} \in X$. The procedure for establishing the existence of a unique solution to (3.7.19) satisfying (3.7.20) starting from the equivalent integral equations

$$x_i(t) = x_{i,0} + \int_0^t f_i(x_1(s), x_2(s), \ldots, x_n(s), s)\, ds \tag{3.7.23}$$

is entirely analogous to the one-dimensional problem and is left for the student to complete.

## EXERCISES

**3.7.1** Let $X = [0, 1]$ and $d(x, y) = |x - y|$. If $f\colon X \longrightarrow X$ is a continuously differentiable function, then show that $f$ is a contraction if and only if $|f'(x)| < 1$. Sketch such a function, and realize the implication of the contraction mapping theorem. Also sketch a function whose derivative is not bounded by 1 but which has a unique fixed point in [0, 1].

**3.7.2** Let $(X, d)$ be a metric space and $f\colon X \longrightarrow X$. Further, let $g\colon X \longrightarrow X$ be continuous and such that $g^{-1}$ exists and is continuous. Show that if the composite function $g(f(g^{-1}))\colon X \longrightarrow X$ is a contraction, $f$ has a unique fixed point. Does $g$ have a fixed point? Is it unique?

---

†A formal definition of a *connected* metric space $(X, d)$ is that it cannot be represented as the union of two disjoint (i.e., nonintersecting) nonempty open (or closed) sets. When $X = \mathcal{R}$, connectedness implies the geometrical interpretation that any two points in $X$ can be connected by a continuous curve entirely in $X$.

**3.7.3** Consider the metric space $X = \mathcal{C}[0, T]$, $d = d_\infty$, and the function $F_\lambda: X \longrightarrow X$, whose action on $\mathbf{f} \in X$ is defined by

$$g(t) = \lambda \int_a^t \kappa(t, s) f(s)\, ds$$

where $\kappa(t, s)$ is a continuous function in $[0, T] \times [0, T]$ and $\lambda$ is a fixed real number.

Use Theorem 3.7.2 to show that for any fixed $\lambda$, $F_\lambda$ has a unique fixed point. Therefore, show that there are no nontrivial solutions of the homogeneous Volterra integral equation

$$f(t) = \lambda \int_0^t \kappa(t, s) f(s)\, ds$$

Further establish that there *exists* a *unique* solution, for any fixed $\lambda$, to the inhomogeneous Volterra integral equation

$$g(t) = f(t) - \lambda \int_0^t \kappa(t, s) f(s)\, ds$$

where $\mathbf{g} \in X$.

## 3.8 Concluding Remarks

Since the elements of a metric space may be "near" or "away" from one another, the subsets in the space can possess geometrical shape. The essential concept of convergence and continuity could be developed in terms of a metric. The classes of subsets such as open, closed, and particularly, compact, play a very important role in subsequent analysis.

The concept of a pseudometric space and the process of forming a quotient metric space consisting of equivalence classes of elements is a very useful one. Approximating discontinuous functions by continuous functions is necessarily in terms of a metric involving the *integral* and if the benefits of the machinery of a metric space must be realized in such cases, the concept of a quotient metric space is indispensable.

Another vital aspect of metric spaces is the property of completeness. Entirely similar to the role of irrational numbers in "completing" the set of all rational numbers are "inclusions" which complete other metric spaces. Thus the quotient metric space derived from the space of integrable functions turns out to be incomplete and the process of its completion necessitates an extended theory of integration (called Lebesgue integration). The importance of completeness was demonstrated, for example, by the application to the fixed-point theorem and its implication to the existence of solutions to equations.

The concept of dense subsets of a metric space is closely associated with the procedure of *approximation*. That every continuous function may be approximated uniformly by polynomials is a consequence of denseness of polynomials in the space of continuous functions. The expansion of arbitrary functions in terms of Fourier sine and cosine series is another important instance of denseness of subspaces that appear somewhat constrained relative to the spaces in which they are dense.

## FURTHER READING

For a lucid treatment of metric spaces punctuated with examples, the reader may consult Chapter 2 of

> *Introductory Real Analysis* by A. N. Kolmogorov and S. V. Fomin, English translation edited by R. A. Silverman, Prentice-Hall, Englewood Cliffs, N.J., 1970.

The foregoing book is also an excellent reference for the reader interested in topological spaces which are more general than metric spaces.

# Introduction to Integration Theory. The Riemann and Lebesgue Integrals

# 4

## 4.0 Foreword

We present here a rudimentary treatment of integration theory. The integral with which the student is normally acquainted is called the Riemann integral. A more general concept of integration due to Lebesgue extends the class of integrable functions. Thus not all Lebesgue-integrable functions are Riemann integrable. On the other hand, Riemann-integrable functions are necessarily Lebesgue integrable, the value of the Lebesgue integral being the same as the Riemann integral. The extended class of integrable functions has the desirable property of completeness, to which reference was made in Chapter 3.

## 4.1 Introduction

The integral is a fundamental notion of the calculus. It involves summing a "very large" (infinite) number of "very small" (infinitesimal) quantities and may be understood precisely only in terms of the *limit* concept. Geometrically, it represents an area or a volume; indeed, other interpretations are possible depending on the nature of the integrand and the integration variable. As representative of the area enclosed by a curve, the concept of an integral dates back to the period of Archimedes, although it was not until the time of Newton and Leibnitz that the notion assumed its rightful place as a rudiment of the calculus. However, a systematic development of what is known today as the Riemann integral evidently did not take place until later. We first outline here the concept of the Riemann integral, then briefly discuss the departures of the Lebesgue integral from the former, and conclude with the implications of Lebesgue integration to analysis.

## 4.2 The Riemann Integral

We will restrict considerations to the Riemann integral of real-valued functions of a single real variable defined on a closed, bounded interval, say $[a, b]$. Let the function to be integrated be denoted by $f(x)$ and as represented in Figure 4.2.1. We begin by partitioning the interval $[a, b]$ into $n$ disjoint intervals $I_1, I_2, \ldots, I_n$, such that

$$I_j = [x_{j-1}, x_j), \qquad x_{j-1} < x_j$$

where $x_0 = a$, $x_n = b$, and $x_j \in (a, b)$, $j = 1, 2, \ldots, n-1$. There need be no restriction either on the location of the points $x_1, x_2, \ldots, x_{n-1}$ or on the number of points $n$. Thus there may be an arbitrarily large number of such points located arbitrarily in the interval $(a, b)$ such that $x_{j-1} < x_j$; that is, one can have "coarse" partitions in which the number of points is relatively small and "fine" partitions comprising a large number of closely spaced points. A measure of "coarseness" of a partition such as the one selected above may be obtained as follows. The partition may be denoted as

$$P = \{a, x_1, x_2, \ldots, x_{n-1}, b\}$$

The coarseness of the foregoing partition, denoted by $|P|$, is measured by

$$|P| = \max\{\Delta x_j : j = 1, 2, \ldots, n\}$$

where $\Delta x_j \equiv (x_j - x_{j-1})$. Arbitrarily fine partitions are obtained by letting $|P|$ be as small as possible. We denote the set of all partitions by $\mathcal{P}$, of which $P$ is a representative element.

We assume that the function $f(x)$ is bounded on $[a, b]$ so that the following numbers exist:

$$\bar{m} \equiv \sup_{x \in [a, b]} f(x), \qquad \underline{m} \equiv \inf_{x \in [a, b]} f(x)$$

Further, for any partition $P \in \mathcal{P}$, we define

$$\bar{m}_j \equiv \sup_{x \in I_j} f(x), \qquad \underline{m}_j \equiv \inf_{x \in I_j} f(x)$$

Clearly,

$$\bar{m} \geq \bar{m}_j \geq \underline{m}_j \geq \underline{m} \tag{4.2.1}$$

We now form the two sums

$$\bar{s}(P) \equiv \sum_{j=1}^{n} \bar{m}_j \Delta x_j, \qquad \underline{s}(P) \equiv \sum_{j=1}^{n} \underline{m}_j \Delta x_j$$

We call $\bar{s}(P)$ the *upper sum* and $\underline{s}(P)$ the *lower sum*. In view of inequality (4.2.1), we have

$$\bar{m}(b-a) \geq \bar{s}(P) \geq \underline{s}(P) \geq \underline{m}(b-a) \qquad \forall \quad P \in \mathcal{P}$$

Theorem 1.8.1 now implies the existence of the following numbers:

$$\bar{S} \equiv \inf\{\bar{s}(P)\colon \ P \in \mathcal{P}\}, \qquad \underline{S} \equiv \sup\{\underline{s}(P)\colon \ P \in \mathcal{P}\}$$

From a geometric point of view the numbers $\bar{s}(P)$ and $\underline{s}(P)$ represent upper and lower estimates, respectively, of the area under the curve (see Figure 4.2.1). Further, it is possible to show (see Exercise 4.2.2) that $\forall P, P' \in \mathcal{P}$, $\bar{s}(P) \geq \underline{s}(P')$. Hence

$$\bar{S} \geq \underline{S}$$

which brings us to the definition of the Riemann integral.

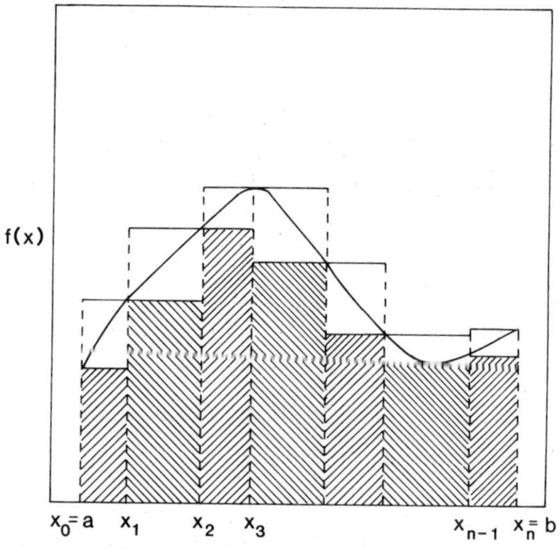


**Figure 4.2.1** Partition $P$ and the upper and lower sums in the definition of the Riemann integral.

***Definition 4.2.1*** A function $f(x)$ defined on $[a, b]$ is said to be *Riemann integrable* if $\bar{S} = \underline{S}$. The common value is the Riemann integral of $f(x)$, denoted $\int_a^b f(x)\,dx$.

We denote the set of all Riemann-integrable functions by $\mathcal{R}[a, b]$. Note that, in practice, the Riemann integral can be calculated by computing $\bar{s}(P_n)$ or $\underline{s}(P_n)$† for a sequence of partitions $\{P_n\}$ of diminishing coarseness

†More generally, one forms the "Riemann sum" $s(P_n) \equiv \sum_{j=1}^{n} f(\xi_j)\,\Delta x_j$, where $\xi_j \in I_j$.

($|P_n| \longrightarrow 0$) and evaluating $\lim \bar{s}(P_n)$ or $\lim \underline{s}(P_n)$. We refer to any standard text on calculus for examples of such calculations.† That boundedness is not sufficient to guarantee that a function belong to $\mathcal{R}[a, b]$ is established by the following example:

$$f(x) = \begin{cases} 1, & x \text{ rational} \\ 0, & x \text{ irrational} \end{cases} \tag{4.2.2}$$

For the integral between the limits $a$ and $b$, every partition $P$ has $\bar{m}_j = 1$, $\underline{m}_j = 0, j = 1, 2, \ldots, n$, since each $I_j$ must contain rational *and* irrational numbers. Thus for each $n$, $\bar{s} = (b - a)$ and $\underline{s} = 0$. Hence $\bar{S} = (b - a)$ and $\underline{S} = 0$, so that $f(x)$ is *not* Riemann integrable. Since $f(x)$ is bounded, it follows that boundedness is a *necessary* but *not* a sufficient condition for a function to be Riemann integrable.‡

Now it is easy to show that functions continuous on $[a, b]$ are Riemann integrable. Thus consider

$$\bar{s} - \underline{s} = \sum_{j=1}^{n} (\bar{m}_j - \underline{m}_j) \, \Delta x_j$$

for a partition $P$ with coarseness $|P|$. If $\varepsilon > 0$ is given, let $\varepsilon' = \varepsilon/(b - a)$. Since $f(x)$ is continuous on $[a, b]$, it is uniformly continuous there: thus $\exists \; \delta_{\varepsilon'} > 0 \ni |x - y| < \delta_{\varepsilon'} \longrightarrow |f(x) - f(y)| < \varepsilon'$, where $\delta_{\varepsilon'}$ is independent of $x$ and $y$.

Now consider $I_j$. Because $f(x)$ is continuous $\exists \; \bar{x}_j, \underline{x}_j \in I_j$, $\bar{m}_j = f(\bar{x}_j)$, and $\underline{m}_j = f(\underline{x}_j)$. If $P$ is chosen such that $|P| < \delta_{\varepsilon'}$,

$$|\bar{x}_j - \underline{x}_j| \leq \Delta x_j < \delta_{\varepsilon'}$$

so that $(\bar{m}_j - \underline{m}_j) < \varepsilon'$, from which

$$(\bar{s} - \underline{s}) < \varepsilon' \sum_{j=1}^{n} \Delta x_j = \varepsilon' \, (b - a) = \varepsilon$$

Thus by making the partition sufficiently fine, $\bar{s}$ may be made arbitrarily close to $\underline{s}$. The conclusion follows that $\bar{S} = \underline{S}$, so that (see Exercise 1.8.7) $f(x)$ is Riemann integrable.

We have thus shown that $\mathcal{C}[a, b] \subset \mathcal{R}[a, b]$. We leave it as an exercise to show that $\mathcal{R}[a, b]$ is a linear space with $\mathcal{R}$ as the associated field for

---

†See, for example, *Advanced Calculus*, by A. E. Taylor, Ginn, Boston, 1955.

‡This does not imply that unbounded functions cannot be integrated. The definition in such a case must be modified somewhat. For example, the function $1/\sqrt{x}$ is unbounded as $x$ approaches zero. However, the integral $\int_0^1 dx/\sqrt{x}$ exists for the reason that $\int_\varepsilon^1 dx/\sqrt{x}$ exists for each $\varepsilon > 0$ since the integral is bounded in $[\varepsilon, 1]$. Thus $\int_0^1 dx/\sqrt{x} \equiv \lim_{\varepsilon \to 0} \int_\varepsilon^1 dx/\sqrt{x}$, if the limit on the right-hand side exists, which in this case does and equals 2.

real-valued functions. For complex-valued functions the associated field should clearly be $\mathcal{C}$. Thus elements of $\mathcal{R}[a, b]$ may be denoted by symbols **f**, **g**, and so on.

The linear space $\mathcal{R}[a, b]$ contains considerably more than continuous functions. For example, it is possible to show that if $f(x)$ is bounded on $[a, b]$ and is continuous everywhere on $[a, b]$ except at some point $c \in [a, b]$, it is Riemann integrable. This may be extended to a finite number of discontinuities; that is, $\mathbf{f} \in \mathcal{R}[a, b]$ if the set of points in $[a, b]$ at which $f(x)$ is discontinuous is finite. Indeed, this can be further extended to functions that have discontinuities at a countably infinite number of points in $[a, b]$, and so on. But we will not be concerned with such questions since our present goal is to touch briefly on Lebesgue-integrable functions which encompass the class of Riemann-integrable functions.

**EXERCISES**

**4.2.1** Let $P, P' \in \mathcal{P}$ be two partitions of $[a, b]$ such that $P'$ is a refinement of $P$, by which it is meant that $P'$ is obtained by adding more points to $P$. Show that $\bar{s}(P) \geq \bar{s}(P')$ and $\underline{s}(P) \leq \underline{s}(P')$.

**4.2.2** Let $P, P' \in \mathcal{P}$ be *any* two partitions of $[a, b]$. Show that $\bar{s}(P) \geq \underline{s}(P')$. (*Hint:* Form partition $P''$, which is a refinement of both $P$ and $P'$. Use the result of Exercise 4.2.1.) Thus conclude that $\bar{S} \geq \underline{S}$.

**4.2.3** Show that $\mathcal{R}[a, b]$ is a linear space with $\mathcal{R}$ as the associated field for real-valued functions.

**4.2.4** Let $\mathbf{f}, \mathbf{g} \in \mathcal{R}[a, b]$. If $f(t) \leq g(t)\ \forall\ t \in [a, b]$, show that

$$\int_a^b f(t)\,dt \leq \int_a^b g(t)\,dt$$

and conclude that

$$\left|\int_a^b f(t)\,dt\right| \leq \int_a^b |f(t)|\,dt$$

where it is presumed that the right-hand side exists.

**4.2.5** Show that the Dirichlet function (4.2.2) is not continuous anywhere.

## 4.3 The Lebesgue Integral

A rigorous treatment of the Lebesgue integral is beyond our scope. However, the essential ideas can be presented without excessive complication. In developing the Riemann integral of a function $f(x)$, $x \in [a, b]$, we recall that approximating sums were constructed by partitioning the interval $[a, b]$. There are several ways of approaching the construction of the Lebesgue integral. One way is to focus on the *interval of values* of $f(x)$ instead of on the domain $[a, b]$, as was the case for the Riemann integral. We present the implements required for developing the Lebesgue integral in a rather informal way.

Let us recall that the bounded function (4.2.2) which vanishes for irrational $x$ does not possess a Riemann integral. On the other hand, it is possible to construct sequences of Riemann-integrable functions $f_n(x)$ which converge (in some sense to be clarified later) to the function $f(x)$ given by (4.2.2). Thus consider on the interval [0, 1] the function

$$f_n(x) = \begin{cases} 1, & x = \dfrac{k}{n}, \quad k = 0, 1, 2, \ldots, n \\ 0, & \text{otherwise} \end{cases} \tag{4.3.1}$$

Since the function above is discontinuous at a finite number of points, it is Riemann integrable on [0, 1] and clearly we must have

$$\int_0^1 f_n(x)\, dx = 0$$

As $n$ is increased indefinitely, *every rational* number between 0 and 1 is covered by the ratio $k/n$ and clearly in the limit we have

$$\lim_{n\to\infty} f_n(x) = f(x) = \begin{cases} 1, & x \text{ rational} \\ 0, & x \text{ irrational} \end{cases}$$

which, as observed earlier, does not possess a Riemann integral. The foregoing convergence occurs pointwise unambiguously for irrational $x$, while for any given rational $x$, as $n$ is increased, $f_n(x)$ fluctuates from 1 to 0 depending, respectively, on whether or not $x$ belongs to the set $\{k/n: k = 0, 1, 2, \ldots, n\}$.† Clearly, one has

$$\lim_{n\to\infty} \int_0^1 f_n(x)\, dx = 0 \tag{4.3.2}$$

so that the net implication of the foregoing is that the limiting process above cannot be carried into the Riemann integral. Notice further that

$$\int_0^1 |f_n(x) - f_m(x)|\, dx = 0 \qquad \forall \quad n, m$$

so that $\{\mathbf{f}_n\}$ is a (trivial) Cauchy sequence in $\mathcal{R}[0, 1]$. However, $\mathbf{f} \notin \mathcal{R}[0, 1]$, so that the incompleteness that we attributed to Riemann-integrable functions comes more to light. We will see presently that the Lebesgue integral of $f(x)$ exists, and is in fact equal to zero.

†This convergence comes within the category of what is defined later as "convergence almost everywhere." Here there is no convergence for rational $x$.

### 4.3.1 The Concept of Measure

As we had once observed, there are many ways of approaching the theory of Lebesgue integration. We take here the route of *measure theory*, which will be expounded most simply and informally. Consider, for example, sets on the real line that may be collections of *points* or *intervals*. The sets may be finite or infinite. A sequence is countably infinite and an interval is uncountably infinite. We associate no length with a point on the real line. (Similarly, no area or volume is associated with a point in two- and three-dimensional spaces, respectively.) On the other hand, an interval has length defined by the absolute value of the difference between the two numbers at the interval extremities (boundary). The total length of nonoverlapping intervals is obtained by summing their individual lengths. If two intervals overlapped partially, their combined length would be obtained by deducting the length of the intersection (the overlapping part) from the sum of the individual lengths. The meausre of a set on the real line is a generalization of the foregoing concept of length. Thus we would be interested in measures of *arbitrary* sets on the real line which are not necessarily made up of discrete points or intervals. In an abstract sense, a measure is a function mapping sets (on the real line in this case) into real numbers (nonnegative for our purposes).

Before we consider the measure of more general sets, it is of interest to identify a special class of sets that have *measure zero*. We *define* a set $E$ to have *measure zero* if the set can be *enclosed* within a family of intervals whose total length (measure) may be made as small as possible. Obviously, a finite number of points makes a set of measure zero. Let us cite more general examples.

*Example 1*

A countably infinite number of points has measure zero because by taking the interval about the $n$th point to be $\varepsilon/2^{n+1}$, where $n = 1, 2, \ldots$, the total length of the intervals is $\varepsilon \sum_{n=1}^{\infty} 1/2^{n+1} = \varepsilon/2$, which is less than $\varepsilon$.

*Example 2*

An ingenious set constructed by Cantor is uncountably infinite and has measure zero. This set, called the Cantor set, is obtained as follows. Consider the interval $[0, 1]$. Delete from the interval the open set $(\frac{1}{3}, \frac{2}{3})$, which is the middle third of $[0, 1]$. From each of the remaining closed intervals $[0, \frac{1}{3}]$ and $[\frac{2}{3}, 1]$, delete the middle-third open intervals. Indefinitely continue the process of deleting the middle third of the remaining intervals. The point set that remains on $[0, 1]$ is called the Cantor set. Denoting this set by $C$, we recognize that

(a) $C$ is not empty since it contains the end points of each deleted interval (it contains other points as well).
(b) $C$ is closed because it is obtained by deleting disjoint open intervals from the closed set [0, 1].
(c) The total length of the deleted sets from [0, 1] is given by

$$\frac{1}{3} + 2\left(\frac{1}{3}\right)^2 + 2^2\left(\frac{1}{3}\right)^3 + \ldots + 2^{n-1}\left(\frac{1}{3}\right)^n + \ldots$$
$$= \frac{1}{3}\left[1 + \frac{2}{3} + \left(\frac{2}{3}\right)^2 + \ldots\right] = \frac{1}{3}\frac{1}{(1 - \frac{2}{3})} = 1$$

Thus the length of set $C$ is $1 - 1 = 0$. Hence the Cantor set has measure zero.

(d) That the set $C$ is uncountably infinite is less apparent. We prefer to simply assert this here and leave the interested reader to other works.†

Hence even an uncountably infinite set may have measure zero. Sets of measure zero are in a sense trivial and inconsequential to many aspects of analysis. Thus if any statement concerning points on an interval, say $[a, b]$, holds everywhere in $[a, b]$ except possibly for points within a set of measure zero, we say that the statement is true *almost everywhere* in $[a, b]$. To cite some examples, we say that a function $f(x)$ is zero almost everywhere in $[a, b]$ if

$$f(x) = 0, \qquad x \notin E$$

where $E \subset [a, b]$ is any set of measure zero. On $E, f(x)$ may take arbitrary nonzero values (e.g., it may even be unbounded if $E$ is infinite). As an immediate corollary, two functions $f$ and $g$ are equal almost everywhere if $f(x) = g(x)$ for all $x \in [a, b]$ except for $x \in E$. Thus functions which are equal to each other almost everywhere form an equivalence class. The student may recall the discussion in Section 3.4 on how a pseudometric space may be converted to a metric space by grouping certain elements of the former together into an equivalence class. It may further be recalled that the integral metric (see Example 2 of Section 3.4) originally yields a pseudometric space with the space of integrable (in the Riemann sense) functions. The formation of an equivalence class then depended on collecting together functions that were equal everywhere barring some "exceptional" points. We are now able

†See, for example, G. Temple, *The Structure of Lebesgue Integration Theory*, Oxford University Press, London, 1971.

to identify these exceptional points as belonging to any set of measure zero. It is this concept that renders the space of Riemann-integrable functions an authentic metric space (with the Riemann integral as the metric) albeit an incomplete one.

Other common statements that hold almost everywhere are, for example, in regard to convergence of a sequence of functions $\{f_n(x)\}$ to a function $f(x)$, and continuity of a function $f(x)$ in an interval. To consider the first, we say that $\{f_n(x)\}$ *converges almost everywhere* to $f(x)$ if the real number sequence $\{f_n(x)\}$ converges to the real number $f(x)$ for *each fixed x except* for those in a set of measure zero. The sequence of functions $\{f_n(x)\}$ given by (4.3.1) converge to the function (4.2.2) almost everywhere in [0, 1] because for any irrational $x, f_n(x) = f(x) = 0$ for each $n$, which implies convergence. But for any rational $x, f_n(x)$ assumes either of the values 0 and 1, depending on whether or not $x = k/n$ for some $k$. Thus no convergence may be expected for rational $x$. Since the rationals are countable, they form a set of measure zero so that almost-everywhere convergence of the said sequence is implied. Let us then conclude that a sequence of Riemann-integrable functions converging almost everywhere does not necessarily do so to a Riemann-integrable function.

Next, it is possible to talk about almost-everywhere continuity. A function is said to be *continuous almost everywhere* if it is continuous at every $x$ except for those that comprise a set of measure zero. Thus $f_n(x)$ in (4.3.1) is continuous almost everywhere. [Note from Exercise 4.2.5 that $f(x)$, as given by (4.2.2), is not continuous anywhere.]

Now it is possible to show that a function is Riemann integrable if and only if it is continuous almost everywhere.† On the other hand, Lebesgue integrability does not require almost-everywhere continuity. For example, the Lebesgue integral of the Dirichlet function (4.2.2) exists.

Let us inquire into the measure of *general* subsets of the bounded interval $[a, b]$. It is in this concept that the generalization of the Riemann integral to that of Lebesgue lies. Denoting the measure of a set $E$ by $\mu E$, we define the measure of any *open interval* $(c, d) \subset [a, b]$ as

$$\mu(c, d) = (d - c)$$

The measure of the *closed interval* $[c, d]$ is defined as

$$\mu[c, d] = (d - c)$$

which would be found subsequently to be consistent with the property that the measure of the union of two disjoint sets is the sum of the individual measures and that the measure of a finite number of points is zero. We shall

†See, for example, F. Riesz and B. Sz. Nagy, *Functional Analysis*, Frederick Ungar, New York, 1955, pp. 23–24.

now let $E$ be an *arbitrary* set in $[a, b]$. It is possible to *cover* the set $E$ by a *set* of open (or closed) intervals in $[a, b]$. (Indeed, $[a, b]$ is itself such a covering.) There are *many* such coverings and we collect *all* such coverings $\{I_j\}$ and note that $\mu \bigcup_j I_j$ is bounded from below (certainly, $\mu \bigcup_j I_j \geq 0$). Hence the infimum of $\mu \bigcup_j I_j$ over all coverings $\{I_j\}$ exists. We define the *outer measure* of $E$, denoted by $\mu_e E$, by

$$\mu_e E = \inf\{\mu \bigcup_j I_j \colon \bigcup_j I_j \supset E\}$$

Similarly, an *inner measure* of $E$, denoted by $\mu_i E$, is defined as

$$\mu_i E = (b - a) - \mu_e E'$$

where $E'$ is the complement of $E$ relative to the universal set $[a, b]$. In general,

$$\mu_e E \geq \mu_i E$$

because $\mu_i E$ arises from the measures of intervals left over by removing coverings of $E'$ from $[a, b]$. We now *define* a set $E$ to be *measurable* if

$$\mu_e E = \mu_i E$$

It is obvious that the application of this criterion to open and closed intervals and unions of them reveals them as measurable sets. Similarly, open and closed sets are measurable because it is possible to show that such sets can be expressed as the union of countable, disjoint intervals. The question then arises as to whether all sets are measurable. It turns out that there indeed are nonmeasurable sets, which in fact are quite difficult to construct.† Our concern is only about measurable sets. Since measurable sets have the same inner and outer measure, we denote the common value by $\mu E$.

Some important properties of measurable sets are given below.

(i) If $E_1$ and $E_2$ are measurable and $E_1 \subset E_2$, $\mu E_1 \leq \mu E_2$.

(ii) If $E_1$ and $E_2$ are measurable, $E_1 \cup E_2$ is also measurable. Further,

$$\mu(E_1 \cup E_2) \leq \mu E_1 + \mu E_2$$

(iii) If $\{E_i\}$ is a countable family of measurable sets such that $E_i \cap E_j = 0$, $i \neq j$ (i.e., the sets are mutually disjoint), then

$$\mu(\bigcup_j E_j) = \sum_j \mu E_j$$

The foregoing properties are obvious for intervals but not so for more general measurable sets.

Finally, we observe that the intersection of measurable sets is measurable.

---

†Note clearly the distinction between a set of measure zero, which is measurable, and a nonmeasurable set.

**EXERCISE**

**4.3.1** Show that the following sets are measurable.
(a) Complement of a measurable set.
(b) $E_1 \cap E_2$, where $E_1$ and $E_2$ are measurable.
[*Hint:* Note that $(E_1 \cap E_2)' = E'_1 \cup E'_2$.]

### 4.3.2 Measurable Functions

The next concept crucial to Lebesgue integration is that of a *measurable function*. Let $f(x)$ be a function defined on a measurable set $E$. Further let $\alpha$ be a real number. Form the set

$$E[f(x) > \alpha] \equiv \{x \in E; f(x) > \alpha\}$$

We *define* a function $f(x)$ as *measurable* if $E[f(x) > \alpha]$ is measurable $\forall\, \alpha \in \mathfrak{R}$. The inequality sign $>$ could just as well have been replaced by $\geq$, $<$, or $\leq$. Furthermore, measurability of $f(x)$ implies that the set $E[\alpha < f(x) < \beta]$ is measurable for arbitrary $\alpha$ and $\beta$. That a continuous function is measurable is evident from Figure 4.3.1. Here $E = [0, 1]$, and for the par-

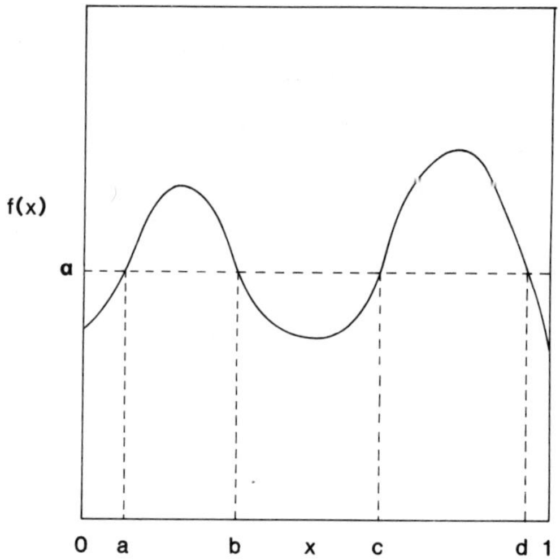


**Figure 4.3.1** Measurability of a continuous function.

ticular $\alpha$ shown in the figure, $E[f(x) > \alpha]$ is the union of the open intervals $(a, b)$ and $(c, d)$ which are measurable. That this result is true for every $\alpha$ is also clear from the figure. (Wildly oscillating continuous functions must similarly produce at most a countable infinity of open intervals for $E[f(x) > \alpha]$ so that they are measurable.) A function, continuous almost everywhere in some measurable set $E$, must be measurable, since the function is continuous on the (measurable) complement of the zero measure set where it is discontinuous. Thus all Riemann-integrable functions are measurable. The class of measurable functions is, however, larger than that of Riemann-integrable functions. Consider, for example, the Dirichlet function (4.2.2). Clearly, we should have

$$E[f(x) > \alpha] = \begin{cases} [0, 1], & \alpha < 0 \\ \text{all rationals in } [0, 1], & 0 \leq \alpha < 1 \\ \text{empty set } 0, & \alpha \geq 1 \end{cases}$$

from which $f(x)$ emerges as a measurable function. Thus the class of measurable functions is larger than the class of Riemann-integrable functions.

It is readily ascertained that if $f_1(x)$ and $f_2(x)$ are measurable functions, the linear combination $\alpha_1 f_1(x) + \alpha_2 f_2(x)$ is measurable. The product of measurable functions is measurable. The reciprocal of a nonvanishing measurable function is also measurable.

A very important result is that a sequence of measurable functions $\{f_n(x)\}$ in $E$, converging almost everywhere in $E$, does so to a measurable function $f(x)$.

Before we proceed to develop the Lebesgue integral, we consider yet another measurable function which is an essential implement of the construction of the Lebesgue integral. Let $E$ be a measurable set. Define the function

$$X_E(x) = \begin{cases} 1, & x \in E \\ 0, & x \notin E \end{cases}$$

The function $X_E(x)$ is usually referred to as the *characteristic* function on the set $E$. That it is measurable follows readily from the definition. When $E$ is an interval, the function $X_E(x)$ is the familiar "step function." The Dirichlet function (4.2.2) is also a characteristic function since $E$ is the set of rationals in [0, 1] that is measurable (it has measure zero).

Suppose now that $E \subset [a, b]$. We represent the Lebesgue integral of any function over the interval $[a, b]$ in the same way as we have done the Riemann integral.† We *define* the Lebesgue integral of $X_E(x)$ by

$$\int_a^b X_E(x)\, dx = \mu E \tag{4.3.3}$$

If $E$ is an interval, then clearly the Lebesgue integral above is the same as the Riemann integral. The Lebesgue integral of the Dirichlet function is therefore zero since the rationals have measure zero.

†This is permissible since, when both the Riemann and Lebesque integrals exist for a function, they have the same value.

Let us consider an arbitrary bounded measurable function defined on $[a, b]$. Since $f(x)$ is bounded, we let

$$A = \inf_{x \in [a,b]} f(x), \qquad B = \sup_{x \in [a,b]} f(x)$$

The range of values of $f(x)$ is covered by the interval $[A, B]$. At the beginning of this section we had observed that in constructing the Lebesgue integral it is the interval of the values of $f(x)$ (the ordinate) that is partitioned. This is precisely what we do at this stage. Consider a partition $P = \{y_0 = A, y_1, y_2, \ldots, y_n = B\}$ of the interval $[A, B]$. Clearly, $y_i < y_{i+1}$. As in the case of the Riemann integral, we describe the coarseness of the partition $P$ by its "norm," denoted $|P|$ and defined as

$$|P| = \max |y_{i+1} - y_i|, \qquad i = 0, 1, 2, \ldots, n-1$$

because $f(x)$ is measurable, the sets

$$E_i = \{x \in [a, b];\ y_{i-1} \leq f(x) < y_i\}, \qquad i = 1, 2, \ldots, n-1$$
$$E_n = \{x \in [a, b];\ y_{n-1} \leq f(x) \leq B\}$$

are measurable. Let us inspect the sets $\{E_j\}_{j=1}^n$ through a specific example. Figure 4.3.2 shows a continuous (and therefore measurable) function which is clearly not monotonic. The sets $E_1, E_2, \ldots, E_n$ are found to be either intervals or unions of intervals. The intervals that have been left unmarked on the figure are actually the union of $E_4, E_5, \ldots, E_{n-1}$. Notice how the sets $\{E_j\}$ are scattered about in the interval $[a, b]$ and that $\bigcup_j E_j = [a, b]$.

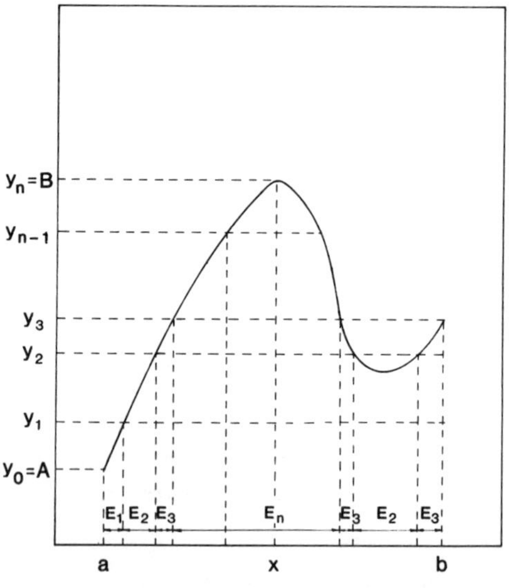


**Figure 4.3.2** Partition for the Lebesgue integral.

Whether or not the function $f(x)$ is continuous, it is always true that $\bigcup_j E_j = [a, b]$. This is because $f(x)$ must necessarily assume the values which it takes between its infimum $A$ and its supremum $B$ at points that together constitute $[a, b]$. With discontinuous functions, it is possible that some $E_j$'s may be empty.

Denoting the characteristic function on $E_j$ by $X_{E_j}$, let us form the two linear combinations

$$\bar{\phi}_n(x) = \sum_{j=1}^{n} y_j X_{E_j}(x), \qquad \underline{\phi}_n(x) = \sum_{j=1}^{n} y_{j-1} X_{E_j}(x)$$

For any $x \in [a, b]$, there is some $i \in \{1, 2, \ldots, n\}$ such that $x \in E_i$ and $\bar{\phi}_n(x) = y_i$, $\underline{\phi}_n(x) = y_{i-1}$. Clearly,

$$\underline{\phi}_n(x) \leq f(x) < \bar{\phi}_n(x) \tag{4.3.4}$$

As in Riemann integration, we form the *upper sum* $\bar{s}(P)$ and the lower sum $\underline{s}(P)$ defined as follows:

$$\bar{s}(P) = \sum_{j=1}^{n} y_j \mu E_j, \qquad \underline{s}(P) = \sum_{j=1}^{n} y_{j-1} \mu E_j$$

Since $\sum_{j=1}^{n} \mu E_j = (b - a)$, we clearly have

$$A(b - a) \leq \underline{s}(P) \leq \bar{s}(P) \leq B(b - a) \tag{4.3.5}$$

which holds for every partition $P$ of $[A, B]$. In view of (4.3.4) the upper and lower sums are, in fact,

$$\bar{s}(P) = \int_a^b \bar{\phi}_n(x)\,dx, \qquad \underline{s}(P) = \int_a^b \underline{\phi}_n(x)\,dx$$

From the inequality (4.3.5), it follows that the following numbers must exist:

$$\bar{S} = \inf_{P \in \mathcal{P}} \bar{s}(P), \qquad \underline{S} = \sup_{P \in \mathcal{P}} \underline{s}(P)$$

where $\mathcal{P}$ is the set of *all* partitions of $[A, B]$. It is possible in a manner entirely analogous to that used in Riemann integration (see Exercises 4.2.1 and 4.2.2) that $\bar{S} \geq \underline{S}$. In fact, by refining the partition such that $|P| < \varepsilon$, we have

$$\bar{s}(P) - \underline{s}(P) < \varepsilon$$

from which it follows that, for a bounded measurable function $f(x)$, we have

$$\bar{S} = \underline{S} \equiv I$$

which is *defined* as the *Lebesgue integral* of $f(x)$.

We have thus defined the Lebesgue integral of a bounded, measurable function. With a slight modification of the definition above, it is possible to accommodate a limited class of unbounded (but measurable) functions among the Lebesgue-integrable functions. We will first define the Lebesgue integral for nonnegative, unbounded functions from which it is possible to extend the definition to functions of mixed signs in a straightforward manner.

Let $f(x)$ be a nonnegative, unbounded, measurable function on $[a, b]$, by which is meant that for arbitrarily large $n$, the set $E_n \equiv \{x \in [a, b]; f(x) > n\}$ is nonempty and measurable. We define the sequence of functions $\{f_n(x)\}$ with

$$f_n(x) = \begin{cases} f(x), & x \in E_n' \\ n, & x \in E_n \end{cases}$$

Obviously, for each $n$, the function $f_n(x)$ is bounded and measurable, and consequently its Lebesgue integral, as defined earlier, exists. The sequence of numbers $\int_a^b f_n(x)\, dx$ may or may not converge. If it does converge, we define

$$\int_a^b f(x)\, dx = \lim_{n\to\infty} \int_a^b f_n(x)\, dx \tag{4.3.6}$$

An unbounded, measurable function for which the limit on the right-hand side of (4.3.6) exists is said to be *summable* or Lebesgue integrable.

For a function $f(x)$ of mixed signs consider the measurable sets $S_+ = \{x \in [a, b]: f(x) \geq 0\}$, $S_- = \{x \in [a, b]: f(x) \leq 0\}$. Define the functions

$$f_+(x) = \begin{cases} f(x), & x \in S^+ \\ 0, & \text{otherwise} \end{cases} \qquad f^-(x) = \begin{cases} -f(x), & x \in S^- \\ 0, & \text{otherwise} \end{cases}$$

Clearly, $f_+(x)$ and $f_-(x)$ are both nonnegative, and

$$f(x) = f_+(x) - f_-(x)$$

We say that the Lebesgue integral of $f(x)$ exists *if* the Lebesgue integrals of $f_+(x)$ and $f_-(x)$ exist, and

$$\int_a^b f(x)\, dx = \int_a^b f_+(x)\, dx - \int_a^b f_-(x)\, dx$$

We have thus defined the Lebesgue integral completely. The Lebesgue integral satisfies all of the known properties of the Riemann integral and more. A particularly important property of the Lebesgue integral, not shared by the Riemann integral, is the content of the Lebesgue dominated convergence theorem, which states that if $\{f_n(x)\}$ are measurable and uniformly bounded, that is,

$$|f_n(x)| \leq K, \qquad K > 0$$

for every $n$ and $x \in [a, b]$, and if the sequence $\{f_n(x)\}$ converges almost everywhere to $f(x)$, then $f(x)$ is Lebesgue integrable and

$$\lim_{n\to\infty} \int_a^b f_n(x)\, dx = \int_a^b f(x)\, dx$$

Another important property of the class of Lebesgue-integrable functions is its completeness, which we shall state later more formally.

The Lebesgue integral can be extended readily to sets in multidimensional spaces such as $\mathcal{R}_n$. Thus area and volume integrals can be defined in an analogous manner. We shall not be concerned with the actual extensions in this book. In what follows we will retain our considerations to the Lebesgue integral over closed, bounded intervals. The functions are assumed to be real-valued, although the extension to complex-valued functions is virtually immediate.

### 4.3.3 Lebesgue Spaces

We have dealt with Lebesgue-integrable functions. The class of Lebesgue-integrable functions forms a linear space and relative to the Lebesgue integral metric

$$d(\mathbf{f}, \mathbf{g}) = \int_a^b |f(t) - g(t)|\, dt$$

yields a complete metric space. This space is referred to as $\mathcal{L}[a, b]$. One may also collect functions $f(t)$, $t \in [a, b]$, such that the Lebesgue integral

$$\int_a^b |f(t)|^p\, dt < \infty \tag{4.3.7}$$

The metric space obtained by letting

$$d_p(\mathbf{f}, \mathbf{g}) \equiv \left\{\int_a^b \left|f(t) - g(t)\right|^p dt\right\}^{1/p}, \qquad p \geq 1 \tag{4.3.8}$$

is again *complete* and is referred to as $\mathcal{L}_p[a, b]$. In particular, the space $\mathcal{L}_2[a, b]$ is an extremely important space in functional analysis. The spaces $\mathcal{L}_p[a, b]$ may be either complex or real, depending on whether complex- or real-valued functions are considered.

Although the class of Lebesgue-integrable functions is larger than Riemann-integrable functions, one can always construct a sequence of Riemann-integrable functions converging with respect to (4.3.8) to any given Lebesgue-integrable function. Alternatively, a Riemann-integrable function can be found arbitrarily close to a given Lebesgue-integrable function. Thus $\mathcal{R}_p[a, b]$, for which (4.3.7) and (4.3.8) must be considered in the sense of Riemann, is *dense* in $\mathcal{L}_p[a, b]$.

This is a very important result which we formally state as the Riesz–Fischer theorem.

***Theorem 4.3.1 (Riesz–Fischer)*** The space $\mathcal{L}_p[a, b]$, $1 < p < \infty$, is complete. Moreover, $\mathcal{R}_p[a, b]$ is dense in $\mathcal{L}_p[a, b]$.

For a proof of this theorem the reader is referred to other works.† It is not difficult to show that $\mathcal{L}_q[a, b] \subset \mathcal{L}_p[a, b]$ when $1 \leq p < q < \infty$. Because

---

†See, for example, *An Introduction to Analysis and Integration Theory*, by E. R. Phillips, Intext, Scranton, Pa., 1971.

$\mathcal{C}[a, b]$ can be shown to be dense in $\mathcal{R}_p[a, b]$, it is also dense in $\mathcal{L}_p[a, b]$ (see Exercise 3.6.3). We consider other dense subsets of $\mathcal{R}_p[a, b]$ (and therefore of $\mathcal{L}_p[a, b]$) at a later stage.

### 4.3.4 Differentiation

We now address the properties of the indefinite Lebesgue integral

$$F(x) = \int_a^x f(y)\, dy, \qquad x \in [a, b] \tag{4.3.9}$$

It is readily shown that $F$ is continuous in $[a, b]$. With a little more work $F$ can be shown to be differentiable *almost everywhere* in $[a, b]$ with $F' = f$. Thus one recovers the integrand by differentiating its indefinite integral. For every integrable function $f$, the function $F(x)$ resulting from (4.3.9) has the property of *absolute continuity*, which is defined below.

***Definition 4.3.1*** Let $F(x)$ be a function defined on $[a, b]$. If $\forall\, \varepsilon > 0\ \exists\ \delta_\varepsilon > 0 \ni$ for *any* finite set of $n$ subintervals of $[a, b]$ such as $[x_j, y_j], j = 1, 2, \ldots, n$, with the property that $\sum_{j=1}^{n} (y_j - x_j) < \delta_\varepsilon$ we have

$$\sum_{j=1}^{n} |F(y_j) - F(x_j)| < \varepsilon$$

then $F(x)$ is said to be *absolutely* continuous on $[a, b]$.

The property of absolute continuity has clearly to do with the rapidity with which the function changes values between neighboring points in the interval. For example, the function $\sin(1/x)$ takes arbitrarily frequent jumps near $x = 0$ so that it is not absolutely continuous on $[0, 1]$; on the other hand, one can show that $x \sin(1/x)$ is absolutely continuous there.

The important conclusion from the discussion above is that $F(x)$ must be absolutely continuous so that $F'(x)$ is integrable. This conclusion is important because of its implications to subsequent discussions on "unbounded" operators whose domains must be restricted so that vectors do not get transformed out of the space of integrable functions (see, e.g., Section 5.5).

### 4.3.5 Dense Subspaces of $\mathcal{L}_2[a, b]$

We conclude this chapter with a discussion of some dense subspaces of interest to us. This interest originates from subsequent necessity to define "unbounded" (differential) operators on dense subspaces of $\mathcal{L}_2[a, b]$. We do not seek to establish rigorous proofs of denseness and will merely point to the intuitive basis of the result.

In the study of differential equations it is common to introduce boundary conditions as additional constraints on the boundary values of the unknown function and its derivatives. When homogeneous, linear boundary conditions are used, the set of all functions satisfying such conditions obviously form a linear subspace. For example, in dealing with second-order, linear differential equations, two homogeneous boundary conditions are specified. We need consider for discussion only one such boundary condition since the arguments leading to the denseness of the subspace will apply in an obvious way to the case of several boundary conditions. Assume that the boundary condition is given by

$$\alpha_1 u(a) + \alpha_2 u'(a) + \alpha_3 u(b) + \alpha_4 u'(b) = 0 \tag{4.3.10}$$

Suppose that we are given an arbitrary element $\mathbf{f} \in \mathfrak{L}_2[a, b]$. Let $D \subset \mathfrak{L}_2[a, b]$ be the subspace of all functions satisfying (4.3.10). Our objective is to establish that for each $\varepsilon > 0$ there exists an element $\mathbf{u} \in D$ such that

$$\int_a^b |u(x) - f(x)|^2 \, dx < \varepsilon \tag{4.3.11}$$

which implies the denseness of $D$ in $\mathfrak{L}_2[a, b]$. Since we know that $\mathfrak{C}^{(1)}[a, b]$ is dense in $\mathfrak{L}_2[a, b]$,† we need only establish the denseness of $D$ in $\mathfrak{C}^{(1)}[a, b]$. Let $\mathbf{v} \in \mathfrak{C}^{(1)}[a, b]$ such that

$$\alpha_1 v(a) + \alpha_2 v'(a) + \alpha_3 v(b) + \alpha_4 v'(b) \equiv \beta$$

Suppose now that we choose $\mathbf{u} \in D$ such that

$$u(x) = v(x), \qquad a + \delta_\varepsilon < x < b \tag{4.3.12}$$

where $\delta_\varepsilon$ is to be determined. Clearly, (4.3.12) implies that $u(b) = v(b)$, $u'(b) = v'(b)$. In the interval $[a, a + \delta_\varepsilon]$ we pick $u'(a) = v'(a)$ so that

$$u(a) = -\frac{1}{\alpha_1}[\beta - \alpha_1 v(a)] = v(a) - \frac{\beta}{\alpha_1} \tag{4.3.13}$$

where we have presumed that $\alpha_1 \neq 0$. (If $\alpha_1 = 0$, a slightly different strategy would be required.) The student may verify that by choosing

$$u(x) = v(x) + \frac{\beta}{\alpha_1}\left[-1 + 3\left(\frac{x-a}{\delta_\varepsilon}\right)^2 - 2\left(\frac{x-a}{\delta_\varepsilon}\right)^3\right] \qquad a \leq x \leq a + \delta_\varepsilon \tag{4.3.14}$$

$\mathbf{u} \in D$. Further,

$$\begin{aligned}\int_a^b [u(x) - v(x)]^2 \, dx &= \int_a^{a+\delta_\varepsilon} [u(x) - v(x)]^2 \, dx \\ &= \int_0^{a+\delta_\varepsilon} \frac{\beta^2}{\alpha_1^2}\left[1 - 3\left(\frac{x-a}{\delta_\varepsilon}\right)^2 + 2\left(\frac{x-a}{\delta_\varepsilon}\right)^3\right]^2 dx \\ &= K\delta_\varepsilon \end{aligned} \tag{4.3.15}$$

†See *Methods in Analysis*, by J. Indritz, Macmillan, New York, 1963, Sec. 5.2. In fact, $\mathfrak{C}^{(\infty)}[a, b]$ is dense in $\mathfrak{L}_2[a, b]$.

where $K \equiv (\beta^2/\alpha_1^2) \int_0^1 (1 - 3y^2 + 2y^3)^2 \, dy$. By taking $\delta_\varepsilon < \varepsilon/K$, we have

$$\int_a^b [u(x) - v(x)]^2 \, dx < \varepsilon$$

so that $D$ is dense in $\mathcal{C}^{(1)}[a, b]$ and hence in $\mathcal{L}_2[a, b]$. We have thus shown that the subspace of functions satisfying homogeneous boundary conditions of the type (4.3.10) is dense in the space $\mathcal{L}_2[a, b]$. This result is extended without difficulty to situations in which several boundary conditions are encountered. The consequence is of great significance to the definition of domains of differential operators which will be encountered in subsequent chapters. Frequently, the domains are restricted to absolutely continuous functions, a constraint not required of functions for square integrability (example: $\{\sin (1/x)\} \in \mathcal{L}_2[0, 1]$). The question then arises as to whether a subspace of absolutely continuous functions (see Exercise 4.3.4) on $[a, b]$ is dense in $\mathcal{L}_2[a, b]$. Absolute continuity is violated only when functions change values with arbitrarily large frequencies. Indeed, they may be approximated in the mean (i.e., in the sense of the metric in $\mathcal{L}_2[a, b]$) by absolutely continuous functions with sufficiently large frequencies.

**EXERCISES**

**4.3.2** Show that the class of measurable functions forms a linear space.

**4.3.3** Verify that $\mathbf{u} = \{u(x)\}$, satisfying (4.3.10), (4.3.12), and (4.3.13), where $\mathbf{v} \in \mathcal{C}^{(1)}[a, b]$ is given, is an element of $\mathcal{C}^{(1)}[a, b]$.

**4.3.4** Show that the set of absolutely continuous functions on $[a, b]$ forms a subspace of $\mathcal{L}_2[a, b]$.

**4.3.5** Show that

$$D = \{\mathbf{u} \in \mathcal{C}^{(1)}[a, b]; \alpha_{i,1}u(a) + \alpha_{i,2}u'(a) + \alpha_{i,3}u(b) + \alpha_{i,4}u'(b) = 0\}, \quad i = 1, 2$$

is dense in $\mathcal{L}_2[a, b]$.

## 4.4 Concluding Remarks

We summarize here the main conclusions of a rather hurried exposure to the ideas of Lebesque integration. Most importantly, it has emerged that the concept of the Riemann integral, normally familiar to engineers, is inadequate in analysis in that the space of Riemann-integrable functions is incomplete as a metric space. The merits of a complete metric space were discussed briefly in Chapter 3 (Sections 3.6 to 3.8).

A further concept that has emerged is that some apparently "small" subspaces such as functions satisfying certain homogeneous boundary conditions, absolute continuity requirements, and so on, can approximate arbitrarily closely (in the sense of Lebesgue-integral metric) any square-integrable function. This result has a tremendous impact on the issue of expanding arbitrary functions in terms of an infinite set of functions satisfying special boundary conditions. Indeed, the concept of Fourier sine and/or cosine series expansion of arbitrary square-integrable functions is one such example. We recommend Section 7.5.1 for a continuation of this discussion.

## FURTHER READING

There are several expositions of the Lebesque integral varying considerably in regard to the details of its development. A highly simplified treatment, which conveys the idea in an elementary analytical way, may be found in

> *An Introduction to the Theory of Linear Spaces* by G. E. Shilov, Prentice-Hall, Englewood Cliffs, N.J., 1961, p. 272.

A conceptually very simple and graphically illustrative approach to Lebesgue integration is available in

> "Theory of Functions of a Real Variable," by S. B. Steckin, Chapter XV of *Mathematics: Its Contents, Methods, and Meaning*, Vol. III. edited by A. D. Alexandrov, A. N. Kolmogorov, and M. A. Lavrentev translated by K. Hirsh, MIT Press, Cambridge, Mass., 1969.

A very readable treatment of integration theory can be found in

> *An Introduction to Analysis and Integration Theory*, by E. R. Phillips, Intext, Scranton, Pa., 1971.

A standard reference to the background of measure theory in Lebesgue integration is

> *Measure Theory*, by P. R. Halmos, D. Van Nostrand, Princeton, N.J., 1960.

A lucid treatment of Lebesgue integration theory with several examples is available in

> *The Structure of Lebesgue Integration Theory*, by G. Temple, Oxford University Press, London, 1971.

# Normed Linear Spaces

# 5

## 5.0 Foreword

In Chapter 2 we expressed our interest in vectors as arising from the fact that the states of engineering systems could be mathematically represented by vectors. Our study of vectors was, however, limited to their algebraic ties. In dealing with numbers, the concept of one being "near" or "away" from another followed readily once the absolute value of a number was defined. This idea is the essence of all *approximations*, one that frequently appears in the verdict "good enough for engineering purposes!" Clearly, therefore, the proximity between vectors is an important feature of the mathematical treatment of engineering systems. The implement for this concept, which was developed abstractly in the treatment of metric spaces (Chapter 3), is readily made available for linear spaces through the definition of the "norm" of a vector (analogous to the absolute value of a number). This norm is an abstract generalization of the "magnitude" of a physical vector in three-dimensional Euclidean space. The concept of angle (and orthogonality) between vectors is postponed until the next chapter in order that the role of a norm may be studied in generality; the inner product spaces of Chapter 6 are also normed linear spaces but possess distinctive properties not common to all normed linear spaces.

Linear transformations on normed linear spaces acquire important properties from the norm of a vector. Thus bounded linear transformations form an important class of operators of interest to engineering applications.

Another class that is of special significance to us in infinite-dimensional spaces is the set of what are called compact operators.

It had transpired that linear transformations could themselves be regarded as vectors but that some of them (the bounded ones) could also be normed is established in this chapter. The "nearness" between transformations is especially significant to the use of approximate methods for solving mathematical equations. Thus, very frequently, approximate solutions to a given equation featuring a linear transformation are obtained by solving "exactly" an approximate equation that has a transformation "near" that which appears in the equation of direct interest.

## 5.1 Introduction

So far, our discussions have been confined to either linear spaces that have only algebraic structure or to metric spaces that have only topological structure. Of course, our examples of metric spaces were essentially derived from linear spaces, but their algebraic structure had been largely unused except for what may have been involved in the definition of the metric. In this chapter we concern ourselves with normed linear spaces, in which vectors or elements of a linear space are assigned a positive-valued "magnitude" called the norm. It will then transpire that this norm naturally generates a metric so that the resulting space has both algebraic and topological structure. Vectors belonging to a normed linear space are more akin to the familiar three-dimensional physical vectors in, say, mechanics, than the vectors from the linear spaces of Chapter 2. The analogy to physical vectors is, however, incomplete since normed linear spaces do not admit the concept of angle between vectors which will be introduced in the next chapter. We now turn to the definition of a normed linear space.

***Definition 5.1.1*** A normed linear space is a pair of objects $(\mathcal{L}, \|\cdot\|)$, where $\mathcal{L}$ is a linear space with an associated field $\mathfrak{F}$ and $\|\cdot\|: \mathcal{L} \longrightarrow \mathcal{R}$ is a *norm* function satisfying the following axioms.

A1 Positivity: $\|\mathbf{x}\| \geq 0$,
A2 Triangular inequality: $\|\mathbf{x} + \mathbf{y}\| \leq \|\mathbf{x}\| + \|\mathbf{y}\|$,
A3 Homogeneity: $\|\alpha\mathbf{x}\| = |\alpha|\,\|\mathbf{x}\|$,
A4 Positive definiteness: $\|\mathbf{x}\| = 0 \leftrightarrow \mathbf{x} = \mathbf{0}$,
where $\mathbf{x}$ and $\mathbf{y}$ are arbitrary elements of $\mathcal{L}$ and $\alpha$ is an arbitrary scalar belonging to $\mathfrak{F}$.†

†Normally, the associated field is the field of complex numbers. However, it may also be extended to an abstract field $\mathfrak{F}$ provided that a mapping $|\cdot|: \mathfrak{F} \longrightarrow \mathcal{R}$ is defined which satisfies axioms A1, A2, and A4 with the vectors replaced by members of $\mathfrak{F}$ and $|\cdot|$ substituted for $\|\cdot\|$.

We denote the normed linear space $(\mathcal{L}, ||\cdot||)$ by a single symbol $\mathfrak{N}$; we will then refer to elements of $\mathcal{L}$ as those of $\mathfrak{N}$. If we define

$$d(\mathbf{x}, \mathbf{y}) = ||\mathbf{x} - \mathbf{y}||$$

then $d$ satisfies all the axioms of a metric in Section 3.1, the proof of which is left to the student. Thus the normed linear space $\mathfrak{N}$ acquires a topological structure from the metric above. The property of completeness had been found to be an important asset to metric spaces, so that normed linear spaces that have this property are of special interest. A complete normed linear space is called a *Banach space*, denoted by $\mathcal{B}$.

We consider some examples of normed linear and Banach spaces.

*Example 1*

$$\mathcal{L} = \mathcal{R}_n\colon ||\mathbf{x}||_p = \{\sum_{i=1}^{n} |x_i|^p\}^{1/p}, \qquad p \geq 1$$

$$||\mathbf{x}||_\infty = \max\{|x_1|, |x_2|, \ldots, |x_n|\}$$

It is readily shown that $||\cdot||_p$ and $||\cdot||_\infty$ satisfy the axioms A1, A3, and A4. The triangular inequality A2 is implied by the Minkowski inequality, to which we have referred in Section 3.1. $\{\mathcal{R}_n, ||\cdot||_p\}$ and $\{\mathcal{R}_n, ||\cdot||_\infty\}$ are normed linear spaces. That they are also Banach spaces follows from the fact that real numbers form a complete system. These statements also hold for the complex linear space $\mathcal{C}_n$ with the absolute value in the norms replaced by the moduli of the complex numbers.

*Example 2*

$$\mathcal{L} = \mathcal{C}[a, b], \qquad ||\mathbf{f}||_p = \left\{\int_a^b |f(t)|^p\,dt\right\}^{1/p}, \qquad p \geq 1$$

The pair $\{\mathcal{L}, ||\cdot||_p\}$ is a normed linear space since $||\cdot||_p$ satisfies the axioms of a norm the proof of which depends on using the properties of the Riemann integral. It is, however, *not* a Banach space (see Example 3 in Section 3.6). If the norm on $\mathcal{C}[a, b]$ is given by

$$||\mathbf{f}||_\infty = \sup_{t \in [a, b]} |f(t)|$$

then $\{\mathcal{C}[a, b], ||\cdot||_\infty\}$ is a Banach space. The results of Chapter 4 show that for a Banach space with $||\cdot||_p$, the latter must be defined in terms of the Lebesgue integral and the linear space must include all functions $f(t)$ such that $|f(t)|^p$ is Lebesgue integrable.

*Example 3*

We consider $\mathcal{L}$ = the subspace of $\mathcal{R}_\infty$ in which $\mathbf{x} = (x_1, x_2, \ldots)$ has the property that

$$\sum_{i=1}^{\infty} |x_i|^p < \infty, \qquad p \geq 1$$

and define the norm by

$$\|\mathbf{x}\|_p = \{\sum_{i=1}^{\infty} |x_i|^p\}^{1/p}$$

which, we show, satisfies the axioms of a norm. Properties A1, A3, and A4 hold obviously. Property A2, the triangular inequality, follows from the fact that if two series of real numbers $\sum_{n=1}^{\infty} a_n$ and $\sum_{n=1}^{\infty} b_n$ converging to $\alpha$ and $\beta$, respectively, are such that $a_n \leq b_n$ for every $n$, then $\alpha \leq \beta$ (why?). This normed linear space, which is referred to as $l_p$ space, is a Banach space.

In general, the results that are true for normed linear spaces are proved more readily for the inner product spaces of Chapter 6, which are specialized normed linear spaces. Since our eventual interest is in inner product spaces, we will spare ourselves the additional technicality required for establishing many useful results for normed linear spaces in which an inner product is not defined.† In particular, the theory of finite-dimensional normed linear spaces is naturally simpler than that of their infinite-dimensional counterparts. When an inner product is defined, the theory of finite-dimensional spaces becomes even simpler.

In Section 3.5 it was pointed out that closed, bounded sets in finite-dimensional spaces are compact, while stricter conditions are required for compactness in infinite-dimensional spaces. It is possible to show that if in a normed linear space $\mathfrak{N}$ the sphere $S = \{\mathbf{x} \in \mathfrak{N}: \|\mathbf{x}\| \leq 1\}$ is compact, then $\mathfrak{N}$ must necessarily be finite-dimensional. The proof of this requires a theorem due to Riesz which we do not cover in this book. Briefly, however, it relates to the existence of vectors *outside* a given proper subspace $M \subset \mathfrak{N}$ at nonzero distances from the subspace.‡ Thus it is readily shown that in infinite-dimensional normed linear spaces, sequences of vectors $\{\mathbf{x}_n\}$ can be found with $\mathbf{x}_n \in S$ and distinctly apart from one another so that no convergent subsequence will exist.

---

†The reader interested in details should consult Sects. 5.1 to 5.10 of *Linear Operator Theory in Engineering and Science*, by A. W. Naylor and G. R. Sell, Holt, Rinehart and Winston, New York, 1971.

‡The distance between a subspace $M$ and a vector $\mathbf{z}$ outside it is defined by $d(\mathbf{z}, M) = \inf\{\|\mathbf{z} - \mathbf{x}\|: \mathbf{x} \in M\}$. We will encounter this again in Chapter 6.

**EXERCISES**

**5.1.1** Show that $(\mathcal{R}_n, \|\cdot\|_p)$ is a normed linear space and, further, that it is a Banach space.

**5.1.2** Show that $|\|\mathbf{x}\| - \|\mathbf{y}\|| \leq \|\mathbf{x} + \mathbf{y}\|$ and infer that the norm is a continuous transformation of $\mathfrak{N}$ into $\mathcal{R}$.

**5.1.3** Show that $l_p$ space is a Banach space.

**5.1.4** Let $\mathcal{L} = \mathcal{C}^{(n)}[a, b]$, which consists of functions with $n$ continuous derivatives in $[a, b]$, and

$$\|\mathbf{f}\| = \left\{\sup_{t\in[a,b]} |f(t)| + \sup_{t\in[a,b]} \left|\frac{df}{dt}\right| + \ldots \sup_{t\in[a,b]} \left|\frac{d^n f}{dt^n}\right|\right\}$$

Show that $\{\mathcal{L}, \|\cdot\|\}$ is a Banach space.

**5.1.5** Let $\mathcal{B}$ be Banach space and $L$ be a subspace. If $L$ is closed, then show that $L$ is also a Banach space.

**5.1.6** Let $\mathfrak{N} = (\mathcal{L}, \|\cdot\|)$ be a normed linear space, where $\mathcal{L}$ is defined over the complex field and $\mathbf{x} \in \mathcal{L}$ such that $\mathbf{x} \neq \mathbf{0}$. Consider the linear manifold $L(\mathbf{x})$ spanned by $\mathbf{x}$. Find $\mathbf{y} \in L(\mathbf{x}) \ni \|\mathbf{y}\| = 1$. Is $\mathbf{y}$ unique?

**5.1.7** Let $\mathfrak{N}$ be a normed linear space. Consider $M = \{\mathbf{x} \in \mathcal{L}: \|\mathbf{x}\| < 1\}$. Show that $M$ is an open set. What is the closure of $M$, denoted $\bar{M}$? Show that $\bar{M}$ is compact if it is finite dimensional. [*Hint:* Use the fact that the Bolzano–Weierstrass theorem (3.5.4) holds for $\mathcal{R}_n$, with any of the norms $\|\cdot\|_p$.]

**5.1.8** Let $\{x_j\}_1^n \subset \mathfrak{N}$ be a linearly independent set. Is the linear manifold $L(\mathbf{x}_1, \mathbf{x}_2, \ldots, \mathbf{x}_n)$ a closed subspace of $\mathfrak{N}$?

**5.1.9** Comment on the nature of closed linear subspaces of an incomplete normed linear space.

### 5.1.1 Direct Sums of Normed Linear Spaces

Given two normed linear spaces $\mathfrak{N}' = (\mathcal{L}', \|\cdot\|')$ and $\mathfrak{N}'' = (\mathcal{L}'', \|\cdot\|'')$, a norm $\|\cdot\|$ can be defined on the linear space of the direct sum $\mathcal{L}' \oplus \mathcal{L}''$ (see Section 2.12) from the individual norms $\|\cdot\|'$ and $\|\cdot\|''$ as follows. Denoting $\mathbf{u} \in \mathcal{L}' \oplus \mathcal{L}''$ by $\mathbf{u} = (\mathbf{x}', \mathbf{x}'')$, where $\mathbf{x}' \in \mathcal{L}'$ and $\mathbf{x}'' \in \mathcal{L}''$,

$$\|\mathbf{u}\|_p = \{\|\mathbf{x}'\|'^p + \|x''\|''^p\}^{1/p}, \qquad p \geq 1$$

or

$$\|\mathbf{u}\|_\infty = \max\,[\|\mathbf{x}'\|', \|\mathbf{x}''\|'']$$

The properties of $\|\cdot\|'$ and $\|\cdot\|''$ are easily used to establish that the norms defined above do satisfy the axiomatic properties of a norm (see Exercise 5.1.10). We denote the normed linear space of the direct sum by $\mathfrak{N}' \oplus \mathfrak{N}''$.

Direct sums can be extended to any number of normed linear spaces and to subspaces of a given normed linear space. If $M$ and $N$ are subspaces of a normed linear space $\mathfrak{N} = \{\mathcal{L}, \|\cdot\|\}$, then the direct sum $M \oplus N$ can be "normed" as follows. For $\mathbf{u} \in M \oplus N$, representing by $\mathbf{u} = (\mathbf{x}, \mathbf{y})$, where $\mathbf{x} \in M$ and $\mathbf{y} \in N$, we have

$$\|\mathbf{u}\|_p = \{\|\mathbf{x}\|^p + \|\mathbf{y}\|^p\}^{1/p}, \qquad 1 \leq p < \infty$$

or

$$\|\mathbf{u}\|_\infty = \max\{\|\mathbf{x}\|, \|\mathbf{y}\|\}$$

Since it is possible to form direct sums of normed linear spaces, the question naturally arises as to whether a norm can be found for the tensor product of two linear spaces $\mathcal{L}'$ and $\mathcal{L}''$ belonging to the normed linear spaces $\mathfrak{N}' = \{\mathcal{L}', \|\cdot\|'\}$ and $\mathfrak{N}'' = \{\mathcal{L}'', \|\cdot\|''\}$, respectively. Indeed, the answer to this question is in the affirmative. But it is not possible to deal with this question in generality, and further discussion is deferred to the next chapter.

**EXERCISES**

**5.1.10** Show that the norms defined in the text for $\mathfrak{N}' \oplus \mathfrak{N}''$ satisfy the axioms for a norm.

**5.1.11** Identify the choice of norms available for the direct sum spaces in Exercise 2.12.1.

**5.1.12** Show that the direct sum of two Banach spaces is a Banach space regardless of the choice of norm among those suggested in the text.

## 5.2 Bases

In dealing with finite-dimensional linear spaces in Chapter 2 we had observed that infinite-dimensional spaces can be defined as linear spaces in which an arbitrarily large number of linearly independent vectors can be identified. Therefore, there could be no finite set of *basis* vectors in terms of which *all* vectors in the space could be expressed as a linear combination. This naturally brings up the question of whether or not infinite-dimensional spaces have bases. Whatever is meant by a basis in spaces of infinite dimension, there is no doubt that there must be an infinite number of vectors in a basis set. There are two problems to be resolved. The *first* is that given that the concept of a basis involves the linear expansion of *every* vector in the space in terms of the basis vectors, the meaning of an infinite sum of vectors must be elucidated. This problem is readily settled since infinite sums derive their meanings from the concept of convergence, a feature that is present in the normed linear spaces under discussion. For example, if we have an equality of the type

$$\mathbf{x} = \sum_{j=1}^{\infty} \alpha_j \mathbf{x}_j, \qquad \mathbf{x}_j \in \mathfrak{N}, \quad \alpha_j \in \mathfrak{F}, \quad j = 1, 2, \ldots$$

it is to be understood as follows. We form the sequence of vectors $\{\mathbf{y}_n\}$ where

$$\mathbf{y}_n = \sum_{j=1}^{n} \alpha_j \mathbf{x}_j$$

and indicate that

$$\mathbf{x} = \lim_{n\to\infty} \mathbf{y}_n$$

by which is meant that

$$\lim_{n\to\infty} ||\mathbf{x} - \mathbf{y}_n|| = 0$$

The *second* problem to be resolved lies in whether the basis set of an infinite-dimensional normed linear space is *countably* infinite or *uncountably* infinite. It turns out that the answer to this question lies in the "type" of space with which are dealing. We shall have more to say on this later. For the present, whether or not a *countable basis* exists in a given normed linear space, we define it as follows.

***Definition 5.2.1*** A *countably infinite* set of vectors $\{\mathbf{x}_j\}$, belonging to a normed linear space $\mathfrak{N}$, is called a (countable) *basis* if $\forall\ \mathbf{x} \in \mathfrak{N}$, $\exists$ a sequence $\{\alpha_n\} \ni$

$$\mathbf{x} = \sum_{j=1}^{\infty} \alpha_j \mathbf{x}_j \qquad (\lim_{n\to\infty} ||\mathbf{x} - \sum_{j=1}^{n} \alpha_j \mathbf{x}_j|| = 0)$$

We now make a few further observations which will establish the circumstances under which a countable basis exists in $\mathfrak{N}$. Consider the linear manifold $L \equiv L(\mathbf{x}_1, \mathbf{x}_2, \ldots)$, which consists of *all finite* linear combinations of vectors in the set $\{\mathbf{x}_j\}$. If the associated field of $\mathfrak{N}$ is real (or complex), then $L$ is an uncountably infinite set, since the coefficients in the linear combination may be uncountable.

Now let $\mathfrak{N}$, defined over the real field, have a countable basis $\{\mathbf{x}_j\}$. Let $\mathbf{x} \in \mathfrak{N}$. Then $\exists$ $\{\alpha_j\} \subset \mathcal{R}$ such that the following is true. For every $\varepsilon > 0$, $\exists$ $n_\varepsilon \ni$ for $n \geq n_\varepsilon$ the vector

$$\mathbf{y}_n \equiv \sum_{j=1}^{n} \alpha_j \mathbf{x}_j$$

satisfies the inequality

$$||\mathbf{x} - \mathbf{y}_n|| < \varepsilon$$

Since $\mathbf{y}_n \in \mathfrak{N}$, we conclude that $L$ is *dense* in $\mathfrak{N}$. Now if we restrict the linear combinations in $L$ to only *rational* numbers and call this manifold $L_r$, then clearly from the denseness of the rationals in the real field, $L_r$ is *dense* in $L$. This leads to the result that $L_r$ is dense in $\mathfrak{N}$ (see Exercise 3.6.3). Now $L_r$ is a countable set (why?), so that we conclude that $\mathfrak{N}$ must be *separable* (see Section 3.6). Thus we have shown that it is *necessary* for $\mathfrak{N}$ to be separable in order to have a countable basis. That it is *also sufficient* follows almost immediately because the integral listing of vectors of the countably dense subset of $\mathfrak{N}$ is then by definition a countable basis in $\mathfrak{N}$.

We will only be interested in separable spaces in this book.† No attention has been paid here to the problem of determining the coefficients $\{\alpha_j\}$ of expansion of a vector $\mathbf{x} \in \mathfrak{N}$ in terms of a given basis $\{\mathbf{x}_j\}$. This is frequently a difficult problem in general, although for certain kinds of bases that arise in more specialized normed linear spaces (discussed in Chapter 6), the problem becomes trivial. Next we consider some examples of bases in selected normed linear spaces.

*Example 1*

Consider the Banach space $l_p$ defined over the real field. The vectors $\{\mathbf{e}_j\}$

$$\mathbf{e}_j = [0, 0, \ldots, 0, 1, 0, \ldots]$$

where the unity occurs in the $j$th place, form a basis in $l_p$. Here the coefficients of expansion of any vector $\mathbf{x} \in l_p$ in terms of $\{\mathbf{e}_j\}$ are obtained readily. Thus $\mathbf{x} \equiv [x,_1\ x_2, \ldots]$ is given by $\mathbf{x} = \sum_{j=1}^{\infty} \mathbf{x}_j \mathbf{e}_j$.

*Example 2*

The Banach space $\{\mathcal{C}[a, b], \|\cdot\|_\infty\}$ has a basis $\{\mathbf{f}_j : f_j(t) = t^{j-1}\}_{j=1}^{\infty}$ since from the Weierstrass theorem it is well known that for every $\mathbf{f} \in \mathcal{C}[a, b]$, $\exists\ \alpha_1, \alpha_2, \ldots$ such that

$$\lim_{n\to\infty} \|\mathbf{f} - \sum_{j=1}^{n} \alpha_j \mathbf{f}_j\|_\infty = 0$$

There are several other examples that are important to us but these will be covered in the next chapter, since they belong to more special normed linear spaces.

### EXERCISES

**5.2.1** Let $\{\mathbf{x}_j\}$ be a sequence in a normed linear space $\mathfrak{N}$ defined over the real field. Show that the linear manifold $L_r(\mathbf{x}_1, \mathbf{x}_2, \ldots)$ consisting of all *finite* linear combinations in terms of *rational* coefficients is a countable set.

**5.2.2** Let $\{\mathbf{x}_j\}$ be a basis of the normed linear space $\mathfrak{N}$ and $L \equiv L(\mathbf{x}_1, \mathbf{x}_2, \ldots)$ be the linear manifold obtained by all finite linear combinations of $\{\mathbf{x}_j\}$. Show that $\bar{L} = \mathfrak{N}$, where $\bar{L}$ is the closure of $L$.

---

†For a discussion of bases in nonseparable spaces, see *Linear Operator Theory in Engineering and Science*, by A. W. Naylor and G. R. Sell, Holt Rinehart and Winston, New York, 1971, p. 314.

## 5.3 Bounded Linear Transformations

Linear transformations are fundamentally associated with linear spaces. Since normed linear spaces inherit a topological structure from the norm, the question arises as to whether or not linear transformations can be continuous. We recall the following definition of continuity.

***Definition 5.3.1*** Let $\mathbf{T}: \mathfrak{N} \longrightarrow \mathfrak{N}$ be a linear transformation. $\mathbf{T}$ is said to be *continuous* at $\mathbf{x}_0$ if $\forall\, \varepsilon > 0 \;\exists\; \delta_\varepsilon > 0 \;\exists\; \|x - x_0\| < \delta_\varepsilon \rightarrow \|\mathbf{Tx} - \mathbf{Tx}_0\| < \varepsilon$.

An interesting feature of continuity of linear transformation is contained in the following theorem.

***Theorem 5.3.1*** If $\mathbf{T}: \mathfrak{N} \longrightarrow \mathfrak{N}$ is continuous at $\mathbf{x}_0$, then it is continuous everywhere.

*Proof*: Since $T$ is continuous at $x_0$, $\forall\, \varepsilon > 0 \;\exists\; \delta_\varepsilon > 0 \ni \|\mathbf{x} - \mathbf{x}_0\| < \delta_\varepsilon \rightarrow \|\mathbf{Tx} - \mathbf{Tx}_0\| < \varepsilon$. Let $\mathbf{y} \in \mathfrak{N}$. Now $\|\mathbf{x} - \mathbf{y}\| = \|\mathbf{x} - \mathbf{y} + \mathbf{x}_0 - \mathbf{x}_0\|$, so that $\|\mathbf{x} - \mathbf{y}\| < \delta_\varepsilon \rightarrow \|(\mathbf{x} - \mathbf{y} + \mathbf{x}_0) - \mathbf{x}_0\| < \delta_\varepsilon \rightarrow \|\mathbf{T}(\mathbf{x} - \mathbf{y} + \mathbf{x}_0) - \mathbf{Tx}_0\| < \varepsilon \rightarrow \|\mathbf{Tx} - \mathbf{Ty}\| < \varepsilon$. Thus $\mathbf{T}$ is continuous at $\mathbf{y}$ and hence continuous everywhere. ▌

We now define an alternative concept associated with linear transformations.

***Definition 5.3.2*** A linear transformation $\mathbf{T}: \mathfrak{N} \longrightarrow \mathfrak{N}$ is said to be *bounded* if $\exists$ a real number $m > 0 \ni \forall\, \mathbf{x} \in \mathfrak{N}$, $\|\mathbf{Tx}\| < m\|\mathbf{x}\|$.

The next theorem establishes that continuity and boundedness of linear transformations are entirely equivalent concepts.

***Theorem 5.3.2*** A linear transformation $\mathbf{T}: \mathfrak{N} \longrightarrow \mathfrak{N}$ is continuous if and only if it is bounded.

*Proof*: To prove the "if" part we assume that $\mathbf{T}$ is bounded. Thus $\exists$ $m > 0 \ni \|\mathbf{Tx}\| \leq m\|\mathbf{x}\| \;\forall\, \mathbf{x} \in \mathfrak{N}$. If for $\varepsilon > 0$ we take $\delta_\varepsilon = \varepsilon/m$, then $\|\mathbf{x}\| < \delta_\varepsilon \rightarrow \|\mathbf{Tx}\| < \varepsilon$, so that $\mathbf{T}$ is continuous at $\mathbf{0}$ from which $\mathbf{T}$ is continuous everywhere. To prove the converse, assume that $\mathbf{T}$ is continuous. Then for $\varepsilon = 1$, $\exists\; \delta_1 > 0 \ni \|\mathbf{x}\| < \delta_1 \rightarrow \|\mathbf{Tx}\| < 1$. Let $\mathbf{y} \in \mathfrak{N}$ such that $\mathbf{y} \neq \mathbf{0}$. Let $\alpha \in \mathfrak{F}$, where $|\alpha| = \delta_1/2\|\mathbf{y}\|$. Then $\alpha\mathbf{y} \in \mathfrak{N}$ has $\|\alpha\mathbf{y}\| = \delta_1/2 < \delta_1$, which implies that $\|\mathbf{T}\alpha\mathbf{y}\| < 1$, so that $\|\mathbf{Ty}\| < 1/|\alpha| = (2/\delta_1)\|\mathbf{y}\|$. Hence $\mathbf{T}$ is bounded. ▌

Thus bounded linear transformations are the *same* as continuous linear transformations. The former term is used commonly in the literature and will be used in this book. An alternative and equivalent way of looking at a bounded linear transformation $\mathbf{T}$ is to regard $||\mathbf{Tx}||$ as bounded on the unit sphere defined by $||\mathbf{x}|| = 1$; that is, $\exists\ m > 0 \ni ||\mathbf{Tx}|| \leq m$ if $||\mathbf{x}|| = 1$.

*Examples of Bounded Linear Transformations*

1. Every linear operator on a finite-dimensional normed linear space is bounded. To prove this we recognize that the linear space $\mathcal{L}$ of dimension $n$ associated with $\mathfrak{N}$ is isomorphic with $\mathcal{R}_n$ or $\mathcal{C}_n$ (see Section 2.11) depending on whether $\mathcal{L}$ is associated with the complex or the real field. Furthermore, the space of linear transformations on $\mathcal{L}$, $\mathfrak{J}(\mathcal{L})$, is isomorphic with the space of $n \times n$ matrices (see Exercise 2.11.6). Thus for any linear operator $\mathbf{T}$, there is a unique matrix operator $\mathbf{A}$ with respect to a given basis in $\mathcal{L}$. Now every matrix operator is a bounded operator. Consider, for instance, $(\mathcal{R}_n, ||\cdot||_1)$ and an $n \times n$ matrix $A$ whose $ij$th coefficient is $a_{ij}$. If $\boldsymbol{\eta} = \mathbf{A}\boldsymbol{\xi}$, then

$$\eta_i = \sum_{i=1}^{n} a_{ij}\xi_j, \qquad i = 1, 2, \ldots, n$$

and

$$\begin{aligned} ||\boldsymbol{\eta}||_1 &= \sum_{i=1}^{n} |\eta_i| = \sum_{i=1}^{n} \left| \sum_{j=1}^{n} a_{ij}\xi_j \right| \\ &\leq \sum_{i=1}^{n} \max\{|a_{ij}|, j = 1, 2, \ldots, n\} \sum_{j=1}^{n} |\xi_j| \\ &\leq k\,||\boldsymbol{\xi}||_1 \end{aligned}$$

where we have used the triangular inequality for real numbers. Thus $\mathbf{A}$ is indeed a bounded operator. It remains, however, to show that the corresponding element $\mathbf{T} \in \mathfrak{J}(\mathcal{L})$ is a bounded operator. Now it will be sufficient to show that there are two numbers $\bar{m}, \underline{m} > 0$ such that $\forall\, \mathbf{x} \in \mathcal{L}$, whose corresponding element is $\boldsymbol{\xi} \in \mathcal{R}_n$, $||\mathbf{x}|| \leq \bar{m}\,||\boldsymbol{\xi}||_1$, and $||\mathbf{x}|| \geq \underline{m}\,||\boldsymbol{\xi}||_1$.

Let $\{\mathbf{z}_j\}_1^n$ be the basis in $\mathcal{L}$. Given $\mathbf{x} \in \mathcal{L}$, $\boldsymbol{\xi} = (\xi_1, \xi_2, \ldots, \xi_n)$ is obtained from the expansion

$$\mathbf{x} = \sum_{j=1}^{n} \xi_j \mathbf{z}_j$$

From the triangular inequality

$$||\mathbf{x}|| \leq \sum_{j=1}^{n} |\xi_j|\,||\mathbf{z}_j|| \leq \bar{m}\,||\boldsymbol{\xi}||_1$$

where $\bar{m} = \max\{||\mathbf{z}_j||, j = 1, 2, \ldots, n\}$. The calculation of $\underline{m}$ is somewhat more involved. Define $f: \mathcal{R}_n \longrightarrow \mathcal{R}$ by

$$f(\boldsymbol{\xi}) = ||\mathbf{x}|| = \left\| \sum_{j=1}^{n} \xi_j \mathbf{z}_j \right\|$$

Indeed, $f$ is continuous since

$$|f(\boldsymbol{\xi}) - f(\boldsymbol{\eta})| = |\,||\mathbf{x}|| - ||\mathbf{y}||\,| \leq ||\mathbf{x} - \mathbf{y}|| \leq \bar{m}\,||\boldsymbol{\xi} - \boldsymbol{\eta}||_1$$

so that $\forall\, \varepsilon > 0$, if $\delta_\varepsilon = \varepsilon/\bar{m}$, then $||\boldsymbol{\xi} - \boldsymbol{\eta}||_1 < \delta_\varepsilon \rightarrow |f(\boldsymbol{\xi}) - f(\boldsymbol{\eta})| < \varepsilon$.

It can be shown that the set $S = \{\boldsymbol{\xi} \in \mathcal{R}_n, ||\boldsymbol{\xi}||_1 = 1\}$ is a compact subset of $\mathcal{R}_n$ (see Exercise 5.1.7). Thus $f$ has a minimum on $S$; let $\underline{m} = \inf\{f(\boldsymbol{\xi}): \boldsymbol{\xi} \in S\}$, where $\underline{m} > 0$ (why?). For $\boldsymbol{\xi} \in \mathcal{R}_n \ni \boldsymbol{\xi} \neq \mathbf{0}$, we have $(\boldsymbol{\xi}/||\boldsymbol{\xi}||_1) \in S$ and

$$f\left(\frac{\boldsymbol{\xi}}{||\boldsymbol{\xi}||_1}\right) = \frac{1}{||\xi||_1}\,||\mathbf{x}|| \geq \underline{m}$$

so that

$$||\mathbf{x}|| \geq \underline{m}\,||\boldsymbol{\xi}||_1$$

We can now establish that the boundedness of $\mathbf{A}$ on $\mathcal{R}_n$, where $\mathbf{A}$ is a representation of $\mathbf{T}$ in terms of the basis $\{\mathbf{z}_j\}_1^n$ in $\mathcal{L}$, implies the boundedness of $\mathbf{T}$. Let $\mathbf{x} \in \mathcal{L}$ for which $\boldsymbol{\xi} \in \mathcal{R}_n$. Then $\mathbf{Tx} \in \mathcal{L}$ corresponds to $\mathbf{A}\boldsymbol{\xi} \in \mathcal{R}_n$. Clearly,

$$||\mathbf{Tx}|| \leq \bar{m}\,||\mathbf{A}\boldsymbol{\xi}||_1 \leq \bar{m}k\,||\boldsymbol{\xi}||_1 \leq \left(\frac{\bar{m}k}{\underline{m}}\right)||\mathbf{x}||$$

which shows that $\mathbf{T}$ is bounded. Thus every linear operator on a finite-dimensional normed linear space is bounded.

In establishing the result above, we have in fact shown that $\mathcal{L}$ and $\mathcal{R}_n$ are not only isomorphic in the algebraic sense but also in the topological sense implied by Definition 3.6.5.

2. The conclusion that all linear operators on finite-dimensional normed linear spaces are bounded is easily shown to be invalid for infinite-dimensional spaces. Thus consider the normed linear space $(\mathcal{C}^{(1)}[0, 1], ||\cdot||_\infty)$ which consists of all functions with continuous first derivatives in $[0, 1]$. The operator $\mathbf{D} \equiv \{d/dt\}$ is unbounded because if $\mathbf{f}_n = \{t^n\}$, then $\mathbf{f}_n \in \mathcal{C}^{(1)}[0, 1]$ and $||\mathbf{f}_n||_\infty = 1$, so that $\mathbf{f}_n$ is on the unit sphere for every $n$. However,

$$||\mathbf{Df}_n|| = \sup_{t \in [0,1]} |nt^{n-1}| = n$$

which can be arbitrarily large so that $\mathbf{D}$ is unbounded.

3. *The continuous kernel*: We consider the Banach space $\mathcal{B} = \{\mathcal{C}[a, b], ||\cdot||_\infty\}$. Define an operator $\mathbf{K}$ by

$$\mathbf{g} = \mathbf{Kf}: \quad g(t) = \int_a^b K(t, s) f(s)\, ds, \qquad a \leq t \leq b$$

where $K(t, s)$ is continuous on $[a, b] \times [a, b]$. Clearly, $\exists\ k > 0 \ni |K(t, s)| \leq k\ \forall\ t, s \in [a, b]$, so that

$$\begin{aligned} |g(t)| &= \left| \int_a^b K(t, s) f(s)\, ds \right| \leq \int_a^b |K(t, s)|\, |f(s)|\, ds \\ &\leq k\, \|\mathbf{f}\|_\infty (b - a) \end{aligned}$$

from which we have

$$\|\mathbf{g}\|_\infty = \|\mathbf{Kf}\|_\infty \leq k(b - a)\, \|\mathbf{f}\|_\infty$$

The foregoing inequality shows that $\mathbf{K}$ is a bounded operator. The operator $\mathbf{K}$ is called a *Fredholm operator*. When the kernel $K(t, s)$ has the property $K(t, s) = 0$, $s \geq t$, $\mathbf{K}$ is called a *Volterra operator*, or alternatively we may write

$$\mathbf{g} = \mathbf{Kf}: \quad g(t) = \int_a^t K(t, s) f(s)\, ds$$

Clearly, the Volterra operator is a bounded operator and has some interesting properties, some of which have been covered in Exercise 3.7.3.

4. *The Hilbert–Schmidt kernel*: The Banach space of interest here is $\mathcal{B} = \{\mathcal{L}_2[a, b], \|\cdot\|_2\}$. The operator $\mathbf{K}$ is represented as before by a kernel $K(t, s)$ which has the property

$$\int_a^b \int_a^b |K(t, s)|^2\, dt\, ds < \infty \tag{5.3.1}$$

where the integral is in the Lebesgue sense. Such a function, when used in defining $\mathbf{K}$ as in the previous example, is called a Hilbert–Schmidt kernel. Note that the continuous kernel of Example 3 is certainly a Hilbert–Schmidt kernel. Obviously, $\mathbf{K}$ is a linear operator. To show that $\mathbf{K}$ is a bounded linear operator, we observe that

$$\begin{aligned} \|\mathbf{Kf}\|_2^2 &= \int_a^b \left| \int_a^b K(t, s) f(s)\, ds \right|^2 dt \\ &\leq \int_a^b \left[ \int_a^b |K(t, s)|^2\, ds \int_a^b |f(s)|^2\, ds \right] dt \\ &= \int_a^b \int_a^b |K(t, s)|^2\, ds\, dt\, \|\mathbf{f}\|_2^2 \end{aligned} \tag{5.3.2}$$

which follows from the Hölder inequality for integrals (see Chapter 4). In view of (5.3.1) and (5.3.2), $\mathbf{K}$ is a bounded linear operator. This demonstration also holds for the case where all quantities involved are complex-valued.

If $\mathbf{T}$ is bounded on $\mathfrak{N}$, $N(\mathbf{T})$ is a closed subspace of $\mathfrak{N}$, since for any subsequence $\{\mathbf{x}_n\} \subset N(\mathbf{T})$ converging to $\mathbf{x}_0 \in \mathfrak{N}$, $\mathbf{0} = \lim_{n\to\infty} \mathbf{T}\mathbf{x}_n = \mathbf{T}(\lim_{n\to\infty} \mathbf{x}_n) = \mathbf{T}\mathbf{x}_0$ so that $\mathbf{x}_0 \in N(\mathbf{T})$. Similarly, $R(\mathbf{T})$ is closed.

### 5.3.1 Domains of Bounded Linear Transformations

In Section 2.7 it was pointed out that some linear operators are *necessarily* restricted to domains that are proper subspaces of a linear space. We are now in a position to elaborate on this point. The derivative operator $\mathbf{D}$ (see Example 2) was necessarily restricted to the normed linear space of continuously differentiable functions with the norm $||\cdot||_\infty$, which is a proper subspace of the Banach space of continuous functions. Indeed, $\mathbf{D}$ cannot be defined on continuous functions that are not differentiable so that the domain of $\mathbf{D}$ cannot be the entire Banach space, say $\{\mathcal{C}[0, 1], ||\cdot||_\infty\}$. More generally, we may say that this is the characteristic of all *unbounded* operators; that is, unbounded operators cannot be defined on *all* the elements of a Banach space. We shall have more to say about this at a subsequent stage.

A bounded linear transformation can always be defined on *every* element of a Banach space. If, for example, a bounded linear transformation $\mathbf{T}$ is defined on a domain $D(\mathbf{T})$ which is *dense* in a Banach space $\mathfrak{B}$, then for every element $\mathbf{x} \in \mathfrak{B}$, there is a sequence $\{\mathbf{x}_n\}$ contained entirely in $D(\mathbf{T})$ which converges to $\mathbf{x}$. We shall show how $\mathbf{T}\mathbf{x}$ may be defined. The sequence $\{\mathbf{x}_n\}$ is, of course, a Cauchy sequence. Form the sequence $\{\mathbf{y}_n = \mathbf{T}\mathbf{x}_n\}$ and note that

$$||\mathbf{y}_n - \mathbf{y}_k|| \leq m\,||\mathbf{x}_n - \mathbf{x}_k||$$

from which $\{\mathbf{y}_n\}$ is a Cauchy sequence in $\mathfrak{B}$ converging to some $\mathbf{y} \in \mathfrak{B}$. Now define $\mathbf{T}\mathbf{x} \equiv \mathbf{y}$. Since there could be *many* sequences that converge to $\mathbf{x}$, whether all such transformed sequences would converge to the same element $\mathbf{y}$ raises some doubt about the uniqueness of the foregoing definition. This is not a problem, however, because the boundedness of $\mathbf{T}$ implies its continuity and we must have $\lim \mathbf{T}\mathbf{x}_n = \mathbf{T}(\lim \mathbf{x}_n)$.

### 5.3.2 Space of Bounded Linear Transformations

In Section 2.11 we had found that the class of linear transformations on linear space $\mathfrak{L}$ formed a linear space denoted $\mathfrak{I}(\mathfrak{L})$. For finite-dimensional normed linear spaces, we have just seen that $\mathfrak{I}(\mathfrak{N})$ is also the class of all bounded linear operators. For infinite-dimensional normed linear spaces, however, the question arises as to whether the class of bounded linear operators forms a linear subspace of $\mathfrak{I}(\mathfrak{N})$. That the answer to this question is in the affirmative is easily established, for let $\mathbf{T}_1$ and $\mathbf{T}_2$ be bounded linear operators on $\mathfrak{N}$. Then $\exists\ m_1, m_2 < 0 \ni \forall\, \mathbf{x} \in \mathfrak{N}, ||\mathbf{T}_1\mathbf{x}|| \leq m_1\,||\mathbf{x}||$,

$\|\mathbf{T}_2\mathbf{x}\| \leq m_2 \|\mathbf{x}\|$. Let $\alpha_1, \alpha_2 \in \mathfrak{F}$ be arbitrary scalars. Consider the transformation $\mathbf{T}$ defined by $\mathbf{T} \equiv \alpha_1\mathbf{T}_1 + \alpha_2\mathbf{T}_2$, which is of course linear. For $\mathbf{x} \in \mathfrak{N}$,

$$\begin{aligned} \|\mathbf{Tx}\| &= \|\alpha_1\mathbf{T}_1\mathbf{x} + \alpha_2\mathbf{T}_2\mathbf{x}\| \\ &\leq |\alpha_1| \, \|\mathbf{T}_1\mathbf{x}\| + |\alpha_2| \, \|\mathbf{T}_2\mathbf{x}\| \\ &\leq [|\alpha_1| m_1 + |\alpha_1| m_2] \, \|\mathbf{x}\| \\ &= m \, \|\mathbf{x}\| \end{aligned}$$

so that $\mathbf{T}$ is a bounded linear transformation. Thus, clearly, the class of bounded linear transformations on $\mathfrak{N}$ is a linear subspace of $\mathfrak{J}(\mathfrak{N})$. We denote this subspace of bounded linear transformations on $\mathfrak{N}$ by $\text{Blt}(\mathfrak{N})$.†

A further question of interest is whether the linear space $\text{Blt}(\mathfrak{N})$ can be "normed"; that is, can one define the norm of a bounded linear operator? To answer this we recognize that for $\mathbf{T} \in \text{Blt}(\mathfrak{N})$,

$$\|\mathbf{Tx}\| \leq m \, \|\mathbf{x}\| \qquad \forall \quad \mathbf{x} \in \mathfrak{N}$$

If $\mathbf{x} \neq \mathbf{0}$, then

$$\frac{\|\mathbf{Tx}\|}{\|\mathbf{x}\|} \leq m$$

Thus the set $\{\|\mathbf{Tx}\|/\|\mathbf{x}\|: \mathbf{x} \neq \mathbf{0}\}$ is bounded on the positive real line and hence possesses a lowest upper bound (see Theorem 1.8.1), which prompts the following definition.

***Definition 5.3.3*** Let $\mathbf{T} \in \text{Blt}(\mathfrak{N})$. Then the *norm* of $\mathbf{T}$ is defined by

$$\|\mathbf{T}\| = \sup \left\{\frac{\|\mathbf{Tx}\|}{\|\mathbf{x}\|}, \ \mathbf{x} \neq \mathbf{0}\right\}$$

An alternative definition of the norm of $\mathbf{T}$ is given by

$$\|\mathbf{T}\| = \sup \{\|\mathbf{Tx}\|: \|\mathbf{x}\| = 1\}$$

It is left as exercises for the student to show that the preceding definition is equivalent to Definition 5.3.3 and that either of them satisfies the axiomatic properties of a norm.

Thus the space of bounded linear transformations $\text{Blt}(\mathfrak{N})$ is a normed linear space. Hence a topological structure is acquired by the space $\text{Blt}(\mathfrak{N})$, which makes it possible to consider, for example, converging sequences of operators, completeness, and so on. Given $\mathbf{T} \in \text{Blt}(\mathfrak{N})$, we may write

$$\|\mathbf{Tx}\| \leq \|\mathbf{T}\| \, \|\mathbf{x}\| \qquad \forall \quad \mathbf{x} \in \mathfrak{N}$$

†A more self-explanatory notation of $\text{Blt}(\mathfrak{N})$ is $\text{Blt}(\mathfrak{N}, \mathfrak{N})$, since mappings from one normed linear space $\mathfrak{N}$ to another, $\mathfrak{N}'$, may then be denoted as $\text{Blt}(\mathfrak{N}, \mathfrak{N}')$. However, the second argument is suppressed for notational brevity with the understanding that when the need arises, the expanded notation may be used.

If $\mathfrak{N}$ is a Banach space $\mathfrak{B}$, then it can be shown that Blt($\mathfrak{B}$) is also a Banach space. Let $\{\mathbf{T}_n\} \subset$ Blt($\mathfrak{B}$) be a Cauchy sequence of bounded linear operators. Let $\mathbf{x} \in \mathfrak{B}$, $\mathbf{x} \neq \mathbf{0}$. Then consider the sequence $\{\mathbf{y}_n\}$ where $\mathbf{y}_n = \mathbf{T}_n\mathbf{x}$. For distinct integers $m$ and $n$ we have

$$\|\mathbf{y}_m - \mathbf{y}_n\| = \|\mathbf{T}_m\mathbf{x} - \mathbf{T}_n\mathbf{x}\| \leq \|\mathbf{T}_m - \mathbf{T}_n\| \, \|\mathbf{x}\|$$

Given $\varepsilon > 0$, if $m$ and $n$ are chosen sufficiently large such that $\|\mathbf{T}_m - \mathbf{T}_n\| < \varepsilon/\|\mathbf{x}\|$, then $\|\mathbf{y}_m - \mathbf{y}_n\| < \varepsilon$, which shows that $\{\mathbf{y}_n\}$ is a Cauchy sequence in $\mathfrak{B}$. Thus $\mathbf{y} = \lim \mathbf{y}_n$ exists and belongs to $\mathfrak{B}$. We may now *define* an operator $\mathbf{T}$ on $\mathfrak{B}$ by

$$\mathbf{Tx} = \lim \mathbf{T}_n\mathbf{x} \qquad \forall \quad \mathbf{x} \in \mathfrak{B}$$

Evidently, $\mathbf{T}$ is a linear operator. That it is bounded is proved as follows. The sequence of real numbers $\{\|\mathbf{T}_n\|\}$ converges since

$$|\|\mathbf{T}_n\| - \|\mathbf{T}_m\|| \leq \|\mathbf{T}_m - \mathbf{T}_n\|$$

so that we may let

$$\alpha = \lim_{n\to\infty} \|\mathbf{T}_n\|$$

Clearly, since

$$\frac{\|\mathbf{y}_n\|}{\|\mathbf{x}\|} \leq \|\mathbf{T}_n\|, \qquad \frac{\|\mathbf{y}\|}{\|\mathbf{x}\|} \leq \alpha$$

or

$$\|\mathbf{Tx}\| \leq \alpha \|\mathbf{x}\|$$

which implies that $\mathbf{T}$ is bounded.† We now show that $\mathbf{T}$ is the limit point of $\{\mathbf{T}_n\}$. Evidently, $\|\mathbf{T}\| \leq \alpha$. Now for $m$ and $n$ sufficiently large, that is, $m, n \geq n_\varepsilon$,

$$\|\mathbf{T}_m\mathbf{x} - \mathbf{T}_n\mathbf{x}\| < \varepsilon \|\mathbf{x}\|$$

Keeping $m$ fixed, the left-hand side of the preceding inequality may be regarded as a sequence with respect to $n$ bounded by $\varepsilon \|\mathbf{x}\|$. By letting $n \longrightarrow \infty$ and using the continuity of the norm (see Exercise 5.1.2), we have

$$\lim_{n\to\infty} \|\mathbf{T}_m\mathbf{x} - \mathbf{T}_n\mathbf{x}\| = \|\mathbf{T}_m\mathbf{x} - \mathbf{Tx}\|$$

Clearly,

$$\|\mathbf{T}_m\mathbf{x} - \mathbf{Tx}\| \leq \varepsilon \|\mathbf{x}\|$$

from which

$$\|\mathbf{T}_m - \mathbf{T}\| \leq \varepsilon$$

which shows that $\lim \mathbf{T}_m = \mathbf{T}$. Thus Blt($\mathfrak{B}$) is a Banach space.

Besides the notion of convergence of bounded linear operators with respect to the natural norm in Blt($\mathfrak{N}$), one can also have a sequence of transformations $\{\mathbf{T}_n\} \subset$ Blt($\mathfrak{N}$) such that $\lim \mathbf{T}_n\mathbf{x} = \mathbf{Tx} \; \forall \; \mathbf{x} \in \mathfrak{N}$, referred to

---

†In establishing this, we have made use of the result of Exercise 1.8.5.

as *strong convergence*, which does not necessarily imply convergence with respect to the norm on $\mathrm{Blt}(\mathfrak{N})$. It is obvious, however, that convergence in $\mathrm{Blt}(\mathfrak{N})$ implies strong convergence (note that this property was used in the preceding discussion).

In Section 2.9 we had observed that the linear space of operators had the additional structure of multiplication of operators and that the product of two linear operators was also a linear operator. If $\mathbf{T}_1, \mathbf{T}_2 \in \mathrm{Blt}(\mathfrak{N})$, then $\forall\, \mathbf{x} \in \mathfrak{N}$,

$$||\mathbf{T}_1\mathbf{T}_2\mathbf{x}|| = ||\mathbf{T}_1(\mathbf{T}_2\mathbf{x})|| \leq ||\mathbf{T}_1||\,||\mathbf{T}_2\mathbf{x}|| \leq ||\mathbf{T}_1||\,||\mathbf{T}_2||\,||\mathbf{x}||$$

so that $\mathbf{T}_1\mathbf{T}_2 \in \mathrm{Blt}(\mathfrak{N})$. It is also evident that $||\mathbf{T}_1\mathbf{T}_2|| \leq ||\mathbf{T}_1||\,||\mathbf{T}_2||$. We consider below the example of a *Volterra operator*, which was the subject of Exercise 3.7.3, in regard to the solution of Volterra equations of the *second kind* by iterative methods. The point of interest is the type of convergence of the successive iterants.

*Example. The Volterra Operator*

Consider the Banach space $\mathcal{B} = (\mathcal{C}\,[0, T], ||\cdot||_\infty)$ and the operator $\mathbf{K}\colon \mathcal{B} \longrightarrow \mathcal{B}$ defined by

$$\mathbf{g} = \mathbf{Kf}; \qquad g(t) = \int_0^t K(t, s) f(s)\, ds$$

where $K(t, s)$ is continuous on $[0, T] \times [0, T]$. The following facts have been established about this operator.

(i) $\mathbf{K}$ is a bounded linear operator (see examples considered earlier in this section).
(ii) $\mathbf{K}$ has *only* the zero element as its fixed point.
(iii) There *exists* a *unique* solution to the inhomogeneous Volterra integral equation of the second kind given by

$$g(t) = f(t) - \lambda \int_0^t K(t, s) f(s)\, ds; \qquad \mathbf{g} = \mathbf{f} - \lambda\mathbf{Kf}$$

where $\lambda$ is any fixed real number (see Exercise 3.7.3).

In regard to (iii) the existence is to be established by showing that the sequence $\{\mathbf{f}_n\}$ formed by the successive iterants $\mathbf{f}_n = \mathbf{g} + \lambda\mathbf{K}\mathbf{f}_{n-1}$ is a Cauchy sequence, the limit of which exists and is the solution to the integral equation. Now let $\mathbf{f}_0 = \mathbf{g}$. Then

$$\begin{aligned}\mathbf{f}_2 &= \mathbf{g} + \lambda\mathbf{K}\mathbf{f}_1 = \mathbf{g} + \lambda\mathbf{K}(\mathbf{g} + \lambda\mathbf{Kg})\\ &= \mathbf{g} + \lambda\mathbf{Kg} + \lambda^2\mathbf{K}^2\mathbf{g}\\ &= (\mathbf{I} + \lambda\mathbf{K} + \lambda^2\mathbf{K}^2)\mathbf{g}\end{aligned}$$

Proceeding along similar lines, we obtain

$$\mathbf{f}_n = \sum_{j=0}^{n} \lambda^j \mathbf{K}^j \mathbf{g}$$

Since the solution of the integral equation is obtained by taking the limit of $\{\mathbf{f}_n\}$ (why?), we have from Exercise 3.7.3

$$\mathbf{f} = \lim_{n\to\infty} \sum_{j=0}^{n} \lambda^j \mathbf{K}^j \mathbf{g} = (\mathbf{I} - \lambda\mathbf{K})^{-1}\mathbf{g}$$

The foregoing equation essentially says that the sequence of operators

$$\left\{\sum_{j=0}^{n} \lambda^j \mathbf{K}^j\right\}$$

converges *strongly* to the operator $(\mathbf{I} - \lambda\mathbf{K})^{-1}$. Indeed, this does not necessarily imply that the convergence of the sequence above also holds with respect to the operator norm. However, we will show that this is, in fact, the case here. We define the operator $\mathbf{T}_n$ by

$$\mathbf{T}_n \equiv \sum_{j=0}^{n} \lambda^j \mathbf{K}^j$$

Since $\mathbf{K}^j \in \text{Blt}(\mathcal{B})$ (why?), we have $\mathbf{T}_n \in \text{Blt}(\mathcal{B})$ for each $n$. Now for $m > n$, we have

$$\|\mathbf{T}_m - \mathbf{T}_n\| \leq \sum_{j=n}^{m} |\lambda|^j \|\mathbf{K}^j\| \tag{5.3.3}$$

At this point we remark that it is possible to apply the conservative bound $\|\mathbf{K}^j\| \leq \|\mathbf{K}\|^j$ and conclude that *if*†

$$0 < \alpha \equiv |\lambda| \|\mathbf{K}\| < 1 \tag{5.3.4}$$

then

$$\|\mathbf{T}_m - \mathbf{T}_n\| \leq \sum_{j=n}^{m} \alpha^j \leq \alpha^n \sum_{j=0}^{m-n} \alpha^j$$

from which

$$\|\mathbf{T}_m - \mathbf{T}_n\| < \varepsilon \tag{5.3.5}$$

by making $n$ sufficiently large. However, the constraint (5.3.4) is unnecessary here because it is easily shown that

$$\|\mathbf{K}^j\| \leq \frac{(kt)^j}{j!} \tag{5.3.6}$$

where $k = \sup\{|K(t, s)| : t, s \in [0, T]\}$. For establishing (5.3.6) the reader is advised to study Example 2 of Section 3.7. Hence (5.3.5) holds regardless

---

†We say this here particularly because it holds for the *Fredholm* operator defined by $\mathbf{g} = \mathbf{Kf}$: $g(t) = \int_0^1 K(t, s) f(s)\, ds$, where

$$\mathbf{K}: \mathcal{B} \longrightarrow \mathcal{B}, \qquad \mathcal{B} = \{\mathcal{C}[0, 1], \|\cdot\|_\infty\}$$

of the value of $\lambda$ because from (5.3.3) and (5.3.6), we have

$$\|\mathbf{T}_m - \mathbf{T}_n\| \leq \sum_{j=n}^{m} \frac{\{|\lambda| kT\}^j}{j!}$$

which can be made as small as we please by making $m$ and $n$ sufficiently large. Thus $\{\mathbf{T}_n\}$ is a Cauchy sequence in Blt($\mathcal{B}$) and must converge to some $\mathbf{T} \in$ Blt($\mathcal{B}$) with respect to the operator norm. We know that $\{\mathbf{T}_n\}$ converges strongly to $(\mathbf{I} - \lambda\mathbf{K})^{-1}$, that is, for $\mathbf{g} \in \mathcal{B}$.

$$\lim_{n\to\infty} \mathbf{T}_n\mathbf{g} = (\mathbf{I} - \lambda\mathbf{K})^{-1}\mathbf{g} \tag{5.3.7}$$

If we let $\mathbf{T}' \equiv (\mathbf{I} - \lambda\mathbf{K})^{-1}$, then the task which remains is to show that $\mathbf{T} = \mathbf{T}'$. Now from (5.3.5), which holds for $m$ and $n$ sufficiently large, we have

$$\|\mathbf{T}_m\mathbf{g} - \mathbf{T}_n\mathbf{g}\|_\infty \leq \varepsilon \|\mathbf{g}\|_\infty \tag{5.3.8}$$

for each $\mathbf{g} \in \mathcal{B}$. Suppose that we hold $n$ fixed and let $m$ go to infinity in (5.3.8); then from the continuity of the norm, we obtain

$$\lim_{m\to\infty} \|\mathbf{T}_m\mathbf{g} - \mathbf{T}_n\mathbf{g}\|_\infty = \|\lim_{m\to\infty} \mathbf{T}_m\mathbf{g} - \mathbf{T}_n\mathbf{g}\|_\infty \leq \varepsilon \|\mathbf{g}\| \tag{5.3.9}$$

Using (5.3.7), the inequality (5.3.9) becomes

$$\|\mathbf{T}'\mathbf{g} - \mathbf{T}_n\mathbf{g}\|_\infty \leq \varepsilon \|\mathbf{g}\|_\infty$$

from which

$$\|\mathbf{T}' - \mathbf{T}_n\| \leq \varepsilon$$

Thus clearly $\{\mathbf{T}_n\}$ converges to $\mathbf{T}'$ with respect to the operator norm so that $\mathbf{T} = \mathbf{T}'$.

We have thus shown that not only is the operator $\mathbf{T}_n$ strongly convergent to the inverse operator $(\mathbf{I} - \lambda\mathbf{K})^{-1}$ but is also convergent with respect to the operator norm. Note particularly that $(\mathbf{I} - \lambda\mathbf{K})^{-1}$ is a bounded operator.

We had observed that a bounded linear operator is always definable on an entire Banach space. It is possible to show that if $\mathbf{T}$ is a linear operator defined on some $D(\mathbf{T}) \subset \mathcal{B}$ and $D(\mathbf{T})$ is of finite dimension, then $\mathbf{T}$ is necessarily bounded. The proof is not difficult (for normed linear spaces) but is considerably simpler for inner product spaces.

## EXERCISES

**5.3.1** Show that the norm defined on Blt($\mathfrak{N}$) satisfies all the axiomatic properties of a norm.

**5.3.2** Let $\{\mathbf{T}_n\} \subset$ Blt($\mathfrak{N}$) such that $\|\mathbf{T}_n\| \leq \alpha$, where $\alpha$ is a fixed positive number. If $\{\mathbf{T}_n\}$ converges strongly to $\mathbf{T} \in$ Blt($\mathfrak{N}$), then show that $\|\mathbf{T}\| \leq \alpha$.

**5.3.3** Let $\mathbf{T} \in$ Blt($\mathfrak{N}$). Define $e^{\mathbf{T}}$ through an appropriate infinite sum. Does this sum converge (a) strongly?; (b) with respect to the operator norm? Is $e^{\mathbf{T}} \in$ Blt($\mathfrak{N}$)?

**5.3.4** Let $\mathbf{T}$ be a linear operator defined on $D(\mathbf{T}) \subset \mathfrak{N}$. $\mathbf{T}$ is said to be *bounded from below* if $\exists\ m > 0 \ni \|\mathbf{Tx}\| \geq m\|\mathbf{x}\| \ \forall\ \mathbf{x} \in D(\mathbf{T})$. Show that an operator bounded from below has a continuous inverse.

**5.3.5** Let $\mathfrak{N}_1 = \{\mathcal{L}_1, \|\cdot\|_1\}$ and $\mathfrak{N}_2 = \{\mathcal{L}_2, \|\cdot\|_2\}$. Show that $\mathbf{T} \in \text{Blt}(\mathfrak{N}_1) \oplus \text{Blt}(\mathfrak{N}_2)$ is a bounded linear operator on $\mathfrak{N}_1 \oplus \mathfrak{N}_2$ in which the norm may be any one of those suggested in Section 5.1.

More generally, establish that a linear transformation $\mathbf{T}$ on $\mathfrak{N}_1 \oplus \mathfrak{N}_2$ is bounded *if and only if* each operator in its matrix form as represented in Section 2.12 is bounded.

**5.3.6** Consider the Banach space $\mathcal{L}_1[0, 1]$, that is, functions $f(t)$ such that

$$\|\mathbf{f}\|_1 \equiv \int_0^1 |f(t)|\,dt$$

where the integral above is in the Lebesgue sense. Let $K(t, s)$ be a function such that the Lebesgue integral

$$\int_0^1 |K(t, s)|\,dt$$

is bounded. Show that the operator $K$ defined by

$$\mathbf{Kf} \equiv \left\{\int_0^1 K(t, s)f(s)\,ds\right\}$$

is bounded.

**5.3.7** Let $\mathcal{B}_1 \equiv \{\mathcal{L}_1[0, 1], \|\cdot\|_1\}$ and $\mathcal{B}_2 \equiv \{\mathcal{L}_2[0, 1], \|\cdot\|_2\}$. If $K(t, s) = |t - s|^{-\alpha}$, $\alpha > 0$, represents an integral operator $\mathbf{K}$, find the range of $\alpha$ for which (a) $\mathbf{K} \in \text{Blt}(\mathcal{B}_1)$; (b) $\mathbf{K} \in \text{Blt}(\mathcal{B}_2)$.

**5.3.8** Let $\Omega$ be the region within and on a closed curve in $\mathcal{R}_2$. Denote by $\mathcal{L}_2(\Omega)$ the linear space of all functions $\mathbf{f} \equiv \{f(x, y)\}$ such that the Lebesgue integral

$$\iint_\Omega |f(x, y)|^2\,dx\,dy$$

exists and is defined as $\|\mathbf{f}\|_2^2$. An integral operator $\mathbf{K}$ on $\{\mathcal{L}_2(\Omega), \|\cdot\|_2\}$ is defined by

$$\mathbf{Kf} = \left\{\iint_\Omega K(x, y; \xi, \eta)f(\xi, \eta)\,d\xi\,d\eta\right\}$$

Show that

$$K(x, y; \xi, \eta) = \tfrac{1}{2}\ln[(x - \xi)^2 + (y - \eta)^2]$$

is a Hilbert-Schmidt kernel.† [*Hint:* Transform to polar coordinates whose origin coincides with $(x, y)$. Then let $x - \xi = \rho\cos\theta$, $y - \eta = \rho\sin\theta$, so that $d\xi\,d\eta = \rho\,d\theta\,d\rho$.]

**5.3.9** Let $\Omega$ be the region within and on a closed surface in $\mathcal{R}_3$. $\mathcal{L}_2(\Omega)$ is the linear space of functions $\mathbf{f} \equiv \{f(x, y, z)\}$ such that the Lebesgue integral

$$\iiint_\Omega |f(x, y, z)|^2\,dx\,dy\,dz$$

---

†Here the square integrability of $K(x, y; \xi, \eta)$ is on the set $\Omega \times \Omega$.

exists and is defined as $\|\mathbf{f}\|_2^2$. The operator $\mathbf{K}$ on $\{\mathcal{L}_2(\Omega), \|\cdot\|_2\}$ is defined by

$$\mathbf{Kf} = \left\{\iiint_{\Omega} K(x, y, z; \xi, \eta, \zeta) f(\xi, \eta, \zeta)\, d\xi\, d\eta\, d\zeta\right\}$$

Show that

$$K(x, y, z; \xi, \eta, \zeta) \equiv [(x - \xi)^2 + (y - \eta)^2 + (z - \zeta)^2]^{-1/2}$$

is a Hilbert–Schmidt kernel. [*Hint:* Transform to spherical polar coordinates with the origin at $(x, y, z)$. Then let $x - \xi = \rho \sin\theta \cos\phi$, $y - \eta = \rho \sin\theta \sin\phi$, $z - \zeta = \rho\cos\theta$, and show that $d\xi\, d\eta\, d\zeta = \rho^2 \sin\theta\, d\rho\, d\theta\, d\phi$.]

We now consider a special class of bounded linear operators, called *compact operators*, which are of central importance to this book. They are also referred to as *completely continuous* operators. It will be established subsequently that compact operators have properties somewhat similar to those of finite-dimensional operators or matrices. Thus in the hierarchy of operators on infinite-dimensional spaces, the compact operator occupies a primary position.

### 5.3.3 Compact Operators

The compact operator, also known as a *completely continuous* operator, is defined as follows.

***Definition 5.3.4*** Let $\mathbf{T}: \mathfrak{N} \longrightarrow \mathfrak{N}$, where $\mathfrak{N}$ is a normed linear space. $\mathbf{T}$ is said to be *compact* or *completely continuous* if for *every bounded sequence* $\{\mathbf{x}_n\} \subset \mathfrak{N}$ (i.e., $\exists\ m > 0 \ni \|\mathbf{x}_n\| \leq m$) the sequence $\{\mathbf{Tx}_n\}$ has a convergent subsequence.

An alternative but equivalent definition of a compact operator is that it maps a bounded set (such as $\{\mathbf{x} \in \mathfrak{N}: \|\mathbf{x}\| \leq 1\}$) *into* a compact set. Some preliminary observations about compact operators are in order.

First, we observe that a compact operator is necessarily bounded for if we have an unbounded operator, say $\mathbf{T}$, there exists a bounded sequence $\{\mathbf{x}_n\} \subset D(\mathbf{T})$ such that $\|\mathbf{Tx}_n\|$ increases without bound; thus no subsequence of $\{\mathbf{Tx}_n\}$ will converge which precludes $\mathbf{T}$ from being compact.

Second, every linear operator on a finite-dimensional normed linear space is compact. This follows from the fact that every linear operator on a finite-dimensional space is bounded so that bounded sets are mapped into bounded sets. Since every bounded set in a finite-dimensional space is a *subset* of a compact set,† it follows from the foregoing that every linear operator on a finite-dimensional space is also compact.

---

†If $A \subset \mathfrak{N}$ is a bounded set, its closure $\bar{A}$ is closed and bounded which, if $\mathfrak{N}$ is finite dimensional, implies compactness.

When $\mathfrak{N}$ is infinite dimensional, even the identity operator is not compact (although it is bounded). This is because closed, bounded sets in infinite-dimensional spaces are not necessarily compact; that is, there may be sequences in these sets that have no convergent subsequences. Thus, if $\mathfrak{N}$ has infinite dimension, the set of compact operators on $\mathfrak{N}$ is a proper subset of Blt($\mathfrak{N}$). We denote the set of all compact operators on $\mathfrak{N}$ by $\mathcal{K}(\mathfrak{N})$. It is left as an exercise to show that $\mathcal{K}(\mathfrak{N})$ is a subspace of Blt($\mathfrak{N}$).

We present below two useful theorems on compact operators.

***Theorem 5.3.3*** Let $\mathbf{T} \in$ Blt($\mathcal{B}$) such that $R(\mathbf{T})$ is finite dimensional. Then $\mathbf{T} \in \mathcal{K}(\mathcal{B})$.

*Proof*: Let $\{\mathbf{x}_k\}$ be a bounded sequence in $\mathfrak{N}$. Since $\mathbf{T}$ is bounded $\{\mathbf{T}x_k\}$ is a bounded sequence in $R(\mathbf{T})$. Because $R(\mathbf{T})$ is finite dimensional, the bounded sequence $\{\mathbf{T}x_k\}$ must have a convergent subsequence (see Exercise 5.1.7). Thus $\mathbf{T}$ is compact. ∎

The next theorem is given by

***Theorem 5.3.4*** $\mathcal{K}(\mathcal{B})$ is a closed linear subspace of Blt($\mathcal{B}$).

*Proof*: (See Exercise 5.3.11 for part of the proof.) Let $\{\mathbf{T}_n\} \subset \mathcal{K}(\mathcal{B})$ be a Cauchy sequence. We have shown before that there exists $\mathbf{T} \in$ Blt($\mathcal{B}$) $\ni \lim_{n\to\infty} ||\mathbf{T}_n - \mathbf{T}|| = 0$. We need to show that $\mathbf{T} \in \mathcal{K}(\mathcal{B})$. Let $\{\mathbf{x}_k\} \subset \mathcal{B}$ be a bounded sequence with $||\mathbf{x}_k|| \leq 1$. Further, let $\{\varepsilon_n\}$ be a sequence of *positive* numbers converging *monotonically* to zero.

For $\varepsilon_1 > 0$, $\exists\ n_{\varepsilon_1} \ni n \geq n_{\varepsilon_1} \rightarrow ||\mathbf{T}_n - \mathbf{T}|| < \frac{1}{3}\varepsilon_1$, so that

$$||\mathbf{T}_n\mathbf{x}_k - \mathbf{T}\mathbf{x}_k|| < \tfrac{1}{3}\varepsilon_1 \qquad \forall\ k$$

*Fix* $n_1 \geq n_{\varepsilon_1}$. $\mathbf{T}_{n_1}$ is compact, so that $\{\mathbf{T}_{n_1}\mathbf{x}_k\}$ has some convergent subsequence. Thus $\exists$ integers $k_1$ and $k_2$ large enough to satisfy $n_{\varepsilon_1} < k_1 < k_2$ and

$$||\mathbf{T}_{n_1}\mathbf{x}_{k_1} - \mathbf{T}_{n_1}\mathbf{x}_{k_2}|| < \tfrac{1}{3}\varepsilon_1$$

Now initiate a subsequence of $\{\mathbf{x}_k\}$ with $\mathbf{x}_{k_1}$ and $\mathbf{x}_{k_2}$.

$$\begin{aligned} ||\mathbf{T}\mathbf{x}_{k_1} - \mathbf{T}\mathbf{x}_{k_2}|| &= ||\mathbf{T}\mathbf{x}_{k_1} - \mathbf{T}_{n_1}\mathbf{x}_{k_1} + \mathbf{T}_{n_1}\mathbf{x}_{k_1} - \mathbf{T}_{n_1}\mathbf{x}_{k_2} + \mathbf{T}_{n_1}\mathbf{x}_{k_2} - \mathbf{T}\mathbf{x}_{k_2}|| \\ &\leq ||\mathbf{T}\mathbf{x}_{k_1} - \mathbf{T}_{n_1}\mathbf{x}_{k_1}|| + ||\mathbf{T}_{n_1}\mathbf{x}_{k_1} - \mathbf{T}_{n_1}\mathbf{x}_{k_2}|| + ||\mathbf{T}_{n_1}\mathbf{x}_{k_2} - \mathbf{T}\mathbf{x}_{k_2}|| \\ &< \varepsilon_1 \end{aligned}$$

Proceed in a similar fashion to add elements $\mathbf{x}_{k_3}$ and $\mathbf{x}_{k_4}$ such that

$$||\mathbf{T}\mathbf{x}_{k_3} - \mathbf{T}\mathbf{x}_{k_4}|| < \varepsilon_2$$

where $k_2 < k_3 < k_4$. The sequence $\{x_{k_n}\}$ thus formed produces a sequence $\{\mathbf{T}\mathbf{x}_{k_n}\}$ such that

$$||\mathbf{T}\mathbf{x}_{k_{n+1}} - \mathbf{T}\mathbf{x}_{k_{n+2}}|| < \varepsilon_n$$

We now leave it for the student to prove that $\{\mathbf{Tx}_{k_n}\}$ is a Cauchy sequence in $\mathcal{B}$ which must converge. Thus $\mathbf{T} \in \mathcal{K}(\mathcal{B})$, so that $\mathcal{K}(\mathcal{B})$ is a closed linear subspace of $\text{Blt}(\mathcal{B})$. ▮

The two theorems just proved have useful applications which are discussed below.

*Examples of Compact Operators*

1. *The degenerate kernel:* Consider the Banach space $\mathcal{B} = \{\mathcal{C}[a, b], \|\cdot\|_\infty\}$. Define an operator $\mathbf{K}$ by

$$\mathbf{g} = \mathbf{Kf}; \qquad g(t) = \int_a^b K(t, s)f(s)\,ds, \qquad a \leq t \leq b$$

where $K(t, s) = \sum_{j=1}^n x_j(t)y_j(s)$, $t, s \in [a, b]$, and $\{\mathbf{x}_j\}_{j=1}^n$ and $\{\mathbf{y}_j\}_{j=1}^n$ are linearly independent sets in $\mathcal{B}$, so that $K(t, s)$ is continuous in $[a, b] \times [a, b]$. Thus $\mathbf{K}$ is a bounded linear operator. Moreover, for $\mathbf{f} \in \mathcal{B}$, we have

$$g(t) = \sum_{j=1}^n x_j(t) \int_a^b y_j(s)f(s)\,ds$$

or

$$\mathbf{g} = \sum_{j=1}^n \alpha_j \mathbf{x}_j, \qquad \alpha_j \equiv \int_a^b y_j(s)f(s)\,ds$$

Clearly, $R(K) = L(\mathbf{x}_1, \mathbf{x}_2, \ldots, \mathbf{x}_n)$ is an $n$-dimensional space so that by Theorem 5.3.3, $\mathbf{K}$ is a compact operator. The function $K(t, s)$ in this example is called a *degenerate kernel.*

2. *The continuous kernel:* We consider again the same Banach space $\mathcal{B}$ as in the previous example and require that $\mathbf{K}$ be represented by a continuous kernel $K(t, s)$ without being degenerate. Clearly, $\mathbf{K} \in \text{Blt}(\mathcal{B})$. Now the Weierstrass approximation theorem (see, e.g., Section 3.6) has established that every continuous function can be *uniformly* approximated by polynomials. This implies that one can find polynomials $\{K_n(t, s)\}$ where

$$K_n(t, s) = \sum_{r=0}^n \sum_{j=0}^n \alpha_{jr}^n t^{r-j} s^j \tag{5.3.10}$$

such that

$$\lim_{n \to \infty} |K_n(t, s) - K(t, s)| = 0 \tag{5.3.11}$$

the foregoing convergence occurring *uniformly*. Let us define operator $\mathbf{K}_n$ by

$$\mathbf{g} = \mathbf{K}_n\mathbf{f}: \quad g(t) = \int_a^b K_n(t, s)f(s)\,ds$$

For any $t, s \in [a, b]$, in view of (5.3.11), $\{K_n(t, s)\}$ is a sequence of real numbers converging to $K(t, s)$. If $\varepsilon > 0$ is given, let $\varepsilon' = \varepsilon/(b - a)$.

Then $\exists\ n_{\varepsilon'}$ (independent of $t$ and $s$) $\ni n \geq n_{\varepsilon'} \rightarrow |K_n(t, s) - K(t, s)| < \varepsilon'$. Let $\mathbf{f} \in \mathcal{B}$,

$$\|(\mathbf{K}_n - \mathbf{K})\mathbf{f}\|_\infty = \sup_{t \in [a,b]} \left| \int_a^b [K_n(t, s) - K(t, s)]f(s)\, ds \right|$$

$$\leq \sup_{t \in [a,b]} \int_a^b |K_n(t, s) - K(t, s)|\,|f(s)|\, ds$$

$$\leq \varepsilon' \|\mathbf{f}\|_\infty (b - a) = \varepsilon \|\mathbf{f}\|_\infty$$

from which it follows that†

$$\|\mathbf{K}_n - \mathbf{K}\| < \varepsilon \tag{5.3.12}$$

Thus $\{\mathbf{K}_n\}$ converges in Blt($\mathcal{B}$) to $\mathbf{K}$. Now in light of (5.3.10), $\mathbf{K}_n(t, s)$ is a degenerate kernel. From the discussion of Example 1, it is clear that $\mathbf{K}_n$ is a compact operator. $\mathbf{K}$ is therefore a limit point of $\mathcal{K}(\mathcal{B})$, the subspace of compact operators. From Theorem 5.3.4, we must have $\mathbf{K} \in \mathcal{K}(\mathcal{B})$, that is, $\mathbf{K}$ is a compact operator.

Before we proceed to the next example, some observations are in order. If we define on $\mathcal{C}[a, b]$, the norm $\|\cdot\|_2$ by

$$\|\mathbf{f}\|_2 = \left\{ \int_a^b |f(t)|^2\, dt \right\}^{1/2} \tag{5.3.13}$$

then as observed in Chapter 3 and earlier in the present chapter, the normed linear space $\mathfrak{N} = \{\mathcal{C}[a, b], \|\cdot\|_2\}$ is not complete. The operator $\mathbf{K}$ considered above is readily seen to be in Blt($\mathfrak{N}$) where its norm is defined by

$$\|\mathbf{K}\| \equiv \sup\{\|\mathbf{Kf}\|_2 : \|\mathbf{f}\| \leq 1\} \tag{5.3.14}$$

The reader must be careful to distinguish the norm of $\mathbf{K}$ in (5.3.14) from that used in (5.3.12). Convergence of a sequence $\{\mathbf{f}_n\} \subset \mathcal{C}[a, b]$ with respect to $\|\cdot\|_\infty$ can be readily shown to imply convergence with respect to $\|\cdot\|_2$. Thus we conclude that $\mathbf{K} \in \mathcal{K}(\mathcal{B}) \rightarrow \mathbf{K} \in \mathcal{K}(\mathfrak{N})$ (why?).‡ Furthermore, using essentially the same reasoning, it is possible to show that convergence in Blt($\mathcal{B}$) implies convergence in Blt($\mathfrak{N}$). Note that convergence in the former is relative to the operator norm in (5.3.12).

3. *The Hilbert–Schmidt kernel:* It was established earlier that the operator $\mathbf{K}$, represented by the Hilbert–Schmidt kernel $K(t, s)$, is a bounded linear operator on the Banach space $\{\mathcal{L}_2[a, b], \|\cdot\|_2\}$.

---

†Obviously, here the norm of operator $\mathbf{K}$ is defined by

$$\|\mathbf{K}\| = \sup\{\|\mathbf{Kf}\|_\infty : \|\mathbf{f}\|_\infty \leq 1\}$$

‡The difference between $\mathcal{K}(\mathcal{B})$ and $\mathcal{K}(\mathfrak{N})$ here lies in that compactness in the former is relative to $\|\cdot\|_\infty$, while that in the latter is measured relative to $\|\cdot\|_2$.

From the Riesz–Fischer theorem of Chapter 4 (Theorem 4.3.1) we know that the subspace of *continuous* functions on $[a, b] \times [a, b]$ is dense in $\mathfrak{L}_2\{[a, b] \times [a, b]\}$, so that for each Hilbert–Schmidt kernel $K(t, s)$, it is possible to find a sequence $\{K_n(t, s)\}$ of continuous functions such that

$$\lim_{n\to\infty} \int_a^b \int_a^b |K_n(t, s) - K(t, s)|^2 \, dt \, ds = 0 \tag{5.3.15}$$

Using arguments identical to those which led to inequality (5.3.2), it is readily shown from (5.3.15) that

$$\lim_{n\to\infty} \|\mathbf{K}_n - \mathbf{K}\| = 0 \tag{5.3.16}$$

where the operator norm is as defined in (5.3.14). Because each operator $\mathbf{K}_n$ is a compact operator (as shown in Example 2) the limit point $\mathbf{K}$, in view of Theorem 5.3.4, is also compact. This is an important result for the purposes of this book.

### EXERCISES

**5.3.10** Repeat Exercise 5.3.5 by substituting "compact" for the word "bounded."

**5.3.11** Show that $\mathcal{K}(\mathfrak{N})$ is a linear subspace of $\text{Blt}(\mathfrak{N})$.

**5.3.12** If $\mathbf{T}_1, \mathbf{T}_2 \in \mathcal{K}(\mathfrak{N})$, show that $\mathbf{T}_1\mathbf{T}_2, \mathbf{T}_2\mathbf{T}_1 \in \mathcal{K}(\mathfrak{N})$.

## 5.4 Bounded Linear Functionals

Linear functionals were defined in Section 2.6 to be linear mappings of a linear space $\mathfrak{L}$ into the associated field $\mathfrak{F}$. If a linear functional is defined on a normed linear space, one may ask if the functional can be continuous. The notion of continuity of a functional $l$ is straightforward: $l$ is continuous at $\mathbf{x}_0$ if $\forall\, \varepsilon > 0 \;\exists\; \delta_\varepsilon > 0 \ni \|\mathbf{x} - \mathbf{x}_0\| < \delta_\varepsilon \rightarrow |l(\mathbf{x}) - l(\mathbf{x}_0)| < \varepsilon$. In exactly the same way as in Theorem 5.3.1 for linear transformations, continuity of a linear functional at a point in $\mathfrak{N}$ implies continuity everywhere. Furthermore, the notion of boundedness of linear transformations such as in Definition 5.3.2 is obviously extendable to linear functionals; that is, a linear functional $l$ is said to be bounded if $\exists\; m > 0 \ni |l(\mathbf{x})| \leq m\|\mathbf{x}\|$ $\forall\, \mathbf{x} \in \mathfrak{N}$.

On the application of the arguments in Theorem 5.3.2, the equivalence of a bounded linear functional with a continuous linear functional becomes evident. The null space $N(l)$ of a bounded linear functional is readily seen to be a closed linear subspace of $\mathfrak{N}$, again in the same way as it is shown that the null space of a bounded linear operator is closed.

It should be evident that all linear functionals on a finite-dimensional normed linear space are bounded. That this is not true of linear functionals on infinite-dimensional normed linear spaces is easily seen by constructing suitable examples. This is left as an exercise.

**EXERCISES**

**5.4.1** Show that

$$l(\mathbf{f}) \equiv \int_0^1 g(t)f(t)\,dt$$

is a bounded linear functional on the Banach space $\{\mathcal{L}_2[0, 1], \|\cdot\|_2\}$ if $\mathbf{g} \in \mathcal{L}_2[0, 1]$.

**5.4.2** Construct an unbounded linear functional on the normed linear space $\{\mathcal{C}^{(1)}[0, 1], \|\cdot\|_\infty\}$.

**5.4.3** Let $l$ be a bounded linear functional on a normed linear space $\mathfrak{N}$. Show that for every bounded sequence $\{\mathbf{x}_n\} \subset \mathfrak{N}$, there exists some subsequence $\{\mathbf{x}_{k_n}\}$ for which $\{l(\mathbf{x}_{k_n})\}$ converges in the associated field of, say, complex numbers. For nonlinear functionals, what property would be required to ensure the property above?

## 5.5 Unbounded Operators

Unbounded transformations necessarily occur on infinite-dimensional normed linear spaces. They are characteristically defined on domains that are proper subspaces of Banach space.† Let us state at the outset that our interest in unbounded operators is entirely with respect to differential operators. We will thus dispense with certain important aspects of the general treatment of unbounded transformations, some of which are concerned more with operators on inner product spaces.

To identify more definitely an unbounded operator, we must associate with it a *formal operation* $T$ (note especially that it is not a boldface symbol because it is not considered as an operator) and a domain $D(T)$ that is a linear subspace of a Banach space $\mathcal{B}$. Thus we denote the unbounded operator $\mathbf{T}$ by

$$\mathbf{T} \equiv \{T, D(T)\} \tag{5.5.1}$$

The reader should carefully study this notation because it contains the essence of what it takes to define an unbounded transformation. The formal

†Differential operators have the important property of being *closed* and cannot have Banach spaces as their domains. This is a result of the "closed graph theorem," for which the interested reader is referred to *Unbounded Linear Operators*, by S. Goldberg, McGraw-Hill, New York, 1966.

operation $T$ may give rise to several different operators for different domains $D(T)$. Thus for two unbounded operators $\mathbf{T} \equiv \{T, D(T)\}$ and $\mathbf{T}' \equiv \{T', D(T')\}$ to be equal, we must have $T = T'$ and $D(T) = D(T')$; that is, *both* the formal operations and their domains must be the same. An important aspect of notation (5.5.1) is that when one writes $\mathbf{Tx}$, it automatically implies that $\mathbf{x} \in D(T)$, so that no such explicit qualification need be made.† Of course, there may be elements $\mathbf{x} \notin D(T)$ for which $T\mathbf{x}$ may make sense. However, we may *not* write $\mathbf{Tx}$ if $\mathbf{x} \notin D(T)$. Alternatively,

$$\mathbf{Tx} \equiv \{T\mathbf{x}, \mathbf{x} \in D(T)\} \tag{5.5.2}$$

The domain $D(T)$ is a proper linear subspace, and for reasons that have to do with applications to inner product spaces, we require that $D(T)$ be a *dense linear* subspace of $\mathcal{B}$. Alternatively, one refers to $\mathbf{T}$ as being densely defined on $\mathcal{B}$. Another constraint that is applied to $D(T)$ is that the range space $R(\mathbf{T})$ is a Banach space, which may either be $\mathcal{B}$ or, more generally, another space $\mathcal{B}'$.

We consider some examples of unbounded operators.

*Example 1*

Let $\mathcal{B} = \{\mathcal{L}_2[0, 1], \|\cdot\|_2\}$. We consider the formal operation $T \equiv d/dt$. For the domain $D(T)$, it would seem natural to choose $\mathcal{C}^{(1)}[0, 1]$, which is of course a dense linear subspace of $\mathcal{L}_2[0, 1]$ (see Section 4.3.5). However, there are functions in $\mathcal{C}^{(1)}[0, 1]$ which when differentiated cannot be square integrable. Thus consider $f(t) = t \sin(1/t)$. This function is differentiable almost everywhere in (0, 1) and is of course square integrable in the interval. However,

$$f'(t) = \sin\frac{1}{t} - \frac{1}{t}\cos\frac{1}{t}$$

which is *not* square integrable in (0, 1). Thus $\mathbf{f}' \notin \mathcal{L}_2[0, 1]$. This clearly does not meet with the goal that $R(\mathbf{T})$ be a Banach space. Hence it is necessary to restrict $D(T)$ further to exclude functions of the kind to which reference was just made, in order that the differentiated function be square integrable.‡

---

†In continuation of the preceding footnote, an operator $\mathbf{T} = \{T, D(T)\}$ is said to be *closed* if the following is true:

Let $\{\mathbf{x}_n\} \subset D(T)$ converge to $\mathbf{x}_0$ and $\{\mathbf{Tx}_n\}$ converge to $\mathbf{y}_0$. *Then* $\mathbf{x}_0 \in D(T)$ and $\mathbf{y}_0 = T\mathbf{x}_0$.

‡In order that $f'(t)$ be square integrable, it turns out that $f(t)$ must be *absolutely continuous* on [0, 1] (see Section 4.3.4). But as a rule we will avoid this detail in the stipulation of the operator domain.

The domain $D(T)$ may then be written as

$$D(T) = \{\mathbf{f} \in \mathcal{C}^{(1)}[0, 1]; T\mathbf{f} \in \mathcal{L}_2[0, 1]\} \tag{5.5.3}$$

The differential operator is then given by the pair $\{T, D(T)\}$. The domain $D(T)$ is a dense linear subspace of $\mathcal{L}_2[0, 1]$. We shall have more to say about this example in the next chapter.

*Example 2*

For our second example, we consider $T \equiv d^2/dx^2$ and define its domain as

$$D(T) = \{u(x): a \leq x \leq b; T\mathbf{u} \in \mathcal{L}_2[a, b]; u(a) = 0, u(b) = 0\} \tag{5.5.4}$$

Notice that in the foregoing, we have included two homogeneous boundary conditions, a feature not present in Example 1. However, $D(T)$ in (5.5.4) is still a dense linear subspace of $\mathcal{L}_2[a, b]$. Similar examples will be encountered again in Chapter 6.

**EXERCISES**

**5.5.1** Define domains as dense linear subspaces of $\mathcal{L}_2[a, b]$ for the following formal operations.

(a) $T \equiv \dfrac{d}{dx}\left[p(x)\dfrac{d}{dx}\right]$.

(b) $T \equiv \sum_{j=0}^{n} a_j(x)\dfrac{d^j}{dx^j}$.

**5.5.2** Let $\mathbf{T} \equiv \{T, D(T)\}$, where $T \equiv d/dx$ and $D(T) = \{u(x): T\mathbf{u} \in \mathcal{L}_2[a, b]; u(a) = 0\}$. If $\mathbf{T}$ has an inverse, find it and comment on its nature. If $\mathbf{T}$ is defined as in Example 1 of this section, does it have an inverse?

## 5.6 Concluding Remarks

It would be fair to say that for the purposes of this book the treatment of compact operators in this chapter forms its most important component. The main results of the next chapter, too, are concerned with compact operators, so that this emphasis calls for some explanation at this stage. Engineering students are normally more familiar with the role of differential equations in engineering models. Their exposure to integral equations is relatively less intense (in fact, frequently meager). Integral equations (barring the singular ones) involve compact operators, and while they are important in their own right in regard to applications, they also derive their significance from the fact that many boundary value problems can be represented as integral equations by "turning them inside out." Thus unbounded operators (which as a general class are not as well behaved as the bounded or compact operators) inherit the powerful properties of their compact inverses when the latter exist.

In Chapter 3 we dealt with the fixed-point theorem for contraction mappings in abstract metric spaces that were complete. In Banach spaces, one has more powerful fixed-point theorems in that the severe constraint of contraction is relaxed for the more general condition of continuity of the nonlinear transformation. For example, we state here Schauder's fixed-point theorem, for which it is necessary to recognize a *convex* set in a normed linear space. A convex set $S$ is defined by the property that $\mathbf{x}, \mathbf{y} \in S \rightarrow \lambda\mathbf{x} + \mu\mathbf{y} \in S$ for every $\lambda, \mu \geq 0$ such that $\lambda + \mu = 1$. Geometrically, the line segment joining any two vectors in the set is contained entirely in the set. The fixed-point theorem due to Schauder states that a *nonlinear* transformation $\mathbf{T}: S \longrightarrow \mathcal{B}$, where $S$ is a convex, compact subset of Banach space $\mathcal{B}$, has at least one fixed point $\mathbf{x} \in S$, provided that $\mathbf{T}$ is continuous in $S$.† This fixed-point theorem is extremely useful in establishing the existence of solutions to equations at the expense of uniqueness. For example, the existence of solutions to the initial value problem consisting of the ordinary differential equation (3.7.8) and the initial condition (3.7.9) may be established under conditions weaker than the Lipschitz condition (3.7.10). However, now uniqueness cannot be established. There are many other applications of the Schauder fixed-point theorem, which are outside the scope of this book. Thus the theorem can be used to show the existence of solutions to many nonlinear boundary value problems‡ which represent complex engineering models.

Another problem of interest in Banach spaces that has powerful applications to engineering problems is the *bifurcation problem.* This is concerned with the investigation of steady-state multiplicity and stability of many physical systems.

## FURTHER READING

Normed linear spaces can display some remarkable differences from inner product spaces. These and other items, including a more detailed treatment of finite-dimensional normed linear spaces, have been dealt with in Chapter 5 of

*Linear Operator Theory in Engineering and Science*, by A. W. Naylor and G. R. Sell, Holt, Rinehart and Winston, New York, 1971.

---

†For a proof of this theorem, the reader is referred to *Functional Analysis in Normed Spaces*, by L. V. Kantorovich and G. P. Akhilov, Macmillan, New York, 1964.

‡See, for example, *Integral Equations and their Applications*, by W. Pogorzelski, Vol. I, Pergamon Press, Oxford, 1966.

For the use of approximate methods in Banach spaces and details concerning fixed-point theorems due to Brouwer, Schauder, LeRay–Schauder, and so on, the following book is recommended.

*Functional Analysis in Normed Spaces*, by L. V. Kantorovich and G. P. Akhilov, Macmillan, New York, 1964.

A more recent book on approximate methods in which the bifurcation problem has been treated is

*Approximate Solution of Operator Equations*, by M. A. Kraisnoselskii, G. M. Vainikko, P. P. Zabreibo, Ya. B. Rutitskii, and V. Ya Stetsenko, Wolters-Noordhoff, Groningen, The Netherlands, 1972.

# 6 Inner Product Spaces

## 6.0 Foreword

In the generalization of the physical vector to a mathematical vector, the one concept that remains is that associated with "direction." Geometrically, direction is measured by *angles* with *given* directions. This chapter presents the extension of the idea to abstract linear spaces.

The idea of expanding a physical vector in terms of components along three mutually perpendicular directions is well known. The components of a vector are readily determined given the unit vectors along the perpendicular directions. This idea is far-reaching and has deep implications to the solution of many linear and nonlinear problems encountered in engineering applications. Indeed, vectors are identified by their components relative to a basis set, and frequently the quest for an unknown vector (the solution to a problem stated as a mathematical equation) becomes equivalent to that for a "suitable" basis set. It turns out that for certain linear problems a most useful (and beautiful!) prescription is available for the suitable basis set. When this basis set has been determined, the solution of the problem follows almost immediately. The student must recognize, therefore, that the present chapter introduces the structural features into linear spaces essential for the purposes of our prime pursuit—the solution of mathematical equations that contain linear operators.

## 6.1 Introduction

The normed linear spaces of Chapter 5 possess both the algebraic features of linear spaces and the topological features of metric spaces. Yet they do not reflect the full geometric structure of, for example, physical space with which we are familiar. Thus the concept of "direction" of a vector, which is not accommodated by normed linear spaces, will be introduced in this chapter.

The student will recall that the *dot product* between two physical vectors **a** and **b** is defined by†

$$\mathbf{a}\cdot\mathbf{b} \equiv ab\cos\theta \tag{6.1.1}$$

where $a$ and $b$ are the magnitudes of **a** and **b**, respectively; $\theta$ is the angle between the two vectors. Alternatively, one could use (6.1.1) to *define* the angle $\theta$ between two vectors by

$$\theta = \cos^{-1}\frac{\mathbf{a}\cdot\mathbf{b}}{ab} \tag{6.1.2}$$

where $(\mathbf{a}\cdot\mathbf{b})$ must be *defined* in some other way consistent with the inequality

$$|\mathbf{a}\cdot\mathbf{b}| \leq ab \tag{6.1.3}$$

for otherwise there would be no meaning to (6.1.2) since the cosine of any angle must lie in $[-1, 1]$.

Clearly, the concept of direction or *relative orientation* between two vectors is tied up with the dot product between two vectors. This dot product is also called a *scalar product*‡ or *inner product*. We will adopt the second of these for our usage since it is the one that is commonly used in the mathematical literature. The linear space in which an inner product is defined is called an *inner product space*. A formal definition follows.

***Definition 6.1.1*** An *inner product space* is a pair of objects $(\mathcal{L}, \langle\cdot,\cdot\rangle)$, where $\mathcal{L}$ is a linear space with an associated field $\mathfrak{F}$ (assumed to be, in general, the field of complex numbers $\mathcal{C}$) and $\langle\cdot,\cdot\rangle\colon (\mathcal{L}\times\mathcal{L}) \to \mathcal{C}$ is a mapping called the *inner product*, satisfying the following axioms.

A1 $\langle\mathbf{x}+\mathbf{y}, \mathbf{z}\rangle = \langle\mathbf{x}, \mathbf{z}\rangle + \langle\mathbf{y}, \mathbf{z}\rangle$,

A2 $\langle\mathbf{x}, \mathbf{y}\rangle = \langle\mathbf{y}, \mathbf{x}\rangle^*$, where * denotes the *complex conjugate*. If $\mathfrak{F} = \mathcal{R}$, then the foregoing condition implies the *symmetry*:

$$\langle\mathbf{x}, \mathbf{y}\rangle = \langle\mathbf{y}, \mathbf{x}\rangle$$

---

†Using the notation of Chapter 5, $a$ and $b$ must be represented by $\|\mathbf{a}\|$ and $\|\mathbf{b}\|$, respectively.

‡The term *scalar product* is used because the result of forming the scalar product of two vectors is a scalar (number).

A3 $\langle\alpha\mathbf{x}, \mathbf{y}\rangle = \alpha\langle\mathbf{x}, \mathbf{y}\rangle$,

A4 $\langle\mathbf{x}, \mathbf{x}\rangle > 0$ if $\mathbf{x} \neq \mathbf{0}$, where $\mathbf{x}$, $\mathbf{y}$, and $\mathbf{z}$ are arbitrary elements of $\mathcal{L}$ and $\alpha$ is an arbitrary complex number. The pair $\{\mathcal{L}, \langle\cdot, \cdot\rangle\}$ is denoted by a single symbol $\mathcal{I}$ and elements of $\mathcal{L}$ will be referred to as elements of $\mathcal{I}$.

From the four properties above, other important relations may be deduced. Thus

$$\langle\mathbf{x}, \mathbf{y} + \mathbf{z}\rangle = \langle\mathbf{y} + \mathbf{z}, \mathbf{x}\rangle^* = \langle\mathbf{y}, \mathbf{x}\rangle^* + \langle\mathbf{z}, \mathbf{x}\rangle^* = \langle\mathbf{x}, \mathbf{y}\rangle + \langle\mathbf{x}, \mathbf{z}\rangle \qquad (6.1.4)$$

where axioms A1 and A2 have been used in the order A2, A1, A2. Another relationship is

$$\langle\mathbf{x}, \alpha\mathbf{y}\rangle = \langle\alpha\mathbf{y}, \mathbf{x}\rangle^* = \alpha^*\langle\mathbf{y}, \mathbf{x}\rangle^* = \alpha^*\langle\mathbf{x}, \mathbf{y}\rangle \qquad (6.1.5)$$

where we have used A2, A3, and A2, in that order. Clearly, axioms A1 and A3 imply that, for a fixed element $\mathbf{y}$, the inner product $\langle\mathbf{x}, \mathbf{y}\rangle$ is a *linear functional* of $\mathbf{x}$ on $\mathcal{L}$ (or $\mathcal{I}$). Indeed, in combination with (6.1.4) and (6.1.5) we conclude that $\langle\mathbf{x}, \mathbf{y}\rangle$ is a bilinear functional of $\mathbf{x}$ and $\mathbf{y}$ (see Section 2.6). Thus it follows that $\langle\mathbf{x}, \mathbf{x}\rangle$ is a *real-valued* quadratic expression (form). Axiom A4 asserts that this quadratic form be *positive definite*. From axiom A3 we have $\langle\mathbf{0}, \mathbf{y}\rangle = 0$ for each $\mathbf{y}$ and $\langle\mathbf{x}, \mathbf{0}\rangle = 0$ for each $\mathbf{x}$ and, of course, $\langle\mathbf{0}, \mathbf{0}\rangle = 0$. In what follows we show that $\mathcal{I}$ generates a normed linear space in which the norm of a vector is defined by

$$\|\mathbf{x}\| \equiv \sqrt{\langle\mathbf{x}, \mathbf{x}\rangle} \qquad (6.1.6)$$

Of course, that the foregoing definition is consistent with the requirements of a norm stipulated in Section 5.1 remains to be established, to which extent the symbol $\|\cdot\|$ in (6.1.6) is presumptuous. Of these, the positive definiteness has just been established. The homogeneity property follows immediately from A3 and (6.1.5), so that the triangular inequality alone remains to be shown. As a prerequisite we must prove a very important inequality, called the *Schwarz inequality*, which states that

$$|\langle\mathbf{x}, \mathbf{y}\rangle| \leq \|\mathbf{x}\|\,\|\mathbf{y}\| \qquad (6.1.7)$$

where $\|\cdot\|$ refers to that defined in (6.1.6). To prove this inequality we let $\alpha \in \mathcal{C}$ be *arbitrary*. For $\mathbf{x}, \mathbf{y} \in \mathcal{I}$, we have

$$0 \leq \langle\mathbf{y} + \alpha\mathbf{x}, \mathbf{y} + \alpha\mathbf{x}\rangle = \|\mathbf{y}\|^2 + \alpha\langle\mathbf{x}, \mathbf{y}\rangle + \alpha^*\langle\mathbf{y}, \mathbf{x}\rangle + |\alpha|^2\|\mathbf{x}\|^2 \qquad (6.1.8.)$$

Since $\langle\mathbf{x}, \mathbf{y}\rangle$ is a complex number, we may represent it by its polar form,

$$\langle\mathbf{x}, \mathbf{y}\rangle = |\langle\mathbf{x}, \mathbf{y}\rangle|\, e^{i\theta} \qquad (6.1.9)$$

We now constrain our $\alpha$ to be given by

$$\alpha = re^{-i\theta} \qquad (6.1.10)$$

where $r$ is an arbitrary real, positive number and $\theta$ is the same angle as that appearing in (6.1.9). When (6.1.9) and (6.1.10) are both incorporated in (6.1.8), one obtains the inequality

$$0 \leq ||\mathbf{y}||^2 + 2r|\langle \mathbf{x}, \mathbf{y}\rangle| + r^2||\mathbf{x}||^2 \tag{6.1.11}$$

which implies that for arbitrarily fixed $\mathbf{x}, \mathbf{y} \in \mathcal{I}$, the quadratic expression

$$q(r) \equiv ||\mathbf{y}||^2 + 2r|\langle \mathbf{x}, \mathbf{y}\rangle| + r^2||\mathbf{x}||$$

cannot possess *two distinct real* roots [for then $q(r)$ would change sign on either side of each root, a situation forbidden by (6.1.11)]. Thus from elementary algebra we conclude that the *discriminant* of the quadratic expression $q(r)$ must be negative, or

$$|\langle \mathbf{x}, \mathbf{y}\rangle|^2 \leq ||\mathbf{x}||^2\,||\mathbf{y}||^2$$

which implies the Schwarz inequality (6.1.7).

We are now in a position to show that the norm defined in (6.1.6) does satisfy the triangular inequality. For $\mathbf{x}, \mathbf{y} \in \mathcal{I}$,

$$\begin{aligned} ||\mathbf{x}+\mathbf{y}||^2 = \langle \mathbf{x}+\mathbf{y}, \mathbf{x}+\mathbf{y}\rangle &= ||\mathbf{x}||^2 + 2\,\mathrm{Re}\langle \mathbf{x}, \mathbf{y}\rangle + ||\mathbf{y}||^2 \\ &\leq ||\mathbf{x}||^2 + 2\,|\langle \mathbf{x}, \mathbf{y}\rangle| + ||\mathbf{y}||^2 \\ &\leq ||\mathbf{x}||^2 + 2\,||\mathbf{x}||\,||\mathbf{y}|| + ||\mathbf{y}||^2 \\ &= (||\mathbf{x}|| + ||\mathbf{y}||)^2 \end{aligned}$$

where $\mathrm{Re}\langle \mathbf{x}, \mathbf{y}\rangle$ represents the real part of $\langle \mathbf{x}, \mathbf{y}\rangle$, from which

$$||\mathbf{x}+\mathbf{y}|| \leq ||\mathbf{x}|| + ||\mathbf{y}||$$

so that the triangular inequality is proved. Thus the inner product space generates a normed linear space in which the norm will be understood to be that given by (6.1.6). It then follows naturally that a metric space is generated by the metric

$$d(\mathbf{x}, \mathbf{y}) \equiv \sqrt{\langle(\mathbf{x}-\mathbf{y}), (\mathbf{x}-\mathbf{y})\rangle} \tag{6.1.12}$$

which assigns *all* the topological properties characteristic of a metric space to the inner product space $\mathcal{I}$. This leads us to a concept of utmost importance to this book, that of a *complete* inner product space, which is called a *Hilbert space*. Indeed, completeness is relative to the metric in (6.1.12). Obviously, a Hilbert space is also a Banach space. Some examples of inner product spaces are presented below.

*Example 1*

Let $\mathcal{L} = \mathcal{R}_n$ and define the inner product by

$$\langle \mathbf{x}, \mathbf{y}\rangle \equiv \sum_{i=1}^{n} x_i y_i \tag{6.1.13}$$

It is readily shown that the definition above satisfies all the axiomatic requirements of a norm. In fact, it is the familiar Euclidean norm. Thus $\{\mathcal{R}_n, \langle \cdot, \cdot \rangle\}$ is an inner product space. Furthermore, the completeness of the real number system implies that the space above is a Hilbert space. If $n$ is increased to $\infty$, then by restricting $\mathcal{L}$ to be the subspace of $\mathcal{R}_\infty$ consisting of vectors $\{\mathbf{x} \equiv (x_1, x_2, \ldots)\}$ such that

$$\sum_{i=1}^{\infty} x_i^2 < \infty$$

an inner product space is obtained by defining the inner product as the extension of the finite sum (6.1.13) to the limit of infinite $n$. Again the resulting space is easily shown to be a Hilbert space and is called *real* $l_2$ space.

*Example 2*

The complex counterparts of Example 1 are obtained by letting $\mathcal{L} = \mathcal{C}_n$ and defining

$$\langle \mathbf{x}, \mathbf{y} \rangle \equiv \sum_{i=1}^{n} x_i y_i^* \tag{6.1.14}$$

which can be easily established as a valid inner product. For the same reasons as in Example 1, the inner product space here is a Hilbert space. Similarly, a *complex* $l_2$ space (a Hilbert space) is obtained from the subspace of $\mathcal{C}_\infty$ comprising vectors $\mathbf{x}$ such that

$$\sum_{i=1}^{\infty} |x_i|^2 < \infty$$

in which the inner product is given by extending the sum (6.1.14) to $n = \infty$.

*Example 3*

Let $\mathcal{L} = \mathcal{C}[a, b]$, where we consider only *real-valued* functions so that the inner product is defined as

$$\langle \mathbf{x}, \mathbf{y} \rangle \equiv \int_a^b x(t)y(t)\, dt \tag{6.1.15}$$

For *complex-valued* functions the foregoing inner product must be replaced by†

$$\langle \mathbf{x}, \mathbf{y} \rangle \equiv \int_a^b x(t)y^*(t)\, dt \tag{6.1.16}$$

In either case, we obtain an inner product space that is not *complete*, so it is not a Hilbert space. The reason for this has been elaborated in the examples in Section 5.1.

---

†Whether we are dealing with real or complex spaces should be evident from the inner product and does not require any further explicit indication to that effect.

*Example 4*

Redefining the inner product (6.1.15) or (6.1.16) as a Lebesgue integral and including in $\mathfrak{L}$ all functions $x(t)$ such that the Lebesgue integral

$$\int_a^b |x(t)|^2 \, dt < \infty$$

(i.e., the integral exists), we obtain the Hilbert space $\{\mathfrak{L}_2[a, b], \langle\cdot, \cdot\rangle\}$, which is normally referred to as just $\mathfrak{L}_2[a, b]$.

In the examples above, we have dealt with rather "straightforward" inner products. There is a considerably large number of different inner products that could be defined which would conform to the axiomatic requirements laid down earlier. Such inner products are extremely important for applications and the reader is urged to go through *all* the exercises below.

**EXERCISES**

**6.1.1** Define the following product on $\mathfrak{C}_n$:

$$\langle \mathbf{x}, \mathbf{y} \rangle \equiv \sum_{i=1}^{n} r_i x_i y_i^*$$

where $r_i > 0$, $i = 1, 2, \ldots, n$. Show that the foregoing is a valid inner product. Is this a Hilbert space?

**6.1.2** As a generalization of Exercise 6.1.1, consider on $\mathfrak{C}_n$ the inner product

$$\langle \mathbf{x}, \mathbf{y} \rangle \equiv \sum_{i=1}^{n} \sum_{j=1}^{n} a_{ij} x_j y_i^*$$

where $\{a_{ij}\}$ contains the coefficients of a *positive-definite* Hermitian matrix for which $a_{ij} = a_{ji}^*$. (Positive definiteness of $\{a_{ij}\}$ implies that $\sum_{i=1}^{n} \sum_{j=1}^{n} a_{ij} x_i x_j^* > 0$ if $\mathbf{x} \neq \mathbf{0}$.) Justify the choice above as an inner product. Is this a Hilbert space?

**6.1.3** Consider the closed interval $[a, b]$ and for functions $f(t)$ and $g(t)$ on $[a, b]$ define the inner product as

$$\langle \mathbf{f}, \mathbf{g} \rangle = \sum_{i=1}^{n} r_i \int_{t_{i-1}}^{t_i} f(t) g^*(t) \, dt$$

where $a = t_0 < t_1 < t_2 < \ldots < t_n = b$ and $r_i > 0$, $i = 1, 2, \ldots, n$. Identify the linear space of interest and establish that the above definition satisfies the stipulations for an inner product. Is the resulting space a Hilbert space?

**6.1.4** Suppose that in a function space on the interval $[a, b]$ we define

$$\langle \mathbf{x}, \mathbf{y} \rangle \equiv \int_a^b r(t) x(t) y^*(t) \, dt$$

where the integral is in the Lebesgue sense. Discuss the restrictions to be placed on the function $r(t)$ so that the integral above is an appropriate inner product. Identify the linear space. Is this a Hilbert space?

**6.1.5** Show that in an inner product space

$$\|\mathbf{x}+\mathbf{y}\|^2 + \|\mathbf{x}-\mathbf{y}\|^2 = 2\|\mathbf{x}\|^2 + 2\|\mathbf{y}\|^2.$$

**6.1.6** Show that

$$\left\langle \sum_{i=1}^{n} \mathbf{x}_i, \sum_{j=1}^{m} \mathbf{y}_j \right\rangle = \sum_{i=1}^{n} \sum_{j=1}^{m} \langle \mathbf{x}_i, \mathbf{y}_j \rangle$$

**6.1.7** Show that for *fixed* $\mathbf{y}$, $\langle \mathbf{x}, \mathbf{y} \rangle$ is a bounded linear functional of $\mathbf{x}$.

### 6.1.1 Orthogonality

We return to the concept of angle between vectors as defined by (6.1.2). The Schwarz inequality (6.1.7) makes it possible to define an angle $\theta$ between any two vectors $\mathbf{x}$ and $\mathbf{y}$ belonging to an inner product space $\mathscr{I}$ in analogy with (6.1.2) by

$$\theta = \cos^{-1} \frac{|\langle \mathbf{x}, \mathbf{y} \rangle|}{\|\mathbf{x}\| \|\mathbf{y}\|} \tag{6.1.17}$$

From (6.1.17), two vectors could be construed as "perpendicular" or *orthogonal* if $\langle \mathbf{x}, \mathbf{y} \rangle = 0$. For orthogonal vectors $\mathbf{x}$ and $\mathbf{y}$ it immediately follows that

$$\|\mathbf{x}+\mathbf{y}\|^2 = \langle \mathbf{x}+\mathbf{y}, \mathbf{x}+\mathbf{y} \rangle = \|\mathbf{x}\|^2 + \|\mathbf{y}\|^2$$

which is the equivalent of the theorem due to Pythagoras in Euclidean geometry. We have already seen that $\mathbf{0}$ is orthogonal to every vector in $\mathscr{I}$.† This result, together with the fact that there is no other vector with the foregoing property, is the content of the following theorem.

***Theorem 6.1.1*** Let $\mathscr{I}$ be an inner product space. Then $\langle \mathbf{x}, \mathbf{y} \rangle = 0 \ \forall\ \mathbf{y} \in \mathscr{I}$ *if and only if* $\mathbf{x} = \mathbf{0}$.

*Proof*: The "if" part is already known to be true. For the "only if" part we recognize that for $\mathbf{y} = \mathbf{x}$, $\langle \mathbf{x}, \mathbf{x} \rangle = 0$ from which $\mathbf{x} = \mathbf{0}$. ∎

This is a familiar result, for no nonzero vector can be perpendicular to an entire space.

It is possible to consider a set of vectors, say $\{\mathbf{x}_1, \mathbf{x}_2, \ldots, \mathbf{x}_n\}$, such that each pair of vectors $\mathbf{x}_i$ and $\mathbf{x}_j$ is orthogonal, that is,

$$\langle \mathbf{x}_i, \mathbf{x}_j \rangle = 0, \qquad i, j = 1, 2, \ldots, n, \quad i \neq j \tag{6.1.18}$$

Such a set of vectors is called an *orthogonal family* of vectors. If in addition to (6.1.18) we also have $\|\mathbf{x}_i\| = 1$, $i = 1, 2, \ldots, n$, then the orthogonal family is called an *orthonormal family*. Note that it is always possible to

---

†This result is also implied by the Schwarz inequality because

$$0 \leq |\langle \mathbf{0}, \mathbf{x} \rangle| \leq \|\mathbf{0}\| \|\mathbf{x}\| = 0 \rightarrow \langle \mathbf{0}, \mathbf{x} \rangle = 0$$

form an orthonormal family $\{\mathbf{x}_i\}_{i=1}^n$ from an orthogonal family, say $\{\mathbf{y}_i\}_{i=1}^n$ by *normalizing* a vector, that is,

$$\mathbf{x}_i = \frac{\mathbf{y}_i}{\|\mathbf{y}_i\|}, \qquad i = 1, 2, \ldots, n$$

so that $\|\mathbf{x}_i\| = 1$ for each $i$.

It is readily shown that an orthonormal family of vectors not including the zero vector is linearly independent. For example, let $\{\mathbf{x}_i\}_{i=1}^n$ be an orthonormal family. Let

$$\mathbf{0} = \sum_{i=1}^n \alpha_i \mathbf{x}_i$$

Thus

$$0 = \langle \mathbf{0}, \mathbf{x}_j \rangle = \sum_{i=1}^n \alpha_i \langle \mathbf{x}_i, \mathbf{x}_j \rangle = \alpha_j, \qquad j = 1, 2, \ldots, n$$

from which it follows that the set $\{\mathbf{x}_i\}_{i=1}^n$ is linearly independent. Obviously, the foregoing result is also valid for an orthogonal family. Indeed, in infinite-dimensional space one may have an infinite number of vectors in an orthogonal family. Next we establish an important inequality via the following theorem.

***Theorem 6.1.2*** Let $\{\mathbf{x}_j\}$ be an orthonormal family of vectors (not necessarily finite in number) in an inner product space $\mathcal{I}$. Then $\forall\, \mathbf{x} \in \mathcal{I}$,

$$\sum_j |\langle \mathbf{x}, \mathbf{x}_j \rangle|^2 \leq \|\mathbf{x}\|^2 \tag{6.1.19}$$

*Proof*: We first prove the inequality for a finite number $n$. Let $\alpha_j \equiv \langle \mathbf{x}, \mathbf{x}_j \rangle$. Then

$$\begin{aligned} \left\| \mathbf{x} - \sum_{j=1}^n \alpha_j \mathbf{x}_j \right\|^2 &= \left\langle \mathbf{x} - \sum_{j=1}^n \alpha_j \mathbf{x}_j \quad \mathbf{x} - \sum_{k=1}^n \alpha_k \mathbf{x}_k \right\rangle \\ &= \|\mathbf{x}\|^2 - \sum_{k=1}^n \alpha_k^* \langle \mathbf{x}, \mathbf{x}_k \rangle - \sum_{j=1}^n \alpha_j \langle \mathbf{x}_j, \mathbf{x} \rangle + \sum_{j=1}^n \sum_{k=1}^n \alpha_j \alpha_k^* \langle \mathbf{x}_j, \mathbf{x}_k \rangle \\ &= \|\mathbf{x}\|^2 - \sum_{j=1}^n |\alpha_j|^2 \geq 0 \end{aligned}$$

Thus $\sum_{j=1}^n |\langle \mathbf{x}, \mathbf{x}_j \rangle|^2 \leq \|\mathbf{x}\|^2$, so that the inequality holds for *every* finite $n$. Since the left-hand side of the inequality is a nondecreasing sequence of real numbers bounded from above, it must converge to a limit which is also bounded by $\|\mathbf{x}\|^2$ (see Theorem 1.8.2). Thus

$$\sum_{j=1}^\infty |\langle \mathbf{x}, \mathbf{x}_j \rangle|^2 \leq \|\mathbf{x}\|^2 \quad \blacksquare$$

The inequality above is known as *Bessel's inequality*. A simple interpretation of this inequality is provided at a later stage.

We had observed earlier that an orthogonal family of vectors is linearly independent. Quite obviously, not every linearly independent set of vectors is an orthogonal family, but it is always possible to produce an orthogonal family from a given set of linearly independent vectors by taking suitable linear combinations of them. This orthogonalization procedure is discussed next.

### 6.1.2 Gram–Schmidt Orthogonalization

Let $\{\mathbf{x}_j\}$ be a linearly independent set of vectors in $\mathcal{I}$. Our objective is to construct an orthogonal family $\{\mathbf{y}_j\}$ by taking appropriate linear combinations of vectors from the set $\{\mathbf{x}_j\}$. In so doing we will also normalize the new set so that

$$\langle \mathbf{y}_i, \mathbf{y}_j \rangle = \delta_{ij} = \begin{cases} 1, & i = j \\ 0, & i \neq j \end{cases}$$

Thus $\{\mathbf{y}_j\}$ will be an orthonormal family. To accomplish this, we first set

$$\mathbf{y}'_1 = \mathbf{x}_1, \qquad \mathbf{y}_1 \equiv \frac{1}{||\mathbf{y}'_1||}\mathbf{y}'_1$$

Clearly, $||\mathbf{y}_1|| = 1$. Next we put $\mathbf{y}'_2 = \mathbf{x}_2 - \alpha\mathbf{y}_1$ and require that $\langle \mathbf{y}'_2, \mathbf{y}_1 \rangle = 0$, so that $\alpha = \langle \mathbf{x}_2, \mathbf{y}_1 \rangle$. Thus

$$\mathbf{y}'_2 = \mathbf{x}_2 - \langle \mathbf{x}_2, \mathbf{y}_1 \rangle \mathbf{y}_1, \qquad \mathbf{y}_2 \equiv \frac{1}{||\mathbf{y}_2||}\mathbf{y}'_2$$

By setting $\mathbf{y}'_3 = \mathbf{x}_3 - \beta\mathbf{y}_1 - \gamma\mathbf{y}_2$ and requiring that $\langle \mathbf{y}'_3, \mathbf{y}_1 \rangle = 0$ and $\langle \mathbf{y}'_3$ $\mathbf{y}_2 \rangle = 0$, we obtain

$$\mathbf{y}'_3 = \mathbf{x}_3 - \langle \mathbf{x}_3, \mathbf{y}_1 \rangle \mathbf{y}_1 - \langle \mathbf{x}_3, \mathbf{y}_2 \rangle \mathbf{y}_2, \qquad \mathbf{y}_3 \equiv \frac{1}{||\mathbf{y}'_3||}\mathbf{y}'_3$$

Proceeding similarly, we obtain

$$\mathbf{y}'_j = \mathbf{x}_j - \sum_{k=1}^{j-1} \langle \mathbf{x}_j, \mathbf{y}_k \rangle \mathbf{y}_k, \qquad \mathbf{y}_j \equiv \frac{1}{||\mathbf{y}'_j||}\mathbf{y}'_j$$

Clearly, the set $\{\mathbf{y}_j\}$ is an orthonormal family. This orthogonalization (or orthonormalization) procedure is a very useful concept to construct orthonormal bases discussed in Section 6.3.

#### EXERCISES

**6.1.8** Let $\mathbf{x}_1, \mathbf{x}_2, \ldots, \mathbf{x}_n$ be $n$ vectors in a complex inner product space $\mathcal{I}$. Show that the matrix $\{a_{ij} \equiv \langle \mathbf{x}_i, \mathbf{x}_j \rangle; i, j = 1, 2, \ldots, n\}$ is nonnegative definite (by which is meant that for any set of complex numbers $\alpha_1, \alpha_2, \ldots, \alpha_n$, $\sum_{i=1}^{n} \sum_{j=1}^{n} a_{ij}\alpha_i^*\alpha_j \geq 0$). Note that the matrix $\{\langle \mathbf{x}_1, \mathbf{x}_j \rangle\}$ is called the *Gram matrix* of the set $\{\mathbf{x}_i\}$. Show further that the Gram matrix of a set $\{\mathbf{x}_i\}_{i=1}^{n}$ is positive definite (i.e., $\sum_{i=1}^{n} \sum_{j=1}^{n} a_{ij}\alpha_i^*\alpha_j > 0$ if not all of the $\alpha_i$'s are zero) *if and only if* the set $\{\mathbf{x}_i\}_{i=1}^{n}$ is linearly independent.

**6.1.9** If $\{\mathbf{x}_j\}_{j=1}^n$ is an orthonormal family, show that

$$\left\|\sum_{j=1}^{n} \alpha_j \mathbf{x}_j\right\|^2 = \sum_{j=1}^{n} |\alpha_j|^2$$

Note that this is an extension of the Pythagorean theorem.

## 6.2 Orthonormal Bases

Since an inner product space generates a normed linear space, the discussion on bases in normed linear spaces in Section 5.2 also applies to inner product spaces. Indeed, the existence of a *countable* basis is therefore restricted to *separable* inner product spaces. All the inner product spaces of interest to us satisfy this requirement of separability, so that countable bases are guaranteed.

Of particular significance to us are bases which are also orthonormal families. Such bases are termed *orthonormal bases.* The advantage of an orthonormal basis lies in the fact that the coefficients of expansion of a vector in terms of the basis can be identified immediately. Consider the finite-dimensional case first. Let $\{\mathbf{x}_j\}_{j=1}^n$ be an orthonormal basis and $\mathbf{x} \in \mathcal{I}$. Then there exist scalars $\{\alpha_j\}_{j=1}^n$ such that

$$\mathbf{x} = \sum_{j=1}^{n} \alpha_j \mathbf{x}_j \tag{6.2.1}$$

Forming the inner product of $\mathbf{x}$ with $\mathbf{x}_k$, from (6.2.1) we obtain

$$\langle \mathbf{x}, \mathbf{x}_k \rangle = \sum_{j=1}^{n} \alpha_j \langle \mathbf{x}_j, \mathbf{x}_k \rangle = \alpha_k \tag{6.2.2}$$

which is because $\langle \mathbf{x}_j, \mathbf{x}_k \rangle = 0$ for $j \neq k$ and $\langle \mathbf{x}_k, \mathbf{x}_k \rangle = 1$. (If $\{\mathbf{x}_j\}_{j=1}^n$ were an *orthogonal* basis, that is, the vectors are not normalized, then clearly $\alpha_k = \langle \mathbf{x}, \mathbf{x}_k \rangle / \langle \mathbf{x}_k, \mathbf{x}_k \rangle$.) The student is quite familiar with such expansions of vectors in three-dimensional space, where the mutually perpendicular unit vectors **i**, **j**, and **k** are chosen along the $x$, $y$, and $z$ axes.

It turns out that the result (6.2.2) holds also for *infinite-dimensional* spaces. Let $\{\mathbf{x}_j\}$ be an orthonormal basis in $\mathcal{I}$ and $\mathbf{x} \in \mathcal{I}$. Then

$$\mathbf{x} = \sum_{j=1}^{\infty} \alpha_j \mathbf{x}_j \tag{6.2.3}$$

by which is meant that

$$\lim_{n\to\infty} \left\| \mathbf{x} - \sum_{j=1}^{n} \alpha_j \mathbf{x}_j \right\| = 0$$

Since for fixed $\mathbf{x}$, $\langle \mathbf{x}, \mathbf{y} \rangle$ is a bounded linear functional of $\mathbf{y}$ (see Exercise 6.1.7), we have

$$\langle \mathbf{x}, \mathbf{x}_k \rangle = \left\langle \lim_{n\to\infty} \sum_{j=1}^{n} \alpha_j \mathbf{x}_j, \mathbf{x}_k \right\rangle = \lim_{n\to\infty} \left\langle \sum_{j=1}^{n} \alpha_j \mathbf{x}_j, \mathbf{x}_k \right\rangle = \alpha_k \tag{6.2.4}$$

so that (6.2.2) is again true.

A concept parallel to that of a basis set is a *maximal* set or a *complete* set.

***Definition 6.2.1*** A set of vectors $\{\mathbf{x}_j\}$ is said to be *maximal* or *complete* in inner product space $\mathcal{I}$ if $\langle \mathbf{x}, \mathbf{x}_j \rangle = 0 \ \forall j \rightarrow \mathbf{x} = \mathbf{0}$.

Note that this means that a vector orthogonal to a maximal set is necessarily zero. Theorem 6.1.1 established that a vector orthogonal to the entire inner product space $\mathcal{I}$ is necessarily zero, which would seem to imply that a maximal set somehow represents the entire inner product space. In fact, the following theorem establishes the equivalence of a maximal orthonormal family to an orthonormal basis in a separable Hilbert space.

***Theorem 6.2.1*** Let $\{\mathbf{x}_j\}$ be an orthonormal family in a Hilbert space $\mathcal{H}$. Then $\{\mathbf{x}_j\}$ is a basis in $\mathcal{H}$ *if and only if* it is a maximal set.

*Proof*: To prove the "only if" part, we assume that $\{\mathbf{x}_j\}$ is an orthonormal basis in $\mathcal{H}$. Then for $\mathbf{x} \in \mathcal{H}$, we have

$$\mathbf{x} = \sum_{j=1}^{\infty} \langle \mathbf{x}, \mathbf{x}_j \rangle \mathbf{x}_j$$

If $\langle \mathbf{x}, \mathbf{x}_j \rangle = 0 \ \forall j$, then $\mathbf{x} = \mathbf{0}$, so that $\{\mathbf{x}_j\}$ is maximal.

To prove the "if" part assume that $\{\mathbf{x}_j\}$ is a maximal orthonormal set. Let $\mathbf{x} \in \mathcal{H}$ be arbitrary. From Bessel's inequality,

$$\|\mathbf{x}\|^2 \geq \sum_{j=1}^{\infty} |\langle \mathbf{x}, \mathbf{x}_j \rangle|^2$$

We define a sequence of vectors $\{\mathbf{y}_n\}$ where $\mathbf{y}_n \equiv \sum_{j=1}^{n} \langle \mathbf{x}, \mathbf{x}_j \rangle \mathbf{x}_j$. From the extension of the Pythagorean theorem (see Exercise 6.1.9),

$$\|\mathbf{y}_n - \mathbf{y}_m\|^2 = \sum_{j=n}^{m} |\langle \mathbf{x}, \mathbf{x}_j \rangle|^2 \tag{6.2.5}$$

Since $\sum_{j=1}^{\infty} |\langle \mathbf{x}, \mathbf{x}_j \rangle|^2$ is convergent, the right-hand side of (6.2.5) can be made as small as possible for sufficiently large $m$ and $n$. Thus $\{\mathbf{y}_n\}$ is a Cauchy sequence and must converge to some $\mathbf{x}' \in \mathcal{H}$ (since $\mathcal{H}$ is complete). Thus

$$\mathbf{x}' = \sum_{j=1}^{\infty} \langle \mathbf{x}, \mathbf{x}_j \rangle \mathbf{x}_j$$

Through arguments identical to those that led to (6.2.4), we obtain

$$\langle \mathbf{x}', \mathbf{x}_k \rangle = \langle \mathbf{x}, \mathbf{x}_k \rangle \qquad \forall \ k$$

or

$$\langle \mathbf{x}' - \mathbf{x}, \mathbf{x}_k \rangle = 0 \qquad \forall \ k$$

From the maximality of the set $\{\mathbf{x}_k\}$, we have $\mathbf{x} = \mathbf{x}'$.

Thus $\{\mathbf{x}_j\}$ form an orthonormal basis in $\mathcal{H}$. ∎

Theorem 6.2.1 has some interesting implications. First, the vector $\mathbf{y}_n \equiv \sum_{j=1}^{n} \langle \mathbf{x}, \mathbf{x}_j \rangle \mathbf{x}_j$ satisfies the relation

$$\|\mathbf{y}_n\|^2 = \sum_{j=1}^{n} |\langle \mathbf{x}, \mathbf{x}_j \rangle|^2$$

The continuity of the inner product implies that

$$\lim_{n \leftarrow \infty} \|\mathbf{y}_n\|^2 = \|\lim_{n \leftarrow \infty} \mathbf{y}_n\|^2 = \|\mathbf{x}\|^2 = \sum_{j=1}^{\infty} |\langle \mathbf{x}, \mathbf{x}_j \rangle|^2 \tag{6.2.6}$$

Thus Bessel's inequality, which holds for any orthonormal family $\{\mathbf{x}_j\}$, becomes an *equality* when $\{\mathbf{x}_j\}$ is an orthonormal basis. The foregoing equality is commonly referred to as *Parseval's relation.* A second implication of Theorem 6.2.1 is contained in the following theorem.

***Theorem 6.2.2*** Every complex (real) separable Hilbert space $\mathcal{H}$ is isometrically equivalent to the complex (real) Hilbert space $l_2$.

*Proof*: First we establish that there is a one-to-one correspondence between elements of $\mathcal{H}$ and those of $l_2$; that is, $\mathcal{H}$ is isomorphic with $l_2$. Since $\mathcal{H}$ is separable, it has a countable basis (see Section 5.2) which may be orthonormalized to obtain $\{\mathbf{x}_j\}$ as an orthonormal basis in $\mathcal{H}$.

Let $\mathbf{x} \in \mathcal{H}$. The complex sequence $\{\langle \mathbf{x}, \mathbf{x}_j \rangle\}$ is such that

$$\sum_{j=1}^{\infty} |\langle \mathbf{x}, \mathbf{x}_j \rangle|^2 = \|\mathbf{x}\|^2 < \infty$$

so that $\{\langle \mathbf{x}, \mathbf{x}_j \rangle\} \in l_2$ corresponds to $\mathbf{x} \in \mathcal{H}$. Similarly, for any element $\{\alpha_j\} \subset l_2$, we have

$$\sum_{j=1}^{\infty} |\alpha_j|^2 < \infty \tag{6.2.7}$$

Form the sequence $\{\mathbf{y}_n\} \subset \mathcal{H}$, where $\mathbf{y}_n = \sum_{j=1}^{n} \alpha_j \mathbf{x}_j$. Now

$$\|\mathbf{y}_m - \mathbf{y}_n\|^2 = \sum_{j=n}^{m} |\alpha_j|^2 \tag{6.2.8}$$

In view of (6.2.7), by making $m$ and $n$ sufficiently large, the right-hand side of (6.2.8) may be made arbitrarily small so that $\{\mathbf{y}_n\}$ is a Cauchy sequence in $\mathcal{H}$. Then $\exists\ \mathbf{x} \in \mathcal{H} \ni \lim_{n \to \infty} \mathbf{y}_n = \mathbf{x}$. We have thus shown that there is a unique correspondence between elements of $\mathcal{H}$ and elements of $l_2$. Furthermore, the correspondence may be represented by the mappings

$$\mathbf{x} \in \mathcal{H} \longrightarrow \{\langle \mathbf{x}, \mathbf{x}_j \rangle\} \in l_2$$

$$\{\alpha_j\} \in l_2 \longrightarrow \sum_{j=1}^{\infty} \alpha_j \mathbf{x}_j \in \mathcal{H} \quad \blacksquare$$

Theorem 6.2.2 represents a rather remarkable property of separable infinite-dimensional inner product spaces. We will consider below some examples of separable Hilbert spaces by citing countable orthonormal bases in them (see Section 5.2 for the equivalence between separability and the existence of a countable basis).

*Example 1*

Every finite-dimensional Hilbert space defined over the complex field is obviously separable because of the existence of a finite basis.

*Example 2.* $\mathfrak{L}_2[-\pi, \pi]$

The separability of this space results from the following. First, the set $\{\boldsymbol{\phi}_n, \boldsymbol{\psi}_n; n = 0, 1, 2, \ldots\}$, where

$$\phi_0(x) \equiv \frac{1}{\sqrt{2\pi}}, \qquad \phi_n(x) \equiv \frac{1}{\sqrt{\pi}} \cos nx, \qquad \psi_n(x) = \frac{1}{\sqrt{\pi}} \sin nx$$

is an orthonormal family because

$$\|\boldsymbol{\phi}_0\|^2 = \langle \boldsymbol{\phi}_0, \boldsymbol{\phi}_0 \rangle = \frac{1}{2\pi} \int_{-\pi}^{\pi} dx = 1$$

$$\langle \boldsymbol{\phi}_0, \boldsymbol{\phi}_n \rangle = \frac{1}{\sqrt{2\pi}} \int_{-\pi}^{\pi} \cos nx \, dx = 0, \qquad n = 1, 2, \ldots$$

$$\langle \boldsymbol{\phi}_0, \boldsymbol{\psi}_n \rangle = \frac{1}{\sqrt{2\pi}} \int_{-\pi}^{\pi} \sin nx \, dx = 0, \qquad n = 1, 2, \ldots$$

$$\langle \boldsymbol{\phi}_n, \boldsymbol{\phi}_m \rangle = \frac{1}{\pi} \int_{-\pi}^{\pi} \cos nx \cos mx \, dx = \begin{cases} 0, & n \neq m \\ 1, & n = m \end{cases}$$

$$\langle \boldsymbol{\phi}_n, \boldsymbol{\psi}_m \rangle = \frac{1}{\pi} \int_{-\pi}^{\pi} \cos nx \sin mx \, dx = 0$$

$$\langle \boldsymbol{\psi}_n, \boldsymbol{\psi}_m \rangle = \frac{1}{\pi} \int_{-\pi}^{\pi} \sin nx \sin mx \, dx = \begin{cases} 0, & n \neq m \\ 1, & n = m \end{cases}$$

Second, it is well known from the theory of Fourier series that any function $\mathbf{f} \in \mathfrak{L}_2[-\pi, \pi]$ can be expressed as

$$f(x) = \frac{a_0}{2} + \sum_{n=1}^{\infty} (a_n \cos nx + b_n \sin nx) \tag{6.2.9}$$

where $a_n = \frac{1}{\pi} \int_{-\pi}^{\pi} f(x) \cos nx \, dx$ and $b_n = \frac{1}{\pi} \int_{-\pi}^{\pi} f(x) \sin nx$. The infinite sum (6.2.9) implies that

$$\lim_{N \to \infty} \int_{-\pi}^{\pi} \left[ f(x) - \frac{a_0}{2} - \sum_{n=1}^{N} (a_n \cos nx + b_n \sin nx) \right]^2 dx = 0 \tag{6.2.10}$$

We will not prove this here but instead refer the reader to any book on Fourier series.† Equation (6.2.10) implies that the countable set $\{\boldsymbol{\phi}_n, \boldsymbol{\psi}_n\}$ is an orthonormal basis in $\mathfrak{L}_2[-\pi, \pi]$, establishing the separability of the space.

---

†See, for example, *Fourier Series and Boundary Value Problems*, 2nd ed., by R. V. Churchill, McGraw-Hill, New York, 1963.

This conclusion of course also extends to $\mathcal{L}_2[a, b]$, where $[a, b]$ is any other bounded interval on the real line.

*Example 3.* $\mathcal{L}_2[0, \infty)$

Consider the set of functions

$$\phi_n(x) = e^{-x/2}L_n(x), \qquad n = 0, 1, 2, \ldots$$

where $\{L_n(x)\}$, called *Laguerre polynomials*, are given by

$$L_n(x) = \frac{e^x}{n!}\frac{d^n}{dx^n}(x^n e^{-x})$$

The functions $\{\phi_n(x)\}$ are referred to as *Laguerre functions* and are orthonormal because

$$\int_0^\infty \phi_n(x)\phi_m(x)\,dx = \begin{cases} 1, & n = m \\ 0, & n \neq m \end{cases}$$

That the foregoing set is a basis in $\mathcal{L}_2[0, \infty)$ may be found in standard textbooks.† Thus $\mathcal{L}_2[0, \infty)$ is seen to be a separable Hilbert space.

*Example 4.* $\mathcal{L}_2(-\infty, \infty)$

This space is separable bacause the following set is an orthonormal basis:

$$\psi_n(x) = \frac{e^{-x^2/2}}{\sqrt{2^n n!\sqrt{\pi}}}H_n(x), \qquad n = 0, 1, 2, \ldots$$

where the functions $\{H_n(x)\}$ are the *Hermite polynomials* given by

$$H_n(x) = (-1)^n e^{x^2}\frac{d^n}{dx^n}(e^{-x^2})$$

The functions $\{\psi_n(x)\}$, known as the *Hermite functions*, satisfy the orthonormality conditions

$$\int_{-\infty}^\infty \psi_n(x)\psi_m(x)\,dx = \begin{cases} 1, & n = m \\ 0, & n \neq m \end{cases}$$

Each of the separable Hilbert spaces above is equivalent to the Hilbert space $l_2$ from Theorem 6.2.2. This is particularly interesting because a vector in a function space is identified by an uncountably infinite number of function values, whereas an element in $l_2$ space is characterized by a countably infinite number of quantities.

---

†See, for example, *Methods of Mathematical Physics*, Vol. 1, by R. Courant and D. Hilbert, Interscience, New York, 1953, p. 95.

**EXERCISES**

**6.2.1** Consider $\mathcal{L}_2[0, 1]$. Show that the set of functions

$$\phi_n(x) = x^n, \qquad n = 0, 1, 2, \ldots$$

is linearly independent. Given that it is also complete, obtain an orthonormal basis in $\mathcal{L}_2[0, 1]$ by Gram–Schmidt orthogonalization.

**6.2.2** The set

$$f(x) = x^n e^{-x/2}, \qquad n = 0, 1, 2, \ldots$$

is complete in $\mathcal{L}_2[0, \infty)$. Obtain by Gram–Schmidt orthogonalization an orthonormal basis $\{\psi_n\}$ in $\mathcal{L}_2[0, \infty)$ and show that the result may be cast in the form $\psi_n(x) = (e^{-x/2}/n!)L_n(x)$, where $L_n(x)$ is the $n$th Laguerre polynomial given by

$$L_n(x) = e^x \frac{d^n}{dx^n}[x^n e^{-x}].$$

**6.2.3** Obtain the Hermite functions as an orthonormal basis in $\mathcal{L}_2(-\infty, \infty)$ by Gram–Schmidt orthogonalization of a suitable set.

## 6.3 Direct Sums and Tensor Products of Inner Product Spaces

Having encountered direct sums of normed linear spaces, it should not come as a surprise that direct sums could also be defined on inner product spaces. The issue of a tensor product is less straightforward than that of the direct sum. We consider the direct sum first.

***Definition 6.3.1*** Let $\mathcal{I}_1 \equiv (\mathcal{L}_1, \langle \cdot, \cdot \rangle_1)$ and $\mathcal{I}_2 \equiv (\mathcal{L}_2, \langle \cdot, \cdot \rangle_2)$ be two inner product spaces where $\mathcal{L}_1$ and $\mathcal{L}_2$ have the same associated field. Then the *direct sum* of $\mathcal{I}_1$ and $\mathcal{I}_2$, denoted $\mathcal{I}_1 \oplus \mathcal{I}_2$, is an inner product space with $\mathcal{L}_1 \oplus \mathcal{L}_2$ as the linear space on which the inner product for any two vectors $(\mathbf{x}_1, \mathbf{x}_2), (\mathbf{y}_1, \mathbf{y}_2) \in \mathcal{L}_1 \oplus \mathcal{L}_2$, where $\mathbf{x}_1, \mathbf{y}_1 \in \mathcal{I}_1, \mathbf{x}_2, \mathbf{y}_2 \in \mathcal{I}_2$, is defined by

$$\langle (\mathbf{x}_1, \mathbf{x}_2), (\mathbf{y}_1, \mathbf{y}_2) \rangle \equiv \alpha_1 \langle \mathbf{x}_1, \mathbf{y}_1 \rangle_1 + \alpha_2 \langle \mathbf{x}_2, \mathbf{y}_2 \rangle_2 \tag{6.3.1}$$

where $\alpha_1$ and $\alpha_2$ are *real, positive* numbers.

It is left as an exercise to show that the inner product as defined in (6.3.1) satisfies the axioms of a norm. It is readily shown further that if $\mathcal{I}_1$ and $\mathcal{I}_2$ are Hilbert spaces, then $\mathcal{I}_1 \oplus \mathcal{I}_2$ is also a Hilbert space.

In applications, it is frequently possible to encounter direct sums of the type $\mathcal{I} \oplus \mathcal{C}$ or $\mathcal{I} \oplus \mathcal{R}$, depending on whether $\mathcal{I}$ is a complex or real inner product space, respectively. For example elements of $\mathcal{I} \oplus \mathcal{R}$ would be of the type $(\mathbf{x}, x_1)$, where $\mathbf{x} \in \mathcal{I}$ and $x_1 \in \mathcal{R}$. The inner product between two such elements $(\mathbf{x}, x_1)$ and $(\mathbf{y}, y_1)$ may then be defined as†

$$[(\mathbf{x}, x_1), (\mathbf{y}, y_1)] \equiv \alpha_1\langle \mathbf{x}, \mathbf{y}\rangle + \alpha_2 x_1 y_1 \tag{6.3.2}$$

where we have used the symbol $[\bullet, \bullet]$ to denote the inner product on $\mathcal{I} \oplus \mathcal{R}$ to distinguish it from the inner product $\langle \bullet, \bullet \rangle$ on $\mathcal{I}$. Again in (6.3.2) $\alpha_1$ and $\alpha_2$ are real numbers, and if $\mathcal{I}$ is a Hilbert space, we conclude that $\mathcal{I} \oplus \mathcal{R}$ is also a Hilbert space.

Indeed, the extension of the direct sum to more than two inner product spaces is trivial.

We next consider the tensor product of two inner product spaces.

***Definition 6.3.2*** Let $\mathcal{I}_1 \equiv (\mathcal{L}_1, \langle \bullet, \bullet \rangle_1)$ and $\mathcal{I}_2 \equiv (\mathcal{L}_2, \langle \bullet, \bullet \rangle_2)$ be two inner product spaces, where $\mathcal{L}_1$ and $\mathcal{L}_2$ are defined over the same field. The *tensor* product of $\mathcal{I}_1$ and $\mathcal{I}_2$, denoted as $\mathcal{I}_1 \otimes \mathcal{I}_2$, is an inner product space with $\mathcal{L}_1 \otimes \mathcal{L}_2$ as the linear space. The inner product for any two vectors of the type $(\mathbf{x}_1 \otimes \mathbf{x}_2)$, $(\mathbf{y}_1 \otimes \mathbf{y}_2) \in \mathcal{L}_1 \otimes \mathcal{L}_2$, where $\mathbf{x}_1, \mathbf{y}_1 \in \mathcal{I}_1, \mathbf{x}_2, \mathbf{y}_2 \in \mathcal{I}_2$ is defined by

$$\langle (\mathbf{x}_1 \otimes \mathbf{x}_2), (\mathbf{y}_1 \otimes \mathbf{y}_2)\rangle \equiv \langle \mathbf{x}_1, \mathbf{y}_1\rangle_1 \langle \mathbf{x}_2, \mathbf{y}_2\rangle_2 \tag{6.3.3}$$

For elements of the more general type $\sum_j \alpha_j(\mathbf{x}_{j,1} \otimes \mathbf{x}_{j,2}) \in \mathcal{L}_1 \otimes \mathcal{L}_2$, where $\mathbf{x}_{j,1} \in \mathcal{I}_1, \mathbf{x}_{j,2} \in \mathcal{I}_2$, the inner product (6.3.3) is naturally extended as

$$\langle \sum_j \alpha_j(\mathbf{x}_{j,1} \otimes \mathbf{x}_{j,2}), \sum_k \beta_k(\mathbf{y}_{k,1} \otimes \mathbf{y}_{k,2})\rangle$$
$$\equiv \sum_j \sum_k \alpha_j \beta_k^* < \mathbf{x}_{j,1}, \mathbf{y}_{k,1}\rangle_1 \langle \mathbf{x}_{j,2}, \mathbf{y}_{k,2}\rangle_2 \tag{6.3.4}$$

Let us show that the inner product above satisfies the axioms A1 to A4 in Section 6.1. The relation (6.3.4) is essentially an assertion of A1, so that it would be adequate to establish the other axioms for (6.3.3). Let $\alpha \in \mathcal{C}$. Then

$$\begin{aligned}\langle \alpha(\mathbf{x}_1 \otimes \mathbf{x}_2), (\mathbf{y}_1 \otimes \mathbf{y}_2)\rangle &= \langle (\alpha\mathbf{x}_1 \otimes \mathbf{x}_2), (\mathbf{y}_1 \otimes \mathbf{y}_2)\rangle \\ &= \langle \alpha\mathbf{x}_1, \mathbf{y}_1\rangle_1 \langle \mathbf{x}_2, \mathbf{y}_2\rangle_2 \\ &= \alpha\langle \mathbf{x}_1, \mathbf{y}_1\rangle_1 \langle \mathbf{x}_2, \mathbf{y}_2\rangle_2 \\ &= \alpha\langle (\mathbf{x}_1 \otimes \mathbf{x}_2), (\mathbf{y}_1 \otimes \mathbf{y}_2)\rangle\end{aligned}$$

The scalar $\alpha$ could also have been associated with $\mathbf{x}_2$ and the result would have been the same. Thus A3 holds. To establish A2 we need only recognize that the complex conjugate of the product of two complex numbers is the product of their conjugates. From (6.3.3), we have

$$\sqrt{\langle (\mathbf{x}_1 \otimes \mathbf{x}_2), (\mathbf{x}_1 \otimes \mathbf{x}_2)\rangle} = \|\mathbf{x}_1\|_1 \|\mathbf{x}_2\|_2 \tag{6.3.5}$$

where we have used $\|\bullet\|_1$ and $\|\bullet\|_2$ to denote the norms in $\mathcal{I}_1$ and $\mathcal{I}_2$ generated

---

†Of course, this is consistent with (6.3.1) if we regard $\mathcal{R}$ or $\mathcal{C}$ as an inner product space with the inner product between two numbers $x$ and $y$ as $xy$ in $\mathcal{R}$ or $xy^*$ in $\mathcal{C}$.

by their respective inner products. From (6.3.5) the norm in $\mathcal{S}_1 \otimes \mathcal{S}_2$ is clearly the product of the norms of the constituent elements in $\mathcal{S}_1$ and $\mathcal{S}_2$. Axiom A4 is readily apparent from (6.3.5) for an element of the type $\mathbf{x}_1 \otimes \mathbf{x}_2$, but not so for elements as in (6.3.4). The student is referred to Exercise 6.3.4. The inner product as defined in (6.3.4), therefore, satisfies all the stipulations required of it.

It is interesting to observe that the inner products $\langle \cdot, \cdot \rangle_1$ and $\langle \cdot, \cdot \rangle_2$ may also be regarded as the following mappings:

$$\langle \cdot, \cdot \rangle_1 : \quad \mathcal{S}_1 \otimes \mathcal{S}_2 \times \mathcal{S}_1 \longrightarrow \mathcal{S}_2$$
$$\langle \cdot, \cdot \rangle_2 : \quad \mathcal{S}_1 \otimes \mathcal{S}_2 \times \mathcal{S}_2 \longrightarrow \mathcal{S}_1$$

More specifically, let $\mathbf{x}_1 \otimes \mathbf{x}_2 \in \mathcal{S}_1 \times \mathcal{S}_2, \mathbf{y}_1 \in \mathcal{S}_1, \mathbf{y}_2 \in \mathcal{S}_2$. Then we have

$$\begin{aligned} \langle \mathbf{x}_1 \otimes \mathbf{x}_2, \mathbf{y}_1 \rangle_1 &= \langle \mathbf{x}_1, \mathbf{y}_1 \rangle_1 \mathbf{x}_2 \\ \langle \mathbf{x}_1 \otimes \mathbf{x}_2, \mathbf{y}_2 \rangle_2 &= \langle \mathbf{x}_2, \mathbf{y}_2 \rangle_2 \mathbf{x}_1 \end{aligned} \tag{6.3.6}$$

The above formulas are readily extended to obtain

$$\begin{aligned} \langle \sum_j \alpha_j (\mathbf{x}_{j,1} \otimes \mathbf{x}_{j,2}), \mathbf{y}_1 \rangle_1 &= \sum_j \alpha_j \langle \mathbf{x}_{j,1}, \mathbf{y}_1 \rangle_1 \mathbf{x}_{j,2} \\ \langle \sum_j \alpha_j (\mathbf{x}_{j,1} \otimes \mathbf{x}_{j,2}), \mathbf{y}_2 \rangle_2 &= \sum_j \alpha_j \langle \mathbf{x}_{j,2}, \mathbf{y}_2 \rangle_2 \mathbf{x}_{j,1} \end{aligned} \tag{6.3.7}$$

where $\mathbf{x}_{j,1} \in \mathcal{S}_1, \mathbf{x}_{j,2} \in \mathcal{S}_2$. Formulas (6.3.7) are especially useful for dealing with elements of $\mathcal{S}_1 \otimes \mathcal{S}_2$ that *cannot be factored* into the form $\mathbf{x}_1 \otimes \mathbf{x}_2$. Thus if $\mathbf{w} \in \mathcal{S}_1 \otimes \mathcal{S}_2$, $\langle \mathbf{w}, \mathbf{y}_1 \rangle_1$ is a vector in $\mathcal{S}_2$ and $\langle \mathbf{w}, \mathbf{y}_2 \rangle_2$ is a vector in $\mathcal{S}_1$. Also, it is possible to write

$$\langle \mathbf{w}, \mathbf{y}_1 \otimes \mathbf{y}_2 \rangle = \langle \langle \mathbf{w}, \mathbf{y}_1 \rangle_1, \mathbf{y}_2 \rangle_2 \tag{6.3.8}$$

We leave it to the exercises to establish that if $\mathcal{H}_1$ and $\mathcal{H}_2$ are Hilbert spaces, $\mathcal{H}_1 \otimes \mathcal{H}_2$ is a Hilbert space. Furthermore, let $\{\mathbf{u}_j\}$ be an orthonormal basis in $\mathcal{H}_1$ and $\{\mathbf{v}_k\}$ be an orthonormal basis in $\mathcal{H}_2$. Then it is readily shown that $\{\mathbf{u}_j \otimes \mathbf{v}_k\}$ is an orthonormal basis in $\mathcal{H}_1 \otimes \mathcal{H}_2$. If $\mathbf{w} \in \mathcal{H}_1 \otimes \mathcal{H}_2$, then we have

$$\mathbf{w} = \sum_j \sum_k \langle \mathbf{w}, \mathbf{u}_j \otimes \mathbf{v}_k \rangle (\mathbf{u}_j \otimes \mathbf{v}_k) \tag{6.3.9}$$

The expansion above is, in fact, the basis of the application of the method of separation of variables to the solution of partial differential boundary value problems.

## EXERCISES

**6.3.1** Show that the inner product defined by (6.3.1) on the direct sum of two Hilbert spaces obtains a Hilbert space. If $\{\mathbf{u}_j\}$ is an orthonormal basis in $\mathcal{H}_1$ and $\{\mathbf{v}_k\}$ an orthonormal basis in $\mathcal{H}_2$, construct an orthonormal basis in $\mathcal{H}_1 \oplus \mathcal{H}_2$ based on the inner product (6.3.1).

**6.3.2** Consider the direct sum space $\mathcal{H} = \mathcal{L}_2[0, 1] \oplus \mathcal{R}$. Let $\{\lambda_n\}$ be the denumerably infinite number of roots of the transcendental equation

$$(\beta - \lambda) = \sqrt{\lambda} \tan \sqrt{\lambda}$$

and

$$\boldsymbol{\phi}_n \equiv \begin{bmatrix} \cos \sqrt{\lambda_n}\, x \\ \cos \sqrt{\lambda_n} \end{bmatrix}$$

Show that $\{\boldsymbol{\phi}_n\}$ is an orthogonal family, with respect to the inner product (6.3.1), using $\alpha_1 = \alpha_2 = 1$. Obtain an orthonormal family of vectors from the set above. Assuming that the family thus obtained is an orthonormal *basis* in $\mathcal{H}$, identify the expansion of an arbitrary vector in $\mathcal{H}$ in terms of the foregoing basis.

**6.3.3** Let $\mathcal{H} = \mathcal{L}_2[0, 1] \oplus \mathcal{L}_2[0, 1]$ in which the inner product is defined as in 6.3.1, with $\alpha_1 = \alpha_2 = 1$. Show that the set of vectors

$$\boldsymbol{\phi}_n \equiv \begin{bmatrix} \sin (n\pi + \theta)x \\ \cos (n\pi + \theta)x \end{bmatrix}, \qquad n = 0, \pm 1, \pm 2, \ldots$$

where $\theta$ is a fixed angle between 0 and $\pi/2$, is an orthonormal family in $\mathcal{H}$. If $\{\boldsymbol{\phi}_n\}$ is assumed to be an orthonormal basis, obtain the expansion formula for an arbitrary vector in $\mathcal{H}$.

**6.3.4** Establish the strict positive-definiteness axiom A4 of Section 6.1 for the inner product defined in (6.3.4) for elements of the type $\sum_j \alpha_j(x_{j,1} \otimes x_{j,2})$ in $\mathcal{G}_1 \otimes \mathcal{G}_2$. [*Hint:* Work with bases in $\mathcal{G}_1$ and $\mathcal{G}_2$ and the result of Exercise 2.12.3(d).]

**6.3.5** Show that the tensor product of two Hilbert spaces is a Hilbert space.

**6.3.6** Find suitable replacements for (6.3.6) if $\langle \cdot, \cdot \rangle_1$ in $\mathcal{G}_1$ and $\langle \cdot, \cdot \rangle_2$ in $\mathcal{G}_2$ are regarded as mappings:

$$\langle \cdot, \cdot \rangle_1 \colon \ \mathcal{G}_1 \times \mathcal{G}_1 \otimes \mathcal{G}_2 \longrightarrow \mathcal{G}_2$$
$$\langle \cdot, \cdot \rangle_2 \colon \ \mathcal{G}_2 \times \mathcal{G}_1 \otimes \mathcal{G}_2 \longrightarrow \mathcal{G}_1$$

**6.3.7** It is common in continuum mechanics to define vectors and higher-order tensors. The stresses at a point in a body are thus represented by a second-order tensor.† Another example of a second-order tensor is the velocity gradient tensor. Show that a second-order tensor may be regarded as an element of $\mathcal{R}_3 \otimes \mathcal{R}_3$. In terms of the notations of Section 6.3, interpret the following quantities

(a) The vector representing the traction force per unit area normal to unit vector **n** at a point.

(b) The rate of work done by the traction force above.

(c) The rate of viscous dissipation per unit volume at a point in a fluid.

---

†See, for example, *Transport Phenomena* by R. B. Bird, W. E. Stewart, and E. N. Lightfoot, Wiley, New York 1960.

**6.3.8** Solve again Exercise 2.12.4. Consider $\mathcal{H}_1 = \{\mathcal{R}_n, \langle\cdot, \cdot\rangle_1\}$, $\mathcal{H}_2 = \{\mathcal{R}_m, \langle\cdot, \cdot\rangle_2\}$ where $\langle\cdot, \cdot\rangle_1$ and $\langle\cdot, \cdot\rangle_2$ are as in (6.1.13). Define suitable inner products on
(a) $\mathcal{R}_{n+m}$ so that the resulting inner product space is isometrically equivalent to the $\mathcal{H}_1 \oplus \mathcal{H}_2$ in which the inner product is as in (6.3.1).
(b) $\mathcal{R}_{nm}$ so that the resulting inner product space is isometrically equivalent to $\mathcal{H}_1 \otimes \mathcal{H}_2$.

**6.3.9** Let $I_1 = \{0 \leq x \leq a\}$, $I_2 = \{0 \leq y \leq b\}$ be intervals on the real line, and $\mathcal{H} \equiv \mathcal{L}_2(I_1 \times I_2)$ be the space of vectors $\mathbf{f} \equiv \{f(x, y): x \in I_1, y \in I_2\}$ such that

$$\int_0^a dx \int_0^b dy\, f^2(x, y) < \infty$$

Further, let $\mathcal{H}_1 \equiv \mathcal{L}_2(I_1)$ and $\mathcal{H}_2 \equiv \mathcal{L}_2(I_2)$. Show that $\mathcal{H}$ and $\mathcal{H}_1 \otimes \mathcal{H}_2$ are isometrically equivalent Hilbert spaces.

## 6.4 Subspaces in Hilbert Space

In our discussion of normed linear spaces, we made only a fleeting reference to subspaces. The concept of a subspace is indeed a very important element of analysis in Hilbert space. Although our discussion is specific to Hilbert spaces, many definitions and theorems apply also to inner product spaces in general. However, we will not bother to make this distinction since our ultimate interest is in the theory of Hilbert spaces. We will be concerned with subsets, subspaces, and especially, *closed* linear subspaces of a Hilbert space $\mathcal{H}$. Since a Hilbert space generates a Banach space, the closedness of subsets is with respect to the metric generated by the norm. We begin with a definition.

***Definition 6.4.1*** Let $M$ be a subset of $\mathcal{H}$. A vector $\mathbf{z} \in \mathcal{H}$ is said to be *perpendicular* to $M$ (written in symbols as $\mathbf{z} \perp M$) if $\langle \mathbf{z}, \mathbf{x} \rangle = 0 \; \forall \; \mathbf{x} \in M$.

The following existence theorem, which we state without proof since it depends on some details not covered earlier, is very important. The validity of the theorem originates from the completeness of the space, as is characteristic of many existence theorems in general.†

†In Chapter 5 we alluded to the shortest distance between a subspace $M$ (of a normed linear space) and a vector $\mathbf{z}$ outside it. Curiously even for a Banach space, it is not possible to assert the existence of a vector in $M$ that is at the shortest distance from $\mathbf{z}$. This anomaly cannot arise in a Hilbert space. See, for example, *Linear Operator Theory in Engineering and Science*, by A. W. Naylor and G. R. Sell, Holt, Rinehart and Winston, New York, 1971. The proof of Theorem 6.4.1 is based on the foregoing existence result. Indeed, every student knows that in ordinary Euclidean geometry it is always possible to *project* a vector outside a plane onto the plane by "dropping a perpendicular"!

***Theorem 6.4.1*** Let $M$ be a proper closed linear subspace of a Hilbert space $\mathcal{H}$. Then there exists a vector $\mathbf{z} \in \mathcal{H}$ but not in $M$ such that $\mathbf{z} \perp M$.

Clearly, $\mathbf{z} \neq \mathbf{0}$, since it would be in $M$ then. It also follows from Theorem 6.4.1 that if $M$ and $N$ are closed linear subspaces of $\mathcal{H}$ such that $M \subset N$ $(M \neq N)$, then there exists a vector $\mathbf{z} \in N \ni \mathbf{z} \perp M$.

Since Theorem 6.4.1 shows that we can always find a vector in a Hilbert space outside of a closed linear (proper) subspace $M$, perpendicular to $M$, the question arises as to whether there are many such vectors. Obviously, any scalar multiple of a vector perpendicular to $M$ also has the same property. There may be others, too, which lead to the following definition.

***Definition 6.4.2*** Let $M$ be a subspace (or more generally subset) of a Hilbert space $\mathcal{H}$. The *orthogonal complement* of $M$, denoted $M^\perp$, is defined by

$$M^\perp = \{\mathbf{x} \in \mathcal{H} \colon \mathbf{x} \perp M\}$$

The orthogonal complement is, in fact, a familiar concept to the student. For example, in three-dimensional space, the $x - y$ plane is a proper subspace. All the vectors along the $z$-axis are perpendicular to the $x - y$ plane so that the $z$-axis is the orthogonal complement of the $x - y$ plane.

The following theorems are useful implements in the development of the theory of Hilbert spaces.

***Theorem 6.4.2*** Let $M$ and $N$ be proper linear subspaces of $\mathcal{H}$.

(i) $M^\perp$ is a closed linear subspace; further, $M \cap M^\perp = \{\mathbf{0}\}$.
(ii) $M \subset M^{\perp\perp}$.
(iii) If $M \subset N \subset \mathcal{H}$, $N^\perp \subset M^\perp$.
(iv) $M^\perp = M^{\perp\perp\perp}$.
(v) If $M$ is closed, $M = M^{\perp\perp}$.

*Proof*: The notation $M^{\perp\perp}$ refers to the orthogonal complement of $M^\perp$, that is, $M^{\perp\perp} = (M^\perp)^\perp$. We prove the results in the order in which they appear above.

(i) Let $\mathbf{u}, \mathbf{v} \in M^\perp$ and $\mathbf{x} \in M$ be arbitrarily chosen. Then

$$\langle \mathbf{u}, \mathbf{x} \rangle = 0, \qquad \langle \mathbf{v}, \mathbf{x} \rangle = 0$$

$$\langle \mathbf{u} + \mathbf{v}, \mathbf{x} \rangle = \langle \mathbf{u}, \mathbf{x} \rangle + \langle \mathbf{v}, \mathbf{x} \rangle = 0$$

Thus

$$(\mathbf{u} + \mathbf{v}) \in M^\perp$$

Similarly, $\langle \alpha\mathbf{u}, \mathbf{x} \rangle = \alpha\langle \mathbf{u}, \mathbf{x} \rangle = 0$, so that $\alpha\mathbf{u} \in M^\perp$, from which $M^\perp$ is clearly a linear subspace. To show that it is *closed*, let $\{\mathbf{u}_n\} \subset M^\perp$ such that it converges to some $\mathbf{u}$, which is a limit point of $M^\perp$. Now $\langle \mathbf{u}_n, \mathbf{x} \rangle = 0$ for each $n$. Hence

$$\lim_{n\to\infty} \langle \mathbf{u}_n, \mathbf{x} \rangle = \langle \lim_{n\to\infty} \mathbf{u}_n, \mathbf{x} \rangle = 0$$

where we have used the continuity of the inner product. Thus $\langle \mathbf{u}, \mathbf{x} \rangle = 0$ and $\mathbf{u} \in M^{\perp}$, so that $M^{\perp}$ is closed.

Finally, let $\mathbf{u} \in M \cap M^{\perp}$. Then $\langle \mathbf{u}, \mathbf{u} \rangle = 0$, so that $\mathbf{u} = 0$. Thus $M \cap M^{\perp} = \{\mathbf{0}\}$.

(ii) To prove $M \subset M^{\perp\perp}$, let $\mathbf{u} \in M$. Then $\forall\, \mathbf{x} \in M^{\perp}$ we have $\langle \mathbf{u}, \mathbf{x} \rangle = 0$. But $M^{\perp\perp}$ contains *all* such elements so that $\mathbf{u} \in M^{\perp\perp}$. Thus $M \subset M^{\perp\perp}$.

Obviously, results (i) and (ii) also hold for subspace $N$.

(iii) Let $M \subset N$ and $\mathbf{u} \in N^{\perp}$. Then $\mathbf{u} \perp N$. Since $M \subset N$, $\mathbf{u} \perp M$, so that $\mathbf{u} \in M^{\perp}$. Thus $N^{\perp} \subset M^{\perp}$.

(iv) Let $N \equiv M^{\perp}$. Using (ii), $N \subset N^{\perp\perp}$ and $M \subset M^{\perp\perp}$ or $M \subset N^{\perp}$. From (iii) $N^{\perp\perp} \subset M^{\perp}$ or $N^{\perp\perp} \subset N$. Thus $N = N^{\perp\perp}$ or $M^{\perp} = M^{\perp\perp\perp}$.

(v) Let $M$ be closed. Now from (ii) we know that $M \subset M^{\perp\perp}$. If $M \neq M^{\perp\perp}$, from Theorem 6.4.1, $\exists\; \mathbf{z} \in M^{\perp\perp} \ni \mathbf{z} \perp M$, so that $\mathbf{z} \in M^{\perp}$. Hence $\mathbf{z} \in M^{\perp} \cap M^{\perp\perp}$. From (i) we obtain $\mathbf{z} = \mathbf{0}$, which is a contradiction. Thus $M = M^{\perp\perp}$. ∎

Most results of Theorem 6.4.2 must appear obvious to the student, at least with respect to three-dimensional space. In the generalization to abstract Hilbert spaces, however, special note must be taken of the role of the completeness property of the space.

We recall now from our discussion of the algebraic features of a linear space in Chapter 2 how a linear space may be decomposed into the algebraic sum of subspaces. In particular, we recall the result of Exercise 2.12.2, which asserts that the decomposition of the algebraic sum $(M + N)$ of subspaces $M$ and $N$ of a linear space is unique if and only if $M \cap N = \{\mathbf{0}\}$. In what follows, we will be concerned with decomposition of a Hilbert space into orthogonal subspaces. Two subspaces $M$ and $N$ are said to be orthogonal $(M \perp N)$ if every element of one of them is perpendicular to the other subspace.

We prove the following theorem first.

***Theorem 6.4.3*** Let $M$ and $N$ be orthogonal closed linear subspaces in $\mathcal{H}$. $M + N$ is closed.

*Proof*: $M \perp N \rightarrow M \cap N = \{\mathbf{0}\} \rightarrow \forall\, \mathbf{u} \in M + N,\ \exists\ \mathbf{x} \in M,\ \mathbf{y} \in N \ni \mathbf{u} = \mathbf{x} + \mathbf{y}$. Let $\mathbf{z}$ be a limit point of $M + N$. Then $\exists\ \{\mathbf{z}_n\} \subset M + N$ such that $\lim_{n\to\infty} \mathbf{z}_n = \mathbf{z}$. Clearly, $\exists\ \{\mathbf{x}_n\} \subset M, \{\mathbf{y}_n\} \subset N \ni \mathbf{z}_n = \mathbf{x}_n + \mathbf{y}_n$. Now

$$(\mathbf{z}_n - \mathbf{z}_m) = (\mathbf{x}_n - \mathbf{x}_m) + (\mathbf{y}_n - \mathbf{y}_m)$$

From the Pythagorean theorem,

$$\|\mathbf{z}_n - \mathbf{z}_m\|^2 = \|\mathbf{x}_n - \mathbf{x}_m\|^2 + \|\mathbf{y}_m - \mathbf{y}_n\|^2$$

Because $\{\mathbf{z}_n\}$ is a converging sequence, it must be a Cauchy sequence so that $\{\mathbf{x}_n\}$ and $\{\mathbf{y}_n\}$ are also Cauchy sequences. Since $M$ and $N$ are closed, they are Hilbert spaces† and $\lim_{n\to\infty} \mathbf{x}_n \equiv \mathbf{x} \in M$ and $\lim_{n\to\infty} \mathbf{y}_n \equiv \mathbf{y} \in N$. Thus

$$\mathbf{z} \equiv \lim_{n\to\infty} \mathbf{z}_n = \lim_{n\to\infty} (\mathbf{x}_n + \mathbf{y}_n) = \lim_{n\to\infty} \mathbf{x}_n + \lim_{n\to\infty} \mathbf{y}_n$$

or $\mathbf{z} = \mathbf{x} + \mathbf{y}$, so that $\mathbf{z} \in M + N$ and $(M + N)$ is closed. ∎

We are just armed to prove the very important *projection theorem*, which follows.

***Theorem 6.4.4 (The Projection Theorem)*** Let $M$ be a closed linear subspace of Hilbert space $\mathcal{H}$. Then $\mathcal{H} = M + M^{\perp}$.

*Proof*: From Theorem 6.4.2(i), $M^{\perp}$ is closed. Since $M$ and $M^{\perp}$ are closed, Theorem 6.4.3 requires that $M + M^{\perp} \equiv N$ be closed. Now $M \subset N$, so that $N^{\perp} \subset M^{\perp}$ from Theorem 6.4.2(iii). Similarly, $N^{\perp} \subset M^{\perp\perp}$. From Theorem 6.4.2(v), $M = M^{\perp\perp}$ so that $N^{\perp} \subset M$. Thus $N^{\perp} \subset M \cap M^{\perp} = \{\mathbf{0}\}$. Since $N$ is closed $N = N^{\perp\perp} = \{\mathbf{0}\}^{\perp} = \mathcal{H}$, so that $\mathcal{H} = M + M^{\perp}$. ∎

The implication of the foregoing theorem is that a Hilbert space can be decomposed into an algebraic sum of *any* closed linear subspace of it and the orthogonal complement of the subspace. For example, in three-dimensional space, the $z$-axis is a closed linear subspace and its orthogonal complement is the $x$-$y$ plane (see Figure 6.4.1). Indeed, it is well known that a vector in three-dimensional space can be expressed as the sum of a vector along the $z$-axis and a vector on the $x$-$y$ plane. Obviously, one may also inquire into the generalization of the process of expressing a three-dimensional vector as the sum of vectors along the $x$, $y$, and $z$ axes. If $M$ and $N$ are two orthogonal closed linear subspaces of $\mathcal{H}$, then $M \cap N = \{0\}$ and from Theorem 6.4.3, $M + N$ is closed and we may use the projection theorem to assert that $\mathcal{H} = M + N + (M + N)^{\perp}$. It is readily shown‡ that $(M + N)^{\perp} = M^{\perp} \cap N^{\perp}$, so that we have the decomposition

$$\mathcal{H} = M + N + M^{\perp} \cap N^{\perp} \tag{6.4.1}$$

The following definition may be used to generalize (6.4.1).

†See, for example, Exercise 5.1.5.

‡To show that $(M + N)^{\perp} = M^{\perp} \cap N^{\perp}$, first observe that $M \subset M + N$, so that $(M + N)^{\perp} \subset M^{\perp}$. Similarly, $(M + N)^{\perp} \subset N^{\perp}$. Thus $(M + N)^{\perp} \subset M^{\perp} \cap N^{\perp}$. Suppose now that $\mathbf{z} \in M^{\perp} \cap N^{\perp}$. Let $\mathbf{w} \in M + N \ni \mathbf{u} \in M, \mathbf{v} \in N \ni \mathbf{w} = \mathbf{u} + \mathbf{v}$. Since $\langle \mathbf{z}, \mathbf{u} \rangle = \langle \mathbf{z}, \mathbf{v} \rangle = 0$, we have $\langle \mathbf{z}, \mathbf{w} \rangle = 0$, from which $\mathbf{z} \in (M + N)^{\perp}$. Hence $M^{\perp} \cap N^{\perp} \subset (M + N)^{\perp}$, thus proving that $(M + N)^{\perp} = M^{\perp} \cap N^{\perp}$.

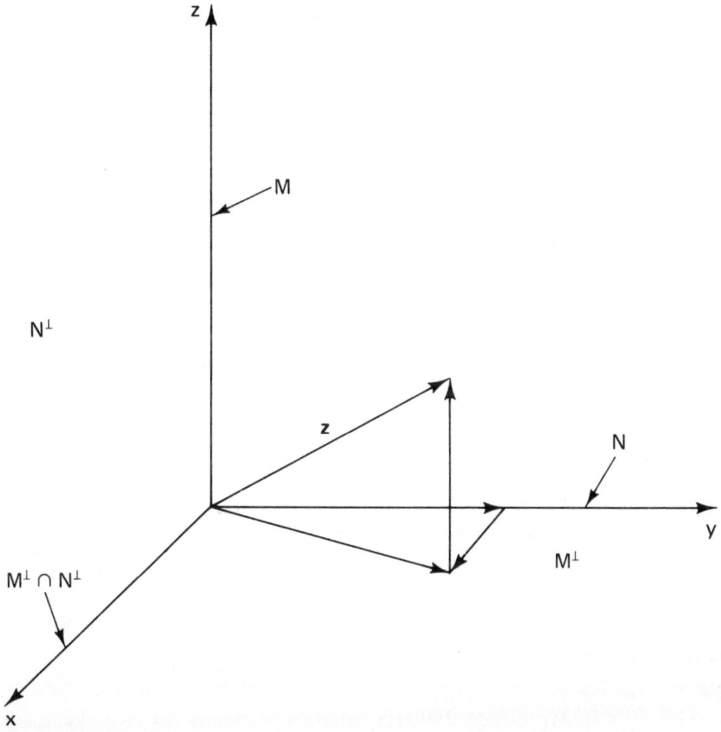


**Figure 6.4.1** Example of decomposition of three-dimensional space as in the projection theorem.

***Definition 6.4.3*** A family of closed linear subspaces $\{M_j\}_{j=1}^n$ of Hilbert space $\mathcal{H}$ is said to be an *orthogonal family* if for $j \neq k$, $M_j \perp M_k$.

The following theorem is readily established and is left as an exercise.

***Theorem 6.4.5*** Let $\{M_j\}_{j=1}^n$ be an orthogonal family of closed linear subspaces of Hilbert space $\mathcal{H}$. Then

$$\mathcal{H} = M_1 + M_2 + \ldots + M_n + \bigcap_{j=1}^{n} M_j^\perp \tag{6.4.2}$$

The decomposition above is readily perceived in three-dimensional space. The vector **z** in Figure 6.4.1 may be expressed of vectors along the $x$-axis (say subspace $M$), the $y$-axis (subspace $N$ perpendicular to $M$), and $z$-axis, which is the *intersection* of $M^\perp$ (the $y$-$z$ plane) and $N^\perp$ (the $x$-$z$ plane). This is precisely the statement of (6.4.1). The extension to higher-dimensional spaces must be evident from (6.4.2) although free from geometrical perception.

## 6.5 Bounded Linear Functionals

The discussion in Section 5.4 on bounded linear functionals on normed linear spaces does apply to inner product spaces. Our interest in reopening this subject matter is spurred by a different consideration. Recall that earlier in this chapter, the inner product of two vectors regarding one of them as fixed was a bounded linear functional of the other (see Exercise 6.1.7). We shall show that for a Hilbert space, *every* bounded linear functional can be written as an inner product. More precisely, given a bounded linear functional $l$ on $\mathcal{H}$, we seek to write

$$l(\cdot) = \langle \cdot, \mathbf{y} \rangle$$

for some $\mathbf{y} \in \mathcal{H}$, called the *representation* of $l$. The existence of a unique such representation is the substance of the following important theorem.

***Theorem 6.5.1 (The Riesz Representation Theorem)*** Let $l$ be a bounded linear functional on $\mathcal{H}$. Then there exists a unique representation $\mathbf{y} \in \mathcal{H}$ for $l$.

*Proof*: We proceed in steps.

If $l \equiv 0$, then the representation is $\mathbf{0}$ since $\langle \mathbf{0}, \mathbf{x} \rangle = 0 \;\forall\; \mathbf{x} \in \mathcal{H}$. The uniqueness is obvious (see Theorem 6.1.1).

If $l \neq 0$, let $M \equiv N(l)$. $M$ is a closed linear subspace (see Section 5.4). Since $N(l)$ is *not* the entire Hilbert space (as when $l = 0$), we know that $\dim M^\perp \geq 1$. We show that the dimension is 1.

Let $\mathbf{x}, \mathbf{y}$ be any two nonzero vectors in $M^\perp$. Clearly, $l(\mathbf{x}) \neq 0$, $l(\mathbf{y}) \neq 0$. The vector $\mathbf{z} = l(\mathbf{y})\mathbf{x} - l(\mathbf{x})\mathbf{y}$ must lie in $M^\perp$. But $l(\mathbf{z}) = l(\mathbf{y})l(\mathbf{x}) - l(\mathbf{x})l(\mathbf{y}) = 0 \rightarrow \mathbf{z} \in M$. Hence $\mathbf{z} = \mathbf{0}$. Thus $\mathbf{x}$ and $\mathbf{y}$ are linearly dependent so that $\dim M^\perp = 1$. Let $\mathbf{u} \in M^\perp \ni \|\mathbf{u}\| = 1$ serve as a basis in $M^\perp$. Now let $\mathbf{x} \in \mathcal{H}$. Then by the projection theorem (Theorem 6.4.4), $\mathcal{H} = M + M^\perp$, so that $\exists\; \mathbf{v} \in M$, $\mathbf{w} \in M^\perp$ such that

$$\mathbf{x} = \mathbf{v} + \mathbf{w} = \mathbf{v} + \lambda\mathbf{u} \qquad (\lambda \text{ a scalar})$$

Now

$$\langle \mathbf{x}, \mathbf{u} \rangle = \langle \mathbf{v}, \mathbf{u} \rangle + \lambda \|\mathbf{u}\|^2 = \lambda$$

Thus

$$\mathbf{x} = \mathbf{v} + \langle \mathbf{x}, \mathbf{u} \rangle \mathbf{u}$$

$$\begin{aligned} l(\mathbf{x}) &= l(\mathbf{v} + \langle \mathbf{x}, \mathbf{u} \rangle \mathbf{u}) = l(\mathbf{v}) + \langle \mathbf{x}, \mathbf{u} \rangle l(\mathbf{u}) \\ &= \langle \mathbf{x}, \mathbf{u} \rangle l(\mathbf{u}) \\ &= \langle \mathbf{x}, l^*(\mathbf{u})\mathbf{u} \rangle \end{aligned}$$

from which the representation of $l$ is clearly $l^*(\mathbf{u})\mathbf{u}$. To show that the representation above is unique, let $\mathbf{y}$ be another representation, that is,

$$l(\mathbf{x}) = \langle \mathbf{x}, \mathbf{y} \rangle \qquad \forall \quad \mathbf{x} \in \mathcal{H}$$

Evidently,

$$0 = \langle \mathbf{x}, \mathbf{y} - l^*(\mathbf{u})\mathbf{u} \rangle \qquad \forall \quad \mathbf{x} \in \mathcal{H}$$

from which

$$\mathbf{y} = l^*(\mathbf{u})\mathbf{u} \quad \blacksquare$$

The importance of this theorem may not seem obvious to the student. However, it will become evident from the next section that the idea of a representation of a bounded linear functional is a crucial one. Note that the Riesz representation theorem is not valid for an incomplete inner product space. The exercises below will establish this result.

## EXERCISES

**6.5.1** In order to show that the representation theorem does not hold for incomplete inner product spaces, consider

$$\mathcal{G} = \{\mathcal{C}[0, 1], \langle \cdot, \cdot \rangle\}$$

where

$$\langle \mathbf{x}, \mathbf{y} \rangle = \int_0^1 x(t)y^*(t)\,dt$$

Define

$$l(\mathbf{x}) \equiv \int_0^{1/2} x(t)\,dt$$

(a) Show that $l$ is a bounded linear functional.

(b) Show that

$$y(t) = \begin{cases} 1, & 0 \leq t < \frac{1}{2} \\ 0, & \frac{1}{2} \leq t \leq 1 \end{cases}$$

is a representation.

(c) Is $\mathbf{y} \equiv \{y(t)\}$ in $\mathcal{C}[0, 1]$?

(d) If you consider the functional $l$ on $\mathcal{H} = \mathcal{L}_2[0, 1]$, how do you reconcile the Riesz representation theorem with the "second" representation of $l$ given by

$$y(t) = \begin{cases} 1, & 0 \leq t \leq \frac{1}{2} \\ 0, & \frac{1}{2} < t \leq 1 \end{cases}$$

**6.5.2** Consider on $\mathcal{L}_2[0, 1]$ the functional

$$l(\mathbf{x}) = x(1)$$

Is this a linear functional? Is it bounded?

## 6.6 Bounded Linear Operators

The subject of linear operators on Hilbert spaces is of focal interest to this book. We will begin with bounded linear operators and examine some of their basic properties. Consequently, we focus attention on a bounded linear transformation $\mathbf{T}$ defined on Hilbert space $\mathcal{H}$.

Let $\mathbf{x}, \mathbf{y} \in \mathcal{H}$. For fixed $\mathbf{y}$, $\langle \mathbf{Tx}, \mathbf{y} \rangle$ is a linear functional on $\mathcal{H}$. Furthermore,

$$\begin{aligned} \|\langle \mathbf{Tx}, \mathbf{y} \rangle\| &\leq \|\mathbf{Tx}\| \|\mathbf{y}\| \\ &\leq \|\mathbf{T}\| \|\mathbf{y}\| \|\mathbf{x}\| \end{aligned}$$

where we have employed the Schwarz inequality. Thus $\langle \mathbf{Tx}, \mathbf{y} \rangle$ is a bounded linear functional of $\mathbf{x}$ for fixed $\mathbf{y}$. Notice that for *each* $\mathbf{y}$, we get a *new* bounded linear functional. The Riesz representation theorem of Section 6.5 guarantees a unique representation, say $\mathbf{y}^*$, for the functional $\langle \mathbf{Tx}, \mathbf{y} \rangle$ for fixed $\mathbf{y}$. Thus we have

$$\langle \mathbf{Tx}, \mathbf{y} \rangle = \langle \mathbf{x}, \mathbf{y}^* \rangle \tag{6.6.1}$$

Since $\mathbf{y}^*$ depends on $\mathbf{y}$, one may define an operator $\mathbf{T}^*$ by†

$$\mathbf{T}^*\mathbf{y} = \mathbf{y}^* \tag{6.6.2}$$

Let $\alpha$ be a complex scalar. The following steps are easily understood. From the above it follows that

$$\mathbf{T}^*(\alpha \mathbf{y}) = \alpha \mathbf{y}^* = \alpha \mathbf{T}^*(\mathbf{y})$$

Similarly, it is readily established that for $\mathbf{y}_1, \mathbf{y}_2 \in \mathcal{H}$,

$$\mathbf{T}^*(\mathbf{y}_1 + \mathbf{y}_2) = \mathbf{T}^*\mathbf{y}_1 + \mathbf{T}^*\mathbf{y}_2$$

so that the operator $\mathbf{T}^*$ is *linear*. Clearly, $\mathbf{T}^*$ depends on $\mathbf{T}$, which prompts the following important definition. We denote the set of bounded linear transformations on $\mathcal{H}$ by Blt($\mathcal{H}$).

***Definition 6.6.1*** Let $\mathbf{T} \in$ Blt($\mathcal{H}$). Then $\mathbf{T}^*$, called its *adjoint* operator, satisfies the relation

$$\langle \mathbf{Tx}, \mathbf{y} \rangle = \langle \mathbf{x}, \mathbf{T}^*\mathbf{y} \rangle \tag{6.6.3}$$

for *every* pair of elements $\mathbf{x}, \mathbf{y} \in \mathcal{H}$.

---

†The asterisk on $\mathbf{T}^*$ may appear to be in conflict with its role in converting scalars to their complex conjugates. However, the two notations will be noted to be indeed harmonious, since switching locations of scalars and operators within the inner product will have the same effect of producing an asterisk on top! $\mathbf{T}^*$ is in a sense a "complex conjugate" of $\mathbf{T}$. See also Exercise 6.6.1.

Note that the existence of $\mathbf{T}^*$ is guaranteed by the Riesz representation theorem and that the definition of (6.6.3) is equivalent to (6.6.2) and (6.6.1). We have seen that $\mathbf{T}^*$ is a linear operator. Also, since $\mathbf{T}^*$ is defined for every $\mathbf{y} \in \mathcal{H}$, $T^*$ is bounded.

Next, we establish that

$$\mathbf{T}^{**} \equiv (\mathbf{T}^*)^* = \mathbf{T}$$

which implies in words that the adjoint of the adjoint of an operator is the operator itself. To show this, let $\mathbf{x}, \mathbf{y} \in \mathcal{H}$ be arbitrary. From Definition 6.6.1 we have

$$\langle \mathbf{T}^*\mathbf{x}, \mathbf{y} \rangle = \langle \mathbf{x}, \mathbf{T}^{**}\mathbf{y} \rangle \tag{6.6.4}$$

But (6.6.3) also implies that

$$\langle \mathbf{T}^*\mathbf{x}, \mathbf{y} \rangle = \langle \mathbf{x}, \mathbf{T}\mathbf{y} \rangle \tag{6.6.5}$$

(why?). Subtracting (6.6.5) from (6.6.4), we obtain

$$0 = \langle \mathbf{x}, (\mathbf{T}^{**} - \mathbf{T})\mathbf{y} \rangle$$

$\forall\, \mathbf{x} \in \mathcal{H}$ so that

$$(\mathbf{T}^{**} - \mathbf{T})\mathbf{y} = \mathbf{0} \qquad \forall \quad \mathbf{y} \in \mathcal{H}$$

which implies that $\mathbf{T}^{**} = \mathbf{T}$. The following theorem establishes that the norm of a bounded linear operator is equal to that of its adjoint.

***Theorem 6.6.1*** Let $\mathbf{T} \in \text{Blt}(\mathcal{H})$. Then $\mathbf{T}^* \in \text{Blt}(\mathcal{H})$ and furthermore, $\|\mathbf{T}\| = \|\mathbf{T}^*\|$.

*Proof*: We have already established that $\mathbf{T}^*$ is linear. Let $\mathbf{x} \in \mathcal{H}$.

$$\begin{aligned}\|\mathbf{T}^*\mathbf{x}\|^2 &= \langle \mathbf{T}^*\mathbf{x}, \mathbf{T}^*\mathbf{x} \rangle = \langle \mathbf{T}\mathbf{T}^*\mathbf{x}, \mathbf{x} \rangle \\ &\le \|\mathbf{T}\mathbf{T}^*\mathbf{x}\|\,\|\mathbf{x}\| \\ &\le \|\mathbf{T}\|\,\|\mathbf{T}^*\mathbf{x}\|\,\|\mathbf{x}\|\end{aligned}$$

Thus

$$\|\mathbf{T}^*\mathbf{x}\| \le \|\mathbf{T}\|\,\|\mathbf{x}\|$$

from which it is clear that

$$\|\mathbf{T}^*\| \le \|\mathbf{T}\|$$

Application of the same inequality to the adjoint of $\mathbf{T}^*$ gives us

$$\|\mathbf{T}\| = \|\mathbf{T}^{**}\| \le \|\mathbf{T}^*\|$$

thus establishing that $\|\mathbf{T}\| = \|\mathbf{T}^*\|$. ∎

We now present some examples of bounded linear operators and their adjoints.

*Example 1*

$$\mathcal{H} = \{\mathcal{C}_n, \langle\cdot, \cdot\rangle\}, \quad \text{where} \quad \langle \mathbf{x}, \mathbf{y}\rangle = \sum_{i=1}^{n} x_i y_i^*.$$

Consider the matrix operator **A** defined by

$$\mathbf{Ax} \equiv \{\sum_{j=1} a_{ij}x_j, i = 1, 2, \ldots, n\}$$

Let $\mathbf{y} \equiv (y_1, y_2, \ldots, y_n) \in \mathcal{H}$.

$$\langle \mathbf{Ax}, \mathbf{y}\rangle = \sum_{i=1}^{n} y_i^* \sum_{j=1}^{n} a_{ij}x_j = \sum_{j=1}^{n} x_j \sum_{i=1}^{n} (a_{ij}^* y_i)^*$$

$$= \sum_{i=1}^{n} x_i \left(\sum_{j=1}^{n} a_{ji}^* y_j\right)^* \equiv \langle \mathbf{x}, \mathbf{A}^*\mathbf{y}\rangle$$

Thus

$$\mathbf{A}^*\mathbf{y} = \left\{\sum_{j=1}^{n} a_{ji}^* y_j, j = 1, 2, \ldots, n\right\}$$

which means that the adjoint of a matrix operator is obtained as a matrix whose element in the $i$th row and $j$th column is the complex conjugate of the element in the $j$th row and the $i$th column of the original matrix. For example, in $\mathcal{C}_2$ the matrix operator $\begin{bmatrix} a_{11} & a_{12} \\ a_{21} & a_{22} \end{bmatrix}$ has as its adjoint the matrix operator $\begin{bmatrix} a_{11}^* & a_{21}^* \\ a_{12}^* & a_{22}^* \end{bmatrix}$. In particular, it should be noticed that the adjoint of a *real* matrix operator is the *transpose* of the matrix.

*Example 2*

Consider $\mathcal{H} = \mathcal{L}_2[a, b]$ and the integral operator **K** defined by

$$\mathbf{Kx} = \left\{\int_a^b K(t, s)x(s)\, ds\right\}$$

Let $\mathbf{y} \in \mathcal{H}$.

$$\langle \mathbf{Kx}, \mathbf{y}\rangle = \int_a^b y^*(t)\left[\int_a^b K(t, s)x(s)\, ds\right] dt$$

$$= \int_a^b K(t, s)y^*(t)\, dt \int_a^b x(s)\, ds$$

$$= \int_a^b \left[\int_a^b K^*(s, t)y(s)\, ds\right]^* x(t)\, dt$$

$$\equiv \langle \mathbf{x}, \mathbf{K}^*\mathbf{y}\rangle$$

where

$$\mathbf{K}^*\mathbf{y} = \left\{\int_a^b K^*(s, t)y(s)\, ds\right\}$$

Thus the adjoint of the integral operator **K** is given by the kernel function obtained by taking the complex conjugate of the original kernel function with its arguments interchanged. To take an example, the operator **K** which has the kernel $K(t, s) = t(s^2 + it)$ has an adjoint operator $\mathbf{K}^*$ whose kernel is given by $s(t^2 - is)$. If the kernel is real valued, then the adjoint operator has simply the arguments interchanged in the kernel. Thus the operator with the kernel $K(t, s) = ts^2$ has an adjoint operator $\mathbf{K}^*$ with the kernel $st^2$.

The concept of an adjoint operator is indeed a very deep one. The operators most important to this book are the class of *self-adjoint* operators. Before we consider them, the following exercises are recommended. The transformations that appear therein are bounded and defined on a Hilbert space $\mathcal{H}$ and no special qualification to that effect will be made.

## EXERCISES

**6.6.1** Show that

(a) $(\alpha\mathbf{T})^* = \alpha^*\mathbf{T}^*$.

(b) $(\mathbf{T}_1 + \mathbf{T}_2)^* = \mathbf{T}_1^* + \mathbf{T}_2^*$.

(c) Every operator **T** may be written as

$$\mathbf{T} = \mathbf{T}_R + i\mathbf{T}_I \qquad (i = \sqrt{-1})$$

by identifying operators $\mathbf{T}_R$ and $\mathbf{T}_I$. Show further that $\mathbf{T}_R^* = \mathbf{T}_R$, $\mathbf{T}_I^* = \mathbf{T}_I$.

(d) $(\mathbf{T}_1\mathbf{T}_2)^* = \mathbf{T}_2^*\mathbf{T}_1^*$; hence $(\mathbf{T}\mathbf{T}^*)^* = \mathbf{T}\mathbf{T}^*$, and $(\mathbf{T}^*)^n = (\mathbf{T}^n)^*$, $n = 1, 2, \ldots$.

(e) If $\mathbf{T}^{-1}$ exists and is bounded, $(\mathbf{T}^*)^{-1}$ exists and equals $(\mathbf{T}^{-1})^*$.

**6.6.2** Consider the Hilbert space of Exercise 6.1.1. Show that the matrix operator **A** with matrix coefficients $\{a_{ij};\ i, j = 1, 2, \ldots, n\}$ has an adjoint operator $\mathbf{A}^*$ with matrix coefficients

$$\left\{a_{ji}^* \frac{r_j}{r_i};\ i, j = 1, 2, \ldots, n\right\}$$

**6.6.3** Consider the Hilbert space of Exercise 6.1.2. Show that the matrix operator **T** with matrix coefficients $\{t_{ij};\ i, j = 1, 2, \ldots, n\}$ has an adjoint operator with matrix coefficients

$$\left\{\sum_{l=1}^{n} a'_{il} \sum_{k=1}^{n} t_{kl}^* a_{kj},\ i, j = 1, 2, \ldots, n\right\}$$

where $a'_{il}$ is the coefficient in the $i$th row and $l$th column of the matrix $\mathbf{A}^{-1}$.

**6.6.4** For the integral operator **K** with a kernel function $K(t,s)$ defined on the Hilbert space of Exercise 6.1.4, identify the kernel of the adjoint operator.

**6.6.5** Refer back to Section 2.11 on matrix representation of linear operators following Theorem 2.11.3. Let **T** be a linear operator on $\mathcal{H}$ and $\{\mathbf{z}_j\}$ be an orthonormal basis in $\mathcal{H}$.

(a) Obtain the matrix representation of **T** relative to the basis above by identifying the element in the $i$th row and $j$th column.

(b) Let **T** be bounded and **T*** be its adjoint. Find the relationship between the matrix representations of **T** and **T*** relative to the given basis.
(c) For $\mathbf{x} \in \mathcal{H}$, obtain the quadratic form $\langle \mathbf{Tx}, \mathbf{x} \rangle$ in terms of the matrix representation of **T** and the vector in $l_2$ which corresponds to **x**. What is your conclusion?

**6.6.6** Let **T** be a compact operator on $\mathcal{H}$.
(a) Show that **TT*** and **T*T** are compact.
(b) Show that **T*** is compact.
[*Hint:* Note that part (a) can be proved without part (b). To prove part (b), let $\{\mathbf{x}_n\}$ be a bounded sequence in $\mathcal{H}$ and $\{\mathbf{x}_{k_n}\}$ be a subsequence for which $\{\mathbf{TT}^*_{x_{k_n}}\}$ converges. Then show that for $k_m$, $k_n$ sufficiently large,

$$\lim |< (\mathbf{x}_{k_m} - \mathbf{x}_{k_n}), \mathbf{TT}^*(\mathbf{x}_{k_m} - \mathbf{x}_{k_n}) > | = 0.]$$

**6.6.7** Consider the Hilbert space $l_2$ and the "shift" operator **T** defined by

$$\mathbf{T}(x_1, x_2, \ldots) = (x_2, x_3, \ldots)$$

Find the adjoint operator **T***.

**6.6.8** Let $\mathcal{H}_1$ and $\mathcal{H}_2$ be two Hilbert spaces, $\mathbf{T}_1 \in \text{Blt}(\mathcal{H}_1)$ and $\mathbf{T}_2 \in \text{Blt}(\mathcal{H}_2)$. Further, let $\mathbf{T} \equiv \mathbf{T}_1 \oplus \mathbf{T}_2 \in \text{Blt}(\mathcal{H}_1) \oplus, \text{Blt}(\mathcal{H}_2)$. Show that $\mathbf{T}^* = \mathbf{T}_1^* \oplus \mathbf{T}_2^*$. Similarly, if $\mathbf{T} \equiv \mathbf{T}_1 \otimes \mathbf{T}_2 \in \text{Blt}(\mathcal{H}_1) \otimes \text{Blt}(\mathcal{H}_2)$, show that $\mathbf{T}^* = \mathbf{T}_1^* \otimes \mathbf{T}_2^*$.

In Theorem 6.6.1 we found that the norm of a bounded linear operator is equal to the norm of its adjoint operator; that is, $||\mathbf{T}|| = ||\mathbf{T}^*||$. Indeed, this equality does not require that **T** and **T*** be the same but the operators for which $\mathbf{T} = \mathbf{T}^*$ form a special class of what are called *self-adjoint* operators.† The subject of prime concern to this work is the study of self-adjoint operators which have powerful properties that can be used to solve equations involving them. It turns out that in many applications the operators involved in the equations describing the physical process are self-adjoint. The behavior of non-self-adjoint operators (i.e., where $\mathbf{T} \neq \mathbf{T}^*$) can be considerably more varied and complicated.

In Chapter 5 we had encountered the compact or completely continuous operator (see Section 5.3). Since compact operators are bounded, their adjoints exist. Thus compact self-adjoint operators will be of special interest to us. The student is bound to feel at this stage that altogether too many different types of operators have been presented at various stages and that some focusing would be of value. We provide a schematic perspective of the various operators that we have covered, emphasizing those that are of direct interest to us at the end of this chapter. For the present, we proceed on to discuss *unbounded* operators for which the discussion of Section 6.6 in regard to the existence of the adjoint will not apply. We strongly recommend the following exercises before progressing to the next section.

†The term "Hermitian" is sometimes used in place of "self-adjoint," although the latter is far more common.

**EXERCISES**

**6.6.9** Let $S \equiv \{\mathbf{T} \in \text{Blt}(\mathcal{H}): \mathbf{T} = \mathbf{T}^*\}$. Is $S$ a subspace of $\text{Blt}(\mathcal{H})$?

**6.6.10** Identify the decomposition of $\mathbf{T}$ in Exercise 6.6.1(c) when $\mathbf{T}$ is self-adjoint.

**6.6.11** Let $f(\lambda)$ be a real-valued analytic function of its real argument $\lambda$, with the series representation

$$f(\lambda) = \sum_{n=0}^{\infty} a_n \lambda^n$$

which converges uniformly for $\lambda$ in any finite interval (example: $f(\lambda) = e^{\lambda}$). Show that if $\mathbf{T} = \mathbf{T}^*$, the operator $f(\mathbf{T})$ is self-adjoint.

**6.6.12** Identify the necessary and sufficient conditions for each of the operators in Exercises 6.6.2 to 6.6.4 to be self-adjoint.

**6.6.13** Show that the direct sum of two self-adjoint operators is self-adjoint. Show also that their tensor product is self-adjoint.

**6.6.14** In Exercise 6.6.8 only operators from $\text{Blt}(\mathcal{H}_1) \oplus \text{Blt}(\mathcal{H}_2)$ and $\text{Blt}(\mathcal{H}_1) \otimes \text{Blt}(\mathcal{H}_2)$ were considered. In Chapter 2 (Section 2.11), however, we learned about operators from $\text{Blt}(\mathcal{H}_1 \oplus \mathcal{H}_2)$.† Let $\mathbf{T} \in \text{Blt}(\mathcal{H}_1 \oplus \mathcal{H}_2)$ be represented by

$$\begin{bmatrix} \mathbf{T}_{1,1} & \mathbf{T}_{1,2} \\ \mathbf{T}_{2,1} & \mathbf{T}_{2,2} \end{bmatrix}$$

Derive the necessary and sufficient conditions for $\mathbf{T}$ to be self-adjoint. Show that the conditions above are sufficient to assure the self-adjointness of $\mathbf{T} \in \text{Blt}(\mathcal{H}_1 \otimes \mathcal{H}_2)$.

## 6.7 Unbounded Operators

Our interest in unbounded operators stems from the importance of differential equations and associated boundary value problems in engineering applications. Some features of unbounded operators have been discussed in Section 5.5. It was observed therein that an unbounded operator can be defined at most only on a dense subspace of a Banach space. Obviously, this constraint must also apply to unbounded operators on Hilbert spaces. Thus the domain on which the operation is defined can at most be a dense subspace of the Hilbert space.

Recall that an unbounded operator $\mathbf{T}$ consists of a formal operation $T$ and a domain $D(T)$ which, in the present context, is a dense subspace of Hilbert space $\mathcal{H}$. Thus recapitulate that the same formal operation $T$ may

†Although the concept of boundedness could not be mentioned then, we did introduce the space of operators $\mathfrak{J}(\mathcal{L}_1 \oplus \mathcal{L}_2)$, where $\mathcal{L}_1$ and $\mathcal{L}_2$ are linear spaces. When applied to Hilbert spaces $\mathcal{H}_1$ and $\mathcal{H}_2$, $\text{Blt}(\mathcal{H}_1 \oplus \mathcal{H}_2)$ is a subspace of $\mathfrak{J}(\mathcal{H}_1 \oplus \mathcal{H}_2)$. The same applies to tensor product spaces.

obtain many different operators **T** depending on the domain of definition $D(T)$. The formal operation $T$ of interest to us is a differential expression such as those considered in Section 5.5 and the domain $D(T)$ is specified by *homogeneous boundary conditions* and certain other conditions that guarantee that the operation $T$ transforms $D(T)$ *onto* the Hilbert space $\mathcal{H}$. The reader is again urged to go through the examples in Section 5.5 and solve the exercises therein.

### 6.7.1 Adjoint of an Unbounded Operator

In defining the adjoint of an unbounded operator, we are unable to employ the Riesz representation theorem to assert its existence since the domain of the operator is not a Hilbert space.† We must therefore take an alternative route to defining the adjoint operator. Indeed, not withstanding how the adjoint operator $T^*$ is defined, we expect it to be unbounded so that we may write

$$\mathbf{T}^* = \{T^*, D(T^*)\}$$

write $T^*$ is the formal operation of the adjoint operator and $D(T^*)$ the domain of operation. Next we define the adjoint operator.

***Definition 6.7.1*** Let $\mathcal{H}$ be a Hilbert space and $\mathbf{T} \equiv \{T, D(T)\}$ be an unbounded operator. The adjoint operator $\mathbf{T}^* \equiv \{T^*, D(T^*)\}$ satisfies the relationship

$$\langle T\mathbf{x}, \mathbf{y}\rangle = \langle \mathbf{x}, T^*\mathbf{y}\rangle \qquad \forall \quad \mathbf{x} \subset D(T), \mathbf{y} \subset D(T^*)$$

or equivalently,

$$\langle \mathbf{Tx}, \mathbf{y}\rangle = \langle \mathbf{x}, \mathbf{T}^*\mathbf{y}\rangle \tag{6.7.1}$$

Notice particularly in (6.7.1) that the use of boldface operator symbols makes it unnecessary to qualify the domains from which **x** and **y** arise.

Having made the foregoing definition, it is desirable to spotlight on some simplifications in our treatment here. First, let us recognize that depending on how $D(T)$ is defined, there could be many adjoint operators since more than one $D(T^*)$ could be identified. We should consider some examples to make this clear.

---

†In this regard, there are two important redeeming features, the detailed discussion of which is outside the scope of this book. These are (1) the *denseness* of the domain $D(T)$ and (2) the *closedness* of the operator (differential operators have this property). The property of closedness received only a footnote allusion in Section 5.5. The interested reader is referred to *Linear Operators, Part II*, by N. Dunford and J. T. Schwartz, Interscience, New York, 1963, pp. 1185–1186, for a more thorough discussion on these aspects.

*Example 1*

Let $\mathcal{H} = \mathcal{L}_2[0, 1]$ and $T \equiv \{T, D(T)\}$, where

$$T \equiv \frac{d}{dt} \quad \text{and} \quad D(T) = [\mathbf{x} \in \mathcal{H}: x(0) = 0, T\mathbf{x} \in \mathcal{H}]$$

We shall assume without proof† that the formal operation of the adjoint is given by $T^* \equiv -d/dt$. It is readily seen that

$$\begin{aligned}\langle T\mathbf{x}, \mathbf{y}\rangle - \langle \mathbf{x}, T^*\mathbf{y}\rangle &= \int_0^1 \left[\frac{dx}{dt}y + x\frac{dy}{dt}\right] dt \\ &= x(1)y(1) - x(0)y(0)\end{aligned}$$

If $\mathbf{x} \in D(T)$, then

$$\langle \mathbf{T}\mathbf{x}, \mathbf{y}\rangle - \langle \mathbf{x}, T^*\mathbf{y}\rangle = x(1)y(1) \tag{6.7.2}$$

The choice of $D(T^*)$ would be with a view to validate (6.7.1) in Definition 6.7.1, which in the present case is clearly accomplished by requiring that $y(1) = 0$. Thus one could define

$$D_1(T^*) = \{\mathbf{y} \in \mathcal{H}: y(1) = 0; T^*\mathbf{y} \in \mathcal{H}\} \tag{6.7.3}$$

which is a dense linear subspace of $\mathcal{L}_2[0, 1]$. On the other hand, one could define a second domain (also a dense subspace) such as

$$D_2(T^*) = \{\mathbf{y} \in \mathcal{H}: y(1) = 0, y(0) = 0, T^*\mathbf{y} \in \mathcal{H}\} \tag{6.7.4}$$

Thus we have produced *two* adjoint operators and clearly there are many more. Notice, however, that $D_2(T^*) \subset D_1(T^*)$, so that the domain $D_1(T^*)$ is the *largest* domain which validates (6.7.1) by allowing the right-hand side of (6.7.2) to vanish.

The foregoing example shows how in defining an adjoint operator the domain $D(T^*)$ could be made unique by picking the largest subspace. No special effort need be made to identify such a domain since it is the one that results naturally without overconstraining functions as in (6.7.4). We consider another example below which shows that the boundary conditions characterizing $D(T^*)$ may not be unique, although the latter itself is unique.

*Example 2*

Let $\mathcal{H} = \mathcal{L}_2[0, 1]$. $\mathbf{T} = \{T, D(T)\}$ where $T \equiv d^2/dt^2$.

$$D(T) = \{\mathbf{x} \in \mathcal{H}: \mathbf{x}(0) = 0; x'(0) = x(1); T\mathbf{y} \in \mathcal{H}\}$$

---

†The formal operations of adjoints of differential operators will be taken up in Chapter 9.

As in Example 1, we assume without proof that the formal operation of the adjoint is $T^* = d^2/dt^2$. (Note that $T$ and $T^*$ are identical.) Now

$$\langle T\mathbf{x}, \mathbf{y}\rangle - \langle \mathbf{x}, T^*\mathbf{y}\rangle = \int_0^1 \left(\frac{d^2x}{dt^2}y - x\frac{d^2y}{dt^2}\right) dt$$

$$= \int_0^1 \frac{d}{dt}\left[\frac{dx}{dt}y - x\frac{dy}{dt}\right] dt = \left[\frac{dx}{dt}y - x\frac{dy}{dt}\right]_0^1$$

$$= x'(1)y(1) - x(1)y'(1) - x'(0)y(0) + x(0)y'(0)$$

$$\rightarrow \langle \mathbf{Tx}, \mathbf{y}\rangle - \langle \mathbf{x}, T^*\mathbf{y}\rangle = x'(1)y(1) - x'(0)[y'(1) + y(0)] \qquad (6.7.5)$$

where consistent with the use of the boldface symbol $\mathbf{T}$ we have put $\mathbf{x} \in D(T)$. The right-hand side of (6.7.5) can be made to vanish for arbitrary $\mathbf{x}$ by requiring that $y(1) = 0$, $y'(1) + y(0) = 0$. The domain

$$D(T^*) \equiv \{\mathbf{y} \in \mathcal{H}\colon y(1) = 0;\, y'(1) + y(0) = 0;\, T^*\mathbf{y} \in \mathcal{H}\} \qquad (6.7.6)$$

is the unique domain of $T^*$ so that the unique adjoint operator is $\mathbf{T}^* = \{T^*, D(T^*)\}$.

An interesting point to note is that the boundary conditions above for $T^*$ are by themselves *not* unique. For example, we could use instead of those in (6.7.6)

$$\begin{aligned} \alpha y(1) + \beta[y'(1) + y(0)] &= 0 \\ \gamma y(1) + \delta[y'(1) + y(0)] &= 0 \end{aligned} \qquad (6.7.7)$$

which also imply the boundary conditions in (6.7.6) as long as $\alpha\delta \neq \beta\gamma$. [Note that the right-hand side of (6.7.5) would not vanish if $\alpha\delta \neq \beta\gamma$.] Although the boundary conditions (6.7.7) appear different, insofar as they imply the boundary conditions in (6.7.6), $D(T^*)$ is unique.

We have thus been able to see how the adjoint operator of an unbounded operator may arise. Note particularly how in Example 2 we had $T = T^*$ but $\mathbf{T} \neq \mathbf{T}^*$ since their domains do not coincide. In this connection we introduce for convenience the notation

$$D_T \equiv D(T), \qquad D_T^* \equiv D(T^*) \qquad (6.7.8)$$

for the case $T = T^*$, so that one may talk about distinct operators $\mathbf{T}$ and $\mathbf{T}^*$ by recognizing that $D_T \neq D_T^*$.

The following definition suggests itself.†

---

†We wish to point to some subtleties here. It is possible to have a formal operation $T$ (whose adjoint operation is also $T$) such that $\langle T\mathbf{x}, \mathbf{y}\rangle = \langle \mathbf{x}, T\mathbf{y}\rangle$ for the *same* boundary conditions in $\mathbf{x}$ and $\mathbf{y}$ except that $\mathbf{x}$ may satisfy additional boundary conditions. In such a case $D(T)$ is a proper subspace of $D(T^*)$, so that $\mathbf{T} \neq \mathbf{T}^*$. Such

***Definition 6.7.2*** An unbounded operator $\mathbf{T} \equiv \{T, D(T)\}$ is said to be *self-adjoint* when $\mathbf{T} = \mathbf{T}^*$, that is, $T = T^*$ and $D(T) = D(T^*)$, or $D_T = D_T^*$. We consider below some examples of self-adjoint operators.

*Example 3*

Let $\mathcal{H} = \mathcal{L}_2[0, 1]$ and $\mathbf{T} = \{T, D(T)\}$, where $T = i(d/dt)$ and $D(T) = \{\mathbf{u} \in \mathcal{H} : u(0) = u(1); T\mathbf{u} \in \mathcal{H}\}$. Again we postpone to the general treatment of differential operators the proof of the fact that $T = T^*$. It is readily verified that

$$\langle T\mathbf{x}, \mathbf{y}\rangle - \langle \mathbf{x}, T\mathbf{y}\rangle = \int_0^1 i\left[\frac{dx}{dt}y + x\frac{dy}{dt}\right] dt$$

$$= i\left[xy\right]_0^1 = i[x(1)y(1) - x(0)y(0) = 0$$

when $\mathbf{x}, \mathbf{y} \in D(T)$. Thus $T$ is self-adjoint. Although $\mathcal{H}$ is a complex Hilbert space, we have used only real elements of $\mathcal{H}$ for the demonstration (as was also the case in Examples 1 and 2). The condition of symmetry above is readily established for complex elements. However, the student must remember the definition of the inner product for complex elements since the differential operator is complex (imaginary) here.

*Example 4*

Another example of a self-adjoint operator in the Hilbert space of Example 3 is given by

$$T \equiv \frac{d^2}{dt^2}, \qquad D(T) = \{\mathbf{x} \in \mathcal{H} : x(0) = x(1); x'(0) = x'(1); T\mathbf{x} \in \mathcal{H}\}$$

---

operators are called *symmetric*. An example of such an operator is

$$\mathbf{T} \equiv \left\{T \equiv \frac{d^2}{dt^2}, D(T)\right\}$$

$$D(T) = \{\mathbf{x} \in \mathcal{L}_2[0, 1]: x(0) = x'(0) = x(1) = 0; T\mathbf{x} \in \mathcal{L}_2[0, 1]\}$$

For the operator above $T^* = T$, and adopt the domain

$$D(T^*) = \{\mathbf{y}: \mathcal{L}_2[0, 1]; y(0) = y(1) = 0\}$$

Clearly, $D(T) \subset D(T^*)$ and $\langle T\mathbf{x}, \mathbf{y}\rangle = \langle \mathbf{x}, T\mathbf{y}\rangle, \forall\, \mathbf{x}, \mathbf{y} \in D(T^*)$. By definition, then, $\mathbf{T}$ is a symmetric operator. In this case one recognizes that the boundary condition $x'(0) = 0$ is irrelevant and may be "dumped." The result is a *self-adjoint extension* of the original operator $\mathbf{T}$ to the self-adjoint operator $\{T, D(T^*)\}$. In the case above it is not unique. (What are the other extensions possible?) As long as we do not consider such "redundant" boundary conditions, we do not have to worry about such subtleties in this book. However, the subtleties themselves are of importance to the development of the abstract theory of unbounded operators.

$$\begin{aligned}\langle \mathbf{Tx}, \mathbf{y}\rangle - \langle \mathbf{x}, \mathbf{Ty}\rangle &= \Big[x'(t)y(t) - x(t)y'(t)\Big]_0^1 \\ &= x'(1)y(1) - x(1)y'(1) - x'(0)y(0) + x(0)y'(0) \\ &= x'(1)[y(1) - y(0)] - x(1)[y'(1) - y'(0)] \\ &= 0\end{aligned}$$

In Chapter 9 we will provide a more complete discussion of differential operators of interest to engineering applications.

**EXERCISE**

**6.7.1** Let $\mathcal{H} = \mathcal{L}_2[0, 1]$, $T = d^2/dt^2$, assuming that $T^* = d^2/dt^2$. Determine which of the following operators are self-adjoint. Where they are not self-adjoint, find the adjoint operator.

The domain $D(T)$ includes $\mathbf{u} \in \mathcal{H}$, $T\mathbf{u} \in \mathcal{H}$, and

(a) $u(0) + u(1) = 0, \quad u'(1) + u'(0) = 0.$
(b) $u(0) = 0, \quad u'(1) + u'(0) = 0.$
(c) $u(0) = u(1), \quad u'(1) + u'(0) = 0.$
(d) $u(0) = \frac{1}{2}u(1), \quad u'(0) = 0.$
(e) $u'(0) = u(1), \quad u'(1) = u(0).$
(f) $u(0) + u(1) + 2u'(0) - 3u'(1) = 0, \quad 2u(0) - 3u(1) + u'(0) + u'(1) = 0.$

## 6.8 Concluding Remarks

In this chapter we have laid out the generalization of ordinary three-dimensional Euclidean space in the abstract setting which was promised earlier. Much of the student's intuitive feeling for geometric interpretations of vector algebra originates from those in Euclidean space, where there is always room for a triplet of rectilinear Cartesian coordinates. The three-dimensional Hilbert space comprising $\mathcal{R}_3$ and the inner product defined by (6.1.13) is exactly the view of ordinary space through a Cartesian frame of reference. In a detailed sense, our generalizations have occurred in many different directions. For example, we have generalized the concept of a *real* space to a *complex* space. In another direction, the real three-dimensional space has been generalized to a finite $n$-dimensional space and on to the infinite dimensional (real $l_2$-space); continuing further we arrived at the infinite-dimensional "function space," although it turned out interestingly enough that the separable spaces from this class were essentially renaming of the $l_2$-space elements. In a third direction, the generalization included *general inner products*,† essentially producing non-Euclidean spaces. The student must

†The inner products presented do not embrace those encountered in dealing with the space-time continuum of the special or general theory of relativity at least superficially.

consider the generalizations in *all* directions extremely important in regard to applications. The scheme of generalizations just recounted above has been presented in Figure 6.8.1.

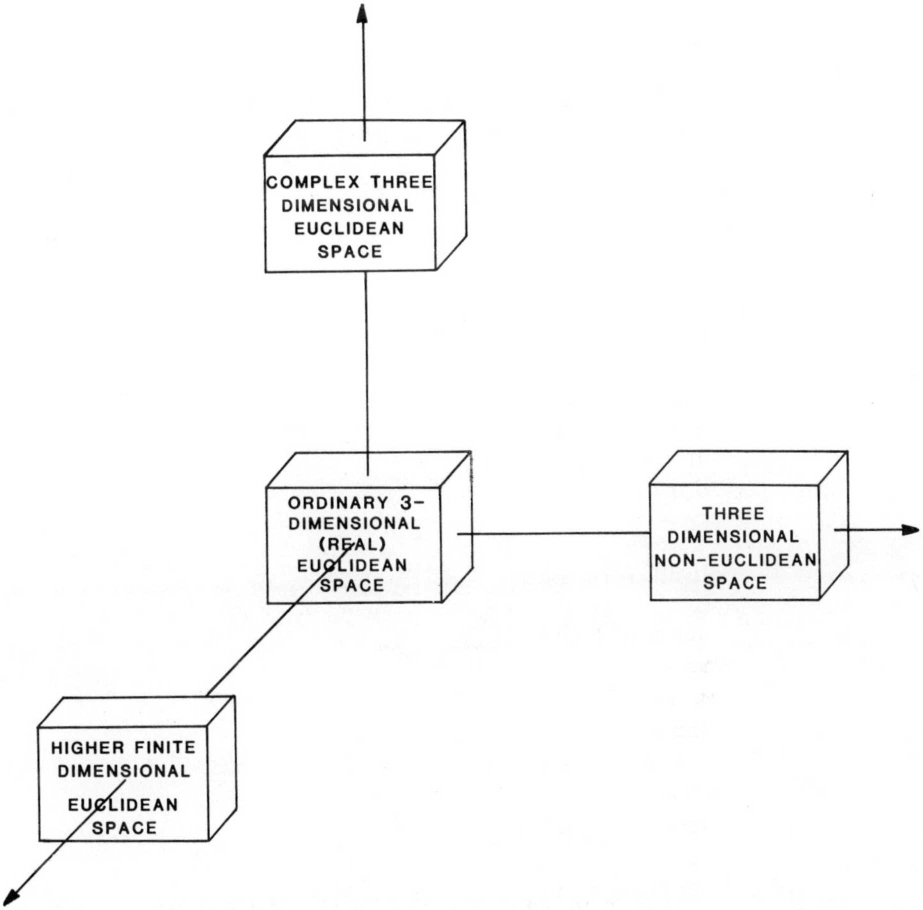


**Figure 6.8.1** Generalizations of inner product spaces.

Next, the concept of "compartmentalizing" space into orthogonal subspaces was introduced for which the projection theorem (Theorem 6.4.4) was the seeding source. How the subspaces are to be selected was not at issue here. That such a "view" of a Hilbert space was possible constituted the main point. The next chapter addresses this problem more deeply.

Finally, the subject of bounded and unbounded operators on a Hilbert space occupied the subsequent portions of this chapter. A crucial concept was that of an adjoint operator. The self-adjoint operator has the most important implication to this book. In fact, that is what it is about and all others are in service of this goal! Self-adjoint operators occur commonly in engineering problems—sometimes in the most recognizable form but frequently in subtle disguise. It is the "unmasking" of those in disguise that requires the generalizations that have occupied us thus far. But why self-adjoint operators? The answer to this is the objective of the next chapter. Here we merely observe that they have powerful properties which can be used to solve equations involving them. We did not refer in the text to the generalization of a self-adjoint operator to a *normal operator* which is defined as that for which $\mathbf{TT^*} = \mathbf{T^*T}$; that is, it commutes with its adjoint. The normal operator has properties almost as powerful as those of a self-adjoint operator. We will have more to say about them in Chapter 7.

In dealing with operators, we have come across several kinds. It will be desirable to take stock of these at this stage and put them in perspective, reflecting the rarity (or abundance) of each kind. Figure 6.8.2 does this schematically. The compact self-adjoint operator in this schematic is the

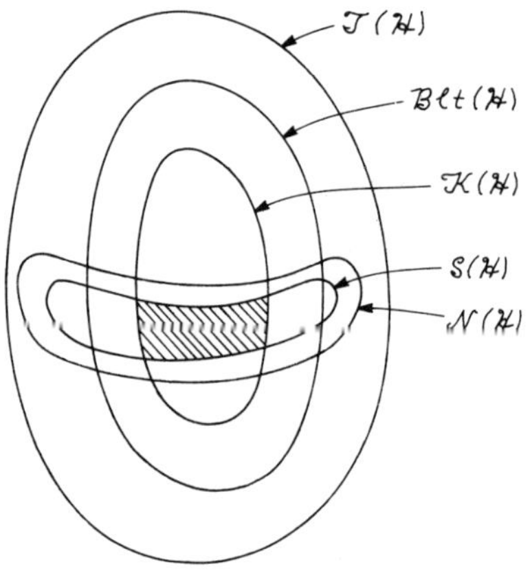


**Figure 6.8.2** Linear operators on Hilbert space.

rarest kind. This is the only operator that we can deal with with some rigor within the scope of this book and its territory appears shaded in Figure 6.8.2. It is fortunate that many important problems fall within this category!

## FURTHER READING

For a slightly more advanced treatment of inner product spaces, the reader is referred to Chapter 5 of

*Linear Operator Theory in Engineering and Science*, by A. W. Naylor and G. R. Sell, Holt, Rinehart and Winston, New York, 1971.

An excellent and extremely comprehensive treatment of linear operators is available in the three-part series of

*Linear Operators*, by N. Dunford and J. T. Schwartz, Wiley-Interscience, New York, Part I, 1958; Part II, 1963; Part III, 1971.

The book is not easily read by engineers but the treatment contains motivating discussions concerning many concepts and definitions and in regard to applications, which some writers find as intruding discontinuities in a logical treatment.

# *Spectral Theory of Self-Adjoint Operators* 7

## 7.0 Foreword

It is our objective here to unearth just what it is that makes a self-adjoint operator so attractive. In Chapter 6 we encountered the important concept of compartmentalizing a Hilbert space into orthogonal subspaces. This provided a certain "view" of the space in much the same way as that of real physical space which one gets, for example, through a Cartesian or any other coordinate frame of reference. Of course, there are infinitely many such views, each providing a different perspective of the space in question. It turns out that, given a self-adjoint operator, its properties generate a unique and most purposeful view of the space on which it acts. Thus the self-adjoint operator permits a compartmentalization of the space into subspaces in which the action of the operator occurs in the simplest manner; that is, the vectors in any subspace are transformed in a very simple way. The nature of this simplicity will become evident from the sections to follow. Now how does this information help solve an equation involving the self-adjoint operator? Suppose that one is interested in solving an equation of the type $\mathbf{Tu} = \mathbf{f}$, where $\mathbf{T}$ is self-adjoint, $\mathbf{f} \in \mathcal{H}$ is known, and $\mathbf{u} \in \mathcal{H}$ is to be determined. In the view of $\mathcal{H}$ (or compartmentalization) which $\mathbf{T}$ generates, $\mathbf{f}$ appears in "pieces" distributed among the subspaces. The piece of $\mathbf{f}$ in each subspace readily leads to the piece of the unknown vector $\mathbf{u}$; the pieces of $\mathbf{u}$ are then put together to obtain $\mathbf{u}$!

## 7.1 Introduction

We begin by laying out the properties of a bounded self-adjoint operator on a Hilbert space $\mathcal{H}$, which will finally lead to the most important theorem of this chapter: the *spectral theorem.* Our emphasis is on the *compact* self-adjoint operator, and the connected theory is presented rigorously. The theory of noncompact self-adjoint operators is considerably more complicated and only a highly simplified treatment is possible within the scope of this book. However, of particular importance to us is a class of unbounded self-adjoint operators whose *inverse* operators are compact and self-adjoint. Consequently, the compact operator theory becomes applicable for this case. Many applications in Chapter 9 fall within this category.

## 7.2 Bounded Self-Adjoint Operators. Some Properties

In the following discussion, it is understood that we are concerned with bounded operators on a complex Hilbert space $\mathcal{H}$. We will establish some preliminary properties that are useful in establishing the spectral theorem. We begin with

***Theorem 7.2.1*** Let $\mathbf{T} \in \text{Blt}(\mathcal{H})$. Then $\mathbf{T} = \mathbf{T}^*$ if and only if $\langle \mathbf{Tx}, \mathbf{x} \rangle$ is *real-valued* $\forall\ \mathbf{x} \in \mathcal{H}$.

*Proof*: First we observe that $\langle \mathbf{Tx}, \mathbf{x} \rangle$ is real-valued if and only if $\langle \mathbf{Tx}, \mathbf{x} \rangle = \langle \mathbf{Tx}, \mathbf{x} \rangle^*$. To prove the *only if* part, we assume that $\mathbf{T} = \mathbf{T}^*$. For $\mathbf{x} \in \mathcal{H}$, $\langle \mathbf{Tx}, \mathbf{x} \rangle = \langle \mathbf{x}, \mathbf{Tx} \rangle = \langle \mathbf{Tx}, \mathbf{x} \rangle^*$, so that $\langle \mathbf{Tx}, \mathbf{x} \rangle$ is real. To prove the *if* part, assume that $\langle \mathbf{Tx}, \mathbf{x} \rangle$ is real for every $\mathbf{x} \in \mathcal{H}$. Let $\mathbf{x}, \mathbf{y} \in \mathcal{H}$. We leave the following identity to be verified by the student.

$$\begin{aligned} 4\langle \mathbf{Tx}, \mathbf{y} \rangle &= \langle \mathbf{T}(\mathbf{x} + \mathbf{y}), (\mathbf{x} + \mathbf{y}) \rangle - \langle \mathbf{T}(\mathbf{x} - \mathbf{y}), (\mathbf{x} - \mathbf{y}) \rangle \\ &\quad + i\langle \mathbf{T}(\mathbf{x} + i\mathbf{y}), (\mathbf{x} + i\mathbf{y}) \rangle - i\langle \mathbf{T}(\mathbf{x} - i\mathbf{y}), (\mathbf{x} - i\mathbf{y}) \rangle \\ &= \langle (\mathbf{x} + \mathbf{y}), \mathbf{T}(\mathbf{x} + \mathbf{y}) \rangle - \langle (\mathbf{x} - \mathbf{y}), \mathbf{T}(\mathbf{x} - \mathbf{y}) \rangle \\ &\quad + i\langle (\mathbf{x} + i\mathbf{y}), \mathbf{T}(\mathbf{x} + i\mathbf{y}) \rangle - i\langle (\mathbf{x} - i\mathbf{y}), \mathbf{T}(\mathbf{x} - i\mathbf{y}) \rangle \\ &= 4\langle \mathbf{x}, \mathbf{Ty} \rangle \end{aligned}$$

so that $\mathbf{T} = \mathbf{T}^*$. ∎

Although the significance of this property may not be very clear at this stage, it will be found to be useful in proving the spectral theorem; in this regard, the next theorem is also an important implement.

***Theorem 7.2.2*** Let $\mathbf{T} \in \text{Blt}(\mathcal{H})$ with $\mathbf{T} = \mathbf{T}^*$. Then

$$\|\mathbf{T}\| = \sup_{\mathbf{x} \in \mathcal{H}} \{|\langle \mathbf{Tx}, \mathbf{x} \rangle : \|\mathbf{x}\| = 1\}$$

*Proof*: Since $\mathbf{T}$ is bounded, $|\langle \mathbf{Tx}, \mathbf{x} \rangle|$ is bounded for $\|\mathbf{x}\| = 1$. Thus the set $S \equiv \{|\langle \mathbf{Tx}, \mathbf{x} \rangle| : \|\mathbf{x}\| = 1\}$ consisting of positive numbers must have a supremum (see Theorem 1.8.2). Let $\alpha = \sup S$. We want to show that $\|\mathbf{T}\| = \alpha$.

From the Schwarz inequality

$$|\langle \mathbf{Tx}, \mathbf{x} \rangle| \leq \|\mathbf{Tx}\| \|\mathbf{x}\| \leq \|\mathbf{T}\| \|\mathbf{x}\|^2 = \|\mathbf{T}\|$$

where we have set $\|\mathbf{x}\| = 1$. Thus

$$\alpha \leq \|\mathbf{T}\| \tag{7.2.1}$$

Next, for any real number $t \neq 0$, it is readily seen that for $\mathbf{x} \in \mathcal{H}$,

$$\begin{aligned} 4\|\mathbf{Tx}\|^2 &= \langle \mathbf{T}(t\mathbf{x} + t^{-1}\mathbf{Tx}), (t\mathbf{x} + t^{-1}\mathbf{Tx}) \rangle \\ &\quad - \langle \mathbf{T}(t\mathbf{x} - t^{-1}\mathbf{Tx}), (t\mathbf{x} - t^{-1}\mathbf{Tx}) \rangle \\ &\leq \alpha \|t\mathbf{x} + t^{-1}\mathbf{Tx}\|^2 + \alpha \|t\mathbf{x} - t^{-1}\mathbf{Tx}\|^2 \qquad \text{(why?)} \end{aligned}$$

Using the parallelogram law (see Exercise 6.1.5), the inequality above becomes

$$4\|\mathbf{Tx}\|^2 \leq 2\alpha \|t\mathbf{x}\|^2 + 2\alpha \|t^{-1}\mathbf{Tx}\|^2$$

If we choose $t^2 = \|\mathbf{Tx}\|/\|\mathbf{x}\|$, we obtain

$$4\|\mathbf{Tx}\|^2 \leq 2\alpha \|\mathbf{Tx}\| \|\mathbf{x}\| + 2\alpha \|\mathbf{Tx}\| \|\mathbf{x}\|$$

or $\|\mathbf{Tx}\| \leq \alpha \|\mathbf{x}\|$, so that $\|\mathbf{T}\| \leq \alpha$, which with (7.2.1) implies that $\alpha = \|\mathbf{T}\| = \sup \{|\langle T\mathbf{x}, \mathbf{x} \rangle| : \|\mathbf{x}\| = 1\}$. ∎

Again Theorem 7.2.2 must be viewed as a prerequisite for the spectral theorem. Our next theorem also falls in this category. Before we state and prove this theorem, the following simple concept is required.

***Definition 7.2.1*** Let $\mathcal{L}$ be a linear space and $M \subset \mathcal{L}$ be a subspace. $M$ is said to be invariant under a linear transformation $\mathbf{T}$ on $\mathcal{L}$ if $\mathbf{x} \in M \rightarrow \mathbf{Tx} \in M$.

The foregoing property of invariance simply means this: If for some reason we restrict the domain of $\mathbf{T}$ to a subspace $M$, the invariance of $M$ under $\mathbf{T}$ restricts the *range* of $\mathbf{T}$ to be in $M$ also. The important concept which germinates here is that under the circumstance of invariance the action of $\mathbf{T}$ on $M$ can be studied without regard to the vectors in the remaining part of the linear space.

***Theorem 7.2.3*** Let $\mathbf{T} \in \text{Blt}(\mathcal{H})$ with $\mathbf{T} = \mathbf{T}^*$, and $M \subset \mathcal{H}$. If $M$ is invariant under $\mathbf{T}$, then $M^\perp$ is invariant under $\mathbf{T}$.

*Proof*: Let $\mathbf{y} \in M^{\perp}$. Now $\mathbf{x} \in M \rightarrow \mathbf{Tx} \in M$, so that $\langle \mathbf{Tx}, \mathbf{y} \rangle = 0$. But

$$\mathbf{T} = \mathbf{T}^* \rightarrow \langle \mathbf{Tx}, \mathbf{y} \rangle = \langle \mathbf{x}, \mathbf{Ty} \rangle$$

so that $\langle \mathbf{x}, \mathbf{Ty} \rangle = 0 \; \forall \; \mathbf{x} \in M$. Thus $\mathbf{Ty} \in M^{\perp}$ and $M^{\perp}$ is invariant under $\mathbf{T}$. ▮

From Theorem 7.2.3 we have been able to ascertain that if the action of a bounded self-adjoint operator can be studied in a subspace $M$ without transferring out of it, then a similar study can be made of its action on $M^{\perp}$. This makes it possible to make a piecewise study of the operator in individual subspaces.

The idea of decomposing a Hilbert space into subspaces has been discussed at some length. In Section 7.0 we had hinted about how a self-adjoint operator affords a decomposition of the Hilbert space into subspaces on which the action of the operator is in some sense "simple." We will alleviate at this stage some of the mystery that surrounds the nature of this simplicity by observing that each of the foregoing subspaces will be invariant under the transformation. Indeed, associated with this decomposition of the Hilbert space generated by a self-adjoint operator $\mathbf{T}$ is a *decomposition of the operator* itself into "simpler" operators which can be linearly combined to yield $\mathbf{T}$. The ensuing discussion will address the details that remain after a preliminary investigation of what are known as orthogonal projections.

### 7.2.1 Orthogonal Projections

We recall from Chapter 2 that a projection operator $\mathbf{P}$ is a linear operator with the property $\mathbf{P}^2 = \mathbf{P}$. The question arises as to whether projection operators are bounded transformations. Although there are unbounded projection operators, those of specific interest to us are bounded.

***Definition 7.2.2*** A projection $\mathbf{P}$ on a Hilbert space $\mathcal{H}$ (or more generally on an inner product space $\mathcal{I}$) is said to be *orthogonal* if $R(\mathbf{P}) \perp N(\mathbf{P})$.

That an orthogonal projection is a bounded operator follows readily because one may write (see Example 5 of Section 2.8)

$$\mathcal{H} = R(\mathbf{P}) + N(\mathbf{P})$$

For any $\mathbf{z} \in \mathcal{H}$, we may write $\mathbf{z} = \mathbf{x} + \mathbf{y}$, where $\mathbf{x} = \mathbf{Pz}$ and from the Pythagorean theorem we have $||\mathbf{z}||^2 = ||\mathbf{x}||^2 + ||\mathbf{y}||^2$. Thus $||\mathbf{x}|| \leq ||\mathbf{z}||$ or $||\mathbf{Pz}|| \leq ||\mathbf{z}||$, so that $\mathbf{P}$ is a bounded operator. Further, we have $||\mathbf{P}|| \leq 1$. Obviously, since for $\mathbf{z} \in R(\mathbf{P})$ we have $\mathbf{Pz} = \mathbf{z}$, we must conclude that $||\mathbf{P}|| = 1$. It is also readily seen that $R(\mathbf{P}) = N(\mathbf{P})^{\perp}$ and $N(\mathbf{P}) = R(\mathbf{P})^{\perp}$.

Since an orthogonal projection is a bounded operator we know that its adjoint $\mathbf{P}^*$ exists and we may inquire into the relationship between $\mathbf{P}$ and $\mathbf{P}^*$. The result is the following theorem.

***Theorem 7.2.4*** Let $\mathbf{P}$ be a bounded projection operator on $\mathcal{H}$. Then $\mathbf{P}$ is orthogonal if and only if $\mathbf{P} = \mathbf{P}^*$.

*Proof*: Assume first that $\mathbf{P}$ is orthogonal to prove the *only if* part. Let $\mathbf{z}_1, \mathbf{z}_2 \in \mathcal{H}$. Then $\exists\ \mathbf{x}_1, \mathbf{x}_2 \in R(\mathbf{P}), \mathbf{y}_1, \mathbf{y}_2 \in N(\mathbf{P}) \ni$

$$\mathbf{z}_1 = \mathbf{x}_1 + \mathbf{y}_1, \quad \mathbf{z}_2 = \mathbf{x}_2 + \mathbf{y}_2$$

with $\mathbf{P}\mathbf{z}_1 = \mathbf{x}_1$ and $\mathbf{P}\mathbf{z}_2 = \mathbf{x}_2$. Now

$$\begin{aligned}\langle \mathbf{P}\mathbf{z}_1, \mathbf{z}_2\rangle &= \langle \mathbf{x}_1, \mathbf{z}_2\rangle = \langle \mathbf{x}_1, \mathbf{x}_2\rangle + \langle \mathbf{x}_1, \mathbf{y}_2\rangle \\ &= \langle \mathbf{x}_1, \mathbf{x}_2\rangle\end{aligned}$$

Also,

$$\begin{aligned}\langle \mathbf{z}_1, \mathbf{P}\mathbf{z}_2\rangle &= \langle \mathbf{z}_1, \mathbf{x}_2\rangle = \langle \mathbf{x}_1, \mathbf{x}_2\rangle + \langle \mathbf{y}_1, \mathbf{x}_2\rangle \\ &= \langle \mathbf{x}_1, \mathbf{x}_2\rangle\end{aligned}$$

Thus $\langle \mathbf{P}\mathbf{z}_1, \mathbf{z}_2\rangle = \langle \mathbf{z}_1, \mathbf{P}\mathbf{z}_2\rangle$, so that $\mathbf{P} = \mathbf{P}^*$. To prove the *if* part we assume that $\mathbf{P} = \mathbf{P}^*$. Let $\mathbf{x} \in R(\mathbf{P})$ and $\mathbf{y} \in N(\mathbf{P})$. Then $\mathbf{P}\mathbf{x} = \mathbf{x}$ and $\mathbf{P}\mathbf{y} = 0$. Now

$$\langle \mathbf{x}, \mathbf{y}\rangle = \langle \mathbf{P}\mathbf{x}, \mathbf{y}\rangle = \langle \mathbf{x}, \mathbf{P}\mathbf{y}\rangle = 0$$

Thus $R(\mathbf{P}) \perp N(\mathbf{P})$ and $\mathbf{P}$ is orthogonal. ∎

Hence orthogonality of a bounded projection and self-adjointness are equivalent concepts. The range and null spaces of a self-adjoint projection are closed linear subspaces of $\mathcal{H}$. As an example we refer to Figure 6.4.1, where the projection shown is clearly an orthogonal projection. Every vector $\mathbf{z}$ is projected on to the $z$-axis so that $R(\mathbf{P})$ is the $z$-axis itself. The null space $N(\mathbf{P})$, on the other hand, is the entire $x$-$y$ plane, since vectors with zero projection on the $z$-axis must necessarily lie on the $\mathbf{x}$-$\mathbf{y}$ plane. It is also obvious that in this case $R(\mathbf{P}) \perp N(\mathbf{P})$.

Thus the projections on the coordinate axes or planes of three-dimensional space with which the student is familiar are examples of orthogonal or self-adjoint projections. From a geometric viewpoint it is clear how given any plane or line, one could construct onto it the projection of any vector in the space. It is also evident that such a projection is a unique thing. Our next theorem is a generalized formulation of this concept.

***Theorem 7.2.5*** Let $M$ be a closed linear subspace of $\mathcal{H}$. Then there exists a *unique* self-adjoint projection with $R(\mathbf{P}) = M$.

*Proof*: From the projection theorem we have $\mathcal{H} = M + M^{\perp}$. Let $\mathbf{z} \in \mathcal{H}$. $\exists$ unique vectors $\mathbf{x} \in M, \mathbf{y} \in M^{\perp} \ni \mathbf{z} = \mathbf{x} + \mathbf{y}$. We define $\mathbf{P}\mathbf{z} = \mathbf{x}$. Then clearly $R(\mathbf{P}) = M$. It is easily seen that $\mathbf{P}$ is linear.

That $\mathbf{P}$ is a projection is established as follows. Let $\mathbf{z} \in \mathcal{H}$ and $\mathbf{z} = \mathbf{x} + \mathbf{y}$, where $\mathbf{x} \in M$, $\mathbf{y} \in M^{\perp}$. Now $\mathbf{Pz} = \mathbf{x}$, so that $\mathbf{P}^2\mathbf{z} = \mathbf{Px}$. Since the unique decomposition of $\mathbf{x}$ is $(\mathbf{x} + \mathbf{0})$, we have $\mathbf{Px} = \mathbf{x}$. Thus $\mathbf{P}^2\mathbf{z} = \mathbf{Pz}$, which implies that $\mathbf{P}$ is a projection.

For $\mathbf{z} \in M^{\perp}$, we have the decomposition $\mathbf{z} = \mathbf{0} + \mathbf{z}$, so that $\mathbf{Pz} = \mathbf{0}$ and $\mathbf{z} \in N(\mathbf{P})$, from which $M^{\perp} \subset N(\mathbf{P})$. On the other hand, let $\mathbf{z} \in N(\mathbf{P})$. $\mathbf{Pz} = \mathbf{0}$ implies the decomposition $\mathbf{z} = \mathbf{0} + \mathbf{z}$ and $\mathbf{z} \in M^{\perp}$. Thus $N(\mathbf{P}) = M^{\perp}$, so that $\mathbf{P}$ is an orthogonal (self-adjoint) projection.

To show that $\mathbf{P}$ is the only such projection, assume a second projection $\mathbf{P}'$ with $R(\mathbf{P}') = M$, $N(\mathbf{P}') = M^{\perp}$. Let $\mathbf{z} \in \mathcal{H}$. $\mathbf{P}'\mathbf{z} \in M$, $(\mathbf{z} - \mathbf{P}'\mathbf{z}) \in N(\mathbf{P}')$. Thus $\mathbf{z} = \mathbf{P}'\mathbf{z} + (\mathbf{z} - \mathbf{P}'\mathbf{z})$, which is a different decomposition of $\mathbf{z}$ with elements in $M$ and $M^{\perp}$ (from $\mathbf{Pz}$ and $\mathbf{z} - \mathbf{Pz}$), which is impossible. Thus $\mathbf{P} = \mathbf{P}'$. ∎

Theorem 7.2.5 has established how to project $\mathcal{H}$ onto a given subspace. Also if subspace $M$ of dimension $n$ has $\{\mathbf{z}_j\}_{j=1}^n$ as an orthonormal basis the self-adjoint projection $\mathbf{P}$ of $\mathcal{H}$ onto $M$ may be represented by

$$\mathbf{Px} = \sum_{j=1}^{n} \langle \mathbf{x}, \mathbf{z}_j \rangle \mathbf{z}_j \qquad \forall\, \mathbf{x} \in \mathcal{H}$$

Suppose now that the Hilbert space is decomposed (compartmentalized) into several orthogonal subspaces, say $\{M_j\}_{j=1}^n$ as in Theorem 6.4.5. Then in each of the subspaces one could define a self-adjoint projection to obtain a family of projections $\{\mathbf{P}_j\}_{j=1}^n$ with the property that $R(\mathbf{P}_i) \perp R(\mathbf{P}_j)$, $i \neq j$. We present a formal definition.

***Definition 7.2.3*** An *orthogonal family* of self-adjoint projections $\{\mathbf{P}_j\}$ (which could be finite, countably, or uncountably infinite) has the property that $\mathbf{P}_i\mathbf{P}_j = \mathbf{0}$ whenever $i \neq j$.

The following theorem presents some relevant properties of a finite family of self-adjoint projections.

***Theorem 7.2.6*** Let $\{\mathbf{P}_j\}_{j=1}^n$ be a finite orthogonal family of self-adjoint projections on $\mathcal{H}$. Then the following are true:

(i) $R(\mathbf{P}_i) \subset N(\mathbf{P}_j)$, $i \neq j$.

(ii) $R(\mathbf{P}_i) \perp R(\mathbf{P}_j)$, $i \neq j$.

(iii) $\mathcal{H} = R(\mathbf{P}_1) + R(\mathbf{P}_2) + \dots + R(\mathbf{P}_n) + N$, where $N = N(\mathbf{P}_1) \cap N(\mathbf{P}_2) \cap \dots \cap N(\mathbf{P}_n)$.

*Proof:*

(i) $\mathbf{z} \in R(\mathbf{P}_i) \rightarrow \mathbf{P}_i\mathbf{z} = \mathbf{z} \rightarrow \mathbf{P}_j\mathbf{P}_i\mathbf{z} = \mathbf{P}_j\mathbf{z}$. Since $\mathbf{P}_i\mathbf{P}_j = \mathbf{0}$ we have $\mathbf{P}_j\mathbf{z} = \mathbf{0}$, so that $\mathbf{z} \in N(\mathbf{P}_j)$ and we have $R(\mathbf{P}_i) \subset N(\mathbf{P}_j)$.

(ii) Let $\mathbf{x} \in R(\mathbf{P}_i)$, $\mathbf{y} \in R(\mathbf{P}_j)$, $i \neq j$, $\mathbf{P}_i\mathbf{x} = \mathbf{x}$, and $\mathbf{P}_j\mathbf{y} = \mathbf{y}$. Then

$$\langle \mathbf{x}, \mathbf{y} \rangle = \langle \mathbf{P}_i\mathbf{x}, \mathbf{P}_j\mathbf{y} \rangle = \langle \mathbf{x}, \mathbf{P}_i\mathbf{P}_j\mathbf{y} \rangle = 0$$

Thus $R(\mathbf{P}_i) \perp R(\mathbf{P}_j)$.

(iii) This follows from Theorem 6.4.5. ▮

In an infinite-dimensional Hilbert space the question arises as to whether Theorem 7.2.6 can accommodate an infinite family of projections in regard to part (iii). [Parts (i) and (ii) are obviously true.] The answer to this is in the affirmative provided that the infinite sums of subspaces is interpreted in terms of the infinite sums of vectors from these spaces. We will have more to say on this when dealing with the spectral theorem. For the present we will investigate the properties of the operator obtained by a linear combination of an orthogonal family of self-adjoint projections, that is, the operator

$$\mathbf{T} = \lambda_1\mathbf{P}_1 + \lambda_2\mathbf{P}_2 + \ldots + \lambda_n\mathbf{P}_n \tag{7.2.2}$$

where $\lambda_1, \lambda_2, \ldots, \lambda_n$ are complex scalars. The next theorem establishes the necessary and sufficient conditions under which $\mathbf{T}$ is self-adjoint.

***Theorem 7.2.7*** Let $\{\mathbf{P}_j\}$ be an orthogonal family of self-adjoint projections on $\mathcal{H}$. The operator defined by (7.2.2) is self-adjoint if and only if $\lambda_1, \lambda_2, \ldots, \lambda_n$ are real.

*Proof*: First we prove the *if* part. Assume that the $\lambda_j$'s are real; that is, $\lambda_j = \lambda_j^*$ for each $j$. Let $\mathbf{x}, \mathbf{y} \in \mathcal{H}$.

$$\begin{aligned} \langle \mathbf{Tx}, \mathbf{y} \rangle &= \langle \sum_{j=1}^{n} \lambda_j\mathbf{P}_j\mathbf{x}, \mathbf{y} \rangle = \sum_{j=1}^{n} \lambda_j\langle \mathbf{P}_j\mathbf{x}, \mathbf{y} \rangle \\ &= \sum_{j=1}^{n} \lambda_j\langle \mathbf{x}, \mathbf{P}_j\mathbf{y} \rangle = \sum_{j=1}^{n} \langle \mathbf{x}, \lambda_j\mathbf{P}_j\mathbf{y} \rangle \\ &= \langle \mathbf{x}, \sum_{j=1}^{n} \lambda_j\mathbf{P}_j\mathbf{y} \rangle = \langle \mathbf{x}, \mathbf{Ty} \rangle \end{aligned}$$

so that $\mathbf{T} = \mathbf{T}^*$. To prove the *only if* condition assume that $\mathbf{T} = \mathbf{T}^*$. Let $\mathbf{x} \in R(\mathbf{P}_i)$. Then from Theorem 7.2.4(i), $\mathbf{x} \in N(\mathbf{P}_j)$, $j \neq i$. Thus

$$\mathbf{Tx} = \sum_{j=1}^{n} \lambda_j\mathbf{P}_j\mathbf{x} = \lambda_i\mathbf{P}_i\mathbf{x} = \lambda_i\mathbf{x}$$

Now

$$\langle \mathbf{Tx}, \mathbf{x} \rangle = \lambda_i \|\mathbf{x}\|^2$$

which from Theorem 7.2.1 must be real-valued. Thus $\lambda_i$ must be real. Similarly, the other $\lambda$'s must be real. ▮

We should observe that if the $\lambda_j$'s are not real in (7.2.2), the operator $\mathbf{T}$ is *normal*, which means that $\mathbf{TT}^* = \mathbf{T}^*\mathbf{T}$ which is left as an exercise. Whether $\mathbf{T}$

is self-adjoint or normal, it is readily seen that

$$N(\mathbf{T}) = N(\mathbf{P}_1) \cap N(\mathbf{P}_2) \ldots \cap N(\mathbf{P}_n) \tag{7.2.3}$$

Denoting the right-hand side of (7.2.3) by $N$, it is obvious that $N \subset N(\mathbf{T})$. To prove that $N(\mathbf{T}) \subset N$, observe that for $\mathbf{x} \in \mathcal{H}$, the set $\{\mathbf{P}_j\mathbf{x}\}_{j=1}^n$ is linearly independent (why?), so that $\mathbf{Tx} = \mathbf{0}$ would imply that $\mathbf{P}_j\mathbf{x} = \mathbf{0}$ for each $j$. Hence (7.2.3) is established. The decomposition of $\mathcal{H}$ represented by part (iii) of Theorem 7.2.4 becomes

$$\mathcal{H} = R(\mathbf{P}_1) + R(\mathbf{P}_2) + \ldots + R(\mathbf{P}_n) + N(\mathbf{T}) \tag{7.2.4}$$

Thus the operator $\mathbf{T}$ defined by (7.2.2) leads to a decomposition of $\mathcal{H}$ of the type (7.2.4). This conclusion is significant to further developments. The crucial question to be pursued is whether, given a self-adjoint operator $\mathbf{T}$, it is possible for us to find an orthogonal family of self-adjoint projections $\{\mathbf{P}_j\}$ and real numbers $\{\lambda_j\}$ such that $\mathbf{T}$ may be represented by (7.2.2).† Then we will have accomplished the compartmentalization of the Hilbert space into subspaces in accordance with (7.2.4) (review Section 7.0 again). We are now in a position to dilate on what we mean by the operator $\mathbf{T}$ acting on the compartment subspaces in a "simple" manner. The compartments in (7.2.4) are the subspaces $\{R(\mathbf{P}_j)\}$ and $N(\mathbf{T})$. Let $\mathbf{x} \in R(\mathbf{P}_j)$; then we have

$$\mathbf{Tx} = \lambda_j \mathbf{x} \tag{7.2.5}$$

which is the simplicity that we had implied. Note that (7.2.5) covers also the behavior on the subspace $N(\mathbf{T})$ by letting $\lambda_j = 0$. In (7.2.5) we have the most important basis for the "purposeful view" which a self-adjoint operator can produce of the Hilbert space. The next section will lay down some important definitions and properties of entities associated with self-adjoint operators.

## EXERCISES

**7.2.1** Let $M$ be an $m$-dimensional subspace of $\mathcal{H}$ and $\mathbf{P}$ be the self-adjoint projection of $\mathcal{H}$ onto $M$. Find the matrix representation of $\mathbf{P}$ with respect to any orthonormal basis in $\mathcal{H}$, say the first $m$ of which are in $M$. If $\mathbf{x} \in \mathcal{H}$, identify the corresponding element of $\mathbf{Px}$ in $l_2$ (see Theorem 6.2.2).

**7.2.2** Let $\mathcal{H}$ be a Hilbert space and $\mathbf{T}$ be an operator of the type (7.2.2). Show that for any integer $k$

$$\mathbf{T}^k = \sum_{j=1}^{n} \lambda_j^k \mathbf{P}_j$$

Suppose that $f(\lambda)$ is an analytic function of $\lambda$ with a series representation

$$f(\lambda) = \sum_{k=0}^{\infty} a_k \lambda^k$$

†More generally, we may accommodate a normal operator.

which has a radius of convergence $R$ such that $|\lambda_j| < R$ for $j = 1, 2, \ldots, n$. Consider the operator

$$\mathbf{f}(\mathbf{T}) = \sum_{k=0}^{\infty} a_k \mathbf{T}^k$$

Show that

$$\mathbf{f}(\mathbf{T}) = \sum_{j=1}^{n} f(\lambda_j)\mathbf{P}_j$$

**7.2.3** Let $\mathcal{H}$ be an $n$-dimensional Hilbert space and $\mathbf{T}$ be given by (7.2.2) such that $N(\mathbf{T}) = \{0\}$. Show that

$$\mathbf{T}^{-1} = \sum_{j=1}^{n} \lambda_j^{-1}\mathbf{P}_j$$

## 7.3 Eigenvalues, Eigenvectors, and Eigenspaces

In the preceding section we observed that given a self-adjoint operator, (7.2.5) is the basis for our inquiry into how a Hilbert space could be decomposed "purposefully." This prompts the following definition, where we drop the qualification "self-adjoint" for the sake of generality.

***Definition 7.3.1*** Let $\mathbf{T} \in \text{Blt}(\mathcal{H})$. A complex number $\lambda$ is an *eigenvalue* (also referred to as *characteristic value*) if there exists a *nonzero* vector $\mathbf{z} \in \mathcal{H}$ such that $\mathbf{Tz} = \lambda\mathbf{z}$. The nonzero vector $\mathbf{z}$ is called an *eigenvector* corresponding to the eigenvalue $\lambda$.†

An alternative way of looking at the eigenvalue $\lambda$ is that it is a complex number for which the operator $(\mathbf{T} - \lambda\mathbf{I})$ is *not* one-to-one, that is, not invertible. Thus the null space $N(\mathbf{T} - \lambda\mathbf{I})$ contains nonzero elements if $\lambda$ is an eigenvalue, and is called the *eigenspace* corresponding to eigenvalue $\lambda$. We will denote the eigenspace by $N_\lambda$. Notice particularly that *every* eigenvector corresponding to eigenvalue $\lambda$ must belong to $N_\lambda$. The eigenvectors corresponding to an eigenvalue are in effect a basis for the eigenspace.

The following theorem concerns self-adjoint operators.

***Theorem 7.3.1*** Let $\mathbf{T} \in \text{Blt}(\mathcal{H}) \ni \mathbf{T} = \mathbf{T}^*$. Then

(i) All the eigenvalues of $\mathbf{T}$ are real.

(ii) If $\lambda$ and $\lambda'$ are two eigenvalues such that $\lambda \neq \lambda'$ $N_\lambda \perp N'_\lambda$; that is, eigenvectors corresponding to distinct eigenvalues are orthogonal. Note that this also means that eigenvectors corresponding to distinct eigenvalues are linearly independent.

---

†The eigenvalue can also assume the value of zero if there exist nonzero vector $\mathbf{x}$ such that $\mathbf{Tx} = \mathbf{0}$.

*Proof*:

(i) Let $\lambda$ be any eigenvalue of $\mathbf{T}$ and $\mathbf{x}$ the corresponding eigenvector. Then $\langle \mathbf{Tx}, \mathbf{x} \rangle = \lambda \|\mathbf{x}\|^2$. Since $\mathbf{T} = \mathbf{T}^*$, Theorem 7.2.1 requires that $\langle \mathbf{Tx}, \mathbf{x} \rangle$ be real-valued. Thus $\lambda$ must be real-valued. Hence all eigenvalues of $\mathbf{T}$ are real.

(ii) Let $\mathbf{x} \in N_\lambda$, $\mathbf{y} \in N'_\lambda$. Then $\mathbf{Tx} = \lambda \mathbf{x}$, $\mathbf{Ty} = \lambda' \mathbf{y}$:

$$\langle \mathbf{Tx}, \mathbf{y} \rangle = \lambda \langle \mathbf{x}, \mathbf{y} \rangle$$

$$\langle \mathbf{x}, \mathbf{Ty} \rangle = \lambda' \langle \mathbf{x}, \mathbf{y} \rangle \qquad \text{(Note: } \lambda \text{ and } \lambda' \text{ are real)}$$

Since $\mathbf{T} = \mathbf{T}^*$, subtracting the second from the first, we have

$$(\lambda - \lambda')\langle \mathbf{x}, \mathbf{y} \rangle = 0 \qquad \text{so that } \langle \mathbf{x}, \mathbf{y} \rangle = 0$$

Thus $N_\lambda \perp N'_\lambda$. ∎

It would seem at first that we have all that we need. The theorem above has shown that all the eigenvalues are real and the eigenspaces are orthogonal to each other. So shouldn't one look for the eigenvalues, the eigenvectors (or the eigenspaces), define projections on each of the eigenspaces, and express the self-adjoint operator in the form (7.2.2)? For several reasons, the answer to the foregoing question is a resounding no. First, we do not know for certain that, given a self-adjoint operator $\mathbf{T}$, eigenvalues *do exist*, that is, that $\mathbf{Tx} = \lambda \mathbf{x}$ has a nonzero solution for any $\lambda$. Second, suppose that eigenvalues do exist and that there are several of them. In a finite-dimensional Hilbert space we can assert forthwith that there can only be a *finite* number of them. (This is because there can only be a finite number of linearly independent eigenvectors in a finite-dimensional space.) On the other hand, for infinite-dimensional Hilbert spaces, there could be an infinite number of eigenvalues; further, we do not know whether there could be a countable or uncountable infinity of eigenvalues. A third source of uncertainty arises as follows. Suppose that all the eigenvectors (or eigenspaces) are computed. We are not at this point able to assert that the algebraic sum of the eigenspaces above will make up the *entire* Hilbert space. Another way of saying the same thing is that we do not know whether the eigenvectors are complete (form a basis) in $\mathcal{H}$. Clearly, the uncertainties recounted above essentially pertain to *existence* and this is about what the spectral theorem is concerned.

Unfortunately, the situation is quite complicated for an arbitrary bounded self-adjoint operator on an infinite-dimensional Hilbert space. In finite-dimensional spaces, only the *existence* of the eigenvalues and the completeness of the eigenvectors need be established, since the finiteness of the eigenvalues (and eigenvectors) is already obvious to us. It is now important to recall from Chapter 5 how *every* linear operator on a finite-dimensional (normed linear) space is a compact operator. It turns out that in infinite-dimensional Hilbert spaces it is the compact self-adjoint operator that possesses the countable infinity of eigenvalues, a fact that the spectral theorem will establish. The next section will present the spectral theorem, first for self-adjoint operators on a finite-dimensional space and then for compact self-adjoint operators on an infinite-dimensional space.

**EXERCISE**

**7.3.1** Let $\mathbf{T} \in \text{Blt}(\mathcal{H})$ and $\lambda$ be an eigenvalue of $\mathbf{T}$. Show that
(a) $N_\lambda$ is invariant under $\mathbf{T}$.
(b) $N_\lambda$ is a closed subspace of $\mathcal{H}$.
(c) if $\mathbf{T} = \mathbf{T}^*$, $N_\lambda^\perp$ is invariant under $\mathbf{T}$ (see Theorem 7.2.3).

## 7.4 Spectral Theorems

We have now all the ammunition required for proving the spectral theorems of great significance to us. We will deal with the finite-dimensional case first.

***Theorem 7.4.1 (Spectral Theorem: Finite-Dimensional Case)*** Let $\mathcal{H}$ be a finite-dimensional Hilbert space and $\mathbf{T} \in \text{Blt}(\mathcal{H}) \ni \mathbf{T} = \mathbf{T}^*$.

(i) There exists at least one eigenvalue $\lambda_1$ of $\mathbf{T}$ such that $|\lambda_1| = \|\mathbf{T}\|$.

(ii) There are at most a finite number of nonzero eigenvalues

$$\lambda_1, \lambda_2, \ldots, \lambda_s \qquad \text{such that } \lambda_j \neq \lambda_k \rightarrow N_{\lambda_j} \perp N_{\lambda_k}$$

If $N(\mathbf{T})$ contains nonzero elements, then zero is an eigenvalue.

(iii) $\mathcal{H} = N_{\lambda_1} + N_{\lambda_2} + \ldots + N_{\lambda_s} + N(\mathbf{T})$.

(iv) $\mathbf{T} = \lambda_1\mathbf{P}_1 + \lambda_2\mathbf{P}_2 + \ldots + \lambda_s\mathbf{P}_s$, where $\mathbf{P}_j$ is the orthogonal projection of $\mathcal{H}$ onto $N_{\lambda_j}$, that is, $R(\mathbf{P}_j) = N_{\lambda_j}$.

*Proof*: Note that since $\mathcal{H}$ is finite dimensional, $\mathbf{T}$ is compact. From Theorem 7.2.2 we know that $\|\mathbf{T}\| = \sup\{|\langle\mathbf{Tx}, \mathbf{x}|: \|\mathbf{x}\| = 1\}$. Consider $S \equiv \{\langle\mathbf{Tx}, \mathbf{x}\rangle: \|\mathbf{x}\| = 1\}$. Since $S$ is a bounded set on the real line, Theorem 1.8.1 provides for the existence of its infimum and supremum. Clearly,

$$\|\mathbf{T}\| = |\inf S| \quad \text{or} \quad |\sup S|$$

(see Exercise 1.8.8). From the Corollary to Theorem 1.8.2 there exist sequences of numbers in $S$ converging either to inf $S$ or sup $S$. More precisely, we can find a sequence $\{\mathbf{x}_n\}$ with $\|\mathbf{x}_n\| = 1$ such that $\lim_{n\to\infty} \langle\mathbf{Tx}_n, \mathbf{x}_n\rangle = \lambda_1$, where $|\lambda_1| = \|\mathbf{T}\|$.† We will now show that $\lambda_1$ is an eigenvalue. The bounded sequence $\{\mathbf{x}_n\}$ is transformed to a bounded sequence $\{\mathbf{Tx}_n\}$ since $\mathbf{T}$ is bounded.

†The existence of the sequence $\{\mathbf{x}_n\}$, $\|\mathbf{x}_n\| = 1$ follows from the fact that all numbers in the set $S$ are obtained from vectors $\mathbf{x}$, with $\|\mathbf{x}\| = 1$, by computing $\langle\mathbf{Tx}, \mathbf{x}\rangle$.

The compactness of operator $\mathbf{T}$ implies that $\{\mathbf{Tx}_n\}$ must have a convergent subsequence $\{\mathbf{Tx}_{k_n}\}$. Let

$$\lim_{n\to\infty} \mathbf{Tx}_{k_n} \equiv \mathbf{y}$$

Now

$$\begin{aligned}\lim \|\mathbf{Tx}_{k_n} - \lambda_1 \mathbf{x}_{k_n}\|^2 &= \lim [\|\mathbf{Tx}_{k_n}\|^2 - 2\lambda_1\langle \mathbf{Tx}_{k_n}, \mathbf{x}_{k_n}\rangle + \lambda_1^2 \|\mathbf{x}_{k_n}\|^2] \\ &= \|\mathbf{y}\|^2 - 2\lambda_1^2 + \lambda_1^2 = \|\mathbf{y}\|^2 - \lambda_1^2 \geq 0\end{aligned}$$

Thus $\|\mathbf{y}\| \geq |\lambda_1| = \|\mathbf{T}\|$. But $\|\mathbf{Tx}_{k_n}\| \leq \|\mathbf{T}\|$, so that $\|\mathbf{y}\| \leq \|\mathbf{T}\|$, from which it follows that

$$\|\mathbf{y}\| = \|\mathbf{T}\|$$

Therefore,

$$\lim_{n\to\infty} \|\mathbf{Tx}_{k_n} - \lambda_1 \mathbf{x}_{k_n}\| = 0$$

or

$$\lim_{n\to\infty} \left\|\mathbf{x}_{k_n} - \frac{1}{\lambda_1}\mathbf{Tx}_{k_n}\right\| = 0$$

Clearly,

$$\left\|\mathbf{x}_{k_n} - \frac{1}{\lambda_1}\mathbf{y}\right\| \leq \left\|\mathbf{x}_{k_n} - \frac{1}{\lambda_1}\mathbf{Tx}_{k_n}\right\| + \frac{1}{|\lambda_1|}\|\mathbf{Tx}_{k_n} - \mathbf{y}\|$$

As $n \longrightarrow \infty$ both the terms on the right-hand side of the foregoing inequality vanish, yielding $\lim_{n\to\infty} \mathbf{x}_{k_n} = (1/\lambda_1)\mathbf{y}$. Since $\mathbf{T}$ is bounded,

$$\mathbf{y} = \lim_{n\to\infty} \mathbf{Tx}_{k_n} = \mathbf{T}(\lim_{n\to\infty} \mathbf{x}_{k_n}) = \mathbf{T}\left(\frac{1}{\lambda_1}\mathbf{y}\right)$$

so that $\mathbf{Ty} = \lambda_1 \mathbf{y}$ and we have shown that $\lambda_1$ is an eigenvalue. Besides, $\mathbf{y}$ is an eigenvector, which we normalize to obtain the eigenvector

$$\mathbf{z}_1 = \frac{\mathbf{y}}{\|\mathbf{y}\|}$$

Note that $\mathbf{Tz}_1 = \lambda_1 \mathbf{z}_1$, $\|\mathbf{z}_1\| = 1$, which proves (i).

From the linear manifold $L(\mathbf{z}_1)$, which has the properties $L(\mathbf{z}_1) \subset N_{\lambda_1}$, $L(\mathbf{z}_1)$ is a closed subspace of $\mathcal{H}$, so that $L(\mathbf{z})$ is invariant under $\mathbf{T}$. From the projection theorem we have

$$\mathcal{H} = L(\mathbf{z}_1) + L(\mathbf{z}_1)^\perp$$

We now restrict $\mathbf{T}$ to $L(\mathbf{z}_1)^\perp$, which is a Hilbert space. Let $\mathbf{T} \neq 0$ on this space. Then by precisely the same arguments as before, we have

$$|\lambda_2| = \sup\{|\langle \mathbf{Tx}, \mathbf{x}\rangle|:\ x \in L(\mathbf{z}_1)^\perp; \|\mathbf{x}\| = 1\}$$

Obviously, $|\lambda_2| \leq |\lambda_1|$. Also, $\exists\ \mathbf{z}_2 \in L(\mathbf{z}_1)^\perp$.

$$\mathbf{Tz}_2 = \lambda_2 \mathbf{z}_2, \qquad \|\mathbf{z}_2\| = 1$$

The following possibilities are obvious.

If $\lambda_2 = \lambda_1$, then $L(\mathbf{z}_1, \mathbf{z}_2) \subset N_{\lambda_1}$; if $\lambda_2 \neq \lambda_1$, then $L(\mathbf{z}_1) = N_{\lambda_1}$, $L(\mathbf{z}_2) \subset N_{\lambda_2}$. In either case $L(\mathbf{z}_1, \mathbf{z}_2)$ is a proper subspace of $\mathcal{H}$ unless $\mathcal{H}$ is two-dimensional.

It is also a closed subspace of $\mathcal{H}$. From the projection theorem we may write

$$\mathcal{H} = L(\mathbf{z}_1, \mathbf{z}_2) + L(\mathbf{z}_1, \mathbf{z}_2)^{\perp}$$

when $\lambda_2 = \lambda_1$ or

$$\mathcal{H} = N_{\lambda_1} + N_{\lambda_1}^{\perp}$$

if $\lambda_2 \neq \lambda_1$. In either case we restrict $\mathbf{T}$ to the Hilbert space $L(\mathbf{z}_1, \mathbf{z}_2)^{\perp}$ which is invariant under $\mathbf{T}$ [since $L(\mathbf{z}_1, \mathbf{z}_2)$ is clearly invariant under $\mathbf{T}$]. As before, we know $\exists\ \lambda_3 \ni$

$$|\lambda_3| \leq |\lambda_2| \leq |\lambda_1|$$

and $\mathbf{z}_3$ such that

$$\mathbf{T}\mathbf{z}_3 = \lambda_3 \mathbf{z}_3, \qquad \|\mathbf{z}_3\| = 1$$

Again various cases are possible. If $\lambda_3 = \lambda_2 = \lambda_1$, then $L(\mathbf{z}_1, \mathbf{z}_2, \mathbf{z}_3) \subset N_{\lambda_1}$; if $\lambda_3 = \lambda_2 \neq \lambda_1$, then $L(\mathbf{z}_1) = N_{\lambda_1}, L(\mathbf{z}_2, \mathbf{z}_3) \subset N_{\lambda_2}$; and if $\lambda_3 \neq \lambda_2 \neq \lambda_1$, then $L(\mathbf{z}_1) = N_{\lambda_1}, L(\mathbf{z}_2) = N_{\lambda_2}, L(\mathbf{z}_3) \subset N_{\lambda_3}$. Regardless of which case is true, we next restrict $\mathbf{T}$ to $L(\mathbf{z}_1, \mathbf{z}_2, \mathbf{z}_3)^{\perp}$.

The procedure of finding $\mathbf{z}_1, \mathbf{z}_2, \ldots, \mathbf{z}_r$ will stop at $\mathbf{z}_r$ if and only if

$$L(\mathbf{z}_1, \mathbf{z}_2, \ldots, \mathbf{z}_r)^{\perp} = N(\mathbf{T}) \tag{7.4.1}$$

This is obvious since if (7.4.1) is true $\langle \mathbf{T}\mathbf{x}, \mathbf{x} \rangle = 0\ \forall\ \mathbf{x} \in L(\mathbf{z}_1, \mathbf{z}_2, \ldots, \mathbf{z}_r)^{\perp}$; if (7.4.1) is not true, then the procedure clearly has not ended because then we could find $\mathbf{z}_{r+1}$, and so on.

Since $\mathcal{H}$ is finite dimensional, the procedure must end if $r = n$, in which case $N(\mathbf{T}) = \{\mathbf{0}\}$ (so that $\mathbf{T}$ has an inverse). If it ends for $r < n$, then (7.4.1) is true and $\lambda = 0$ may be considered an eigenvalue. We could then denote $N(\mathbf{T})$ by $N_0$. At this stage we must elaborate on some important *notational changes*. We denote the *distinct* nonzero eigenvalues as $\lambda_1, \lambda_2, \ldots, \lambda_s$. Call the number of times $\lambda_i$ has occurred the *degeneracy* of $\lambda_i$ and denote it by $r_i$. Clearly,

$$\sum_{i=1}^{s} r_i = r$$

$$\begin{aligned} N_{\lambda_1} &= L(\mathbf{z}_1, \mathbf{z}_2, \ldots, \mathbf{z}_{r_1}) \\ N_{\lambda_2} &= L(\mathbf{z}_{r_1+1}, \mathbf{z}_{r_1+2}, \ldots, \mathbf{z}_{r_1+r_2}) \\ &\ \ \vdots \\ N_{\lambda_s} &= L(\mathbf{z}_{r_1+r_2+\ldots+r_{s-1}}, \mathbf{z}_{r_1+r_2+\ldots+r_{s-1}+1}, \ldots, \mathbf{z}_r) \end{aligned} \tag{7.4.2}$$

Since

$$\mathcal{H} = L(\mathbf{z}_1, \mathbf{z}_2, \ldots, \mathbf{z}_r) + L(\mathbf{z}_1, \mathbf{z}_2, \ldots, \mathbf{z}_r)^{\perp} \tag{7.4.3}$$

we may write from (7.4.2)

$$\mathcal{H} = N_{\lambda_1} + N_{\lambda_2} + \ldots + N_{\lambda_s} + N_0$$

which establishes (ii).

For (iii) let $\mathbf{P}_j$ define the unique self-adjoint projection of $\mathcal{H}$ onto $N_\lambda$. For $\mathbf{x} \in \mathcal{H}$, $\exists$ $\mathbf{x}_j \in N_{\lambda_j}, j = 1, 2, \ldots, s$ and $\mathbf{y} \in N_0 \ni$

$$\mathbf{x} = \mathbf{x}_1 + \mathbf{x}_2 + \ldots + \mathbf{x}_s + \mathbf{y}$$

Clearly,

$$\mathbf{Tx} = \lambda_1 \mathbf{x}_1 + \lambda_2 \mathbf{x}_2 + \ldots + \lambda_s \mathbf{x}_s \tag{7.4.4}$$

and

$$\mathbf{P}_j \mathbf{x} = \mathbf{P}_j(\mathbf{x}_1 + \mathbf{x}_2 + \ldots + \mathbf{x}_s + \mathbf{y}) = \mathbf{P}_j \mathbf{x}_j = \mathbf{x}_j, \qquad j = 1, 2, \ldots, s$$

Thus (7.4.4) becomes

$$\mathbf{Tx} = \sum_{j=1}^{s} \lambda_j \mathbf{P}_j \mathbf{x} \qquad \forall \quad \mathbf{x} \in \mathcal{H}$$

so that

$$\mathbf{T} = \sum_{j=1}^{s} \lambda_j \mathbf{P}_j \quad \blacksquare$$

We have thus established the spectral theorem for a self-adjoint operator on a finite-dimensional space. Clearly, the eigenvectors $\mathbf{z}_1, \mathbf{z}_2, \ldots, \mathbf{z}_r$ form an orthonormal basis in $R(\mathbf{T})$. The space $N(\mathbf{T})$ must have dimension $(n - r)$, where $n$ is the dimension of $\mathcal{H}$, which implies that there are $(n - r)$ linearly independent eigenvectors corresponding to the zero eigenvalue; if we pick these to be orthonormal and denote them by $\mathbf{z}_{r+1}, \mathbf{z}_{r+2}, \ldots, \mathbf{z}_n$, then the set $\{\mathbf{z}_j\}_{j=1}^{n}$ forms an orthonormal basis in $\mathcal{H}$. This is an extremely important result.

We will now be concerned with the spectral theorem for compact self-adjoint operators on an infinite-dimensional Hilbert space. The set of compact operators on $\mathcal{H}$ is denoted $\mathcal{K}(\mathcal{H})$.

***Theorem 7.4.2 (Spectral Theorem: Infinite-Dimensional Case)*** Let $\mathcal{H}$ be an infinite-dimensional separable Hilbert space and $\mathbf{T} \in \mathcal{K}(\mathcal{H}) \ni \mathbf{T} = \mathbf{T}^*$. Then

(i) There exists a sequence of eigenvalues $\lambda_1, \lambda_2, \ldots$ which terminate at a finite number *or* $\lambda_n \longrightarrow 0$. In the first case, zero will be an eigenvalue with $N(\mathbf{T})$ infinite dimensional. In the second case zero is an accumulation point of the set of eigenvalues.

(ii) Each eigenspace $N_{\lambda_j}$ is finite dimensional for $\lambda_j \neq 0$. Also $\lambda_j \neq \lambda_k$ $\rightarrow N_{\lambda_j} \perp N_{\lambda_k}$.

(iii) If $\mathbf{P}_j$ is the self-adjoint projection onto $N_{\lambda_j}$, then

$$\mathbf{T} = \sum_{j=1}^{\infty} \lambda_j \mathbf{P}_j$$

where the convergence is in the operator norm (thus also in the strong sense). Alternatively, the eigenvectors $\{\mathbf{z}_j\}$ form an orthonormal basis in $\mathcal{H}$.

*Proof*:

(i) The existence of eigenvalues is established exactly as in Theorem 7.4.1, since the operator $\mathbf{T}$ is compact. Thus we must have $\lambda_1, \lambda_2, \ldots$ such that $|\lambda_k| \geq |\lambda_{k+1}|$. If it terminates after a finite number $s$ of eigenvalues with eigenvectors $\mathbf{z}_1, \mathbf{z}_2, \ldots, \mathbf{z}_r$ $(r \geq s)$, then $N(\mathbf{T}) = L(\mathbf{z}_1, \mathbf{z}_2, \ldots, \mathbf{z}_r)^{\perp}$. Thus zero is an eigenvalue and $N(\mathbf{T})$ the corresponding eigenspace is infinite dimensional because $\mathcal{H}$ is of infinite dimension. On the other hand, $R(\mathbf{T})$ is finite dimensional.

Suppose that the sequence of eigenvalues does not terminate at a finite number. Since $|\lambda_n|$ must decrease progressively, if $\lambda_n$ does not tend to zero, let $\lim_{n\to\infty} |\lambda_n| = \varepsilon > 0$. Then $\{\mathbf{z}_n/\lambda_n\}$ is a bounded sequence since $||\mathbf{z}_n/\lambda_n|| = 1/|\lambda_n| < 1/\varepsilon$. Now $\mathbf{T}(\mathbf{z}_n/\lambda_n) = (1/\lambda_n)\mathbf{T}\mathbf{z}_n = \mathbf{z}_n$. Since $\mathbf{T}$ is compact, the sequence $\{T(\mathbf{z}_n/\lambda_n)\}$ or equivalently $\{\mathbf{z}_n\}$ must have a convergent subsequence. But this is impossible because for any $n$ and $m$,

$$||\mathbf{z}_n - \mathbf{z}_m||^2 = ||\mathbf{z}_n||^2 + ||\mathbf{z}_m||^2 = 2$$

Thus the contradiction on hand implies that $\lim_{n\to\infty} \lambda_n = 0$.

(ii) To show that $N_{\lambda_j}$ is finite dimensional, we observe that since $\lambda_n \longrightarrow 0$, no eigenvalue can repeat infinitely often. This implies that $N_{\lambda_j}$ is finite dimensional. The orthogonality of the subspaces follows from Theorem 7.4.1.

(iii) We retain the notation in Theorem 7.4.1 and observe that $s$ distinct eigenvalues generate $r(r \geq s)$ orthonormal eigenvectors. Also, if $\mathbf{x} \in \mathcal{H}$, then

$$\mathbf{P}_j\mathbf{x} = \sum_{i=r_1+r_2+\ldots+r_{j-1}+1}^{i=r_1+r_2+\ldots+r_j} \langle \mathbf{x}, \mathbf{z}_i \rangle \mathbf{z}_i \tag{7.4.5}$$

and

$$\mathbf{T}\mathbf{P}_j\mathbf{x} = \sum_{i=r_1+r_2+\ldots+r_{j-1}+1}^{i=r_1+r_2+\ldots+r_j} \lambda_i \langle \mathbf{x}, \mathbf{z}_i \rangle \mathbf{z}_i \tag{7.4.6}$$

Moreover, from (7.4.3) we may write

$$\mathbf{x} = \sum_{j=1}^{s} \mathbf{P}_j\mathbf{x} + \mathbf{y}_r \tag{7.4.7}$$

where $\mathbf{y}_r \in L(\mathbf{z}_1, \mathbf{z}_2, \ldots, \mathbf{z}_r)^{\perp}$. From (7.4.5) to (7.4.7) we may write

$$\mathbf{T}\mathbf{x} - \sum_{j=1}^{s} \lambda_j \mathbf{P}_j\mathbf{x} = \mathbf{T}\mathbf{y}_r \tag{7.4.8}$$

and note that as $s$ is increased arbitrarily $r$ also increases arbitrarily since $s \leq r$. From (7.4.7) we obtain

$$||\mathbf{x}||^2 = \left\| \sum_{j=1}^{s} \mathbf{P}_j\mathbf{x} \right\|^2 + ||\mathbf{y}_r||^2$$

so that

$$||\mathbf{y}_r|| \leq ||\mathbf{x}|| \tag{7.4.9}$$

Clearly,

$$\left\|\left(\mathbf{T} - \sum_{j=1}^{s} \lambda_j \mathbf{P}_j\right)\mathbf{x}\right\| = \|\mathbf{T}\mathbf{y}_r\| \leq |\lambda_{s+1}| \|\mathbf{y}_r\|$$
$$\leq |\lambda_{s+1}| \|\mathbf{x}\| \tag{7.4.10}$$

where we have used (7.4.9) and the fact that

$$|\lambda_{s+1}| = \sup\{|\langle \mathbf{T}\mathbf{y}, \mathbf{y}\rangle|: \mathbf{y} \in L(\mathbf{z}_1, \mathbf{z}_2, \ldots, \mathbf{z}_r)^{\perp}; \|\mathbf{y}\| = 1\}$$

which is also the norm of $\mathbf{T}$ restricted to $L(\mathbf{z}_1, \mathbf{z}_2, \ldots, \mathbf{z}_r)^{\perp}$ (also see Theorem 7.2.2). Inequality (7.4.10) implies that

$$\left\|\mathbf{T} - \sum_{j=1}^{s} \lambda_j \mathbf{P}_j\right\| \leq |\lambda_{s+1}| \tag{7.4.11}$$

We have already seen that $|\lambda_{s+1}| \longrightarrow 0$ as $s \longrightarrow \infty$ so that (7.4.11) yields the result

$$\lim_{n \to \infty} \|\mathbf{T} - \sum_{j} \lambda_j \mathbf{P}_j\| = 0$$

or equivalently,

$$\mathbf{T} = \sum_{j=1}^{\infty} \lambda_j \mathbf{P}_j \quad \blacksquare \tag{7.4.12}$$

In dealing with the spectral theorem above we have emphasized the conceptual aspect of how the compact self-adjoint operator may be expressed as a real linear combination of self-adjoint projections onto subspaces into which the Hilbert space is decomposed. From a computational viewpoint, it is more purposeful to deal directly with the eigenvalues and eigenvectors. We have seen from spectral theorem 7.4.1 how the eigenvectors corresponding to nonzero eigenvalues form an orthonormal basis in $R(\mathbf{T})$. This conclusion of course holds here. When zero is not an eigenvalue, that is, $N(\mathbf{T}) = \{\mathbf{0}\}$, then $\lambda_n \longrightarrow 0$ as $n \longrightarrow \infty$ and the countably infinite set of eigenvectors $\{\mathbf{z}_j\}$ form an orthonormal basis in $\mathcal{H}$.

It is also interesting to note that the operator

$$\sum_{j=1}^{s} \mathbf{P}_j$$

acts as an *identity operator* on $L(\mathbf{z}_1, \mathbf{z}_2, \ldots, \mathbf{z}_r)$. If $\lambda = 0$ is not an eigenvalue, the infinite sum of projections converge in the *strong sense* to the identity operator on $\mathcal{H}$. Thus

$$\sum_{j=1}^{\infty} \mathbf{P}_j = \mathbf{I} \tag{7.4.12a}$$

where the right-hand side is the identity operator on $\mathcal{H}$. The self-adjoint projections $\{\mathbf{P}_j\}$ are referred to as a *resolution of the identity* spurred by the self-adjoint operator $\mathbf{T}$. Equation (7.4.12) may be referred to as the *spectral representation* of $\mathbf{T}$.

We have thus carried out the most important task on hand. The compartmentalization of the Hilbert space associated with the self-adjoint operator

**T**, which was mentioned in Section 7.0, lies in the eigenspaces $\{N_{\lambda_j}\}$. This simplicity of the behavior of **T** on $N_{\lambda_j}$ is expressed by (7.2.5). We are now in a position to see how some of the heuristic remarks in Section 7.0 pertaining to the solution of the equation

$$\mathbf{Tu} = \mathbf{f}, \qquad \mathbf{f} \in \mathcal{H} \tag{7.4.13}$$

can be interpreted more rigorously. It will be presumed that $N(\mathbf{T}) = \{\mathbf{0}\}$, so that there is a unique solution **u** to (7.4.13) where **T** is a compact self-adjoint operator. We will solve (7.4.13) in two different but essentially equivalent ways.

First we form the inner product of (7.4.13) with the $j$th eigenvector $\mathbf{z}_j$ and use the self-adjointness of **T** to get

$$\langle \mathbf{Tu}, \mathbf{z}_j \rangle = \langle \mathbf{u}, \mathbf{Tz}_j \rangle = \langle \mathbf{f}, \mathbf{z}_j \rangle \tag{7.4.14}$$

Now $\mathbf{Tz}_j = \lambda_j \mathbf{z}_j$.† Equation (7.4.14) then becomes

$$\langle \mathbf{u}, \mathbf{z}_j \rangle = \frac{1}{\lambda_j} \langle \mathbf{f}, \mathbf{z}_j \rangle \tag{7.4.15}$$

Since $\{\mathbf{z}_j\}$ is an orthonormal basis in $\mathcal{H}$, we have the solution to (7.4.13) as

$$\mathbf{u} = \sum_{j=1}^{\infty} \langle \mathbf{u}, \mathbf{z}_j \rangle \mathbf{z}_j = \sum_{j=1}^{\infty} \lambda_j^{-1} \langle \mathbf{f}, \mathbf{z}_j \rangle \mathbf{z}_j \tag{7.4.16}$$

Alternatively, we premultiply (7.4.13) by $\mathbf{P}_j$:

$$\mathbf{P}_j \mathbf{Tu} = \mathbf{P}_j \mathbf{f}$$

and make use of the readily proven fact that $\mathbf{P}_j \mathbf{T} = \mathbf{TP}_j = \lambda_j \mathbf{P}_j$ (see Exercise 7.4.1) to obtain

$$\mathbf{P}_j \mathbf{u} = \lambda_j^{-1} \mathbf{P}_j \mathbf{f} \tag{7.4.17}$$

Equation (7.4.17) is the one that clarifies how the solution **u** to (7.4.13) is obtained by solving for its "pieces" $\{\mathbf{P}_j \mathbf{u}\}$ in the respective subspaces $\{N_{\lambda_j}\}$ from the corresponding pieces of **f**, $\{\mathbf{P}_j \mathbf{f}\}$. The final procedure of putting together the pieces $\{\mathbf{P}_j \mathbf{u}\}$ is represented by

$$\mathbf{u} = \sum_{j=1}^{\infty} \lambda_j^{-1} \mathbf{P}_j \mathbf{f} \tag{7.4.18}$$

From a conceptual viewpoint (7.4.18) represents a convenient way of expressing the solution, although the form (7.4.16) (also obtained by using (7.4.5) on (7.4.18) with the alteration of indicial notation on the eigenvalues

---

†At this point it is necessary to be conscious of an irksome notational inconsistency. With the notation of eigenvalues in Theorem 7.4.1, it is conceivable that the $j$th eigenvector may belong to an eigenspace for some eigenvalue $\lambda_i$ $(i < j)$. In dealing with eigenvalues and eigenvectors directly, it is much more convenient to associate the same index with the eigenvalue and eigenvector. In this notation, eigenvalues with *different* indices are *not* necessarily distinct.

as in the footnote below) provides the direct route to computation once the eigenvalues and eigenvectors have been determined.

The problem of determining the eigenvalues and eigenvectors of a self-adjoint operator will be discussed at a later stage. For the present, we observe that the procedure used in the spectral theorem to establish the existence of the eigenvalues provides the basis for what is referred to as the variational method for the estimation of eigenvalues. Exercise 7.4.6 goes into some details.

Although (7.4.16) or (7.4.18) represents the solution to (7.4.13), it is important to recognize some subtleties arising from the nature of the inverse of a compact operator. Since $\mathbf{T}$ is compact and self-adjoint, the eigenvalues $\{\lambda_j\}$ eventually diminish to zero (zero itself is not an eigenvalue since $\mathbf{T}$ has an inverse). The sequence $\{\lambda_j^{-1}\}$ consequently increases without bound so that (7.4.18) may become unbounded for some vectors $\mathbf{f}$ in $\mathcal{H}$ and is therefore not defined for all vectors in $\mathcal{H}$. However, (7.4.18) is defined if *restricted* to a subspace of vectors which would by definition be $D(\mathbf{T}^{-1})$, the domain of $\mathbf{T}^{-1}$, or equivalently $R(\mathbf{T})$, the range space of $\mathbf{T}$. The implication of constraining vectors $\mathbf{f}$ in (7.4.18) is that $D(\mathbf{T}^{-1})$ is not all of $\mathcal{H}$. It is readily shown that $R(\mathbf{T})$ is dense in $\mathcal{H}$.† Thus it transpires that $\mathbf{T}^{-1}$ is an *unbounded* operator, its domain being the subspace of all vectors for which (7.4.16) or (7.4.18) is defined; it must be noted, however, that although (7.4.16) is a solution to (7.4.13), no computational efficacy is implied by it. Equation (7.4.13) is normally referred to as *Fredholm equation* of the *first kind*. We will not consider this equation any further but move on to Fredholm equations of the *second kind*, exemplified by

$$\mathbf{Tx} - \lambda\mathbf{x} = \mathbf{y} \tag{7.4.19}$$

where $\lambda$ is a real parameter and $\mathbf{T}$ will be restricted here to be compact and self-adjoint. At this stage, it is useful to recall that an eigenvalue of an operator $\mathbf{T}$ is a number $\lambda$ for which the operator $(\mathbf{T} - \lambda\mathbf{I})$ does *not* have an inverse. On the other hand, if $\lambda$ is not an eigenvalue for $\mathbf{T} \in \mathcal{K}(\mathcal{H})$, $(\mathbf{T} - \lambda\mathbf{I})^{-1}$ exists, so that (7.4.19) is soluble.‡ Let $\mathbf{T} = \mathbf{T}^*$, and $\{\lambda_j\}$ and $\{\mathbf{z}_j\}$ be the eigen-

†To show that $R(\mathbf{T})$ is dense in $\mathcal{H}$, let $\mathbf{x} \in \mathcal{H}$. We should find vectors in $R(\mathbf{T})$ arbitrarily close to $\mathbf{x}$. The eigenvectors must clearly belong to $R(\mathbf{T})$. Because the eigenvectors are an orthonormal basis in $\mathcal{H}$, there must be linear combinations of them [which are in $R(\mathbf{T})$] that approach $\mathbf{x}$ arbitrarily closely. This establishes that $R(\mathbf{T})$ or $D(\mathbf{T}^{-1})$ is dense in $\mathcal{H}$.

‡This is one of the so-called *Fredholm alternatives* for solving (7.4.19) with $\mathbf{T} \in \mathcal{K}(\mathcal{H})$. The alternatives are as follows. Either for $\mathbf{y} = \mathbf{0}$, the homogeneous equation has *only* the solution $\mathbf{x} = \mathbf{0}$, in which case the inhomogeneous equation has a *unique* solution for each $\mathbf{y} \in \mathcal{H}$, *or* the homogeneous equation has nonzero solutions in which case, the inhomogeneous equation can be solved only for $\mathbf{y} \perp N(T^* - \lambda\mathbf{I})$. The student should review these alternatives for linear algebraic equations.

values and eigenvectors of $\mathbf{T}$, respectively. If $\lambda$ is real, the operator $(\mathbf{T} - \lambda\mathbf{I})$ is self-adjoint, since $\mathbf{T} = \mathbf{T}^*$. Furthermore,

$$(\mathbf{T} - \lambda\mathbf{I})\mathbf{z}_j = (\lambda_j - \lambda)\mathbf{z}_j$$

so that the eigenvalues of $(\mathbf{T} - \lambda\mathbf{I})$ are $\{\lambda_j - \lambda\}$ and the eigenvectors are the same as those of $\mathbf{T}$. [$\mathbf{T}$ and $(\mathbf{T} - \lambda\mathbf{I})$ are said to have a common spectral resolution.] Using (7.4.12), we may write

$$(\mathbf{T} - \lambda\mathbf{I}) = \sum_{j=1}^{\infty} (\lambda_j - \lambda)\mathbf{P}_j$$

Indeed, $(\mathbf{T} - \lambda\mathbf{I})$ is a bounded operator (it is not compact if $\mathcal{H}$ is infinite dimensional) and since $\lambda \neq \lambda_j$ for any $j$, its inverse may be identified as

$$(\mathbf{T} - \lambda\mathbf{I})^{-1} = \sum_{j=1}^{\infty} (\lambda_j - \lambda)^{-1}\mathbf{P}_j \tag{7.4.20}$$

Note that unlike $\mathbf{T}^{-1}$, $(\mathbf{T} - \lambda\mathbf{I})^{-1}$ for $\lambda \neq 0$ is a *bounded* operator since $|\lambda_j - \lambda|^{-1}$ is bounded. Hence the solution to (7.4.19) can be written as

$$\mathbf{x} = \sum_{j=1}^{\infty} (\lambda_j - \lambda)^{-1}\mathbf{P}_j\mathbf{y} \tag{7.4.21}$$

From a computational viewpoint however, the solution (7.4.21) is not as suitable as the following alternative form obtained by rewriting (7.4.20) as

$$\begin{aligned}(\mathbf{T} - \lambda\mathbf{I})^{-1} &= \sum_{j=1}^{\infty} (\lambda_j - \lambda)^{-1}\mathbf{P}_j + \lambda^{-1}\mathbf{I} - \lambda^{-1}\mathbf{I} \\ &= \sum_{j=1}^{\infty} [(\lambda_j - \lambda)^{-1} + \lambda^{-1}]\mathbf{P}_j - \lambda^{-1}\mathbf{I} \\ &= \sum_{j=1}^{\infty} \lambda_j(\lambda_j - \lambda)^{-1}\mathbf{P}_j - \lambda^{-1}\mathbf{I}\end{aligned} \tag{7.4.22}$$

where we have again used (7.4.12a). The form (7.4.22) is more convenient than (7.4.20) because the eigenvalues $\{\lambda_j\}$ diminish progressively to zero. Thus an alternative form of the solution to (7.4.19) is given by

$$\mathbf{x} = \sum_{j=1}^{\infty} \lambda_j(\lambda_j - \lambda)^{-1}\mathbf{P}_j\mathbf{y} - \lambda^{-1}\mathbf{y} \tag{7.4.23}$$

which is computationally superior to (7.4.21). Note again that (7.4.23) is applicable for $\lambda \neq 0$. If $\lambda = 0$, (7.4.21) is still meaningful although only for vectors in $D(\mathbf{T}^{-1})$.

We have seen that the inverse of a compact operator, when it exists, is an unbounded operator. We have dealt with unbounded operators and how they may yield self-adjoint operators. The spectral representation of unbounded operators can be considerably more complicated than that of compact operators. However, there is a class of unbounded operators that is extremely important to us which is suggested by the considerations above in regard to the inverse $\mathbf{T}^{-1}$ of the compact operator $\mathbf{T}$. Indeed, the inverse of $\mathbf{T}^{-1}$ would be $\mathbf{T}$ itself and the question naturally arises as to which of the spectral properties of $\mathbf{T}$, $\mathbf{T}^{-1}$ could inherit. The next theorem addresses this issue and is central to many of the applications to boundary value problems.

***Theorem 7.4.3*** Let $\mathcal{H}$ be an infinite-dimensional Hilbert space and $\mathbf{T} = \{T, D(T)\}$ be an unbounded self-adjoint operator. Suppose that $\mathbf{T}^{-1}: \mathcal{H} \longrightarrow D(T)$ exists and is compact.

(i) $\mathbf{T}^{-1}$ is self-adjoint with the consequence that it has eigenvalues $\{\mu_j\}$, and eigenvectors $\{\mathbf{z}_j\}$ which form an orthonormal basis in $\mathcal{H}$.

(ii) $\mathbf{T}$ has eigenvalues $\{\lambda_j\}$ and eigenvectors $\{\mathbf{z}_j\}$ where $\lambda_j = 1/\mu_j$, so that $\lim_{n\to\infty} |\lambda_n| = \infty$.

*Proof*: To prove (i), let $\mathbf{x}, \mathbf{y} \in \mathcal{H}$. $\exists$ $\mathbf{u}, \mathbf{v} \in D(T) \ni \mathbf{Tu} = \mathbf{x}, \mathbf{Tv} = \mathbf{y}$. Also, $\mathbf{u} = \mathbf{T}^{-1}\mathbf{x}$, $\mathbf{v} = \mathbf{T}^{-1}\mathbf{y}$. Since $\mathbf{T}$ is self-adjoint, we have

$$\langle \mathbf{Tu}, \mathbf{v} \rangle = \langle \mathbf{u}, \mathbf{Tv} \rangle$$

or

$$\langle \mathbf{x}, \mathbf{T}^{-1}\mathbf{y} \rangle = \langle \mathbf{T}^{-1}\mathbf{x}, \mathbf{y} \rangle$$

so that $\mathbf{T}^{-1}$ is self-adjoint. From Theorem 7.4.2, $\mathbf{T}^{-1}$ has eigenvalues $\{\mu_j\}$ and eigenvectors $\{\mathbf{z}_j\}$ which are an orthonormal basis in $\mathcal{H}$.

The proof of (ii) is almost immediate.

$$\mathbf{T}^{-1}\mathbf{z}_j = \mu_j \mathbf{z}_j \rightarrow \mathbf{z}_j = \mu_j \mathbf{T}\mathbf{z}_j$$

or

$$\mathbf{T}\mathbf{z}_j = \lambda_j \mathbf{z}_j \qquad \text{where } \lambda_j \equiv \frac{1}{\mu_j}$$

Clearly, $\{\lambda_j\}$ are *all* of the eigenvalues of $\mathbf{T}$, with $\mathbf{z}_j$ as the eigenvector corresponding to $\lambda_j$, because every eigenvector of $\mathbf{T}$ must be an eigenvector of $\mathbf{T}^{-1}$ and must belong in the orthonormal basis $\{\mathbf{z}_j\}$. Since as $n \longrightarrow \infty$, $|\mu_n| \longrightarrow 0$, we have $|\lambda_n| \longrightarrow \infty$. ▮

Theorem 7.4.3 has a straightforward corollary to include the case where the unbounded operator $\mathbf{T}$ may not have an inverse but for some $\lambda_0$, the inverse $(\mathbf{T} - \lambda_0\mathbf{I})^{-1}$ may exist and may be compact.

***Corollary*** Suppose that the unbounded self-adjoint operator $\mathbf{T}$ in Theorem 7.4.3 does not have an inverse but that for some $\lambda_0 \neq 0$, $(\mathbf{T} - \lambda_0\mathbf{I})^{-1}$ exists and is compact. Again $\mathbf{T}$ has a countably infinite set of eigenvalues $\{\lambda_j\}$ and eigenvectors $\{\mathbf{z}_j\}$ which form an orthonormal basis in $\mathcal{H}$. As before, $\lim_{n\to\infty} |\lambda_n| = \infty$.

*Proof*: The proof of this corollary is obvious since Theorem 7.4.3 may be applied to the operator $(\mathbf{T} - \lambda_0 \mathbf{I})$, and then recognizing that if $\mathbf{z}_j$ is an eigenvector of $(\mathbf{T} - \lambda_0 \mathbf{I})$ with eigenvalue, say $\lambda'_j$, then it is also an eigenvector of $\mathbf{T}$ with $\lambda'_j + \lambda_0$ as the corresponding eigenvalue. ∎

It will turn out subsequently that many boundary value problems involve differential operators that have compact inverses or satisfy the stipulations in the corollary to Theorem 7.4.3. The solution to such boundary value problems is then obtained in terms of the eigenvalues and the eigenvectors of the unbounded self-adjoint operator.

If $\{\mathbf{P}_j\}$ are the self-adjoint projections of the unbounded operator $\mathbf{T}$, then denoting the eigenvalues by $\{\lambda_j\}$, the representation†

$$\mathbf{T} = \sum_{j=1}^{\infty} \lambda_j \mathbf{P}_j$$

is defined for $\mathbf{x} \in D(\mathbf{T})$ but *not* for *all* elements in $\mathcal{H}$. Note that its compact inverse $\mathbf{T}^{-1}$ has the representation

$$\mathbf{T}^{-1} = \sum_{j} \lambda_j^{-1} \mathbf{P}_j \tag{7.4.24}$$

The solution (7.4.18) to equations of the type (7.4.13) is therefore applicable here also. In fact, these applications are very common in engineering. For example, steady-state situations in the analysis of transport processes lead to boundary value problems in temperature, mass concentration, or other dependent variables that satisfy certain differential equations and boundary conditions. These boundary value problems are exactly of the type (7.4.13) where $\mathbf{T}$ is an unbounded (and frequently self-adjoint) operator with a compact inverse.‡ We refer to the example in regard to the temperature distribution in a peripherally cooled sphere in Section 2.12 and in particular to (2.12.3), which is of the form (7.4.13). More generally, we will encounter partial differential equations of the type (7.4.13) in which the partial differential expression together with homogeneous boundary conditions yields an unbounded self-adjoint operator.

---

†This representation must not be interpreted as implying convergence with respect to the operator norm, since such convergence may be expected only when $\mathbf{T}$ is compact as, for example, in (7.4.24).

‡In Section 5.3 we observed that integral operators of certain types give rise to compact operators, while in Sections 5.5 and 6.7 we found that differential operators are examples of unbounded operators. It should not come as a surprise, therefore, that invertible compact operators on an infinite-dimensional Hilbert space have unbounded inverses. Similarly, differential operators could possess compact inverses for the same reason.

### 7.4.1 Functions of a Self-Adjoint Operator

In discussing the solution of (7.4.13) through the spectral resolution of the self-adjoint operator $\mathbf{T}$, we were in effect using the inverse as given by (7.4.24). The question then arises whether other functions of the operator $\mathbf{T}$ (which are operators themselves) can also be represented via the spectral resolution of $\mathbf{T}$. We have, in fact, laid the foundation for the answer to this question through Exercises 6.6.9 and 7.2.2. If $\lambda$ is an eigenvalue of $\mathbf{T}$ with eigenvector $\mathbf{z}$, it is immediately evident that for a positive integer $n$, $\lambda^n$ is an eigenvalue of $\mathbf{T}^n$ with $\mathbf{z}$ as the eigenvector (if $\mathbf{T}^{-1}$ exists, then negative integers are automatically included in this discussion). Thus for polynomials of an operator $\mathbf{T}$ with eigenvalues $\{\lambda_j\}$ and self-adjoint spectral resolution $\{\mathbf{P}_j\}$, we have

$$\mathbf{p}(\mathbf{T}) = \sum_{j=1}^{\infty} p(\lambda_j)\mathbf{P}_j \tag{7.4.25}$$

where $p(\cdot)$ is a polynomial of some finite degree. The operator $\mathbf{p}(\mathbf{T})$ has the same spectral resolution as $\mathbf{T}$. In regard to more general functions, say $f(\cdot)$, an appropriate definition is itself an important issue.† We assume that the function in question is defined over the field of complex numbers in such a way that $f(\lambda_j)$ exists for each (real) eigenvalue $\lambda_j$ of the self-adjoint operator $\mathbf{T}$. Then we may *define*

$$\mathbf{f}(\mathbf{T}) \equiv \sum_{j=1}^{\infty} f(\lambda_j)\mathbf{P}_j \tag{7.4.26}$$

Indeed, $\mathbf{f}(\mathbf{T})$ is linear, and self-adjoint as long as $f(\lambda)$ is real-valued for real arguments $\lambda$. Moreover, $\mathbf{f}(\mathbf{T})$ has eigenvalues $\{f(\lambda_j)\}$ and eigenvectors $\{\mathbf{z}_j\}$. If $\{f(\lambda_j)\}$ is a bounded sequence, then $\mathbf{f}(\mathbf{T})$ is a *bounded* operator. If, on the other hand, $\{f(\lambda_j)\}$ is an *unbounded* sequence, $\mathbf{f}(\mathbf{T})$ is an *unbounded* operator defined at most on a dense domain of $\mathcal{H}$. If $\lim_{j\to\infty} f(\lambda_j) = 0$, it follows that $\mathbf{f}(\mathbf{T})$ is compact. The category to which $\mathbf{f}(\mathbf{T})$ belongs would depend on the signs of the eigenvalues $\{\lambda_j\}$. The following definition is essential.

***Definition 7.4.1*** Let $\mathbf{T}$ be a self-adjoint operator on $\mathcal{H}$ (or a dense subspace of $\mathcal{H}$). $\mathbf{T}$ is said to be *positive definite* or simply positive, if $\forall\, \mathbf{x} \in D(\mathbf{T}) \ni \mathbf{x} \neq 0$,

$$\langle \mathbf{Tx}, \mathbf{x} \rangle > 0$$

When the inequality above is reversed, then the operator is *negative definite* (or simply negative). If, instead, we have

$$\langle \mathbf{Tx}, \mathbf{x} \rangle \geq 0 \tag{7.4.27}$$

---

†The book, *Linear Operators, Part I*, by N. Dunford and J. T. Schwartz, Wiley-Interscience, New York, 1958, Sec. VII, goes about the details in regard to the function of an operator in a proper and rigorous manner.

then $\mathbf{T}$ is said to be *nonnegative* (definite) and with the inequality above reversed, $\mathbf{T}$ is *nonpositive*. When both signs are possible, $\mathbf{T}$ is said to be *indefinite*.

If $\mathbf{T}$ is positive (negative), it is immediately evident that all of its eigenvalues are strictly positive (negative) because $\langle \mathbf{T}\mathbf{z}_j, \mathbf{z}_j \rangle = \lambda_j$ for the $j$th normalized eigenvector. For a nonnegative (nonpositive) operator, the eigenvalues are nonnegative (nonpositive), implying that zero is an eigenvalue in either case (see also Exercise 7.4.5). In what follows we assume that the self-adjoint operator $\mathbf{T}$ has a countable set of eigenvalues $\{\lambda_j\}$ and a corresponding resolution of the identity given by the family of self-adjoint projections $\{\mathbf{P}_j\}$.

The exponential function of a negative operator $\mathbf{T}$ is a bounded operator given by

$$\mathbf{e}^{\mathbf{T}} = \sum_{j=1}^{\infty} e^{\lambda_j} \mathbf{P}_j \tag{7.4.28}$$

because $\{e^{\lambda_j}\}$ is a bounded sequence for negative numbers $\{\lambda_j\}$. Similarly, other functions could be considered. For example, the *square root* of a positive (or nonnegative) operator is defined as

$$\mathbf{T}^{1/2} = \sum_{j=1}^{\infty} \lambda_j^{1/2} \mathbf{P}_j$$

The concept of a function of an operator is a very useful one in the solution of various types of equations. We take this up in the next section.

## EXERCISES

**7.4.1** Let $\mathbf{T} \in \mathcal{K}(\mathcal{H}) \ni \mathbf{T} = \mathbf{T}^*$ and $\{\lambda_j\}$ be the eigenvalues and $\{\mathbf{P}_j\}$ be the associated resolution of the identity. Show that

$$\mathbf{P}_j\mathbf{T} = \mathbf{T}\mathbf{P}_j = \lambda_j \mathbf{P}_j$$

**7.4.2** Let $\{\mathbf{P}_j\}$ be the resolution of identity belonging to some self-adjoint operator on $\mathcal{H}$. Consider the operator $\mathbf{T}$ defined by

$$\mathbf{T} = \sum_{j=1}^{\infty} \alpha_j \mathbf{P}_j$$

for any complex sequence $\{\alpha_j\}$. Show that $\mathbf{T}$ is a bounded linear operator if and only if the sequence $\{\alpha_j\}$ is bounded. [*Hint:* Use Parseval's relation (6.2.6).]

**7.4.3** Study Theorem 6.2.2 again. Show that the matrix representation (see, e.g., Section 2.11) of a self-adjoint operator $\mathbf{T}$ with respect to any orthonormal basis in $\mathcal{H}$ is a *Hermitian* matrix, that is, a matrix of coefficients $\{t_{ij}\}$ for which $t_{ij} = t_{ji}^*$.

(a) Show that $\mathbf{T}$ and $\{t_{ij}\}$ have the same eigenvalues and that the $l_2$-representative of eigenvector $\mathbf{z}_j$ of $\mathbf{T}$ (corresponding to eigenvalue $\lambda_j$) is the eigenvector of $\{t_{ij}\}$.

(b) What is the matrix representation of $\mathbf{T}$ relative to the eigenvectors of $\mathbf{T}$?

**7.4.4** Let $\mathbf{T} \in \mathcal{K}(\mathcal{H}) \ni \mathbf{T} = \mathbf{T}^*$, and $M$ be a subspace of $\mathcal{H}$ invariant under $\mathbf{T}$. If $M$ is of dimension $n$, then show that $\mathbf{T}$ has $n$ eigenvectors in $M$.

**7.4.5** Let $\mathbf{T}$ be a compact self-adjoint operator on an infinite-dimensional Hilbert space $\mathcal{H}$ and $\{\boldsymbol{\phi}_j\}$ be an orthonormal basis in $\mathcal{H}$. Let $L_n$ be the subspace given by the manifold

$$L_n \equiv L(\boldsymbol{\phi}_1, \boldsymbol{\phi}_2, \ldots, \boldsymbol{\phi}_n)$$

and $\mathbf{P}_{(n)}$ be the orthogonal projection of $\mathcal{H}$ onto $L_n$. Denote the operator $\mathbf{P}_{(n)}\mathbf{T}\mathbf{P}_{(n)}$ by $\mathbf{T}_n$.

(a) Show that $L_n$ is invariant under $\mathbf{T}_n$ and that $\|\mathbf{T}_n\| \leq \|\mathbf{T}\|$.

(b) Show that $\mathbf{T}_n$ is self-adjoint. Obtain the matrix representation of $\mathbf{T}_n$ with respect to the set $\{\boldsymbol{\phi}_j\}_{j=1}^n$. How would one determine the eigenvalues of $\mathbf{T}_n$? [see Exercise 7.4.3(a)].

(c) Show that $\mathbf{T}_n$ converges strongly to $\mathbf{T}$ and thus that as $n$ becomes large, the eigenvalues of $\mathbf{T}_n$ approach the eigenvalues of $\mathbf{T}$.

**7.4.6** Show that a compact self-adjoint operator $\mathbf{T}$ is positive (negative) if and only if all its eigenvalues are positive (negative).

**7.4.7** Let $\mathbf{T}$ be a compact, positive self-adjoint operator on $\mathcal{H}$. From Theorem 7.4.1, the largest (positive) eigenvalue $\lambda_1$ is given by

$$\lambda_1 = \sup\{\langle \mathbf{T}\mathbf{x}, \mathbf{x}\rangle : \mathbf{x} \in \mathcal{H}, \|\mathbf{x}\| = 1\}$$

(a) Let $\mathbf{T}_n$ be the operator defined in Exercise 7.4.4. Show that $\mathbf{T}_n$ is positive and that for $\mathbf{x} \in L_n$,

$$\langle \mathbf{T}_n\mathbf{x}, \mathbf{x}\rangle = \langle \mathbf{T}\mathbf{x}, \mathbf{x}\rangle$$

Then an estimate for $\lambda_1$ is the largest eigenvalue $\lambda_1^{(n)}$ of $\mathbf{T}_n$. Also, the estimate $\mathbf{z}_1^{(n)}$ for the first eigenvector $\mathbf{z}_1$ of $\mathbf{T}$ is in $L_n$. Since

$$\lambda_2 = \sup\{\langle \mathbf{T}\mathbf{x}, \mathbf{x}\rangle : \mathbf{x} \in L(\mathbf{z}_1)^\perp, \|\mathbf{x}\| = 1\}$$

which is the next largest eigenvalue of $\mathbf{T}$, an estimate for this may again be obtained as

$$\lambda_2^{(n)} = \sup\{\langle \mathbf{T}_n\mathbf{x}, \mathbf{x}\rangle : \mathbf{x} \in L(\mathbf{z}_1^{(n)})^\perp \cap L_n, \|\mathbf{x}\| = 1\}$$

Show that the foregoing procedure of the first $n$ eigenvalues is exactly equivalent to the method outlined in Exercise 7.4.5(c).

(b) To understand the limitation of this method, we let $\|\mathbf{z}_1^{(n)} - \mathbf{z}_1\| \equiv \varepsilon$, which is the "error" in the estimated eigenvector. Note that $\|\mathbf{z}_1^{(n)}\| = \|\mathbf{z}_1\| = 1$; that is, the eigenvectors are normalized. Show that

$$\lambda_1^{(n)} - \lambda_1 = \langle \mathbf{T}(\mathbf{z}_1^{(n)} - \mathbf{z}_1), (\mathbf{z}_1^{(n)} - \mathbf{z}_1)\rangle - \lambda_1\varepsilon^2 = (\alpha - \lambda_1)\varepsilon^2$$

where $\alpha < \lambda_1$. Thus the "error" in eigenvalue is of order $\varepsilon^2$ in contrast with $\varepsilon$ for the order of error in the eigenfunction. The eigenvalue is estimated more accurately by this method.

**7.4.8** An operator $\mathbf{T}$ is said to be *skew self-adjoint* if $\mathbf{T} = -\mathbf{T}^*$. Let $\mathbf{T}$ be a compact skew self-adjoint operator. Show that the operator $i\mathbf{T}$ (where $i = \sqrt{-1}$) is self-adjoint and hence that $\mathbf{T}$ has an uncountable set of *imaginary* eigenvalues eventually becoming arbitrarily small. Furthermore, the eigenvectors of $\mathbf{T}$ form an orthonormal basis of the Hilbert space $\mathcal{H}$ on which $\mathbf{T}$ acts.

**7.4.9** Let $\mathbf{T} \in \mathcal{K}(\mathcal{H})$. Show that the operators $\mathbf{T}^*\mathbf{T}$ and $\mathbf{T}\mathbf{T}^*$ qualify for spectral theorem 7.4.2.

(a) Further show that $\mathbf{T}^*\mathbf{T}$ and $\mathbf{T}\mathbf{T}^*$ are nonnegative.

(b) Consider the Hilbert space $\mathcal{H} \oplus \mathcal{H}$. Show that the operator

$$\mathbf{A} \equiv \begin{bmatrix} 0 & \mathbf{T}^* \\ \mathbf{T} & 0 \end{bmatrix}$$

is compact and self-adjoint, so that Theorem 7.4.2 is applicable.

(1) Show that if $\mu$ is an eigenvalue of $\mathbf{A}$ with $(\mathbf{u}, \mathbf{v})$ as an eigenvector, then $-\mu$ is also an eigenvalue with $(\mathbf{u}, -\mathbf{v})$ as an eigenvector.

(2) Show that $\mu^2$ is an eigenvalue of both $\mathbf{T}^*\mathbf{T}$ and $\mathbf{T}\mathbf{T}^*$ with the corresponding eigenvectors $\mathbf{u}$ and $\mathbf{v}$, respectively.

(3) Let $\mathbf{x} \in \mathcal{H}$. Show that

$$\mathbf{T}\mathbf{x} = \sum_{j=1}^{\infty} \mu_j \langle \mathbf{x}, \mathbf{u}_j \rangle \mathbf{v}_j$$

$$\mathbf{T}^*\mathbf{x} = \sum_{j=1}^{\infty} \mu_j \langle \mathbf{x}, \mathbf{v}_j \rangle \mathbf{u}_j$$

**7.4.10** Let $\mathbf{T} \in \mathcal{K}(\mathcal{H}) \ni \mathbf{T}\mathbf{T}^* = \mathbf{T}^*\mathbf{T}$. ($\mathbf{T}$ is then called a normal operator.)

(a) Show that for any complex number $\alpha$, and $\mathbf{x} \in \mathcal{H}$,

$$\|\mathbf{T}\mathbf{x} - \alpha\mathbf{x}\| = \|\mathbf{T}^*\mathbf{x} - \alpha^*\mathbf{x}\|$$

(b) If $\mathbf{z}$ is an eigenvector of $\mathbf{T}$ with eigenvalue $\lambda_1$, show that it is also an eigenvector of $\mathbf{T}^*$. What is the corresponding eigenvalue of $\mathbf{T}^*$?

(c) Let $\lambda \equiv (\mu + i\nu)$ be an eigenvalue of $\mathbf{T}$ with eigenvector $\mathbf{z}$. Recall the decomposition $\mathbf{T} \equiv \mathbf{T}_R + i\mathbf{T}_I$ from Exercise 6.6.1. Show that $\mathbf{z}$ is an eigenvector of $\mathbf{T}_R$ and $\mathbf{T}_I$ with $\mu$ and $\nu$ as the respectve eigenvalues.

**7.4.11** Let $\mathbf{T}_1$ and $\mathbf{T}_2$ be compact self-adjoint operators such that they commute, that is, $\mathbf{T}_1\mathbf{T}_2 = \mathbf{T}_2\mathbf{T}_1$.

(a) Show that $R(\mathbf{T}_1)$ and $N(\mathbf{T}_1)$ are invariant under $\mathbf{T}_2$. (Clearly, the statement above holds also when one inserts $\mathbf{T}_2$ for $\mathbf{T}_1$ and $\mathbf{T}_1$ for $\mathbf{T}_2$.)

(b) Let $\lambda$ be an eigenvalue of $\mathbf{T}_1$, and $N_\lambda$ be the corresponding eigenspace of dimension, say $n$ (why should this be finite?). Show that $\mathbf{T}_2$ has $n$ eigenvectors in $N_\lambda$. [*Hint:* Use the result of Exercise 7.4.4 with part (a) of this exercise.]

Thus establish that $\mathbf{T}_1$ and $\mathbf{T}_2$ have the same spectral resolution although different eigenvalues; that is, they may be expressed

$$\mathbf{T}_1 = \sum_{j=1}^{\infty} \lambda_{j_1}\mathbf{P}_j, \qquad \mathbf{T}_2 = \sum_{j=1}^{\infty} \lambda_{j_2}\mathbf{P}_j$$

where $\{\mathbf{P}_j\}$ represents the common spectral resolution of the identity. Obtain the spectral representations of $(\mathbf{T}_1 + \mathbf{T}_2)$ and $\mathbf{T}_1\mathbf{T}_2$. If $(\mathbf{T}_1 + \mathbf{T}_2)^{-1}$ exists, what is its representation? If $\mathbf{T}_1^{-1}$ and $\mathbf{T}_2^{-1}$ exist, represent $(\mathbf{T}_1\mathbf{T}_2)^{-1}$.

**7.4.12** Let $\mathbf{T}$ be a compact normal operator. Show that $\mathbf{T}_R$ and $\mathbf{T}_I$ commute so that from Exercise 7.4.11 they have a common spectral resolution, denoted $\{\mathbf{P}_j\}$ with corresponding real eigenvalues $\{\mu_j\}$ and $\{\nu_j\}$, respectively.

Now infer the generalization of the spectral Theorem 7.4.2 to that for a normal operator. What is the spectral resolution of **T**? What is that of **TT***?

A normal operator is called *unitary* if $\mathbf{TT}^* = \mathbf{I}$. Locate the eigenvalues of a unitary operator on the complex plane.

## 7.5 Some Common Equations

In demonstration of the utility of the spectral representation of an operator, we were concerned with the solution of (7.4.13) and (7.4.19). Here we will provide other types of equations that also occur commonly and show how they can be solved using the spectral representation of the operator involved. Since in succeeding chapters the application of the types of equations encountered here will be treated with details of the physical problem, we will be content with an abstract setup in this section; besides, the abstract treatment establishes the generality of the methods involved. Accordingly, we stipulate that **T** is a self-adjoint operator with a countable set of eigenvalues $\{\lambda_j\}$ and associated family of self-adjoint projections $\{\mathbf{P}_j\}$; the eigenvectors of **T** will be denoted by $\{\mathbf{z}_j\}$. We will consider different examples, each involving the operator **T** with some relevant additional details.

Basically, we will be concerned with a vector function $\mathbf{u}(t)$, where $t$ is a real parameter, which in physical applications may be a time or spatial coordinate. For each $t$, $\mathbf{u}(t)$ belongs to a Hilbert space $\mathcal{H}$. Furthermore, $\mathbf{u}(t)$ is assumed to be differentiable as many times as is required in the application. We recall the definition of the derivative as

$$\frac{d}{dt}\mathbf{u}(t) = \lim_{h\to 0}\frac{1}{h}[\mathbf{u}(t+h) - \mathbf{u}(t)] \tag{7.5.1}$$

The limiting procedure in (7.5.1) implies that $d\mathbf{u}(t)/dt$ is a vector in $\mathcal{H}$ such that

$$\lim_{h\to 0}\left\|\frac{d}{dt}\mathbf{u}(t) - \frac{1}{h}[\mathbf{u}(t+h) - \mathbf{u}(t)]\right\| = 0 \tag{7.5.2}$$

In what follows, it is useful to record the following property:

$$\frac{d}{dt}\langle \mathbf{u}(t), \mathbf{v}(t)\rangle = \left\langle \frac{d}{dt}\mathbf{u}(t), \mathbf{v}(t)\right\rangle + \left\langle \mathbf{u}(t), \frac{d}{dt}\mathbf{v}(t)\right\rangle \tag{7.5.3}$$

The proof of (7.5.3) is left as an exercise. Next we consider what are known as initial value problems.

### 7.5.1 Initial Value Problems

Unsteady-state analysis of a large variety of engineering problems may be represented by an equation of the form

$$-\frac{d}{dt}\mathbf{u}(t) = \mathbf{T}\mathbf{u}(t) - \mathbf{f}(t), \qquad \mathbf{u}(0) = \mathbf{u}_0 \tag{7.5.4}$$

where $\mathbf{T}$ is a self-adjoint operator with the properties stipulated earlier,† $\mathbf{u}_0 \in \mathcal{H}$ and $\mathbf{f}(t) \in \mathcal{H}$ for each $t$. We shall present two essentially equivalent methods of solving (7.5.4) for $\mathbf{u}(t)$, which satisfies the initial value $\mathbf{u}_0$. Taking the inner product of (7.5.4) with $\mathbf{z}_j$, we have, using (7.5.3),‡

$$-\frac{d}{dt}\langle \mathbf{u}(t), \mathbf{z}_j \rangle = \langle \mathbf{Tu}(t), \mathbf{z}_j \rangle - \langle \mathbf{f}(t), \mathbf{z}_j \rangle = \lambda_j \langle \mathbf{u}(t), \mathbf{z}_j \rangle - \langle \mathbf{f}(t), \mathbf{z}_j \rangle \tag{7.5.5}$$

Equation (7.5.5) is a differential equation in the scalar dependent variable $\langle \mathbf{u}(t), \mathbf{z}_j \rangle$ satisfying the initial condition

$$\langle \mathbf{u}(0), \mathbf{z}_j \rangle = \langle \mathbf{u}_0, \mathbf{z}_j \rangle \tag{7.5.6}$$

Equation (7.5.5) is readily solved subject to (7.5.6) to obtain

$$\langle \mathbf{u}(t), \mathbf{z}_j \rangle = \langle \mathbf{u}_0, \mathbf{z}_j \rangle e^{-\lambda_j t} + \int_0^t \langle \mathbf{f}(t'), \mathbf{z}_j \rangle e^{-\lambda_j (t-t')}\, dt'$$

and since $\{\mathbf{z}_j\}$ is an orthonormal basis in $\mathcal{H}$, we write the solution as

$$\mathbf{u}(t) = \sum_{j=1}^{\infty} \left\{ \langle \mathbf{u}_0 \mathbf{z}_j \rangle e^{-\lambda_j t} + \int_0^t \langle \mathbf{f}(t'), \mathbf{z}_j \rangle e^{-\lambda_j (t-t')}\, dt' \right\} \mathbf{z}_j \tag{7.5.7}$$

An alternative approach is to solve (7.5.4) directly by viewing $\mathbf{T}$ as a scalar quantity and to subsequently interpret appropriately the operator function that appears in the solution. Thus we write

$$\mathbf{u}(t) = \mathbf{e}^{-t\mathbf{T}}\mathbf{u}_0 + \int_0^t \mathbf{e}^{-(t-t')\mathbf{T}}\mathbf{f}(t')\, dt' \tag{7.5.8}$$

The terms $e^{-t\mathbf{T}}$ and $e^{-(t-t')\mathbf{T}}$ represent operators and the spectral representation of $\mathbf{T}$ may be used to obtain their representations as in Section 7.4. Thus we may write for the solution

$$\mathbf{u}(t) = \sum_{j=1}^{\infty} e^{-\lambda_j t}[\mathbf{P}_j \mathbf{u}_0 + \int_0^t e^{\lambda_j t'} \mathbf{P}_j \mathbf{f}(t')\, dt'] \tag{7.5.9}$$

The solutions (7.5.9) and (7.5.7) are exactly equivalent, and are meaningful only for a nonnegative operator $\mathbf{T}$.

Next we consider problems of the type

$$-\frac{d^2}{dt^2}\mathbf{u}(t) = \mathbf{Tu}(t), \qquad \mathbf{u}(0) = \mathbf{u}_0, \qquad \frac{d}{dt}\mathbf{u}(0) = \mathbf{u}'_0 \tag{7.5.10}$$

†Generally, in application to boundary value problems $\mathbf{T}$ originates from a differential or partial differential expression together with homogeneous boundary conditions. Thus $\mathbf{T}$ is an unbounded operator, but it will become evident from Chapter 9 that $\mathbf{T}$ satisfies the requirements of Theorem 7.4.3 for the assumed spectral resolution.

‡This operation is equivalent to operating on either side of (7.5.4) with projection $P_j$. Then (7.5.5) is obtained when one uses $P_j T = \lambda_j P_j$.

where $\mathbf{u}_0, \mathbf{u}'_0 \in \mathcal{H}$. Problems of this type occur commonly in vibrations.† Generally, the operator $\mathbf{T}$ in such applications is nonnegative. If $\mathbf{T}$ is positive, we can solve this problem directly by replacing $\mathbf{T}$ in (7.5.10) by a positive number and obtain the solution in a manner similar to that which led to the solution (7.5.9) of (7.5.4). We leave this as an exercise, which also shows how the nonnegative case can be accommodated by a limiting procedure. Here we use the first method used for solving (7.5.4), which does not depend on the constraint that $\mathbf{T}$ be positive. Thus we form the inner product of (7.5.10) with each of the eigenvectors of $\mathbf{T}$. We assume for specificity that $\lambda_1 = 0$ and all other eigenvalues to be positive. The inner product of (7.5.10) with $\mathbf{z}_1$ yields

$$-\frac{d^2}{dt^2}\langle \mathbf{u}(t), \mathbf{z}_1\rangle = \langle \mathbf{T}\mathbf{u}(t), \mathbf{z}_1\rangle = \langle \mathbf{u}(t), \mathbf{T}\mathbf{z}_1\rangle = 0$$

Solving, we obtain

$$\langle \mathbf{u}(t), \mathbf{z}_1\rangle = a_1 + b_1 t$$

where $a_1$ and $b_1$ are constants. Substituting the initial values, we obtain

$$a_1 = \langle \mathbf{u}_0, \mathbf{z}_1\rangle$$
$$b_1 = \langle \mathbf{u}'_0, \mathbf{z}_1\rangle$$

The inner product of (7.5.10) with $\mathbf{z}_j$ yields

$$-\frac{d^2}{dt^2}\langle \mathbf{u}(t), \mathbf{z}_j\rangle = \lambda_j\langle \mathbf{u}(t), \mathbf{z}_j\rangle$$

where $\lambda_j$ is positive. The solution is clearly

$$\langle \mathbf{u}(t), \mathbf{z}_j\rangle = a_j \sin\sqrt{\lambda_j}t + b_j \cos\sqrt{\lambda_j}t$$

Substitution of the initial values yields

$$\langle \mathbf{u}_0, \mathbf{z}_j\rangle = b_j$$
$$\frac{1}{\sqrt{\lambda_j}}\langle \mathbf{u}'_0, \mathbf{z}_j\rangle = a_j$$

The final solution emerges as

$$\mathbf{u}(t) = \{\langle \mathbf{u}_0, \mathbf{z}_1\rangle + \langle \mathbf{u}'_0, \mathbf{z}_1\rangle t\}\mathbf{z}_1 + \sum_{j=2}^{\infty}\left\{\frac{1}{\sqrt{\lambda_j}}\langle \mathbf{u}'_0, \mathbf{z}_j\rangle \sin\sqrt{\lambda_j}t + \langle \mathbf{u}_0, \mathbf{z}_j\rangle \cos\sqrt{\lambda_j}t\right\}\mathbf{z}_j \tag{7.5.11}$$

We also recommend this strategy of solution for other types of equations, which are presented in the exercises. The method is clearly applicable to situations, where instead of "initial" values (at $t = 0$), "boundary" values are

†In a vibration problem, the specification of the initial values of $\mathbf{u}(t)$ and $d\mathbf{u}/dt$ is equivalent to specifying the initial displacement and rate of displacement of the vibrating medium.

specified for $\mathbf{u}(t)$ (and its derivatives) at end points of the interval in which $t$ may be constrained to vary.

In applying the method to boundary value problems (see Section 9.2.3) it may frequently happen that the dependent variable in (7.5.4) [or for that matter any other equation such as (7.5.10)] does not belong to the domain of the operator. Calling the dependent variable $\mathbf{v}(t)$ [instead of $\mathbf{u}(t)$], we rewrite (7.5.4) as

$$-\frac{d}{dt}\mathbf{v}(t) = T\mathbf{v}(t) - \mathbf{f}(t), \qquad \mathbf{v}(0) = \mathbf{v}_0 \tag{7.5.12}$$

which, unlike (7.5.4), features the formal operation $T$ instead of the operator $\mathbf{T}$. Since $\mathbf{v}(t) \notin D(T)$, (7.5.5) does not hold, so that the solution, which is derived therefrom, will not apply for (7.5.12). However, we envisage only situations in which

$$\mathbf{v}(t) = \mathbf{u}(t) + \mathbf{g}(t) \tag{7.5.13}$$

where $\mathbf{u}(t) \in D(T)$, and $\mathbf{g}(t)$ is known with $T\mathbf{g}(t) \in \mathcal{H}$. Making the substitution of (7.5.13) in (7.5.12) it is readily seen that

$$-\frac{d}{dt}\mathbf{u}(t) = \mathbf{T}\mathbf{u}(t) - \mathbf{h}(t), \qquad \mathbf{u}(0) = \mathbf{u}_0 \tag{7.5.14}$$

where $\mathbf{h}(t) \equiv \mathbf{f}(t) - d\mathbf{g}(t)/dt - T\mathbf{g}(t)$ and $\mathbf{u}_0 = \mathbf{v}_0 - \mathbf{g}(0)$. This equation is now the same as (7.5.4) except for $\mathbf{h}(t)$, which replaces $\mathbf{f}(t)$. Thus the solution may be expressed either by (7.5.7) or (7.5.9) with $\mathbf{h}(t)$ in place of $\mathbf{f}(t)$. We write

$$\mathbf{v}(t) = \mathbf{g}(t) + \sum_{j=1}^{\infty} e^{-\lambda_j t}\left[\mathbf{P}_j\{\mathbf{v}_0 - \mathbf{g}(0)\} + \int_0^t e^{\lambda_j t'}\mathbf{P}_j\left\{\mathbf{f}(t') - \frac{d}{dt}\mathbf{g}(t') - T\mathbf{g}(t')\right\} dt'\right]$$

which, on integration by parts, yields the solution

$$\mathbf{v}(t) = \sum_{j=1}^{\infty} e^{-\lambda_j t}\left[\mathbf{P}_j\mathbf{v}_0 + \int_0^t e^{\lambda_j t'}\mathbf{P}_j\{\mathbf{f}(t') - T\mathbf{g}(t') + \lambda_j\mathbf{g}(t')\}\, dt'\right] \tag{7.5.15}$$

The student is referred to Section 9.2.3 for the application of this method to solve boundary value problems of engineering interest.

## EXERCISES

**7.5.1** Establish the formula (7.5.3).

**7.5.2** Assuming that $\mathbf{T}$ is positive in (7.5.10), solve the equation directly to obtain

$$\mathbf{u}(t) = (\sqrt{\mathbf{T}})^{-1} \sin (t\sqrt{\mathbf{T}})\mathbf{u}_0' + \cos (t\sqrt{\mathbf{T}})\mathbf{u}_0$$

Express the solution in terms of the spectral resolution of the identity generated by $\mathbf{T}$. Show also that the case of nonnegative $\mathbf{T}$ (with $\lambda_1 = 0$) can be accommodated by taking the limit of the solution above as $\lambda_1 \longrightarrow 0$ to obtain the same solution as (7.5.11).

**7.5.3** Solve

$$-\frac{d^2}{dt^2}\mathbf{u}(t) = \mathbf{T}\mathbf{u}(t) - \mathbf{f}(t), \qquad \mathbf{u}(0) = \mathbf{u}_0, \qquad \frac{d}{dt}\mathbf{u}(0) = \mathbf{u}'_0$$

where $\mathbf{T}$ is the operator of Exercise 7.5.2, by both methods outlined in Section 7.5.

**7.5.4** Solve

$$\frac{d^2}{dt^2}\mathbf{u}(t) = \mathbf{T}\mathbf{u}(t) + \mathbf{f}(t), \qquad a < t < b$$

subject to the boundary conditions

$$\mathbf{u}(a) = \mathbf{u}(b)$$

$$\frac{d}{dt}\mathbf{u}(t)\bigg|_{t=a} = \frac{d}{dt}\mathbf{u}(t)\bigg|_{t=b}$$

Assume $\mathbf{T}$ to be nonnegative.

**7.5.5** Solve

$$\frac{d^2}{dt^2}\mathbf{u}(t) - \frac{d}{dt}\mathbf{u}(t) - \mathbf{T}\mathbf{u}(t) = \mathbf{0}, \qquad 0 < t < 1$$

where $\mathbf{T}$ is nonnegative and $\mathbf{u}(t)$ satisfies the boundary conditions

$$-\frac{d}{dt}\mathbf{u}(0) + \mathbf{u}(0) = \mathbf{u}_0, \qquad \frac{d}{dt}\mathbf{u}(1) = \mathbf{0}$$

## 7.6 The Spectra of Bounded and Unbounded Self-Adjoint Operators

Our discussion has so far been confined to self-adjoint operators which are either compact or from which compact inverses may be obtained. But the admission of arbitrary, even bounded operators introduce tremendous complexities in the analysis. We will not go into the details of the general spectral theory but instead provide some informal generalizations. First, however, some prior concepts must be expanded. We will assume $\mathbf{T}$ to be bounded.

In introducing the concept of an eigenvalue of an operator $\mathbf{T}$, we had observed it to be a number $\lambda$ for which the operator $\mathbf{T}_\lambda \equiv (\mathbf{T} - \lambda\mathbf{I})$ does not have an inverse. Alternatively, there are nonzero vectors $\mathbf{x}$ such that $\mathbf{T}_\lambda\mathbf{x} = \mathbf{0}$. (We referred to $\mathbf{x}$ as an eigenvector.)

Let us now see what happens when no nonzero vector $\mathbf{x}$ exists such that $\mathbf{T}_\lambda\mathbf{x} = \mathbf{0}$; that is, $\lambda$ is not an eigenvalue. Thus $\mathbf{T}_\lambda\mathbf{x} = \mathbf{0} \rightarrow \mathbf{x} = \mathbf{0}$, so that $\mathbf{T}_\lambda$ maps $\mathcal{H}$ onto its range space $R_\lambda \equiv R(\mathbf{T}_\lambda)$ in a one-to-one manner. Thus $\mathbf{T}_\lambda^{-1}$ is defined, which obtains a unique vector in $\mathcal{H}$ for each vector in $R_\lambda$.† There are three possible situations.

†Recall how in the case of a compact, self-adjoint operator we had obtained in Section 7.4 the representation $\mathbf{T}_\lambda^{-1} = \sum_{j=1}^{\infty} (\lambda_j - \lambda)^{-1}\mathbf{P}_j$ [Eq. (7.4.20)] from which we concluded $\mathbf{T}_\lambda^{-1}$ to be *bounded.*

1. $R_\lambda = \mathcal{H}$.
2. $R_\lambda \neq \mathcal{H}$ but is *dense* in $\mathcal{H}$.
3. $R_\lambda \neq \mathcal{H}$ but is *not* dense in $\mathcal{H}$.

In case 1, $\mathbf{T}_\lambda^{-1}$ is bounded, since if it is unbounded, its domain $R_\lambda$ cannot be the entire space $\mathcal{H}$. We refer to the set of all $\lambda$'s for which case 1 is true as the *resolvent set*, $\rho(\mathbf{T})$. More formally, we write

$$\rho(\mathbf{T}) = \{\lambda \in \mathbb{C}\colon N(\mathbf{T} - \lambda\mathbf{I}) = \{\mathbf{0}\};\ R(\mathbf{T} - \lambda\mathbf{I}) = \mathcal{H}\}$$

For a *compact, self-adjoint* operator, the *entire complex plane* with the *exception of the eigenvalues* (on the real line) constitutes the *resolvent set*.

Next we introduce the concept of the *spectrum of an operator* $\mathbf{T}$, denoted $\sigma(\mathbf{T})$, as

$$\sigma(\mathbf{T}) = \{\lambda \in \mathbb{C}\colon \lambda \notin \rho(\mathbf{T})\}$$

Alternatively, $\sigma(T)$ is the complement of the resolvent set in the complex plane. Indeed, the spectrum of a compact self-adjoint operator comprises the eigenvalues. The set of eigenvalues is normally referred to as the point spectrum [denoted $P\sigma(\mathbf{T})$] or *discrete spectrum*, reflecting the fact that the eigenvalues are points on the complex plane. More generally, however, $\sigma(\mathbf{T})$ may also include the numbers, which are not eigenvalues but for which case 2 or 3 may hold.

Let us consider case 2. We know that $\mathbf{T}_\lambda^{-1}$ must be unbounded for this case. Thus there must exist at least one bounded sequence $\{\mathbf{y}_n\} \subset R_\lambda \ni \|\mathbf{y}_n\| = 1$ and

$$\|\mathbf{T}_\lambda^{-1}\mathbf{y}_n\| \longrightarrow \infty$$

Consider

$$\mathbf{x}_n = \frac{\mathbf{T}_\lambda^{-1}\mathbf{y}_n}{\|\mathbf{T}_\lambda^{-1}\mathbf{y}_n\|}$$

Then

$$\|\mathbf{x}_n\| = 1 \quad \text{and} \quad \mathbf{T}_\lambda\mathbf{x}_n = \frac{\mathbf{y}_n}{\|\mathbf{T}_\lambda^{-1}\mathbf{y}_n\|}$$

Clearly,

$$\|\mathbf{T}_\lambda\mathbf{x}_n\| = \frac{1}{\|\mathbf{T}_\lambda^{-1}\mathbf{y}_n\|}$$

so that

$$\|\mathbf{T}_\lambda\mathbf{x}_n\| \longrightarrow 0 \tag{7.6.1}$$

We have thus produced a bounded sequence $\{\mathbf{x}_n\}$ such that $\mathbf{T}_\lambda\mathbf{x}_n \longrightarrow \mathbf{0}$. It is important to recognize that the sequence $\{\mathbf{x}_n\}$ itself does *not* converge to any vector in $\mathcal{H}$ for if it did ($\mathbf{T}_\lambda$ being bounded), $\lambda$ would be an eigenvalue, which it is not. We refer to the set of $\lambda$'s for which case 2 holds as the *continuous spectrum*, denoted $C\sigma(\mathbf{T})$, for reasons that will become apparent later. For the present let us observe that the point spectrum $P\sigma(\mathbf{T})$ is *disjoint* from the continuous spectrum $C\sigma(\mathbf{T})$. Alternatively, if $\lambda$ is an eigenvalue, it cannot be in the continuous spectrum.

Even when $\lambda$ is an eigenvalue, it is possible to find sequences converging to the eigenvector such that (7.6.1) is true. A terminology that is used to describe the set of all $\lambda$'s for which a sequence $\{\mathbf{x}_n\}$ may be found such that (7.6.1) is true is the *approximate point spectrum*, denoted $\mathrm{AP}\sigma(\mathbf{T})$. Obviously, the approximate point spectrum includes both the point spectrum and the continuous spectrum.

The reason for the terminology of the continuous spectrum in case 2 lies in the fact that $C\sigma(\mathbf{T})$ consists of intervals rather than discrete points. Before we demonstrate this, we consider case 3.

The set of numbers $\lambda$ for which case 3 holds is called the *residual spectrum* and denoted $R\sigma(\mathbf{T})$. Thus the spectrum of an operator consists of the point spectrum, continuous spectrum, and the residual spectrum, all of which are mutually disjoint. In symbols, we write

$$\sigma(\mathbf{T}) = P\sigma(\mathbf{T}) \cup C\sigma(\mathbf{T}) \cup R\sigma(\mathbf{T})$$

We show below that self-adjoint operators do not have residual spectra. (We say the residual spectrum is empty.) Let $\lambda \in R\sigma(\mathbf{T})$. Since $R_\lambda$ is a proper subspace of $\mathcal{H}$ and is not dense, $\exists\ \mathbf{y} \in \mathcal{H} \ni \mathbf{y} \perp R_\lambda$. Thus $\forall\ \mathbf{x} \in \mathcal{H}$, we have

$$\begin{aligned} 0 = \langle \mathbf{T}_\lambda \mathbf{x}, \mathbf{y} \rangle &= \langle \mathbf{x}, \mathbf{T}\mathbf{y} \rangle - \langle \mathbf{x}, \lambda^* \mathbf{y} \rangle \\ &= \langle \mathbf{x}, \mathbf{T}\mathbf{y} - \lambda^* \mathbf{y} \rangle \end{aligned}$$

from which

$$\mathbf{T}\mathbf{y} = \lambda^* \mathbf{y}$$

Thus $\lambda^*$ is an eigenvalue! But since we know that the eigenvalues of $\mathbf{T}$ are real, we must have $\lambda = \lambda^*$ as an eigenvalue. We have thus obtained a contradiction because the residual spectrum and the point spectrum are disjoint. Hence there is no residual spectrum for a self-adjoint operator.

In extending the concept of a spectrum for self-adjoint operators beyond that for a compact operator, we have in effect produced the possibility of a continuous spectrum in addition to the point spectrum (the eigenvalues). Of course, there are bounded self-adjoint operators which have (1) only point (or discrete) spectra, (2) only continuous spectra, or (3) both point and continuous spectra. Regardless of these possibilities, it is readily shown that the *spectrum of a self-adjoint operator must lie on the real line*. Since the eigenvalues are known to be real, we need only show that the continuous spectrum is real. To prove this, we let $\lambda \in C\sigma(\mathbf{T})$ and $\{\mathbf{x}_n\}$ be the sequence with $\|\mathbf{x}_n\| = 1$ such that $\lim_{n\to\infty} \mathbf{T}_\lambda \mathbf{x}_n = 0$. It is then immediately evident that

$$\lim_{n\to\infty} |\langle \mathbf{T}\mathbf{x}_n, \mathbf{x}_n \rangle - \lambda| = 0 \tag{7.6.2}$$

But the sequence $\{\langle \mathbf{T}\mathbf{x}_n, \mathbf{x}_n \rangle\}$ is real and clearly its limit point $\lambda$ must also be real.

We shall now show an example in which a continuous spectrum arises. Consider the Hilbert space $\mathcal{H} = \mathcal{L}_2[0, 1]$, and the operator $\mathbf{T}$ defined by

$$\mathbf{Tx} \equiv \{tx(t)\} \tag{7.6.3}$$

It is readily seen that $\mathbf{T}$ is linear, bounded, and self-adjoint. Our objective is to investigate the spectrum of $\mathbf{T}$.

Let us first look for eigenvalues.

$$\mathbf{Tx} = \lambda \mathbf{x} \longrightarrow tx(t) = \lambda x(t)$$
$$(t - \lambda)x(t) = 0 \tag{7.6.4}$$

It is impossible to find for any real number $\lambda$ a function $x(t)$ which is not zero such that (7.6.4) is true.† Thus there cannot be any eigenvalues.

Now let $\lambda$ be any number between 0 and 1. Consider the sequence of functions $\{x_n(t)\}$ where‡

$$x_n(t) = \begin{cases} \sqrt{\dfrac{n}{\alpha}}, & |t - \lambda| < \dfrac{\alpha}{2n} \\ 0, & \text{otherwise} \end{cases} \tag{7.6.5}$$

where $\alpha \equiv \min\{\lambda, (1 - \lambda)\}$. The sequence is also portrayed in Figure 7.6.1.

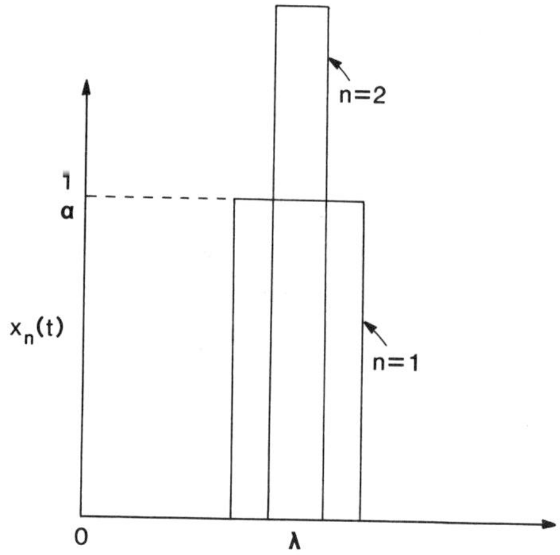


**Figure 7.6.1** Vector sequence $\{\mathbf{x}_n\}$ such that $\langle \mathbf{Tx}_n, \mathbf{x}_n \rangle$ converges to an element in the continuous spectrum of $\mathbf{T}$.

---

†This result is obvious for $\lambda$ outside the interval [0, 1]. But for $\lambda$ in the interval [0, 1], (7.6.4) requires that $x(t)$ vanish for all $t$ except at $t = \lambda$, which is a set of measure zero. Thus $x(t)$ is zero almost everywhere so that $\mathbf{x} = \mathbf{0}$.

‡If $\lambda$ is at either of the endpoints 0 and 1, this general construction must clearly be altered so that interval for which $x_n(t)$ does not vanish must extend only on one side of $\lambda$. This sequence may be recognized as leading to a function of the "Dirac $\delta$ type."

Clearly,

$$\|\mathbf{x}_n\| = \sqrt{\int_0^1 |x(t)|^2\, dt} = \sqrt{\frac{n}{\alpha}\frac{\alpha}{n}} = 1$$

Thus $\{\mathbf{x}_n\}$ is a bounded sequence. Now

$$\|\mathbf{T}_\lambda \mathbf{x}_n\|^2 = \int_{\lambda-\alpha/2n}^{\lambda+\alpha/2n} (t-\lambda)^2 \frac{n}{\alpha}\, dt = \frac{\alpha^2}{24n^2}$$

or

$$\|\mathbf{T}_\lambda \mathbf{x}_n\| = \frac{\alpha}{2\sqrt{6}\,n} \tag{7.6.6}$$

which becomes arbitrarily small. Thus $\lambda$ is in the continuous spectrum. For $\lambda$ not in [0, 1], $(t-\lambda)^{-1}$ is defined and bounded for $t$ in [0, 1]. Thus

$$\mathbf{T}_\lambda^{-1}\mathbf{x} = \{(t-\lambda)^{-1}x(t)\} \tag{7.6.7}$$

so that $\lambda$ is in the resolvent set. Thus we conclude that

$$\sigma(\mathbf{T}) = C\sigma(\mathbf{T}) = [0, 1]$$

The resolvent set $\rho(\mathbf{T})$ is the complex plane with [0, 1] sliced off the real axis.

### 7.6.1 Spectral Resolution

The question of a spectral resolution remains when a continuous spectrum appears. The general treatment of continuous spectral resolutions involves considerable detail and will not occupy us here. Instead, we continue our discussion of the operator (7.6.3).

Consider $\lambda \in C\sigma(T)$ and the operator $\mathbf{P}(\lambda)$ defined on $\mathfrak{L}_2[0, 1]$ by

$$\mathbf{y} = \mathbf{P}(\lambda)\mathbf{x}, \qquad y(t) = \begin{cases} 0, & t > \lambda \\ x(t), & t < \lambda \end{cases} \tag{7.6.8}$$

Figure 7.6.2(a) shows the effect of the transformation. It is obvious that $\mathbf{P}(\lambda)$ is a projection operator. Furthermore, it is an orthogonal projection that we leave for the student to verify.

If $\lambda > \mu$, the operator $\mathbf{P}(\lambda) - \mathbf{P}(\mu)$ is a self-adjoint projection also. It is defined by

$$\mathbf{y} = [\mathbf{P}(\lambda) - \mathbf{P}(\mu)]\mathbf{x}, \qquad y(t) = \begin{cases} x(t), & \mu < t < \lambda \\ 0, & \text{otherwise} \end{cases} \tag{7.6.9}$$

Figure 7.6.2(b) portrays the transformation. When $(\lambda - \mu)$ is very small, then the action of $\mathbf{T}$ on the range space $R[\mathbf{P}(\lambda) - \mathbf{P}(\mu)]$, which consists of functions vanishing outside the interval $(\mu, \lambda)$ is very nearly that of $\lambda[\mathbf{P}(\lambda) - \mathbf{P}(\mu)]$.

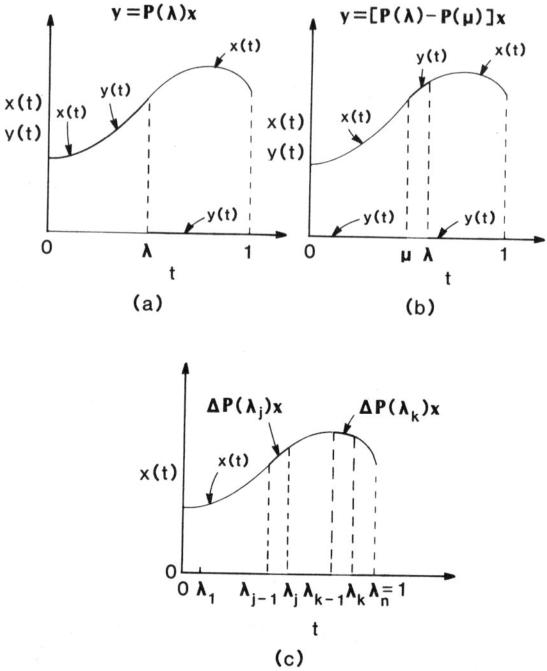


**Figure 7.6.2** Continuous orthogonal family of self-adjoint projections for the operator $\mathbf{Tx} = \{tx(t)\}$, $\mathbf{x} \in \mathfrak{L}_2[0, 1]$.

Clearly, the interval [0, 1] may be partitioned into intervals, using

$$0 = \lambda_0 < \lambda_1 < \ldots < \lambda_n = 1$$

and we may define self-adjoint projections $\Delta\mathbf{P}(\lambda_j) \equiv \mathbf{P}(\lambda_j) - \mathbf{P}(\lambda_{j-1})$, $j = 1, 2, \ldots, n$ [see also Figure 7.6.2(c)]. It is readily shown that they are an orthogonal family (see Exercise 7.6.3). The operator sum $\sum_{j=1}^{n} \Delta\mathbf{P}(\lambda_j) = \mathbf{P}(\lambda_n)$ is the identity operator on $\mathfrak{L}_2[0, 1]$ since $\lambda_n = 1$. Thus we write

$$\sum_{j=1}^{n} \Delta\mathbf{P}(\lambda_j) = \mathbf{I} \tag{7.6.10}$$

which is true for *all* partitions. In the limit of infinitesimally fine partitions we may write

$$\int_0^1 d\mathbf{P}(\lambda) = \mathbf{I} \tag{7.6.11}$$

The integral in (7.6.11) is a Stieltjes integral.† Next we consider the

†In defining the Stieltjes integral, it is necessary to impose a one-sided continuity to avoid ambiguity. For example, we let $\mathbf{P}(\lambda)$ be continuous from the left in that $\lim_{\varepsilon \to 0} \mathbf{P}(\lambda - \varepsilon) = \mathbf{P}(\lambda)$.

operator sum

$$\sum_{j=1}^{\infty} \lambda'_j \, \Delta\mathbf{P}(\lambda_j) \tag{7.6.12}$$

where $\lambda_{j-1} \leq \lambda'_j \leq \lambda_j$. The sum (7.6.12) is called a Riemann–Stieltjes sum. Let $\mathbf{x} \in \mathfrak{L}_2[0, 1]$, and $\mathbf{x}_j \equiv \Delta\mathbf{P}(\lambda_j)\mathbf{x}$. The family $\{\mathbf{x}_j\}_{j=1}^n$ is orthogonal and from (7.6.10) we have $\mathbf{x} = \sum_{j=1}^n \mathbf{x}_j$. The Pythagorean theorem yields

$$||\mathbf{x}||^2 = \sum_{j=1}^{n} ||\mathbf{x}_j||^2 \tag{7.6.13}$$

Since $\mathbf{x}_j \equiv \{x_j(t)\}$ vanishes outside the interval $(\lambda_{j-1}, \lambda_j)$, we must have

$$||\mathbf{T}\mathbf{x}_j - \lambda'_j\mathbf{x}_j|| < (\lambda_j - \lambda_{j-1})||\mathbf{x}_j|| \tag{7.6.14}$$

If we let $(\lambda_j - \lambda_{j-1}) < \varepsilon$ for each $j$, where $\varepsilon$ is a small positive number, then (7.6.14) leads to

$$||\mathbf{T}\mathbf{x}_j - \lambda'_j\mathbf{x}_j|| < \varepsilon\,||\mathbf{x}_j|| \tag{7.6.15}$$

Since $\{\mathbf{T}\mathbf{x}_j\}_{j=1}^n$ also forms an orthogonal family, $\{\mathbf{T}\mathbf{x}_j - \lambda'\mathbf{x}_j\}_{j=1}^n$ is an orthogonal set and

$$\begin{aligned}\left\|\left[\mathbf{T} - \sum_{j=1}^{n} \lambda'_j \, \Delta\mathbf{P}(\lambda_j)\right]\mathbf{x}\right\|^2 &= \left\|\sum_{j=1}^{n} [\mathbf{T}\mathbf{x}_j - \lambda'_j\mathbf{x}_j]\right\|^2 \\ &= \sum_{j=1}^{n} ||\mathbf{T}\mathbf{x}_j - \lambda'_j\mathbf{x}_j||^2 \\ &< \varepsilon \sum_{j=1}^{n} ||\mathbf{x}_j||^2 = \varepsilon\,||\mathbf{x}||^2 \end{aligned} \tag{7.6.16}$$

where we have again used the Pythagorean theorem, (7.6.15), and (7.6.13), in that order. The inequality (7.6.16) clearly implies that

$$||[\mathbf{T} - \sum \lambda'_j \, \Delta\mathbf{P}(\lambda_j)]|| < \varepsilon \tag{7.6.17}$$

In the limit of vanishing partition size we may represent $\mathbf{T}$ as a Stieltjes integral:

$$\mathbf{T} = \int_0^1 \lambda \, d\mathbf{P}(\lambda) \tag{7.6.18}$$

Equation (7.6.18) is then the required *spectral representation* of the bounded self-adjoint operator $\mathbf{T}$ defined by (7.6.3). Having established the foregoing result for the specific operator $\mathbf{T}$, we will foresake a general treatment to make the assertion that an *arbitrary bounded self-adjoint* $\mathbf{T}$ has the representation

$$\mathbf{T} = \int_{\lambda \in \sigma(\mathbf{T})} \lambda \, d\mathbf{P}(\lambda) \tag{7.6.19}$$

The interval of integration has been indicated as the spectrum of the operator. For a bounded operator, this is a bounded interval. Now for every $\lambda \in \sigma(\mathbf{T})$, there exists at least one sequence $\{\mathbf{x}_n\}$, with $\|\mathbf{x}_n\| = 1$, such that

$$\lim_{n\to\infty} \langle \mathbf{Tx}_n, \mathbf{x}_n \rangle = \lambda$$

which was left as an exercise (see Exercise 7.6.1). From Theorem 7.2.2 we know that $\|\mathbf{T}\| = \sup\{|\langle \mathbf{Tx}, \mathbf{x}\rangle| : \|\mathbf{x}\| = 1\}$. Clearly, then, we must have $|\lambda| \leq \|\mathbf{T}\|$, so that the limits of integration in (7.6.19) will not go beyond the interval $-\|\mathbf{T}\| \leq \lambda \leq \|\mathbf{T}\|$.

It is important to recognize that (7.6.19) accommodates the point spectrum also. We do not elaborate on this point but simply indicate how in the presence of a point and continuous spectrum (7.6.19) may be rewritten as

$$\mathbf{T} = \sum_{\lambda_j \in P\sigma(T)} \lambda_j \mathbf{P}_j + \int_{\lambda \in C\sigma(\mathbf{T})} \lambda \, d\mathbf{P}(\lambda) \tag{7.6.20}$$

It is beyond our scope to discuss the details of the spectral resolution of unbounded operators, but barring the details, the conclusion for self-adjoint operators leads to the representation (7.6.20), with the only difference that at least one of the limits of integration in the continuous spectral resolution is infinity. The student would do well to remember that it is *not essential* that bounded or unbounded operators *must* have a continuous spectrum.†

Although we have asserted the existence of a continuous family of self-adjoint projections in the case of a continuous spectrum, we have not discussed any method for *finding* them. This matter is beyond the scope of this book.

We now turn to the spectral resolution of functions of bounded or unbounded self-adjoint operators. Without much ado we will state that for a function $\mathbf{f}(\mathbf{T})$ similar to those in Section 7.4, we write

$$\mathbf{f}(\mathbf{T}) = \sum_{\lambda_j \in P\sigma(\mathbf{T})} f(\lambda_j)\mathbf{P}_j + \int_{\lambda \in C\sigma(\mathbf{T})} f(\lambda) \, d\mathbf{P}(\lambda) \tag{7.6.21}$$

The expression (7.6.21) may then be used to solve the various types of equations considered in Section 7.5. As an example, the solution to the initial value problem

$$-\frac{d}{dt}\mathbf{u}(t) = \mathbf{Tu}(t), \qquad \mathbf{u}(0) = \mathbf{u}_0$$

†This remark is a needless reminder to the alert student who recalls that a linear combination of the orthogonal family of self-adjoint projections $\{\mathbf{P}_j\}$ of the type $\sum_j \alpha_j \mathbf{P}_j$ (with $\alpha_j$ bounded for each $j$) is a bounded operator with only a point spectrum. Again unbounded self-adjoint operators with compact inverses have no continuous spectrum.

is given by

$$\mathbf{u}(t) = \sum_{\lambda \in C\sigma(\mathbf{T})} e^{-\lambda_j t}\mathbf{P}_j\mathbf{u}_0 + \int_{\lambda \in C\sigma(\mathbf{T})} e^{-\lambda t}\, d\mathbf{P}(\lambda)\mathbf{u}_0 \tag{7.6.22}$$

The other equations may be solved in an entirely analogous manner.

**EXERCISES**

**7.6.1** Let $\mathbf{T} \in \text{Blt}(\mathcal{H}) \ni \mathbf{T} = \mathbf{T}^*$. Show that if $\lambda \in C\sigma(\mathbf{T})$ $\exists$ a sequence $\{\mathbf{x}_n\}$, with $\|\mathbf{x}_n\| = 1$,

$$\lambda = \lim_{n \to \infty} \langle \mathbf{T}\mathbf{x}_n, \mathbf{x}_n \rangle$$

**7.6.2** Let $\mathbf{T} \in \text{Blt}(\mathcal{H})$. If $\lambda \in R\sigma(\mathbf{T})$, show that $\lambda^* \in P\sigma(\mathbf{T})$.

**7.6.3** For the operator $\mathbf{T}$ given by (7.6.3), show that the operator $\mathbf{P}(\lambda)$ defined by (7.6.8) is a self-adjoint projection. Further establish that the family $\{\Delta\mathbf{P}(\lambda_j) \equiv \mathbf{P}(\lambda_j) - \mathbf{P}(\lambda_{j-1})\}_{j=1}^{n}$ is an orthogonal family of self-adjoint projections.

## 7.7 Another View of Spectral Representation

We shall present here a very heuristic discussion of the spectral representation of a self-adjoint operator $\mathbf{T}$ from another viewpoint. Recall the familiar Cauchy integral formula† for an analytic function $f(z)$ of complex variable $z$, given by

$$f(z) = \frac{1}{2\pi i}\oint_C \frac{f(\lambda)}{z - \lambda}\, d\lambda \tag{7.7.1}$$

where $c$ is a closed (Jordan) curve enclosing $z$, which is the only singularity of $(z - \lambda)^{-1}$. In particular, (7.7.1) also implies that

$$1 = \frac{1}{2\pi i}\oint_C \frac{d\lambda}{z - \lambda} \tag{7.7.2}$$

$$z = \frac{1}{2\pi i}\oint_C \frac{\lambda}{z - \lambda}\, dy \tag{7.7.3}$$

Spectral theory of an operator $\mathbf{T}$ may be viewed as the attempt to replace $z$ in formulas (7.7.1) to (7.7.4) by the operator $\mathbf{T}$.‡ Then we may write for (7.7.1)

$$\mathbf{f}(\mathbf{T}) = \frac{1}{2\pi i}\oint_C f(\lambda)(\mathbf{T} - \lambda\mathbf{I})^{-1}\, d\lambda \tag{7.7.4}$$

†See *Complex Variables and Applications*, 3rd ed., by R. V. Churchill, J. W. Brown, and R. F. Verhey, McGraw-Hill, New York, 1976.

‡In regard to the likeness of operators to numbers, see Exercise 6.6.1.

Now the spectrum of the operator **T** may be interpreted as the "singularities" of $(\mathbf{T} - \lambda\mathbf{I})^{-1}$ [.ie., values of $\lambda$ for which $(\mathbf{T} - \lambda\mathbf{I})$ does not have a bounded inverse]. Discrete spectra or eigenvalues will correspond to poles and continuous spectra correspond to branch singularities. For self-adjoint operators singularities can occur only along the real line (if $z$ is real, the singularity of $(z - \lambda)^{-1}$ in (7.7.1) is obviously real, which again associates the self-adjoint operator with the real number).

The residue theorem in complex contour integration, which leads to the equality in (7.7.1) in a trivial manner, can be used to construct the right-hand side of (7.7.4) in terms of the residues at the singularities. For a self-adjoint operator with a discrete spectrum, we have

$$\mathbf{f}(\mathbf{T}) = \sum_{j=1}^{\infty} \frac{1}{(r_j - 1)!} \frac{d^{r_j - 1}}{d\lambda^{r_j - 1}} [f(\lambda)(\mathbf{T} - \lambda\mathbf{I})^{-1}]_{\lambda = \lambda_j} \tag{7.7.5}$$

which is obtained by letting $\lambda = \lambda_j$ be a multiple pole of order $r_j$ (corresponding to an eigenvalue $\lambda_j$ with the associated eigenspace $N_{\lambda_j}$ of dimension $r_j$). Thus the resolution of the identity **I** may be written as

$$\mathbf{I} = \sum_{j=1}^{\infty} \frac{1}{(r_j - 1)!} \frac{d^{r_j - 1}}{d\lambda^{r_j - 1}} [(\mathbf{T} - \lambda\mathbf{I})^{-1}]_{\lambda = \lambda_j}$$

Clearly, the $j$th projection $\mathbf{P}_j$ is given by

$$\mathbf{P}_j = \frac{1}{(r_j - 1)!} \frac{d^{r_j - 1}}{d\lambda^{r_j - 1}} [(\mathbf{T} - \lambda\mathbf{I})^{-1}]_{\lambda = \lambda_j}$$

A continuous spectral representation is similarly obtained with branch singularities.

The foregoing discussion, although entirely heuristic, provides a very interesting perspective of the development of spectral theory. We shall have occasion to recall this development in Chapters 9 and 11 in comparing spectral solutions to boundary-initial value problems to those obtained by the method of Laplace transforms.

## 7.8 Concluding Remarks

In this chapter we have dealt with self-adjoint operators, the properties that make them attractive, and how equations involving them can be solved using their spectral representations. The compact self-adjoint operator received special attention in this regard. We found that such an operator **T** afforded a compartmentalization of the Hilbert space into eigenspaces, in each of which the action of $\mathbf{T}$ on a vector was that of a scalar multiple (the eigenvalue). It was then possible to view every vector as being composed of pieces in each of the eigenspaces (or as an expansion in terms of the eigenvectors of $\mathbf{T}$). Given a vector, one could identify its pieces (projections on eigenspaces), or given the pieces of a vector, one could put them together to identify the vector. The powerful feature of the compartmentalization of the Hilbert space which $\mathbf{T}$ produces is that it is also shared by *functions of* $\mathbf{T}$ (in regard to the simplicity of their action on vectors in the compartment subspaces). Alternatively, we say that the self-adjoint operator and functions of it have the same set of eigenvectors or yield the same spectral resolution. Thus the solution of equations depended on the strategy of computing the unknown vector by obtaining its pieces in the eigenspaces. Section 7.5 is particularly important in this regard.

In Chapter 8 we consider applications of spectral theorem 7.4.1. Spectral theorem 7.4.3 is central to the applications to differential equations that constitute Chapter 9. Since differential operators are unbounded operators, it is opportune to recall here that an unbounded operator $\mathbf{T}$ can at most be defined on a dense subspace of the Hilbert space $\mathcal{H}$, say $D(T)$. We have avoided the issue of what it means to talk about functions of unbounded operators except through their interpretation via their spectral representations (7.4.26). The difficulty with unbounded operators, for example, may be demonstrated by considering $\mathbf{T}^2$. $\mathbf{Tx}$ is defined only for $\mathbf{x} \in D(T)$ but since $\mathbf{T}^2\mathbf{x} = \mathbf{T}(\mathbf{Tx})$, and $\mathbf{Tx}$ is not necessarily in $D(T)$, we cannot consider $\mathbf{T}^2$ unless its domain is given by $D(T^2) = \{\mathbf{x} \in D(T) : \mathbf{Tx} \in D(T)\}$. Indeed, while such constraints may be imposed, the entire problem is elegantly handled by von Neumann.† It is possible, however, to depend on (7.4.26) for the solution of problems which are dealt with in this book.

The extension of the spectral theory to bounded and unbounded self-adjoint operators has been admittedly sketchy. When the continuous spectrum arises, the partitioning of the Hilbert space must be viewed as "infinitesimal slicings" of the space into an uncountably infinite number of partitions.

We have not expended any effort in this chapter on the methods of calculation of eigenvalues and eigenvectors of operators. The student must attempt the exercises in Section 7.4 in this regard, which show an approximation technique via matrix representations of the operator. For many applications related to differential operators, however, we will encounter direct methods of computing the spectra in Chapter 9.

---

†See, for example, an article by E. R. Lorch, "The Spectral Theorem," in *Studies in Mathematics*, Vol. I, Math Association of America, 1962. von Neumann's idea is in defining an invertible transformation of the unbounded operator to a bounded normal (unitary see Exercise 7.4.12) operator whose functions are well-defined.

## FURTHER READING

As in Chapters 5 and 6, much of the material of this chapter is discussed at a somewhat higher level in

> *Linear Operator Theory in Engineering and Science* by A. W. Naylor and G. R. Sell, Holt, Rinehart and Winston, New York, 1971.

An interesting account of the basic ideas of spectral theory is contained in an elementary exposition contained in

> "The Spectral Theorem," by E. R. Lorch, in *Studies in Mathematics*, Vol. 1, Mathematical Association of America, Washington, D.C., 1962, pp. 89–137.

A very compact treatment of spectral theory is available in

> *Introduction to Hilbert Space and the Theory of Spectral Multiplicity*, by P. R. Halmos, Chelsea, New York, 1951.

# Applications in Finite-Dimensional Space

# 8

## 8.0 Foreword

Our objective in this chapter is to show that spectral theorem 7.4.1 can be exploited to solve numerous problems of engineering interest. The solutions of the equations will draw on the techniques of Section 7.5. As the Hilbert spaces of interest are finite dimensional, the linear operators involved are compact. Since matrices represent linear operators on finite-dimensional spaces, the numerous applications of matrices to chemical engineering problems come under the purview of this chapter. Many of these have been covered by others, for example, by Amundson† and by Jenson and Jeffreys;‡ we will therefore be selective in our choice, the basis for which is the utility of concepts that were introduced in Chapters 6 and 7 in that they simplify the treatment of problems of interest. Thus we exclude problems that involve only the use of conventional matrix algebra.

It will frequently be found that the treatment in this and later chapters depends on the results of certain key exercises in previous chapters. We advise the student to rework these exercises at the stage they are encountered.

---

†*Mathematical Methods in Chemical Engineering: Matrices and Their Application*, by N. R. Amundson, Prentice-Hall, Englewood Cliffs, N.J., 1966.

‡*Mathematical Methods in Chemical Engineering*, by V. G. Jenson and G. V. Jeffreys, Academic Press, London, 1977.

## 8.1 Introduction

The applications that concern us here are those that pertain to self-adjoint operators. With respect to the ordinary inner product in $\mathcal{R}_n$ [see, e.g., (6.1.13)], real symmetric matrices are self-adjoint. Our pursuit here, however, is of those matrix operators that are nonsymmetric (and hence apparently non-self-adjoint) but may be rendered self-adjoint by a redefinition of the inner product. Of course, this is not always possible, but the question is one of recognizing those situations where it is indeed a possibility. In Chapter 6 we found (see Exercises 6.6.2, 6.6.3, and 6.6.10) that a matrix operator **T** on $\mathcal{R}_n$ would be *self-adjoint* with respect to the inner product

$$\langle \mathbf{x}, \mathbf{y} \rangle \equiv \sum_{i=1}^{n} \sum_{j=1}^{n} a_{ij} x_i y_j \tag{8.1.1}$$

where $\{a_{ij}\}$ are the coefficients of a symmetric, positive-definite matrix **A** (see Exercise 6.1.2) *if and only if*

$$t_{ij} = \sum_{l=1}^{n} a'_{il} \sum_{k=1}^{n} t_{kl} a_{kj} \tag{8.1.2}$$

In (8.1.2) $\{a'_{il}\}$, $i, l = 1, 2, \ldots, n$, are the coefficients of the matrix $\mathbf{A}^{-1}$. In matrix notation, (8.1.2) may be rewritten as

$$\mathbf{AT} = \mathbf{T}^*\mathbf{A} = (\mathbf{AT})^* \tag{8.1.3}$$

which lays down the condition for self-adjointness of **T** with respect to the inner product (8.1.1) as self-adjointness of the matrix **AT** with respect to the usual inner product.

It is also useful to recognize the simpler situation where **A** is a diagonal matrix with (necessarily) positive coefficients $\{r_j\}$ for which

$$a_{kj} = r_j \delta_{kj}, \qquad a'_{il} = r_i^{-1} \delta_{il}$$

so that (8.1.2) becomes

$$r_i t_{ij} = t_{ji} r_j \tag{8.1.4}$$

A particularly interesting area of application is the very elegant treatment of first-order reaction systems by Wei and Prater.† We also consider applications to problems in multicomponent diffusion, an example in multicomponent rectification and transient analysis of some stagewise operations, all of which fit rather well into our method of approach.

†"The Structure and Analysis of Complex Reaction Systems," by J. Wei and C. D. Prater, Chapter 5 of *Advances in Catalysis*, Vol. 13, Academic Press, New York, 1962.

## 8.2 First-Order Reaction Systems

We are concerned here with $n$ chemical species $A_1, A_2, \ldots, A_n$, each of which undergoes first-order chemical transformations to every other species in accordance with the reaction equation

$$A_i \underset{k_{i,j}}{\overset{k_{j,i}}{\rightleftharpoons}} A_j, \qquad i, j = 1, 2, \ldots, n, \quad i \neq j$$

where $k_{j,i}$ is the rate constant for the transformation of $A_i$ to $A_j$. The problem of interest is the determination of the rate constants $\{k_{i,j};\ i, j = 1, 2, \ldots, n,\ i \neq j\}$ at a selected temperature from observations made on suitably designed batch experiments.

Since the total molar concentration is an invariant in a batch reacting system, the reaction rates may be represented in terms of mole fractions instead of concentrations. Thus for species $A_i$ we write

$$\begin{aligned}\frac{dx_i}{dt} &= k_{i,1}x_1 + k_{i,2}x_2 + \ldots + k_{i,i-1}x_{i-1} - \sum_{j=1,i}^{n} k_{j,i}x_i \\ &\quad + k_{i,i+1}x_{i+1} + \ldots + k_{i,n}x_n, \qquad i = 1, 2, \ldots, n\end{aligned} \tag{8.2.1}$$

In (8.2.1) the coefficient of $x_i$ is a summation over index $j$ which runs from 1 to $n$, excluding $i$.

If we let the mole fraction vector $\mathbf{x} \equiv (x_1, x_2, \ldots, x_n) \in \mathcal{R}_n$, the $n$ equations in (8.2.1) may be condensed into the vector equation

$$\frac{d}{dt}\mathbf{x} = \mathbf{K}\mathbf{x} \tag{8.2.2}$$

where $\mathbf{K}$ is the matrix

$$\begin{bmatrix} -\sum_{j=2}^{n} k_{j,1} & k_{1,2} & k_{1,3} & \cdots & k_{1,n} \\ k_{2,1} & -\sum_{j=1,2}^{n} k_{j,2} & k_{2,3} & \cdots & k_{2,n} \\ \cdot & \cdot & \cdot & & \cdot \\ \cdot & \cdot & \cdot & & \cdot \\ \cdot & \cdot & \cdot & & \cdot \\ k_{n,1} & k_{n,2} & k_{n,3} & & -\sum_{j=1}^{n-1} k_{j,n} \end{bmatrix} \tag{8.2.3}$$

Clearly, the matrix $\mathbf{K}$ is nonsymmetric. By summing elements of each column, one generates a row of zeros so that $\mathbf{K}$ is a singular matrix. We assume that the rank of $\mathbf{K}$ is $(n - 1)$. This assumes a *unique* equilibrium composition vector $\mathbf{a}$, which is the solution of the homogeneous equation

$$\mathbf{K}\mathbf{x} = \mathbf{0} \tag{8.2.4}$$

satisfying the additional constraint

$$\sum_{j=1}^{n} x_j = 1 \tag{8.2.5}$$

as all mole fractions do. It is evident that a single batch experiment carried through to equilibrium will yield the vector **a**.

A feature crucial to the entire analysis that follows is the *principle of detailed balancing or microscopic reversibility* requiring that the rate constants satisfy

$$k_{j,i}a_i = k_{i,j}a_j$$

or alternatively,

$$\frac{1}{a_j}k_{j,i} = \frac{1}{a_i}k_{i,j} \tag{8.2.6}$$

which is reminiscent of the relation (8.1.4) since each equilibrium mole fraction $a_j$ is strictly positive.† Thus although **K** is a real, nonsymmetric matrix, we may define on the linear space $\mathcal{R}_n$ the inner product

$$\langle \mathbf{x}, \mathbf{y} \rangle = \sum_{i=1}^{n} \frac{1}{a_i} x_i y_i \tag{8.2.7}$$

with respect to which the operator $\mathbf{K}: \mathcal{R}_n \longrightarrow \mathcal{R}_n$ defined by

$$\mathbf{y} = \mathbf{Kx}: \quad \mathbf{y}_i = \sum_{j=1}^{n} k_{i,j} x_j$$

is *self-adjoint*, that is, we have

$$\langle \mathbf{Kx}, \mathbf{y} \rangle = \langle \mathbf{x}, \mathbf{Ky} \rangle \qquad \forall \quad \mathbf{x}, \mathbf{y} \in \mathcal{H}, \quad \mathcal{H} \equiv \{\mathcal{R}_n, \langle \cdot, \cdot \rangle\}$$

where here and henceforth the inner product symbol will refer to that in (8.2.7). Further, the norm of a vector **x** is given by

$$\|\mathbf{x}\| = \sqrt{\sum_{i=1}^{n} \frac{1}{a_i} x_i^2} \tag{8.2.8}$$

From spectral theorem 7.4.1, **K** has $n$ real eigenvalues $\lambda_0, \lambda_1, \lambda_2, \ldots, \lambda_{n-1}$. In view of (8.2.4) and (8.2.5), it is clear that

$$\mathbf{Ka} = \mathbf{0}$$

so that 0 is an eigenvalue of **K**, which we identify to be $\lambda_0$. Since **K** has rank $(n - 1)$, $\lambda_0 = 0$ is not a repeated eigenvalue. We also assume that the nonzero eigenvalues $\lambda_1, \lambda_2, \ldots, \lambda_{n-1}$ are all distinct. Thus the eigenspaces $N_0 \equiv N(\mathbf{K})$, $N_{\lambda_j} \equiv N(\mathbf{K} - \lambda_j \mathbf{I})$, $j = 1, 2, \ldots, (n - 1)$, are all subspaces, each of dimension 1.

†Although this is an obvious physical fact, it also follows from the strict positivity of the rate constants and (8.2.5) and (8.2.6).

Before proceeding further, let us observe again that the objective is one of determining the unknown matrix **K** from a series of batch experiments. The strategy of Wei and Prater may be broadly stated as recovering **K** from its spectral resolution. Thus the eigenvalues $\lambda_1, \lambda_2, \ldots, \lambda_{n-1}$ and the corresponding eigenvectors $\mathbf{z}_1, \mathbf{z}_2, \ldots, \mathbf{z}_{n-1}$ are to be obtained from experiments. Every batch experiment leads to the equilibrium mole fraction vector **a** ($\equiv \mathbf{z}_0$), which is the eigenvector corresponding to the eigenvalue 0 ($\equiv \lambda_0$). The vectors $\mathbf{z}_0, \mathbf{z}_1, \mathbf{z}_2, \ldots, \mathbf{z}_{n-1}$ are regarded as normalized with respect to the norm given by (8.2.8). Since we have taken $\mathbf{z}_0 = \mathbf{a}$, let us ascertain that it is normalized.

$$||\mathbf{z}_0|| = \sqrt{\sum_{i=1}^{n} \frac{1}{a_i} a_i^2} = \sqrt{\sum_{i=1}^{n} a_i} = 1$$

The last step follows from the fact that the $a_i$'s are mole fractions. Since the eigenvectors $\{\mathbf{z}_j\}_{j=0}^{n-1}$ are an orthonormal basis in $\mathcal{R}_n$, we have for any $\mathbf{x} \in \mathcal{R}_n$,

$$\mathbf{x} = \sum_{i=0}^{n-1} \langle \mathbf{x}, \mathbf{z}_i \rangle \mathbf{z}_i$$

In particular, we define rate constant vectors $\{\mathbf{k}_j\}_{j=1}^{n}$ by

$$\mathbf{k}_j = (k_{1,j}, k_{2,j}, \ldots, k_{n,j})$$

where $k_{j,j} \equiv -\sum_{i=1,j}^{n} k_{i,j}$. It is left as an exercise to show that $\mathbf{k}_j \perp N(\mathbf{K})$ for each $j$. Thus we have

$$\mathbf{k}_j = \sum_{r=1}^{n-1} \langle \mathbf{k}_j, \mathbf{z}_r \rangle \mathbf{z}_r \tag{8.2.9}$$

If we let $\mathbf{z}_r \equiv (z_{1,r}, z_{2,r}, \ldots, z_{n,r})$, then the inner product $\langle \mathbf{k}_j, \mathbf{z}_r \rangle$ is given by

$$\langle \mathbf{k}_j, \mathbf{z}_r \rangle = \sum_{s=1}^{n} \frac{1}{a_s} k_{s,j} z_{s,r}$$

The principle of microscopic reversibility (8.2.6) yields from the foregoing equation

$$\langle \mathbf{k}_j, \mathbf{z}_r \rangle = \frac{1}{a_j} \sum_{s=1}^{n} k_{j,s} z_{s,r} \tag{8.2.10}$$

In view of $\mathbf{K}\mathbf{z}_r = \lambda_r \mathbf{z}_r$, (8.2.10) becomes

$$\langle \mathbf{k}_j, \mathbf{z}_r \rangle = \frac{1}{a_j} \lambda_r z_{j,r}$$

so that (8.2.9) becomes

$$\mathbf{k}_j = \frac{1}{a_j} \sum_{r=1}^{n-1} \lambda_r z_{j,r} \mathbf{z}_r$$

from which the rate constant $k_{i,j}$ is obtained as

$$k_{i,j} = \frac{1}{a_j} \sum_{r=1}^{n-1} \lambda_r z_{j,r} z_{i,r} \tag{8.2.11}$$

Equation (8.2.11) clearly reflects the principle of microscopic reversibility and shows how the rate constants may be obtained if all the nonzero eigenvalues and their corresponding eigenvectors are known. Following Wei and Prater, we lay out the strategy of determining the eigenvalues and eigenvectors through suitable batch experiments. In doing so we proceed by establishing a number of intermediate results as theorems. The physical implications of each theorem are discussed following the proof of that theorem.

***Theorem 8.2.1*** The operator **K** is nonpositive; that is, the nonzero eigenvalues $\lambda_1, \lambda_2, \ldots, \lambda_{n-1}$ are strictly negative.

*Proof*: Let $\mathbf{x} \in \mathcal{R}_n$.

$$\begin{aligned}\langle \mathbf{Kx}, \mathbf{x}\rangle &= \sum_{i=1}^{n} \frac{1}{a_i} x_i \sum_{j=1}^{n} k_{i,j} x_j \\ &= \sum\sum_{i \neq j} \frac{k_{i,j} x_i x_j}{a_i} - \sum\sum_{i \neq j} \frac{k_{j,i}}{a_i} x_i^2 \\ &= \sum\sum_{i \neq j} \left[\sqrt{\frac{k_{j,i}}{a_i}}\, x_i \sqrt{\frac{k_{i,j}}{a_j}}\, x_j - \left(\sqrt{\frac{k_{j,i}}{a_i}}\, x_i\right)^2\right]\end{aligned}$$

By an interchange of indices,

$$\langle \mathbf{Kx}, \mathbf{x}\rangle = \sum\sum_{i \neq j} \left[\sqrt{\frac{k_{i,j}}{a_j}}\, x_j \sqrt{\frac{k_{j,i}}{a_i}}\, x_i - \left(\sqrt{\frac{k_{i,j}}{a_j}}\, x_j\right)^2\right]$$

Adding the two results gives us

$$\langle \mathbf{Kx}, \mathbf{x}\rangle = -\frac{1}{2} \sum\sum_{i \neq j} \left(\sqrt{\frac{k_{i,j}}{a_j}}\, x_j - \sqrt{\frac{k_{j,i}}{a_i}}\, x_i\right)^2 \leq 0$$

so that $\lambda_1, \lambda_2, \ldots, \lambda_{n-1}$ are all negative. ∎

We assume now that the ordering of the eigenvalues is given by $0 > \lambda_1 > \lambda_2 > \ldots \lambda_{n-1}$.

It is convenient to define the following hyperplanes:†

$$H = \{\mathbf{x} = \mathbf{a} + \mathbf{z};\, \mathbf{z} \in N(\mathbf{K})^{\perp}\}$$

$$H_j = \{\mathbf{x} = \mathbf{a} + \mathbf{z};\, \mathbf{z} \in N_{\lambda_j}\}, \qquad j = 1, 2, \ldots, n-1$$

Clearly, $H_j \subset H$. Further we denote by $H^+$ and $H_j^+$ as subsets of $H$ and $H_j$, respectively, in which $x_i \geq 0$, $i = 1, 2, \ldots, n$. It is left as an exercise to show that all mole fraction vectors must lie in $H^+$. Next we define a batch reaction path.

---

†See Section 2.5 for the definition of a hyperplane. $H_j$ is, in fact, a straight line and the term "hyperplane" applied to $H_j$ may sound geometrically inappropriate but no more than its use in higher-dimensional spaces.

***Definition 8.2.1*** A *batch reaction path* of the first-order reaction system is a curve in $\mathcal{R}_n$ with the parametric representation $\mathbf{x} = \mathbf{x}(t)$, where $\mathbf{x}(t)$ is the solution of the differential equation $d\mathbf{x}/dt = \mathbf{K}\mathbf{x}$ subject to some initial composition $\mathbf{x}_0 \in H^+$.

The following theorem concerns reaction paths.

***Theorem 8.2.2*** Every batch reaction path lies entirely on the set $H^+$.

*Proof*: In view of Exercise 8.2.3 and the fact that mole fractions are positive and must add to unity, this theorem must be trivially true. However, since reaction paths are given by the preceding definition, we prove this result.

$$\frac{d}{dt}\langle \mathbf{x}, \mathbf{a} \rangle = \langle \mathbf{K}\mathbf{x}, \mathbf{a} \rangle = \langle \mathbf{x}, \mathbf{K}\mathbf{a} \rangle = 0$$

Thus

$$\langle \mathbf{x}, \mathbf{a} \rangle = \langle \mathbf{x}_0, \mathbf{a} \rangle = 1$$

from which $\sum_{i=1}^{n} x_i(t) = 1$. Since $x_i(0) \geq 0$, $i = 1, 2, \ldots, n$, it is enough to show that $dx_i/dt \geq 0$ whenever $x_i = 0$ to prove that $x_i(t) \geq 0$. This is left as an exercise. Thus $\mathbf{x}(t) \in H^+$ for all $t$. ∎

Theorem 8.2.2 simply asserts that things are in order with the description of the system insofar as mole fractions must always add to 1. Figure 8.2.1 shows the hyperplane $H$ for a three-component system. Next we establish an important result.

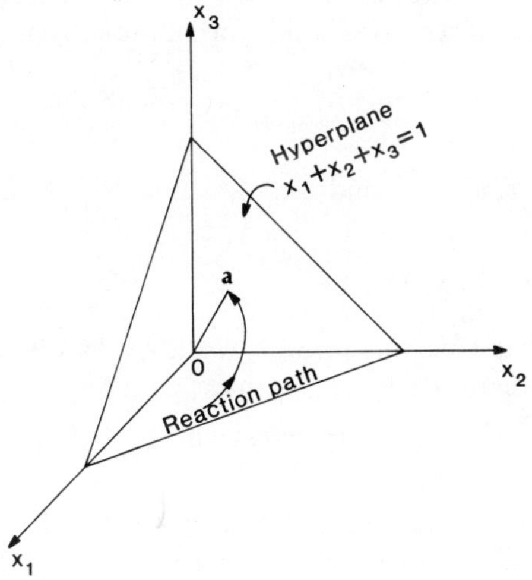


**Figure 8.2.1** Three-component reaction system.

***Theorem 8.2.3*** Let $\mathbf{x}(t)$ be a reaction path with $\mathbf{x}(0) = \mathbf{x}_0$. If $\mathbf{x}_0 \in H_j^+$, then $\mathbf{x}(t) \in H_j^+$ for all $t$. Thus $\mathbf{x}(t)$ is a straight-line path.

*Proof*: Let $\mathbf{x}_0 = \mathbf{a} + \alpha\mathbf{z}_j$. The inner product of differential equation (8.2.2) with $\mathbf{z}_k$ gives

$$\frac{d}{dt}\langle \mathbf{x}, \mathbf{z}_k \rangle = \lambda_k \langle \mathbf{x}, \mathbf{z}_k \rangle$$

which is solved to obtain

$$\langle \mathbf{x}, \mathbf{z}_k \rangle = \langle \mathbf{x}_0, \mathbf{z}_k \rangle e^{\lambda_k t} = \langle \mathbf{a} + \alpha\mathbf{z}_j, \mathbf{z}_k \rangle e^{\lambda_k t} = \alpha e^{\lambda_j t}\delta_{jk}$$

where $\delta_{jk}$ is the Kroenecker delta function. From Theorem 8.2.2 we have $\mathbf{x} = \mathbf{a} + \mathbf{z}$, where $\mathbf{z} \in N(T)^{\perp}$. Now

$$\langle \mathbf{x}, \mathbf{z}_k \rangle = \langle \mathbf{z}, \mathbf{z}_k \rangle = \alpha e^{\lambda_j t}\delta_{jk}$$

so that $\mathbf{z} = \alpha e^{\lambda_j t}\mathbf{z}_j$ and $\mathbf{x} = \mathbf{a} + \alpha e^{\lambda_j t}\mathbf{z}_j$, which implies that $\mathbf{x}(t) \in H_j^+$ for all $t$. From Exercise 8.2.4 it follows that $\mathbf{x}(t)$ is a straight-line path. ▮

The foregoing theorem shows that if in some manner one can obtain an initial composition vector $\mathbf{x}_0$ in the $j$th set $H_j^+$, then a straight line is obtained for the reaction path. Furthermore, this straight line is given by

$$\frac{x_i(t) - a_i}{\alpha z_{i,j}} = e^{\lambda_j t} \quad \text{(independent of } i\text{)}$$

so that the *direction numbers* $\{\alpha z_{i,j}\}$ when normalized with respect to the norm (8.2.8) will yield the eigenvector $\mathbf{z}_j$. Furthermore, a plot of $\ln(x_i - a_i)$ versus $t$ should produce a straight line with the eigenvalue $\lambda_j$ as the slope. Thus the crucial step lies in identifying an approproite initial composition. Notice that the straight-line reaction path leads to the equilibrium mole fraction vector $\mathbf{a}$. The next theorem asserts this for all initial compositions.

***Theorem 8.2.4*** Every reaction path leads to equilibrium, that is, $\forall\, \mathbf{x}_0 \in H^+$ $\lim_{t\to\infty} \mathbf{x}(t) = \mathbf{a}$.

*Proof*: Since $\langle \mathbf{x}, \mathbf{a} \rangle = 1$ and $\langle \mathbf{x}, \mathbf{z}_j \rangle = \langle \mathbf{x}_0, \mathbf{z}_j \rangle e^{\lambda_j t}$, a reaction path is given by

$$\mathbf{x}(t) = \mathbf{a} + \sum_{j=1}^{n-1} \langle \mathbf{x}_0, \mathbf{z}_j \rangle e^{\lambda_j t}\mathbf{z}_j$$

which follows from the fact that $\{\mathbf{a}, \mathbf{z}_1, \mathbf{z}_2, \ldots, \mathbf{z}_{n-1}\}$ are an orthonormal basis in $\mathcal{R}_n$. From Theorem 8.2.1 it follows that

$$\lim_{t\to\infty} \mathbf{x}(t) = \mathbf{a} \quad ▮$$

The implication of Theorem 8.2.4 is obvious. Before stating the next theorem we define the hyperplane

$$H_{(j)} = \{\mathbf{x} = \mathbf{a} + \mathbf{z};\, \mathbf{z} \in N_{\lambda_j} + N_{\lambda_{j+1}} + \ldots + N_{\lambda_{n-1}}\}$$

and confine as before attention to the subset $H^+_{(j)}$ in which $x_i \geq 0$, $i = 1, 2, \ldots, n$. Note that $H = H_{(1)}$ and $H^+ = H^+_{(1)}$. Further, let $\mathbf{P}_j$ be the orthogonal projection of $\mathcal{H}$ onto $N_{\lambda_j}$. In what follows it is convenient to denote a reaction path $\mathbf{x}(t)$ with $\mathbf{x}(0) = \mathbf{x}_0$ by $\mathbf{x}(t; \mathbf{x}_0)$.

***Definition 8.2.2*** Let $\mathbf{x}(t; \mathbf{x}_0)$ be a reaction path. Then $\mathbf{x}(t; \mathbf{a} + \mathbf{P}_j\mathbf{x}_0)$ is called the *projection* of reaction path $\mathbf{x}(t; \mathbf{x}_0)$ on $H_j$.

Note that the projections of a reaction path on hyperplanes $\{H_j\}$ are also reaction paths. The following theorem is significant.

***Theorem 8.2.5*** Let $x(t; x_0)$ be a reaction path with $\mathbf{x}_0 \in H^+_{(j)}$. Then $\lim_{t\to\infty} e^{-\lambda_j t}\|\mathbf{x}(t; \mathbf{x}_0) - \mathbf{x}(\mathbf{t}; \mathbf{a} + \mathbf{P}_j\mathbf{x}_0)\| = 0$; that is, the reaction path of interest asymptotically approaches its projection on a hyperplane, which corresponds to the maximum of the eigenvalues represented in $H_{(j)}$.

*Proof*: Since

$$\mathbf{x}(t; \mathbf{a} + \mathbf{P}_j\mathbf{x}_0) = \mathbf{a} + \langle \mathbf{P}_j\mathbf{x}_0, \mathbf{z}_j\rangle e^{\lambda_j t}\mathbf{z}_j$$

$$= \mathbf{a} + \langle \mathbf{x}_0, \mathbf{z}_j\rangle e^{\lambda_j t}\mathbf{z}_j$$

$$\mathbf{x}(t; \mathbf{x}_0) - \mathbf{x}(t; \mathbf{a} + \mathbf{P}_j\mathbf{x}_0) = \sum_{k=j+1}^{n-1} \langle \mathbf{x}_0, \mathbf{z}_k\rangle e^{\lambda_k t}\mathbf{z}_k$$

$$e^{-\lambda_j t}[\mathbf{x}(t; \mathbf{x}_0) - \mathbf{x}(t; \mathbf{a} + \mathbf{P}_j\mathbf{x}_0)] = \sum_{k=j+1}^{n-1} \langle \mathbf{x}_0, \mathbf{z}_k\rangle e^{(\lambda_k - \lambda_j)t}\mathbf{z}_k$$

In view of $(\lambda_k - \lambda_j) < 0$ for $k = j + 1, j + 2, \ldots, n$, we have

$$\lim_{t\to\infty} e^{-\lambda_j t}\|\mathbf{x}(t; \mathbf{x}_0) - \mathbf{x}(t; \mathbf{a} + \mathbf{P}_j\mathbf{x}_0)\| = 0 \quad \blacksquare$$

Clearly, $\mathbf{x}(t; \mathbf{a} + \mathbf{P}_j\mathbf{x}_0)$ is a straight-line reaction path so that the preceding theorem implies that all reaction paths are asymptotically straight. The straight-line region of a reaction path $\mathbf{x}(t; \mathbf{x}_0)$, $\mathbf{x}_0 \in H^+_{(j)}$ would depend on the spread of the eigenvalues, $\lambda_j, \lambda_{j+1}, \ldots, \lambda_{n-1}$. In particular, a larger value of $(\lambda_j - \lambda_{j+1})$ would produce a more pronounced straight-line region. The straight-line portion of the reaction path is characterized by

$$\mathbf{x}(t) \simeq \mathbf{a} + \langle \mathbf{x}_0, \mathbf{z}_j\rangle e^{\lambda_j t}\mathbf{z}_j$$

so that

$$\frac{x_i - a_i}{x_k - a_k} \simeq \frac{z_{i,j}}{z_{k,j}} \equiv b_{i,k} \quad \text{(suppressing } j\text{)} \tag{8.2.12}$$

which implies that a plot of $x_i$ versus $x_k$ would yield a straight line of slope $b_{i,k}$ for each $i$. Note that the values of $b_{i,k}$ for *fixed* $k$ provide an estimate of the $j$th eigenvector $\mathbf{z}_j$ (at this stage of the experiments it is presumed that accurate estimates are available of the eigenvectors $\mathbf{a}, \mathbf{z}_1, \mathbf{z}_2, \ldots, \mathbf{z}_{j-1}$) since

$$\mathbf{z}_j = \frac{1}{\|\mathbf{b}_k\|}\mathbf{b}_k \tag{8.2.13}$$

where $\mathbf{b}_k \equiv (b_{1,k}, b_{2,k}, \ldots, b_{n,k})$. Again we remind the reader that the dependence of $b_k$ on $j$ has been suppressed. Summing (8.2.12) over $i$, it is obvious that $\sum_{i=1}^{n} b_{i,k} = 0$ so that $b_{i,k}$, $i = 1, 2, \ldots, n$, for fixed $k$ are of mixed signs. We now show that there exists at least one $k$ for which

$$(a_i - b_{i,k}a_k) \geq 0 \tag{8.2.14}$$

Equation (8.2.12) implies that along the *extension* of the straight-line portion of the reaction path, *away from equilibrium*, we have $dx_i = b_{i,k}\, dx_k$. Since $b_{i,k}$ is of mixed signs for varying $i$, some compositions must decrease along the rectilinear path while others must increase. Furthermore, the compositions that *increase* (*decrease*) in the direction *toward* equilibrium must *decrease* (*increase*) away from equilibrium. Thus in the extension (or extrapolation) of the straight-line portion of the reaction path in $H_{(j)}^{+}$ *backward*, that is, away from equilibrium some compositions decrease and others increase. Since *all* mole fractions must be positive, in the interest of maximizing the straight-line portion, the backward extrapolation can only be carried on until one of the mole fractions, which we denote by $x_k$, becomes zero. At this point, the other mole fractions are nonnegative and are given by

$$x_i = a_i - b_{i,k}a_k = a_i - \frac{a_k}{z_{k,j}} z_{i,j} \tag{8.2.15}$$

Equation (8.2.15) is obtained from (8.2.12) by putting $x_k = 0$. Thus (8.2.14) must be true.

The Wei–Prater strategy calls for a new batch experiment with

$$\mathbf{x}(0) = \left(\mathbf{a} - \frac{a_k}{z_{k,j}}\mathbf{z}_j\right) - \sum_{r=1}^{j-1} \mathbf{P}_r\left(\mathbf{a} - \frac{a_k}{z_{k,j}}\mathbf{z}_j\right) \tag{8.2.16}$$

The second term on the right side of (8.2.16) may seem redundant, since it should in fact be $\mathbf{0}$. But it is not unlikely that the estimate of $\mathbf{z}_j$ available from the straight-line portion of the previous experiment via (8.2.13) is inaccurate enough to include components of $\mathbf{x}(0)$ in the subspaces $N_{\lambda_1}, N_{\lambda_2}, \ldots, N_{\lambda_{j-1}}$. The implication of this is that $\mathbf{P}_r(\mathbf{a} - (a_k/z_{k,j})\mathbf{z}_j)$ does not vanish exactly for $r = 1, 2, \ldots, j-1$ (why?). Subtracting the sum of the foregoing terms from the vector $\mathbf{a} - (a_k/z_{k,j})\mathbf{z}_j$ as in (8.2.16) produces an initial composition $\mathbf{x}(0) \in H_j^{+}$. Here one must interpret $H_j^{+}$ as approximately obtained to the extent that it derives from the initial estimate of $\mathbf{z}_j$ from (8.2.13). A batch experiment with this initial composition would produce a generally rectilinear reaction path permitting a better evaluation of $\mathbf{z}_j$.

We have thus determined how given $\mathbf{a}, \mathbf{z}_1, \mathbf{z}_2, \ldots, \mathbf{z}_{j-1}$, the eigenvector $\mathbf{z}_j$ can be determined. The methodology is obviously to be initiated after first determining $\mathbf{a}$ to obtain $\mathbf{z}_1$ and used progessively to identify all the eigenvectors and the corresponding eigenvalues. That the method works well is most ably demonstrated by Wei and Prater. The rate constants emerge from the formula (8.2.11).

**EXERCISES**

**8.2.1** Show that each of the rate constant vectors $\{\mathbf{k}_i\}_1^n$ is orthogonal to the equilibrium mole fraction vector **a**.

**8.2.2** Let $\mathbf{z}_j \equiv (z_{1,j}, z_{2,j}, \ldots, z_{n,j})$ be the $j$th eigenvector of $\mathbf{K}$, where $j = 1, 2, \ldots, n-1$. Show that

$$\sum_{i=1}^{n-1} z_{i,j} = 0$$

**8.2.3** Show that all mole fraction vectors lie in the subset $H^+$ of the hyperplane $H$ defined in the development above.

**8.2.4** Show that $H_j$, as defined in the text, represents a straight-line in $n$-dimensional space.

## 8.3 Multicomponent Rectification

We consider here an example of multicomponent rectification which has been discussed before by Amundson.† Rather than repeat here the details of the physical problem, we shall be content with presenting its bare features and converge on the relevant equations forthwith. The multicomponent rectification problem consists in calculating the liquid and vapor compositions in any given plate in the rectifying section (the section above the feed plate in a continuously fed distillation column), based on a knowledge of the composition of the product. The calculation itself requires no special strategy, but the objective is to circumvent a tedious stepwise procedure to obtain a formula for, say, the $n$th plate composition without having to go through the intermediate plate calculations. What we discuss here is merely a recasting of the method already detailed by Amundson.

The multicomponent mixture consists of $m$ species; the composition of the $i$th species in the $n$th plate is denoted $x_{n,i}$ for the liquid and $y_{n,i}$ for the vapor. Based on constant molal overflow of liquid with a downflow rate $L$ and vapor with an upward flow rate $V$, the mass balance for the $i$th species in the rectifying section above the $n$th plate leads to the equation

$$y_{n+1,i} = \frac{R}{R+1}x_{n,i} + \frac{x_{0,i}}{R+1}, \qquad i = 1, 2, \ldots, m \qquad (8.3.1)$$

where $R \equiv L/(V-L)$ is the reflux ratio and $x_{0,i}$ is the mole fraction of the $i$th species in the product which is known. The standard assumption of equilibrium between the liquid and the vapor departing from the plate is expressed

†See *Mathematical Methods in Chemical Engineering: Matrices and Their Applications*, by N. R. Amundson, Prentice-Hall, Englewood Cliffs, N.J., 1966.

by a relationship of the form

$$x_{n,i} = \frac{p_i y_{n,i}}{\sum_{j=1}^{m} p_j y_{n,j}}, \qquad p_i > 0 \tag{8.3.2}$$

where $p_i$, $i = 1, 2, \ldots, n$, are constants. Note that the mole fractions $x_{n,i}$ in (8.3.2) satisfy the requirement that

$$\sum_{i=1}^{m} x_{n,i} = 1$$

A similar constraint applies to the vapor compositions and is readily apparent from (8.3.1). The total condensation of the vapor product in the condenser leads to the relationship

$$y_{1,i} = x_{0,i} \tag{8.3.3}$$

the right-hand side of which is known. The problem is one of computing $x_{n,i}$ and $y_{n,i}$ for the $m$ species and the various plates. The obvious strategy is using (8.3.3) in (8.3.2) written for $n = 1$, producing $\{x_{1,i}\}_{i=1}^{m}$. The procedure is then one of successively using (8.3.1) and (8.3.2) to calculate the compositions in all the plates. The tedium to which we referred is having to calculate *all* the intermediate plate compositions just in order to compute the $n$th-plate compositions. (This procedure would be especially valuable to calculate the composition just above the feed plate.)

Following Amundson, we define new variables $\{X_{n,i}\}$ iteratively by

$$\begin{aligned} X_{n,i} &= Rp_i X_{n-1,i} + p_i \sum_{j=1}^{m} X_{n-1,j} x_{0,j} \\ X_{0,i} &= p_i, \qquad i = 1, 2, \ldots, m \end{aligned} \tag{8.3.4}$$

The compositions $x_{n,i}$ are then recovered from

$$x_{n,i} = \frac{X_{n,i} x_{0,i}}{\sum_{j=1}^{m} X_{n,j} x_{0,j}} \tag{8.3.5}$$

We have now reset our task as solving the system (8.3.4) so that the answers we seek are directly obtainable from (8.3.5). We define the vector $\mathbf{X}_n \equiv (X_{n,1}, X_{n,2}, \ldots, X_{n,m})$ which inhabits $\mathcal{R}_m$ and the matrix operator $\mathbf{A}$ by

$$\mathbf{Y} = \mathbf{AX}: \quad Y_i = \sum_{j=1}^{m} a_{ij} X_j \tag{8.3.6}$$

where

$$a_{ij} \equiv Rp_i \delta_{ij} + p_i x_{0,j} \tag{8.3.7}$$

Clearly, $\mathbf{A}$ is not a symmetric matrix and is therefore not self-adjoint with respect to the normal inner product in $\mathcal{R}_m$. It would now be of interest to see whether an inner product may be found with positive weights $\{r_j\}_{j=1}^m$ so that **A** becomes self-adjoint. Since for this (8.1.4) must be satisfied, we obtain

$$r_i p_i x_{0,j} = r_j p_j x_{0,i}$$

which yields

$$\frac{r_i p_i}{x_{0,i}} = \frac{r_j p_j}{x_{0,j}}$$

The independence of indices $i$ and $j$ requires that the terms on either side above be a constant which we assume to be unity. Thus

$$r_i = \frac{x_{0,i}}{p_i} > 0, \qquad i = 1, 2, \ldots, m$$

so that we define on $\mathcal{R}_m$ the inner product

$$\langle \mathbf{X}, \mathbf{Y} \rangle = \sum_{i=1}^{m} \frac{x_{0,i}}{p_i} X_i Y_i \tag{8.3.8}$$

Thus we have the Hilbert space $\mathcal{H} = \mathcal{R}_m, \{\langle \cdot, \cdot \rangle\}$, in which **A** is self-adjoint; that is, we have for $\mathbf{X}, \mathbf{Y} \in \mathcal{H}$,

$$\langle \mathbf{AX}, \mathbf{Y} \rangle = \langle \mathbf{X}, \mathbf{AY} \rangle$$

Spectral theorem 7.4.1 then assures the existence of eigenvalues $\lambda_1, \lambda_2, \ldots, \lambda_m$ and the corresponding set of eigenvectors $\mathbf{Z}_1, \mathbf{Z}_2, \ldots, \mathbf{Z}_m$ such that $\langle \mathbf{Z}_i, \mathbf{Z}_j \rangle = \delta_{ij}$. The matrix operator **A** may be expressed by

$$\mathbf{A} = \sum_{j=1}^{m} \lambda_j \mathbf{P}_j \tag{8.3.9}$$

where $\mathbf{P}_j$ is the orthogonal projection of $\mathcal{H}$ onto the eigenspace $N_{\lambda_j} \equiv N(\mathbf{A} - \lambda_j \mathbf{I})$.

To solve problem (8.3.4) we rewrite it as

$$\mathbf{X}_n = \mathbf{AX}_{n-1}, \qquad \mathbf{X}_0 = \mathbf{p}$$

where $\mathbf{p} \equiv (p_1, p_2, \ldots, p_m) \in \mathcal{R}_m$. Successive substitution produces

$$\mathbf{X}_n = \mathbf{A}^n \mathbf{p}$$

The representation (8.3.9) immediately leads to the solution

$$\mathbf{X}_n = \sum_{j=1}^{m} \lambda_j^n \mathbf{P}_j \mathbf{p} \tag{8.3.10}$$

Section 7.4 discusses the details pertinent to the representation of the powers of an operator. In terms of the eigenvectors, the solution (8.3.10) may also be written as

$$\mathbf{X}_n = \sum_{j=1}^{m} \lambda_j^n \langle \mathbf{p}, \mathbf{Z}_j \rangle \mathbf{Z}_j \tag{8.3.11}$$

It turns out that the eigenvalues of **A** can be computed in an interesting manner. $\mathbf{AZ} = \lambda\mathbf{Z}$ and

$$\lambda Z_i = Rp_i Z_i + p_i \sum_{j=1}^{m} x_{0,j} Z_j, \qquad i = 1, 2, \ldots, m$$

Rearranging the equation above, we have

$$Z_i = \frac{p_i}{\lambda - Rp_i} \sum_{j=1}^{m} x_{0,j} Z_j \tag{8.3.12}$$

Multiplying by $x_{0,i}$ and summing over $i$, one obtains

$$1 = \sum_{i=1}^{m} \frac{p_i x_{0,i}}{\lambda - Rp_i} \tag{8.3.13}$$

which is the *characteristic equation* for the eigenvalues of **A**. This equation is readily solved by numerical methods. Amundson demonstrates graphically the location of the roots of (8.3.13), which we will not repeat here. The eigenvector **Z** which corresponds to eigenvalue $\lambda$ is instantly obtained via (8.3.12). The $i$th component of **Z** is given by

$$Z_i = \frac{p_i(\lambda - Rp_i)^{-1}}{\left\{\sum_{k=1}^{m} p_k x_{0,k}(\lambda - Rp_k)^{-2}\right\}^{1/2}} \tag{8.3.14}$$

which is simply derived by imposing the normalization $\|\mathbf{Z}\| = 1$.

## EXERCISE

**8.3.1** The stripping section of an $m$-component distillation column contains $N$ plates including the reboiler (which may be regarded as an equilibrium plate) from which the product withdrawn has known composition $x_{N,i}$, $i = 1, 2, \ldots, m$. The mass balance equation for the $n$th plate is given by

$$x_{n-1,i} = \frac{R'}{R'+1} y_{n,i} + \frac{1}{R'+1} x_{N,i}, \qquad i = 1, 2, \ldots, m$$

based on constant molal vapor upflow $V'$ and liquid downflow $L'$ ($> V'$) and $R' \equiv V'/(L' - V')$. The equilibrium relationship may be assumed to be

$$y_{n,i} = \frac{\alpha_i x_{n,i}}{\sum_{j=1}^{m} \alpha_j x_{n,j}}$$

which is an "inside-out" version of (8.3.5) since $\alpha_i \equiv 1/p_i$. Show that by redefining variables recursively

$$Y_{n-1,i} = R'\alpha_i Y_{n,i} + \alpha_i \sum_{j=1}^{m} Y_{n,j} x_{N,j}, \; Y_{N,i} = \alpha_i, \qquad i = 1, 2, \ldots, m$$

the old variables are related to the new ones by

$$y_{n,i} = \frac{Y_{n,i} x_{N,i}}{\sum_{j=1}^{m} Y_{n,j} x_{N,j}}$$

(Use mathematical induction.) Proceed in an entirely analogous fashion as in the text to arrive at a formula for the $n$th-plate composition as

$$\mathbf{Y}_n = (\mathbf{A}')^{N-n}\boldsymbol{\alpha}$$

where $\boldsymbol{\alpha} \equiv (\alpha_1, \alpha_2, \ldots, \alpha_m)$, by identifying the matrix operator $\mathbf{A}'$. Proceed further to obtain a solution to $\mathbf{Y}_n$ similar to (8.3.10) or (8.3.11) by identifying

(a) A Hilbert space $\mathcal{H}$ by defining a suitable inner product on $\mathcal{R}_m$ such that $\mathbf{A}'$ is self-adjoint.

(b) The characteristic equation for the eigenvalues of $\mathbf{A}'$ and a formula for its eigenvectors.

## 8.4 Jacobi Matrices and Stagewise Operations

There are several stagewise operations in chemical engineering which lead to algebraic equations. When the equations are cast in a vector form, the matrix involved is very frequently of the tridiagonal type. For example, we may write

$$\mathbf{A} \equiv \begin{bmatrix} \beta_1 & \gamma_1 & 0 & 0 & \cdot & \cdot & \cdot & 0 & 0 & 0 \\ \alpha_2 & \beta_2 & \gamma_2 & 0 & & & & & & \\ 0 & \alpha_3 & \beta_3 & \gamma_3 & & & & \cdot & \cdot & \cdot \\ \cdot & & & & & & & & & \\ \cdot & & & & & & & & & \\ \cdot & & & & & & & & & \\ 0 & & & & & 0 & \alpha_{n-1} & \beta_{n-1} & \gamma_{n-1} & \\ 0 & \cdot & \cdot & \cdot & \cdot & \cdot & 0 & 0 & \alpha_n & \beta_n \end{bmatrix} \tag{8.4.1}$$

Such matrices are called *Jacobi matrices*. The Jacobi matrices commonly encountered in chemical engineering are such that $\alpha_2, \alpha_3, \ldots, \alpha_n > 0, \gamma_1, \gamma_2, \ldots, \gamma_{n-1} > 0$. The matrix coefficients may be written as

$$a_{ij} = \alpha_i \delta_{i-1,j} + \beta_i \delta_{ij} + \gamma_i \delta_{i,j-1} \tag{8.4.2}$$

In general the matrix above is not symmetric. It is of interest, however, to see whether the matrix operator $\mathbf{A}$ on $\mathcal{R}_n$ may be rendered self-adjoint using a suitable inner product. Accordingly, we look for positive weights $\{r_j\}$ such that (8.1.4) may be true. Thus $r_i a_{ij} = r_j a_{ji}$ leads to the equations

$$\frac{r_{j+1}}{r_j} = \frac{\gamma_j}{\alpha_{j+1}} > 0 \tag{8.4.3}$$

Letting $r_1 = 1$, we obtain

$$r_j = \prod_{k=1}^{j-1} \frac{\gamma_k}{\alpha_{k+1}}, \qquad j = 2, \ldots, n$$

Thus we define a Hilbert space $\mathcal{H} = \{\mathcal{R}_n, \langle \cdot, \cdot \rangle\}$ where the inner product is defined by

$$\langle \mathbf{x}, \mathbf{y} \rangle \equiv x_1 y_1 + \sum_{j=2}^{n} \prod_{k=1}^{j-1} \frac{\gamma_k}{\alpha_{k+1}} x_j y_j \tag{8.4.4}$$

Indeed, **A** is self-adjoint with respect to the inner product above. Hence Theorem 7.4.1 guarantees the existence of real eigenvalues $\lambda_1, \lambda_2, \ldots, \lambda_n$ and a corresponding set of eigenvectors $\mathbf{z}_1, \mathbf{z}_2, \ldots, \mathbf{z}_n$ which form an orthogonal basis in $\mathcal{H}$.

In the applications of interest to us, the matrix coefficients in (8.4.1) usually satisfy the following equations:

$$\alpha_{i+1} + \beta_i + \gamma_{i-1} = 0, \qquad i = 1, 2, \ldots, n \tag{8.4.5}$$

Equation (8.4.5) generally arises from mass balances. Note that although $\gamma_0$ and $\alpha_{n+1}$ appear here they are not present among the matrix coefficients in (8.4.1). For this case we will show below that the operator $A$ is nonpositive. In establishing this, we will be liberal with the subscripted indices on $\alpha$, $\beta$, and $\gamma$ in that we will not bother to exclude nonexisting terms such as $\alpha_1$, $\gamma_n$, and $r_{n+1}$. When they appear in the manipulations, they must be assumed to be zero. Thus consider

$$\begin{aligned}
\langle \mathbf{Ax}, \mathbf{x} \rangle &= \sum_{i=1}^{n} r_i x_i \sum_{j=1}^{n} [\alpha_i \delta_{i-1,j} + \beta_i \delta_{ij} + \gamma_i \delta_{i,j-1}] x_j \\
&= \sum_{i=1}^{n} r_i x_i [\alpha_i x_{i-1} + \beta_i x_i + \gamma_i x_{i+1}] \\
&= \sum_{i=1}^{n} [r_{i+1} \alpha_{i+1} x_i x_{i+1} + \beta_i r_i x_i^2 + \gamma_i r_i x_i x_{i+1}]
\end{aligned}$$

Using (8.4.3) and (8.4.5), the foregoing becomes

$$\begin{aligned}
\langle \mathbf{Ax}, \mathbf{x} \rangle &= \sum_{i=1}^{n} [2\alpha_{i+1} r_{i+1} x_i x_{i+1} - r_i(\alpha_{i+1} + \gamma_{i-1}) x_i^2] \\
&= \sum_{i=1}^{n} [2\alpha_{i+1} r_{i+1} x_i x_{i+1} - \alpha_{i+1} r_i x_i^2 - \gamma_i r_{i+1} x_{i+1}^2] \\
&= \sum_{i=1}^{n} \alpha_{i+1} \left[ 2r_{i+1} x_i x_{i+1} - r_i x_i^2 - \frac{\gamma_i}{\alpha_{i+1}} r_{i+1} x_{i+1}^2 \right] \\
&= \sum_{i=1}^{n} \frac{\alpha_{i+1}}{r_i} [2 r_i r_{i+1} x_i x_{i+1} - r_i^2 x_i^2 - r_{i+1}^2 x_{i+1}^2] \\
&= -\sum_{i=1}^{n} \frac{\alpha_{i+1}}{r_i} [r_i x_i - r_{i+1} x_{i+1}]^2 \leq 0
\end{aligned}$$

which proves that the nonzero eigenvalues are negative. We shall consider next an application of the matrix just considered to a continuous countercurrent extraction process.

*Example. Continuous Countercurrent Staged Extraction Process*

We consider a sequence of $n$ equilibrium extraction stages which are countercurrently fed with the extracting solvent phase entering stage 1 and the raffinate stream entering stage $n$ (see Figure 8.4.1). The solute-free extract

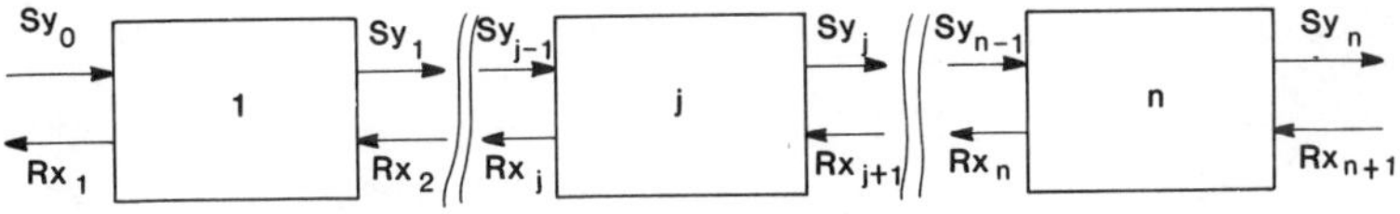


**Figure 8.4.1** Countercurrent extraction process.

and raffinate flow rates are $S$ and $R$, respectively. The solute concentration is measured in mole ratios and denoted in the $j$th stage by $Y_j$ in the extract phase and $X_j$ in the raffinate. The total holdup in the $j$th stage referred to the raffinate phase is denoted $h_j$.† The equilibrium relationship between $Y$ and $X$ is assumed to be given by $Y = KX$. The unsteady-state mass balance for solute in the $j$th stage is given by

$$h_j \frac{dX_j}{dt} = SKX_{j-1} - (SK + R)X_j + RX_{j+1}, \qquad j = 2, 3, \ldots, n-1 \tag{8.4.6}$$

For stages 1 and $n$ we write

$$h_1 \frac{dX_1}{dt} = SY_0 - (SK + R)X_1 + RX_2 \tag{8.4.7}$$

$$h_n \frac{dX_n}{dt} = SKX_{n-1} - (SK + R)X_n + RX_{n+1} \tag{8.4.8}$$

The concentration $Y_0$ in the entering solvent and $X_{n+1}$ in the entering raffinate are assumed to be known. Defining new variables by $x_j \equiv h_j X_j$, (8.4.6) to (8.4.8) are readily written in vectorial form as

$$\frac{d\mathbf{x}}{dt} = \mathbf{Ax} + \mathbf{b} \tag{8.4.9}$$

where $\mathbf{x} \equiv (x_1, x_2, \ldots, x_n)$, $\mathbf{b} \equiv (SY_0, 0, 0, \ldots, RX_{n+1}) \in \mathcal{R}_n$, and $\mathbf{A}$ is the matrix in (8.4.1) provided that

$$\alpha_j \equiv \frac{SK}{h_{j-1}}, \qquad \beta_j \equiv -\frac{SK + R}{h_j}, \qquad \gamma_j \equiv \frac{R}{h_{j+1}}$$

It is readily verified that the coefficients above satisfy (8.4.5). Thus $\mathbf{A}$ is self-

†The total holdup $h_j$, referred to the raffinate phase, is given by the raffinate holdup plus the product of the solvent holdup and the equilibrium constant $K$.

adjoint with respect to the inner product

$$\langle \mathbf{x}, \mathbf{y} \rangle = x_1 y_1 + \sum_{j=2}^{n} \left(\frac{R}{SK}\right)^{j-1} \prod_{k=1}^{j-1} \frac{h_k}{h_{k+1}} x_j y_j \tag{8.4.10}$$

We have also established that $\mathbf{A}$ is nonpositive. If $\{\lambda_j\}_{j=1}^n$ and $\{\mathbf{z}_j\}_{j=1}^n$ are the eigenvalues and eigenvectors of $\mathbf{A}$, then following the methods of Section 7.5, the solution of (8.4.9) subject to the initial condition $\mathbf{x}(0) = \mathbf{x}_0$ is given by

$$\mathbf{x}(t) = \sum_{j=1}^{n} \left[ \langle \mathbf{x}_0, \mathbf{z}_j \rangle e^{\lambda_j t} + \int_0^t \langle \mathbf{b}, \mathbf{z}_j \rangle e^{\lambda_j (t-t')} \, dt' \right] \mathbf{z}_j \tag{8.4.11}$$

The eigenvalues may be determined in an entirely standard manner from the matrix $\mathbf{A}$. The eigenvectors must, however, be normalized with respect to the norm generated by the inner product (8.4.10). When the holdup does not vary from stage to stage, the eigenvalues may be determined in a very convenient way, as demonstrated by Amundson.† For this case the eigenvalues are shown to be

$$\lambda_j = -\frac{1}{h}\left[ SK + R + 2\sqrt{SRK} \cos \frac{\pi j}{n} \right], \qquad j = 1, 2, \ldots, n$$

where $h$ is the holdup common for all the stages. The normalized eigenvector $\mathbf{z}_j \equiv (z_{1,j}, z_{2,j}, \ldots, z_{n,j})$ corresponding to eigenvalue $\lambda_j$ is given by

$$z_{i,j} = \frac{(-1)^{i-1}\left(\frac{SK}{R}\right)^{(i-1)/2} \sin\,[\pi j(n-i+1)/(n+1)]}{\left\{\sum_{r=1}^{n} \sin^2\,[\pi j(n-r+1)/(n+1)]\right\}^{1/2}}, \qquad i = 1, 2, \ldots, n \tag{8.4.12}$$

Thus all quantities connected with the solution (8.4.11) for the case of uniform holdup are known explicitly.

**EXERCISE**

**8.4.1** Consider the sequence of reversible first-order reactions.

$$A_1 \underset{k_1'}{\overset{k_1}{\rightleftharpoons}} A_2 \underset{k_2'}{\overset{k_2}{\rightleftharpoons}} \cdots \underset{k_{n-1}'}{\overset{k_{n-1}}{\rightleftharpoons}} A_n$$

in a well-stirred batch reactor. Denoting the mole fraction vector by $\mathbf{x} \equiv (x_1,$

---

†See *Mathematical Methods in Chemical Engineering: Matrices and Their Applications*, by N. R. Amundson, Prentice-Hall, Englewood Cliffs, N.J., 1966, pp. 162–163.

$x_2, \ldots, x_n$), identify the matrix **K** which yields the batch reactor equation

$$\frac{d\mathbf{x}}{dt} = \mathbf{K}\mathbf{x}$$

Find the inner product on $\mathcal{R}_n$ with respect to which **K** becomes self-adjoint. Is **K** negative or nonpositive? Obtain the solution to the differential equation above subject to the initial condition $\mathbf{x}(0) = \mathbf{x}_0$.

Also solve the transient continuous stirred tank reactor equation

$$\frac{d\mathbf{x}}{dt} = \frac{1}{\Theta}(\mathbf{x}_f - \mathbf{x}) + \mathbf{K}\mathbf{x}, \qquad \mathbf{x}(0) = \mathbf{x}_0$$

where $\mathbf{x}_f$ is the feed mole fraction vector and $\Theta$ is the holding time.

## 8.5 Multicomponent Diffusion

We are concerned here with a multicomponent system containing $n$ chemical species in an $(n + 1)$st "solvent" species, whose nonuniform concentration fields lead to interdiffusional movement. The description of such diffusional transport is subject to the flexibility of a reference frame velocity, the diffusion coefficients, and concentration variables depending on the choice of this frame.† The treatment, however, is essentially equivalent for all frames and in the following we will consider the diffusive fluxes relative to the volume-averaged velocity. Furthermore, we need only consider the relative fluxes of the solute, since the flux of the solvent would be automatically fixed by the values of the others. Noting that the flux of the $i$th species is a physical vector identified by an arrow on top, we let

$$\vec{\mathbf{j}} \equiv (\vec{j}_1, \vec{j}_2, \ldots, \vec{j}_n)$$

Fick's law is represented by

$$\vec{\mathbf{j}} = -\mathbf{D}\,\vec{\nabla}\mathbf{c} \tag{8.5.1}$$

where **D** is the diffusion coefficient matrix, $\vec{\nabla}$ is the spatial gradient operator,‡ and $\mathbf{c} \equiv (c_1, c_2, \ldots, c_n)$ is the solute concentration vector, which depends on

†There are several sources that the reader may consult on this subject: for example, J. G. Kirkwood, R. L. Baldwin, P. S. Dunlop, L. J. Gosting, and G. Kegeles, *J. Chem. Phys.*, **33**, 1505, 1960; R. L. Baldwin, P. S. Dunlop, and L. J. Gosting, *J. Am. Chem. Soc.*, **77**, 5235, 1955; S. R. DeGroot and P. Mazur, *Non-equilibrium Thermodynamics*, North-Holland, Amsterdam, 1962; H. L. Toor, *AIChE J.*, **8**, 561, 1962; H. T. Cullinan, *Ind. Eng. Chem. Fundam.*, *4*, 133, 1965; and E. L. Cussler, *Multicomponent Diffusion*, Elsevier, New York, 1976.

‡For a Cartesian coordinate frame with $\vec{i}, \vec{j}$, and $\vec{k}$ as unit vectors along the $x$, $y$, and $z$ axes, respectively,

$$\vec{\nabla} = \vec{i}\frac{\partial}{\partial x} + \vec{j}\frac{\partial}{\partial y} + \vec{k}\frac{\partial}{\partial z}$$

the position vector $\vec{r}$. The matrix $\mathbf{D}$ given by

$$\mathbf{D} \equiv \begin{bmatrix} D_{11} & D_{12} & \cdots & D_{1n} \\ D_{21} & D_{22} & \cdots & D_{2n} \\ \cdot & & & \\ \cdot & & & \\ \cdot & & & \\ D_{n1} & D_{n2} & \cdots & D_{nn} \end{bmatrix} \tag{8.5.2}$$

is generally unsymmetric. Although the diffusion coefficients $\{D_{ij}\}$ are concentration dependent, they may be regarded as constant at suitably averaged values. Our objective here is to inquire into the solution of certain multicomponent diffusion problems that arise when the flux equations (8.5.1) are combined with the conservation of species mass equation.† The unsteady-state species conservation equation may be written as

$$-\vec{\nabla}\cdot(\vec{j}_i + \vec{v}c_i) = \frac{\partial c_i}{\partial t}$$

or equivalently,

$$-\vec{\nabla}\cdot(\vec{\mathbf{j}} + \vec{v}\mathbf{c}) = \frac{\partial \mathbf{c}}{\partial t} \tag{8.5.3}$$

where $\vec{v}$ must be regarded as the volume-averaged velocity. We will assume further that $\vec{v} \simeq \vec{0}$, so that we have "free" diffusion. Equation (8.5.1) then converts (8.5.3) into

$$\mathbf{D}\nabla^2\mathbf{c} = \frac{\partial \mathbf{c}}{\partial t} \tag{8.5.4}$$

where the expression $\nabla^2$ represents $(\partial^2/\partial x^2 + \partial^2/\partial y^2 + \partial^2/\partial z^2)$ in Cartesian coordinates. Equation (8.5.4) must be satisfied in some spatial region which we denote by $\Omega$ enclosed by a boundary represented by $\partial\Omega$. For a complete specification of the multicomponent diffusion problem the concentration vector must satisfy *boundary* and *initial* conditions. Here, it is possible to be somewhat general, but to be within the scope of the present chapter, we will assume that

$$\mathbf{c}(\vec{r}, t) = \mathbf{0}, \qquad \vec{r} \in \partial\Omega, \quad t > 0 \tag{8.5.5}$$

and

$$\mathbf{c}(\vec{r}, 0) = \mathbf{c}_0, \qquad \vec{r} \in \Omega \tag{8.5.6}$$

where $\mathbf{c}_0 \in \mathcal{R}_n$ is a vector of *constants*. It is our objective to obtain the

---

†See, for example, *Transport Phenomena*, by R. B. Bird, W. E. Stewart, and E. N. Lightfoot, Wiley, New York, 1960.

concentration vector $\mathbf{c}(\vec{r}, t)$. To this end, we consider the equivalent *binary diffusion* problem which arises when a *single* solute is present in the solvent. We let the binary diffusion coefficient be $D$. The concentration of the solute, $c(\vec{r}, t)$, must under conditions identical to those for the multicomponent mixture satisfy the differential equation

$$D\nabla^2 c = \frac{\partial c}{\partial t}, \qquad \vec{r} \in \Omega \tag{8.5.7}$$

the boundary condition

$$c(\vec{r}, t) = 0, \qquad \vec{r} \in \partial\Omega, \quad t > 0 \tag{8.5.8}$$

and the initial condition

$$c(\vec{r}, 0) = c_0, \qquad \vec{r} \in \Omega \tag{8.5.9}$$

Our strategy for the multicomponent problem is to derive its solution from that of the binary diffusion problem. Notice that the boundary condition (8.5.8) and the initial condition (8.5.9) which belong to the binary diffusion problem are the same as (8.5.5) and (8.5.6), respectively. In subsequent chapters we will develop the expertise to solve the binary diffusion problem. For the present we assume that the solution to (8.5.7) subject to (8.5.8) and (8.5.9) may be represented by

$$c(\vec{r}, t) = f(\vec{r}, t; D)c_0 \tag{8.5.10}$$

The function $f(\vec{r}, t; D)$, which is assumed to be known, satisfies both (8.5.7) and (8.5.8); its initial value, however, is unity.

We expect that the solution to the multicomponent problem may be written as

$$\mathbf{c}(\vec{r}, t) = \mathbf{f}(\vec{r}, t; \mathbf{D})\mathbf{c}_0 \tag{8.5.11}$$

and we seek to represent $\mathbf{f}(\vec{r}, t; \mathbf{D})$ suitably.

First we observe that $\mathbf{D}$, being a real nonsymmetric matrix, is non-self-adjoint with respect to the normal inner product on $\mathcal{R}_n$. However, from basic irreversible thermodynamics the matrix $\mathbf{D}$ may be written as†

$$\mathbf{D} = \mathbf{LG} \tag{8.5.12}$$

where $\mathbf{L} \equiv \{L_{ij}\}$ is the matrix of Onsager coefficients which is known to be symmetric and positive definite. The matrix $\mathbf{G}$ has coefficients $\{G_{ij}\}$ given by

$$G_{ij} = \frac{\partial^2 G}{\partial c_i \, \partial c_j}$$

†See, for example, H. T. Cullinan, *Ind. Eng. Chem. Fundam.*, *4*, 133, 1965; and W. E. Stewart and R. Prober, *Ind. Eng. Chem. Fundam.*, *3*, 224, 1964.

where $G$ is volume-specific free energy. Clearly, $G_{ij} = G_{ji}$, so that **G** is a symmetric matrix. Furthermore, thermodynamic stability considerations require that **G** be positive definite. From (8.5.12) we have $\mathbf{L}^{-1}\mathbf{D}$ (= **G**) as a symmetric matrix, so that the discussion in Section 8.1 leads to the fact that **D** is a self-adjoint operator with respect to the inner product $\langle \cdot, \cdot \rangle$ on $\mathcal{R}_n$ defined by

$$\langle \mathbf{x}, \mathbf{y} \rangle = (\mathbf{L}^{-1}\mathbf{x}, \mathbf{y}) \tag{8.5.13}$$

the symbol $(\cdot, \cdot)$ signifying the normal inner product on $\mathcal{R}_n$ given by

$$(\mathbf{x}, \mathbf{y}) = \sum_{i=1}^{n} x_i y_i$$

The self-adjointness of **D** relative to the inner product (8.5.13) follows from

$$\langle \mathbf{Dx}, \mathbf{y} \rangle = (\mathbf{Gx}, \mathbf{y}) = (\mathbf{x}, \mathbf{Gy}) = \langle \mathbf{x}, \mathbf{Dy} \rangle$$

Furthermore, the positive definiteness of **G** and

$$\langle \mathbf{Dx}, \mathbf{x} \rangle = (\mathbf{Gx}, \mathbf{x})$$

establish that **D** is positive definite. Thus **D** has positive eigenvalues $\{\lambda_j\}_{j=1}^n$, and eigenvectors $\{\mathbf{u}_j\}_{j=1}^n$ whose orthonormality is with respect to the inner product (8.5.13). We are also able to write

$$\mathbf{D} = \sum_{j=1}^{n} \lambda_j \mathbf{P}_j \tag{8.5.14}$$

where $\mathbf{P}_j$ is the orthogonal projection of $\mathcal{R}_n$ onto $N_{\lambda_j} \equiv N(\mathbf{D} - \lambda_j \mathbf{I})$ defined by

$$\mathbf{P}_j \mathbf{x} = \langle \mathbf{x}, \mathbf{u}_j \rangle \mathbf{u}_j$$

The representation of $\mathbf{f}(\vec{r}, t; \mathbf{D})$ is now obvious.†

$$\mathbf{f}(\vec{r}, t; \mathbf{D}) = \sum_{j=1}^{n} f(\vec{r}, t; \lambda_j) \mathbf{P}_j \tag{8.5.15}$$

so that the solution to the multicomponent diffusion problem is given by

$$\mathbf{c}(\vec{r}, t) = \sum_{j=1}^{n} f(\vec{r}, t; \lambda_j) \langle \mathbf{c}_0, \mathbf{u}_j \rangle \mathbf{u}_j \tag{8.5.16}$$

In view of (8.5.13), the solution (8.5.16) requires knowledge of the Onsager coefficient matrix **L**, which is frequently not available. If the matrix **G** is known, then **L** is obtained as $\mathbf{G}^{-1}\mathbf{D}$. The difficulty with this lies in the

---

†It is important to reiterate here that because **D** is self-adjoint, regardless of the number of times any of its eigenvalues are repeated, the eigenvectors always form a complete set. (For example, if $\lambda_j$ occurs $r_j$ times, $N_{\lambda_j}$ has dimension $r_j$.)

general unavailability of **G**.† However, in view of the fact that we have established that **D** is self-adjoint with respect to the inner product, the diagonability of $D$ is not in question. Moreover, transient concentration profiles will be a combination of exponential functions in time regardless of the number of times any of the eigenvalues is repeated.

Next we consider the problem of multicomponent diffusion with chemical reaction.

**EXERCISE**

**8.5.1** Show that the multicomponent diffusion matrix **D** is also self-adjoint with respect to the inner product

$$\langle \mathbf{u}, \mathbf{v} \rangle \equiv (\mathbf{G}\mathbf{u}, \mathbf{v})$$

where **G** is the thermodynamic matrix in (8.5.12), and $(\cdot, \cdot)$ is the regular inner product on $\mathcal{R}_n$.

### 8.5.1 Multicomponent Diffusion with Chemical Reaction

We will address the problem of diffusion in a catalyst pore on the surface of which the first-order reactions considered in Section 8.2 occur among $n$-chemical species. The different chemical species commute between the pore interior and exterior by Knudsen diffusion. The gas phase is well mixed and the concentration vector there is $\mathbf{c}_0$ (see Figure 8.5.1). If the mass transfer resistance at the pore mouth is negligible, the concentration there also is $\mathbf{c}_0$. Knudsen diffusion may be characterized by a *diagonal* diffusion coefficient matrix **D** given by

$$\mathbf{D} \equiv \begin{bmatrix} D_1 & 0 & \cdots & 0 \\ 0 & D_2 & \cdots & 0 \\ \cdot & & & \\ \cdot & & & \\ \cdot & & & \\ 0 & 0 & \cdots & D_n \end{bmatrix} \tag{8.5.17}$$

†It is possible to show that for any real nonsymmetric matrix **D** which has real eigenvalues $\{\lambda_j\}_{j=1}^n$ and a complete set of eigenvectors $\{\mathbf{u}_j\}_{j=1}^n$, there is a *family* of inner products with respect to which **D** is self-adjoint. This family is obtained as follows. Let $\mathbf{D}^*$, the adjoint or equivalently the transpose of **D** (which has the same eigenvalues as **D**) have $\{\mathbf{v}_j\}_{j=1}^n$ for its eigenvectors, where each $\mathbf{v}_j$ is determined only up to an arbitrary multiplicative constant. Let **V** be the modal matrix of $\mathbf{D}^*$ formed from the vectors $\{\mathbf{v}_j\}_{j=1}^n$ and $\mathbf{\Delta}$ be an *arbitrary* diagonal matrix of real numbers. Then $\langle \mathbf{x}, \mathbf{y} \rangle \equiv (\mathbf{V}\mathbf{\Delta}^2\mathbf{V}^*\mathbf{x}, \mathbf{y})$ represents a valid inner product with respect to which **D** is self-adjoint. In fact, we must have for some $\mathbf{\Delta}$, $\mathbf{L}^{-1} = \mathbf{V}\mathbf{\Delta}^2\mathbf{V}^*$.

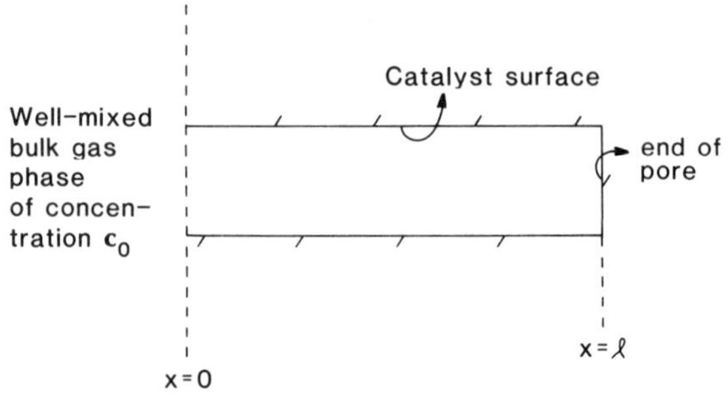


**Figure 8.5.1** First-order reaction system in a single catalyst pore.

The rate constant matrix $\hat{\mathbf{K}}$ is identical to that in Section 8.2 except for it now being defined per unit catalytic surface area. If we let $A$ represent the pore cross section and $P$ denote the surface area per unit length, the steady-state one-dimensional transport equation may be written

$$\mathbf{D}\frac{d^2\mathbf{c}}{dx^2} + \frac{P}{A}\hat{\mathbf{K}}\mathbf{c} = \mathbf{0}, \qquad 0 < x < l \tag{8.5.18}$$

Equation (8.5.18) must be supplemented by the boundary conditions

$$x = 0, \quad \mathbf{c} = \mathbf{c}_0; \qquad x = l, \quad \frac{d\mathbf{c}}{dx} = \mathbf{0} \tag{8.5.19}$$

The objective of the analysis is to describe the *reaction rate* in the pore in terms of the external gas phase concentration $\mathbf{c}_0$. Thus a *matrix effectiveness* factor $\mathbf{E}$ is to be defined such that

$$\text{reaction rate in pore} = \mathbf{E}\hat{\mathbf{K}}\mathbf{c}_0(Pl) \tag{8.5.20}$$

$\mathbf{E}$ accounts for the correction resulting from assuming the entire pore surface to be at the concentration $\mathbf{c}_0$. The reaction rate will become available as soon as (8.5.18) is solved subject to the boundary conditions (8.5.19), and the resulting solution is substituted into

$$\text{reaction rate in pore} = \text{rate of diffusion into the pore} = \mathbf{D}\frac{d\mathbf{c}}{dx}\bigg|_{x=0} A$$

Before we proceed to solve (8.5.18), it is convenient to rewrite it as

$$\frac{d^2\mathbf{c}}{d\xi^2} - \mathbf{\Phi}^2\mathbf{c} = \mathbf{0}, \qquad 0 < \xi < 1 \tag{8.5.21}$$

where we have let

$$-\frac{Pl^2}{A}\mathbf{D}^{-1}\hat{\mathbf{K}} \equiv \mathbf{\Phi}^2, \qquad \xi \equiv x/l$$

in anticipation of some results to be established presently. The matrix $\mathbf{D}^{-1}\hat{\mathbf{K}}$ is real, self-adjoint with respect to an inner product to be identified and *negative definite*, so that $-\mathbf{D}^{-1}\hat{\mathbf{K}}$ is positive and the (dimensionless) matrix abbreviation $\mathbf{\Phi}^2$ may then be regarded as appropriate.

Let us consider the operator $\mathbf{D}^{-1}\hat{\mathbf{K}}$ on $\mathcal{R}_n$. We know that $\hat{\mathbf{K}}$ is self-adjoint with respect to the inner product (8.2.7). Interestingly, the matrix operator $\mathbf{D}^{-1}$ given by

$$\mathbf{D}^{-1} = \begin{bmatrix} \frac{1}{D_1} & 0 & \cdots & 0 \\ 0 & \frac{1}{D_2} & \cdots & 0 \\ \cdot & & & \\ \cdot & & & \\ \cdot & & & \\ 0 & 0 & \cdots & \frac{1}{D_n} \end{bmatrix}$$

is also self-adjoint relative to the inner product (8.2.7), for let $\mathbf{x}, \mathbf{y} \in \mathcal{R}_n$.

$$\begin{aligned} \langle \mathbf{D}^{-1}\mathbf{x}, \mathbf{y} \rangle &= \sum_{i=1}^{n} \frac{1}{a_i} \frac{x_i}{D_i} y_i = \sum_{i=1}^{n} \frac{1}{a_i} x_i \frac{y_i}{D_i} \\ &= \langle \mathbf{x}, \mathbf{D}^{-1}\mathbf{y} \rangle \end{aligned}$$

Since $\mathbf{D}^{-1}$ is clearly positive definite (its eigenvalues are $1/D_1, 1/D_2, \ldots, 1/D_n$), the result of Exercises 6.6.2 and 6.6.10 leads to a new inner product $[\cdot, \cdot]$ on $\mathcal{R}_n$ defined by

$$[\mathbf{x}, \mathbf{y}] \equiv \langle \mathbf{D}\mathbf{x}, \mathbf{y} \rangle \tag{8.5.22}$$

with respect to which the operator $\mathbf{D}^{-1}\hat{\mathbf{K}}$ is self-adjoint. Note that the inner product symbol on the right-hand side of (8.5.22) refers to that in (8.2.7). Furthermore, $\mathbf{D}^{-1}\hat{\mathbf{K}}$ is *nonpositive* because for $\mathbf{x} \in \mathcal{R}_n$,

$$[\mathbf{D}^{-1}\hat{\mathbf{K}}\mathbf{x}, \mathbf{x}] = \langle \mathbf{D}\mathbf{D}^{-1}\hat{\mathbf{K}}\mathbf{x}, \mathbf{x} \rangle = \langle \hat{\mathbf{K}}\mathbf{x}, \mathbf{x} \rangle \leq 0$$

the last of which follows from Theorem 8.2.1. Thus $-\mathbf{D}^{-1}\hat{\mathbf{K}}$ is a *nonnegative* operator; the nonnegativity is clearly inherited by the operator $\mathbf{\Phi}^2$ whose (nonnegative) eigenvalues will be denoted by $\{\mu_j^2\}_{j=1}^n$. Since $\hat{\mathbf{K}}$ is singular, $\mathbf{\Phi}^2$ is also singular, so that one of its eigenvalues is zero. The normalized eigenvectors of $\mathbf{\Phi}^2$ are denoted by $\{\mathbf{u}_j\}_{j=1}^n$, so that the corresponding orthogonal family of self-adjoint projections $\{\mathbf{P}_j\}_{j=1}^n$ are defined by

$$\mathbf{P}_j\mathbf{x} = [\mathbf{x}, \mathbf{u}_j]\mathbf{u}_j, \qquad \mathbf{x} \in \mathcal{R}_n$$

The operator $\mathbf{\Phi}^2$ has the spectral representation

$$\mathbf{\Phi}^2 = \sum_{j=1}^{n} \mu_j^2 \mathbf{P}_j \tag{8.5.23}$$

The determination of the eigenvalues and eigenvectors would proceed exactly as with standard matrices. The normalization of the eigenvector must, however, be with respect to the norm generated by the inner product (8.5.22).

We are now in a position to address the solution of the differential equation (8.5.21) subject to the boundary conditions (8.5.19). Following the methods outlined in Section 7.5, the scalar version of (8.5.21) is

$$\frac{d^2 c}{d\xi^2} = \phi^2 c, \qquad 0 < \xi < 1$$

while the boundary conditions corresponding to (8.5.19) are given by

$$c(0) = c_0, \qquad \frac{dc(1)}{d\xi} = 0$$

The solution to this problem

$$c = (\cosh \phi)^{-1} \cosh \phi(1 - \xi)c_0$$

from which the solution to the multicomponent problem may be written as

$$\mathbf{c} = (\cosh \boldsymbol{\Phi})^{-1} \cosh \boldsymbol{\Phi}(1 - \xi)\mathbf{c}_0 \tag{8.5.24}$$

Since the quantity of eventual interest is the reaction rate in the pore, we obtain for it

$$\begin{aligned} \mathbf{D}\frac{d\mathbf{c}}{dx}\bigg|_{x=0} A &= \mathbf{D}l^{-1}\frac{d\mathbf{c}}{d\xi}\bigg|_{\xi=0} A \\ &= -Al^{-1}\mathbf{D}\,(\cosh \phi)^{-1}\, \boldsymbol{\Phi} \sinh \boldsymbol{\Phi}\mathbf{c}_0 \\ &= -\frac{A}{Pl^2}\mathbf{D\Phi} \tanh \boldsymbol{\Phi}\mathbf{c}_0(Pl) \end{aligned} \tag{8.5.25}$$

Comparing (8.5.25) with (8.5.20), we obtain

$$\mathbf{E\hat{K}} = -\left(\frac{A}{Pl^2}\right)\mathbf{D\Phi} \tanh \boldsymbol{\Phi} \tag{8.5.26}$$

It is interesting to note at this point that although $\boldsymbol{\Phi}$ is a singular matrix, it makes sense to talk about $\boldsymbol{\Phi}^{-1} \tanh \boldsymbol{\Phi}$ if we define it by

$$\boldsymbol{\Phi}^{-1} \tanh \boldsymbol{\Phi} = \sum_{j=1}^{n} (\mu_j^{-1} \tanh \mu_j)\mathbf{P}_j \tag{8.5.27}$$

If $\mu_1 = 0$ and $\mu_j \neq 0$ for $j = 2, 3, \ldots, n$, then (8.5.27) becomes

$$\boldsymbol{\Phi}^{-1} \tanh \boldsymbol{\Phi} = \mathbf{P}_1 + \sum_{j=2}^{n} (\mu_j^{-1} \tanh \mu_j)\mathbf{P}_j \tag{8.5.28}$$

where $\mathbf{P}_1$ is the orthogonal projection onto the null space of $\boldsymbol{\Phi}$ (which coincides with that of $\mathbf{\hat{K}}$). Thus we may rewrite (8.5.26) as

$$\mathbf{E\hat{K}} = \mathbf{\hat{K}}\boldsymbol{\Phi}^{-1} \tanh \boldsymbol{\Phi} \tag{8.5.29}$$

which should be considered in combination with (8.5.28). Clearly, $\mathbf{\hat{K}P}_1 = \mathbf{0}$, so that

$$\mathbf{E\hat{K}} = \sum_{j=2}^{n} (\mu_j^{-1} \tanh \mu_j)\mathbf{\hat{K}P}_j$$

Equation (8.5.29) suggests that we could have defined a matrix effectiveness factor $\mathbf{E}'$ such that $\hat{\mathbf{K}}\mathbf{E}' \equiv \mathbf{E}\hat{\mathbf{K}}$; then we obtain

$$\mathbf{E}' = \mathbf{\Phi}^{-1} \tanh \mathbf{\Phi}$$

which is precisely the matrix version of the effectiveness factor that is obtained for a single first-order reaction. Note that $\mathbf{E}'$ modifies the concentration at the pore mouth rather than the rate.

### 8.5.2 The First-Order Reaction System in a Tubular Reactor

The analysis here applies to the multicomponent first-order reaction mixture in a tubular reactor which is either coated on the wall with the reaction-promoting catalyst or packed with catalyst particles. The mixture consists of $n$ reacting species and an $(n + 1)$st inert fluid in sufficient excess as to maintain a uniform bulk velocity $v$, undisturbed by the effects of diffusion. The individual species disperse axially about the mean velocity such that the dispersive flux of the $i$th species is given by $-D_i(dc_i/dx)$ independently of the other species; $c_i$ represents the molar concentration of $A_i$. The area-specific rate constant matrix $\hat{\mathbf{K}}$ is as defined in the previous example.

The steady-state concentration vector $c$ is readily shown to statisfy the matrix differential equation

$$\mathbf{D}\frac{d^2\mathbf{c}}{dx^2} - v\frac{d\mathbf{c}}{dx} + \alpha\hat{\mathbf{K}}\mathbf{c} = 0, \qquad 0 < x < l \tag{8.5.30}$$

where $l$ is the reactor length and $\alpha$ is the catalytic reaction surface available per unit reactor volume. The matrix of dispersion coefficients is taken to be diagonal. The boundary conditions that must be satisfied are

$$\begin{aligned} -\mathbf{D}\frac{d\mathbf{c}}{dx} + v\mathbf{c} &= v\mathbf{c}_f, \qquad x = 0 \\ \frac{d\mathbf{c}}{dx} &= 0, \qquad x = l \end{aligned} \tag{8.5.31}$$

where $\mathbf{c}_f$ is the feed concentration vector.

Before we proceed to solve the problem above, some further observations are useful. From Section 8.2 we learned that $\hat{\mathbf{K}}$ was self-adjoint with respect to the inner product (8.2.7) and about its consequent spectral representation. For convenience here we denote the eigenvalues of $(-\alpha\hat{\mathbf{K}})$ by $\mu_0, \mu_1, \mu_2, \ldots, \mu_{n-1}$, where $\mu_0 = 0, \mu_j > 0, j = 1, 2, \ldots, n-1$. The corresponding orthogonal projections will be denoted by $\{\mathbf{P}_j\}_0^{n-1}$. The spectral representation of $(-\alpha\hat{\mathbf{K}})$ is given by

$$-\alpha\hat{\mathbf{K}} = \sum_{j=1}^{n-1} \mu_j\mathbf{P}_j \tag{8.5.32}$$

The operator $\mathbf{I}_{n-1}$ defined by

$$\mathbf{I}_{n-1} = \sum_{j=1}^{n-1} \mathbf{P}_j$$

is a self-adjoint projection of $\mathcal{R}_n$ onto $R(\hat{\mathbf{K}})$. On $R(\hat{\mathbf{K}})$, $\hat{\mathbf{I}}_{n-1}$ acts as the identity operator. It also has the interesting property that

$$(\alpha\hat{\mathbf{K}})\mathbf{I}_{n-1} = (\mathbf{I}_{n-1}\alpha\hat{\mathbf{K}}) = \alpha\hat{\mathbf{K}} \tag{8.5.33}$$

a result that was established in Exercise 7.4.1. We may rewrite (8.5.30) as

$$\mathbf{D}\frac{d^2\mathbf{c}}{dx^2} - v\frac{d\mathbf{c}}{dx} + \alpha\hat{\mathbf{K}}\mathbf{I}_{n-1}\mathbf{c} = \mathbf{0} \tag{8.5.34}$$

It is this equation that we will propose to solve. As in the previous examples, our strategy is to start from the scalar versions of (8.5.30) and (8.5.31). The solution of the scalar version of (8.5.30) proceeds via the auxiliary equation

$$D\lambda^2 - v\lambda + \alpha\hat{K} = 0$$

which is solved for the two values of $\lambda$, say $\lambda_1$ and $\lambda_2$, and expressing the solution as the linear combination $u_1e^{\lambda_1 x} + u_2e^{\lambda_2 x}$. The two boundary conditions then lead to the identification of the constant coefficients $u_1$ and $u_2$.

A similar strategy with (8.5.30) or its more convenient substitute (8.5.34) leads to the equation†

$$(\mathbf{D}\lambda^2 - v\mathbf{I}\lambda + \alpha\hat{\mathbf{K}}\mathbf{I}_{n-1})\mathbf{u} = 0 \tag{8.5.35}$$

where $\mathbf{u}$ is a constant vector in $\mathcal{R}_n$ such that $\mathbf{u}e^{\lambda x}$ is a solution to (8.5.34). Equation (8.5.35) leads to the *determinantal* equation

$$|\mathbf{D}\lambda^2 - v\mathbf{I}\lambda + \alpha\hat{\mathbf{K}}\mathbf{I}_{n-1}| = 0 \tag{8.5.36}$$

Equation (8.5.36) could conceivably lead to complex values of $\lambda$. For example, a pair of complex-conjugate roots would lead to spatial oscillations or concentration patterns. We show, however, that for the case on hand, the foregoing situation is impossible.

Our objective now is to recast (8.5.35) into a *pair* of equations in which the parameter $\lambda$ appears linearly. The procedure is reminiscent of (and indeed equivalent to) decomposing a second-order differential equation into a pair of first-order equations. The motivation for this decomposition is to produce a *new eigenvalue* problem in which the parameter $\lambda$ is the eigenvalue. One would then investigate the operator that arises in the problem and seek to render it self-adjoint by an appropriate choice of Hilbert space. If we are successful, we can infer that the parameter $\lambda$ from (8.5.36) can assume only real values.

---

†See, for example, *Elementary Matrices*, by R. A. Frazer, W. J. Duncan, and A. R. Collar, Cambridge University Press, Cambridge, 1952.

We write (8.5.35) as

$$\lambda(-\mathbf{D}\lambda\mathbf{u} + v\mathbf{u}) = \alpha\hat{\mathbf{K}}\mathbf{I}_{n-1}\mathbf{u}$$

We denote the vector within the parentheses above by $-\alpha\hat{\mathbf{K}}\mathbf{w}$, where $\mathbf{w} \in \mathcal{R}_n$; then (8.5.35) is equivalent to the following pair of equations:

$$-\mathbf{D}\lambda\mathbf{u} + v\mathbf{u} = -\alpha\hat{\mathbf{K}}\mathbf{w}$$

$$\lambda\alpha\hat{\mathbf{K}}\mathbf{w} = \alpha\hat{\mathbf{K}}\mathbf{I}_{n-1}\mathbf{u}$$

If $\mathbf{w}$ is restricted to $R(\hat{\mathbf{K}})$, the second equation above may be written as $-\lambda\mathbf{w} = \mathbf{I}_{n-1}\mathbf{u}$. Thus we may rewrite the foregoing pair of equations as

$$v\mathbf{D}^{-1}\mathbf{u} + \mathbf{D}^{-1}(\alpha\hat{\mathbf{K}})\mathbf{w} = \lambda\mathbf{u}$$

$$-\mathbf{I}_{n-1}\mathbf{u} = \lambda\mathbf{w}$$

Using the concept of a partitioned matrix (see also Section 2.11 on operators defined on direct sums of linear spaces), the foregoing pair may be succinctly represented by

$$\mathbf{TX} = \lambda\mathbf{X} \tag{8.5.37}$$

where

$$\mathbf{T} \equiv \begin{bmatrix} v\mathbf{D}^{-1} & \mathbf{D}^{-1}(\alpha\hat{\mathbf{K}}) \\ -\mathbf{I}_{n-1} & \mathbf{0} \end{bmatrix}, \qquad \mathbf{X} \equiv \begin{bmatrix} \mathbf{u} \\ \mathbf{w} \end{bmatrix}$$

The focus is now on the operator $\mathbf{T}$ and it is of interest to see if it is self-adjoint in some Hilbert space. On perusing the operator $\mathbf{T}$, one finds two attractive features.† First, the off-diagonal operator $\mathbf{D}^{-1}(\alpha\hat{\mathbf{K}})$ consists of $\mathbf{D}^{-1}$ and $\alpha\hat{\mathbf{K}}$, both of which are self-adjoint with respect to the inner product (8.2.7). The other is that $\mathbf{D}^{-1}$ is a positive operator, while $(-\alpha\hat{\mathbf{K}})$ is nonnegative. The one disconcerting feature is that $\hat{\mathbf{K}}$ is singular, so that $(-\alpha\hat{\mathbf{K}})$ is not strictly positive.

Since $(-\alpha\hat{\mathbf{K}})$ is strictly positive on $R(\hat{\mathbf{K}})$, and the operator $\mathbf{T}$ in (8.5.37) shows that $\alpha\hat{\mathbf{K}}$ operates only on $\mathbf{w}$, the restriction of $\mathbf{w}$ to $R(\hat{\mathbf{K}})$ has now realized its most important basis.

We consider a linear space $\mathcal{L} = \mathcal{R}_n \oplus R(\hat{\mathbf{K}})$. Observe first that for $\hat{\mathbf{X}} \in \mathcal{L}$,

$$\mathbf{TX} = \begin{bmatrix} v\mathbf{D}^{-1}\mathbf{u} + \mathbf{D}^{-1}\alpha\hat{\mathbf{K}}\mathbf{w} \\ -\mathbf{I}_{n-1}\mathbf{u} \end{bmatrix}$$

†The reason these features are attractive is that a $(2 \times 2)$ matrix of *real* coefficients with both off-diagonal elements of the same sign is an example of the Jacobi matrix of Section 2.4, which can be made self-adjoint. When corresponding matrices of operators are involved, real positive coefficients must be replaced by positive self-adjoint operators. This analogy must, however, be used with care!

Since $(v\mathbf{D}^{-1}\mathbf{u} + \mathbf{D}^{-1}\alpha\hat{\mathbf{K}}\mathbf{w}) \in \mathcal{R}_n$ and $\mathbf{I}_{n-1}\mathbf{u} \in R(\hat{\mathbf{K}})$, we have $\mathbf{TX} \in \mathcal{L}$. Thus $\mathcal{L}$ is invariant under $\mathbf{T}$, the implication being that we can safely work within this space with the operator $\mathbf{T}$ without transforming out of it.

Next we define the inner product on $\mathcal{L} \times \mathcal{L}$ as follows. Let

$$\mathbf{X}_1 \equiv \begin{bmatrix} \mathbf{u}_1 \\ \mathbf{w}_1 \end{bmatrix}, \qquad \mathbf{X}_2 \equiv \begin{bmatrix} \mathbf{u}_2 \\ \mathbf{w}_2 \end{bmatrix} \in \mathcal{L}$$

$$[\mathbf{X}_1, \mathbf{X}_2] \equiv \langle \mathbf{D}\mathbf{u}_1, \mathbf{u}_2 \rangle + \langle -(\alpha\hat{\mathbf{K}})\mathbf{w}_1, \mathbf{w}_2 \rangle \tag{8.5.38}$$

where the inner product symbol $\langle \cdot, \cdot \rangle$ is defined by (8.2.7). The positive definiteness of $\mathbf{D}$, and of $(-\alpha\hat{\mathbf{K}})$ on $R(\hat{\mathbf{K}})$, make (8.5.38) a valid inner product. We denote the Hilbert space $\{\mathcal{L}, [\cdot, \cdot]\}$ by $\mathcal{H}$. The norm of a vector $\mathbf{X} \in \mathcal{H}$ is given by

$$\|\mathbf{X}\| = \sqrt{[\mathbf{X}, \mathbf{X}]}$$

We now show that $\mathbf{T}$ is self-adjoint in $\mathcal{H}$. Let $\mathbf{X}_1, \mathbf{X}_2 \in \mathcal{H}$. Then

$$[\mathbf{TX}_1, \mathbf{X}_2] = v\langle \mathbf{u}_1, \mathbf{u}_2 \rangle + \langle \alpha\hat{\mathbf{K}}\mathbf{w}_1, \mathbf{u}_2 \rangle + \langle \alpha\hat{\mathbf{K}}\mathbf{u}_1, \mathbf{w}_2 \rangle$$
$$[\mathbf{X}_1, \mathbf{TX}_2] = v\langle \mathbf{u}_1, \mathbf{u}_2 \rangle + \langle \mathbf{w}_1, \alpha\hat{\mathbf{K}}\mathbf{u}_2 \rangle + \langle \mathbf{u}_1\, \alpha\hat{\mathbf{K}}\mathbf{w}_2 \rangle$$

from which we have

$$[\mathbf{TX}_1, \mathbf{X}_2] = [\mathbf{X}_1, \mathbf{TX}_2]$$

so that $\mathbf{T}$ is established to be self-adjoint. Thus all its eigenvalues are real. Their evaluation must proceed by solving the determinantal equation (8.5.36). There are $(2n - 1)$ eigenvalues $\lambda_1, \lambda_2, \ldots, \lambda_{2n-1}$ and corresponding eigenvectors $\mathbf{X}_1, \mathbf{X}_2, \ldots, \mathbf{X}_{2n-1}$, which are othonormal with respect to the inner product (8.5.38) and a basis in $\mathcal{H}$. In particular, $\mathbf{TX} = \mathbf{0}$ and

$$v\mathbf{D}^{-1}\mathbf{u} + \mathbf{D}^{-1}(\alpha\hat{\mathbf{K}})\mathbf{w} = \mathbf{0}$$
$$\mathbf{I}_{n-1}\mathbf{u} = \mathbf{0}$$

The second equation above implies that $\mathbf{u} \in N(\hat{\mathbf{K}})$, that is, $\mathbf{u} = \gamma\mathbf{a}$, where $\gamma$ is a constant. However, from the first equation above we find that $\mathbf{u} \in R(\hat{\mathbf{K}})$, so that $\mathbf{u} = \mathbf{0}$, from which it also follows that $\mathbf{w} = \mathbf{0}$. Thus $\lambda = 0$ is not an eigenvalue. Therefore, none of the eigenvalues $\lambda_1, \lambda_2, \ldots, \lambda_{2n-1}$ can be zero. In fact, if $\lambda$ is an eigenvalue with $\mathbf{X} \equiv \begin{bmatrix} \mathbf{u} \\ \mathbf{w} \end{bmatrix}$ as the corresponding eigenvector, then it is easy to show that

$$\lambda = \frac{v\langle \mathbf{u}, \mathbf{u} \rangle + 2\langle \alpha\hat{\mathbf{K}}\mathbf{w}, \mathbf{u} \rangle}{\langle \mathbf{D}\mathbf{u}, \mathbf{u} \rangle + \langle (-\alpha\hat{\mathbf{K}})\mathbf{w}, \mathbf{w} \rangle}$$

Since $\alpha\hat{\mathbf{K}}$ is negative definite on $R(\hat{\mathbf{K}})$ and $\mathbf{D}$ is positive definite on $\mathcal{R}_n$, the numerator can assume either positive or negative values. Thus $\mathbf{T}$ could have both positive and negative eigenvalues.

We now obtain the solution to the tubular reactor problem using the eigenvalues and eigenvectors of **T**. To accomplish this we must express (8.5.34) in terms of the operator **T**. In preparation we take the inner product (8.2.7) of (8.5.34) with vector **a** to obtain

$$\frac{d}{dx}\left\langle \mathbf{D}\frac{d\mathbf{c}}{dx} - v\mathbf{c}, \quad \mathbf{a}\right\rangle = 0 \tag{8.5.39}$$

The equilibrium concentration vector $\mathbf{c}^*$, given by

$$\mathbf{c}^* = \mathbf{a}\sum_{i=1}^{n} c_{i,f}$$

is attained where the axial derivatives vanish so that (8.5.39) yields

$$\left\langle \mathbf{D}\frac{d\mathbf{c}}{dx} - v(\mathbf{c} - \mathbf{c}^*), \quad \mathbf{a}\right\rangle = 0$$

from which it is clear that the vector $\mathbf{D}(d\mathbf{c}/dx) - v(\mathbf{c} - \mathbf{c}^*)$ belongs to $R(\hat{\mathbf{K}})$. Thus define a vector $\mathbf{m}(x)$ such that

$$\mathbf{D}\frac{d\mathbf{c}}{dx} - v(\mathbf{c} - \mathbf{c}^*) = \alpha\hat{\mathbf{K}}\mathbf{m} \tag{8.5.40}$$

In view of (8.5.34) we must have

$$\frac{d\mathbf{m}}{dx} = -\mathbf{I}_{n-1}\mathbf{c} \tag{8.5.41}$$

Indeed, (8.5.40) and (8.5.41) together yield the equation

$$\frac{d}{dx}\mathbf{Y} = \mathbf{T}\mathbf{Y} \tag{8.5.42}$$

where

$$\mathbf{Y} \equiv \begin{bmatrix} \mathbf{c} - \mathbf{c}^* \\ \mathbf{m} \end{bmatrix}$$

The solution of (8.5.42) is written immediately by following the methods of Section 7.5.

$$\mathbf{Y} = \sum_{j=1}^{2n-1} \beta_j e^{\lambda_j x}\mathbf{X}_j, \qquad \mathbf{X}_j \equiv \begin{bmatrix} \mathbf{u}_j \\ \mathbf{w}_j \end{bmatrix} \tag{8.5.43}$$

where $\beta_j \equiv [\mathbf{Y}(0), \mathbf{X}_j]$ is given by

$$\beta_j = \langle \mathbf{D}\{\mathbf{c}(0) - \mathbf{c}^*\}, \mathbf{u}_j\rangle + \langle -\alpha\hat{\mathbf{K}}\mathbf{m}(0), \mathbf{w}_j\rangle$$

Since $\mathbf{c}(0)$ is unknown, constants $\{\beta_j\}_{j=1}^{2n-1}$ are also unknown. The boundary condition (8.5.31) at $x = 0$, when substituted into (8.5.40), yields

$$v\mathbf{c}_f - v\mathbf{c}^* = -\alpha\hat{\mathbf{K}}\mathbf{m}(0)$$

so that

$$\beta_j = \langle \mathbf{D}\mathbf{c}(0), \mathbf{u}_j\rangle - \langle \mathbf{D}\mathbf{c}^*, \mathbf{u}_j\rangle + v\langle \mathbf{c}_f, \mathbf{w}_j\rangle \tag{8.5.44}$$

The exit boundary condition at $x = l$ yields

$$\sum_{j=1}^{2n-1} \beta_j \lambda_j e^{\lambda_j l} \mathbf{u}_j = \mathbf{0} \tag{8.5.45}$$

The unknown vector $\mathbf{c}(0)$ may be expanded in terms of the eigenvectors of $\hat{\mathbf{K}}$. If we recall the eigenvectors $\{\mathbf{z}_0 \equiv \mathbf{a}, \mathbf{z}_1, \mathbf{z}_2, \ldots, \mathbf{z}_{n-1}\}$ of $\hat{\mathbf{K}}$ from Section 8.2, we obtain

$$\mathbf{c}(0) = \sum_{k=0}^{n-1} \langle \mathbf{c}(0), \mathbf{z}_k \rangle \mathbf{z}_k$$

The unknowns to be determined are the constants $\{\langle \mathbf{c}(0), \mathbf{z}_k \rangle\}_{k=0}^{n-1}$, the equations for which are obtained as follows. Clearly,

$$\langle \mathbf{Dc}(0), \mathbf{u}_j \rangle = \sum_{k=0}^{n-1} \langle \mathbf{c}(0), \mathbf{z}_k \rangle \langle \mathbf{Dz}_k, \mathbf{u}_j \rangle \tag{8.5.46}$$

which on plugging into (8.5.44) and combining with (8.5.45) leads to the equations

$$\sum_{k=1}^{n} \gamma_{ik} \langle \mathbf{c}(0), \mathbf{z}_k \rangle = \delta_i \tag{8.5.47}$$

where

$$\gamma_{ik} \equiv \sum_{j=1}^{2n-1} \lambda_j e^{\lambda_j l} \langle \mathbf{Dz}_i, \mathbf{u}_j \rangle \langle \mathbf{Dz}_k, \mathbf{u}_j \rangle$$

$$\delta_i \equiv \sum_{j=1}^{2n-1} \{\langle v\mathbf{c}_f, \mathbf{w}_j \rangle + \langle \mathbf{Dc}^*, \mathbf{u}_j \rangle\} \langle \mathbf{Dz}_i, \mathbf{u}_j \rangle \lambda_j e^{\lambda_j l}$$

the details of which are left as an exercise. Notice that $\{\gamma_{ik}\}$ is a symmetric matrix. The vector $\boldsymbol{\delta} \equiv (\delta_1, \delta_2, \ldots, \delta_{n-1})$ is known on specification of $\mathbf{c}_f$. The solution to the tubular reactor problem is thus reduced to solving the simultaneous equations (8.5.47). When the constants $\{\langle \mathbf{c}(0), \mathbf{z}_k \rangle\}_{k=1}^{n}$ are available, the $\beta_j$'s are obtained via (8.5.44). The concentration profile in the reactor is then obtained as

$$\mathbf{c} = \sum_{j=1}^{2n-1} \beta_j e^{\lambda_j x} \mathbf{u}_j + \mathbf{c}^* \tag{8.5.48}$$

Since the eigenvalues of $\mathbf{T}$ can be both positive and negative, it is not clear from the equation above how $\mathbf{c}$ approaches $\mathbf{c}^*$ for a long reactor. We do not prove here that $\beta_j e^{\lambda_j l} \longrightarrow 0$ for both positive and negative $\lambda_j$'s as $l \longrightarrow \infty$. But the result is true nevertheless, so that $\mathbf{c} \longrightarrow \mathbf{c}^*$ for an infinitely long reactor. In fact, for sufficiently long reactors it is possible to show that

$$\mathbf{c} = \sum_j \beta_j^- e^{\lambda_j^- x} \mathbf{u}_j + \mathbf{c}^*$$

where $\{\lambda_j^-\}$ are only the negative eigenvalues and $\beta_j^-$ the corresponding coefficients.

## 8.6 Concluding Remarks

In this chapter we have shown how the spectral theorem can be effectively used to solve several finite-dimensional problems that appear in engineering analysis. In all examples, the crucial idea was to fit the available physical information into the abstract framework of the theory of self-adjoint operators. Of course, the caution that we have sounded before bears repetition in regard to the fact that not all operators can be made self-adjoint by the methods presented here. Indeed, there are essentially non-self-adjoint operators that will not submit to *any* "symmetrization" procedure. The emphasis here has been on operators that are frequently encountered and can be symmetrized (or rendered self-adjoint).

The reader should study all the discussed examples carefully and develop familiarity with the features that can lead to a self-adjoint structure.

In the ensuing chapters, we will deal with applications in infinite-dimensional spaces.

### FURTHER READING

For a classical background of matrix theory, see

*Elementary Matrices*, by R. A. Frazer, W. J. Duncan, and A. R. Collar, Cambridge University Press, Cambridge, 1952.

A number of problems discussed in this chapter arise as modifications of original versions appearing in

*Mathematical Methods in Chemical Engineering: Matrices and Their Application*, by N. R. Amundson, Prentice-Hall, Englewood Cliffs, N.J., 1966.

# Ordinary Linear Differential Operators 9

## 9.0 Foreword

In this chapter we are concerned with applications of the spectral theorem for operators in infinite-dimensional spaces. Following Chapter 8, which dealt with applications of spectral theorem 7.4.1 for operators in finite-dimensional Hilbert spaces, the next logical step would be to consider compact operators in infinite-dimensional spaces. Hence in such an itinerary of things, the application of integral equations should occupy us next. There indeed are many engineering situations in which integral equations arise, but for two reasons we will begin with differential (and hence unbounded) operators. First, the applications of differential equations in engineering analysis occur with considerably more frequency. Second, in a good many situations integral equations arise in fact from reformulations via inversion of differential operators.

Thus our focus in this chapter (and in the next) is on the applications of spectral theorem 7.4.3, whose concern is with linear unbounded operators that have compact, self-adjoint inverses. Accordingly, we will look at ordinary differential operators in detail and the solution of equations involving them. As in Chapter 8, the methods of Section 7.5 would be the basis of obtaining solutions to the equations once the spectral resolutions of the operators concerned have been computed. The student will do well at this stage to review Sections 5.5 and 6.7, which are concerned with general features of unbounded operators.

## 9.1 Introduction

In view of the preponderance of second-order differential operators in the applications of interest to us, we will consider them first. Furthermore, the treatment of second-order operators is considerably simpler and familiarity with it will be an asset to the subsequent discussion of arbitrary $n$th-order operators.

Recall that an unbounded operator could be defined at most on a dense linear subspace of a Hilbert space. We had represented an unbounded operator $\mathbf{T}$ jointly by a formal operation $T$ and a domain $D(T)$ that was a dense subspace of the Hilbert space. Similarly, $\mathbf{T}^*$, the adjoint operator of $\mathbf{T}$, was associated with a formal adjoint operation $T^*$ and an associated domain $D(T^*)$. In Section 6.7 we considered some examples of differential operators where $T^*$ had been specified without an explanation. We will show here how the formal operation $T^*$ comes about for an operation $T$. Since our eventual interest is in self-adjoint operators, we will seek to identify operations $T$ for which $T^* = T$. The formal operation $T$ will apply for the present to some subspace of $\mathcal{L}_2[a, b]$. Further, we will restrict ourselves to operations $T$ which transform real-valued functions into real-valued functions. The most general linear second-order differential operation may be represented by

$$T \equiv p_0(x)\frac{d^2}{dx^2} + p_1(x)\frac{d}{dx} + p_2(x) \tag{9.1.1}$$

where $p_0(x)$, $p_1(x)$, and $p_2(x)$ are real-valued functions in the interval $[a, b]$ such that $\mathbf{p}_0 \in \mathcal{C}^2[a, b]$, $\mathbf{p}_1 \in \mathcal{C}^1[a, b]$, and $\mathbf{p}_2 \in \mathcal{C}[a, b]$. More generally, however, the foregoing continuity and differentiability may occur in a piecewise manner; that is, the interval $[a, b]$ may be divided into subintervals in each of which the continuity and differentiability properties held in the *interior*, with of course one-sided properties at the end points. We will consider this situation later.

In defining an adjoint operation $T^*$, our objective is to render the expression

$$\langle T\mathbf{u}, \mathbf{v}\rangle - \langle \mathbf{u}, T^*\mathbf{v}\rangle \tag{9.1.2}$$

dependent *entirely* on the *boundary values* of functions $u(x)$ and $v(x)$ and at most their first derivatives. In (9.1.2) the implication is obvious that $\mathbf{u}$ and $\mathbf{v}$ should be such that $T$ and $T^*$ can operate on them. Also, the inner product in (9.1.2) is the one in $\mathcal{L}_2[a, b]$ defined by

$$\langle \mathbf{u}, \mathbf{v}\rangle = \int_a^b u(x)v(x)\, dx \tag{9.1.3}$$

Having stated the objective, the route to finding $T^*$ is provided by integration by parts of the integral†

$$\langle T\mathbf{u}, \mathbf{v}\rangle = \int_a^b v[p_0u'' + p_1u' + p_2u]\,dx \tag{9.1.4}$$
$$= \Big[(p_0v)u' + (p_1v)u - (p_0v)'u\Big]_a^b$$
$$+ \int_a^b u[(p_0v)'' - (p_1v)' + p_2v]\,dx$$

The first term on the right-hand side of (9.1.4) has been integrated by parts twice over. In the expression following (9.1.4) the notation of the square brackets with superscript $b$ and subscript $a$ is used to denote the difference between the values of the expression within the brackets at $b$ and $a$. Thus the expression within the foregoing set of brackets involves only the boundary values of $u$ and $v$ and their first derivatives. The differential expression within the integral term which operates on $v$ can now be used to define the adjoint operation $T^*$. Hence we obtain

$$\langle T\mathbf{u}, \mathbf{v}\rangle - \langle \mathbf{u}, T^*\mathbf{v}\rangle = \Big[p_0(u'v - uv') - (p_0' - p_1)uv\Big]_a^b \tag{9.1.5}$$

where

$$T^* \equiv p_0\frac{d^2}{dx^2} + (2p_0' - p_1)\frac{d}{dx} + (p_0'' - p_1' + p_2) \tag{9.1.6}$$

which is of the same form as (9.1.1), except for the change in coefficients.

Before we proceed any further, let us observe that if the operation $T$ in (9.1.1) is replaced by

$$T \equiv \frac{1}{r(x)}\Big[p_0(x)\frac{d^2}{dx^2} + p_1(x)\frac{d}{dx} + p_2(x)\Big] \tag{9.1.7}$$

where $r(x) > 0$, then by considering the Hilbert space $\mathcal{H}$ of elements $\mathbf{f}$ such that‡

$$\int_a^b r(x)f^2(x)\,dx < \infty$$

in which the inner product $\langle\cdot, \cdot\rangle$ is defined by

$$\langle \mathbf{u}, \mathbf{v}\rangle = \int_a^b r(x)u(x)v(x)\,dx \tag{9.1.8}$$

---

†We use primes to denote differentiation, and drop the arguments of functions frequently to avoid needless repetition. For example, $u(x)$, the "component" of $\mathbf{u} \in \mathfrak{L}_2[a, b]$, may at times be written as $u$.

‡We again refer the reader to Exercise 6.1.4 in this connection.

the adjoint operation $T^*$ is given by

$$T^* \equiv \frac{1}{r(x)}\left[p_0 \frac{d^2}{dx^2} + (2p_0' - p_1)\frac{d}{dx} + (p_0'' - p_1' + p_2)\right] \tag{9.1.9}$$

Again relation (9.1.5) holds if the inner product on the left-hand side is interpreted as that in (9.1.8). Since operations of the type (9.1.7) encompass (9.1.1) when $r(x) \equiv 1$, henceforth we will be concerned with the more general operation $T$ given by (9.1.7) with its adjoint as in (9.1.9).

We may recall at this stage from Section 6.7 how one could find domains $D(T)$ and $D(T^*)$ so that the operator $\mathbf{T} \equiv \{T, D(T)\}$ has, as its adjoint operator, $\mathbf{T}^* \equiv \{T^*, D(T^*)\}$. Clearly, the domains must be such that whenever $\mathbf{u} \in D(T)$ and $\mathbf{v} \in D(T^*)$, the right-hand side of (9.1.5) must vanish. In defining either of the domains $D(T)$ and $D(T^*)$, we must first make certain that the operations $T$ or $T^*$ do not transform elements out of $\mathfrak{L}_2[a, b]$. The reader is referred to Sections 4.3 and 6.7 for some discussion on these details. For example, it is necessary to constrain $D(T)$ to vectors $\mathbf{u} \in \mathfrak{L}_2[a, b]$ such that $u'$ exists and $p_0 u'$ is absolutely continuous. (In Section 4.3 we learned that the derivative of a function absolutely continuous in an interval is Lebesgue-integrable over that interval.) As in Section 6.7, we will not bother to make such detailed qualifications on $D(T)$ or $D(T^*)$ but instead only require that $T\mathbf{u} \in \mathfrak{L}_2[a, b]$ or $T^*\mathbf{u} \in \mathfrak{L}_2[a, b]$. More important, however, the domains must be defined by *boundary conditions*. Moreover, the boundary conditions must be *homogeneous* in order that the implied domains be subspaces of $\mathfrak{L}_2[a, b]$.† The most general boundary conditions for $D(T)$ are given by

$$\begin{aligned} \alpha_{11}u(a) + \alpha_{12}u'(a) + \beta_{11}u(b) + \beta_{12}u'(b) &= 0 \\ \alpha_{21}u(a) + \alpha_{22}u'(a) + \beta_{21}u(b) + \beta_{22}u'(b) &= 0 \end{aligned} \tag{9.1.10}$$

where $\alpha_{ij}$ and $\beta_{ij}$ are real coefficients for $i, j = 1, 2$. In order that the boundary conditions be linearly independent, the $2 \times 4$ matrix of coefficients

$$\begin{bmatrix} \alpha_{11} & \alpha_{12} & \beta_{11} & \beta_{12} \\ \alpha_{21} & \alpha_{22} & \beta_{21} & \beta_{22} \end{bmatrix}$$

must have rank 2. The boundary conditions (9.1.10) may be stated more succinctly by introducing a little terminology. Given a differentiable function $u(x)$, we may form its *Wronskian vector*, denoted $\mathbf{w}(u(x))$ and defined by

$$\mathbf{w}(u(x)) = \begin{bmatrix} u(x) \\ u'(x) \end{bmatrix}$$

†We will not demonstrate here how the homogeneous boundary conditions do assure us a *dense* linear subspace of $\mathfrak{L}_2[a, b]$, since these issues have been addressed in Section 4.3.5. But this fact in itself is important in defining $T$ and $T^*$. In this regard, the reader is persuaded to read Section 6.7.

Note that $\forall\, x \in [a, b]$ $\mathbf{w}(u(x)) \in \mathcal{R}_2$. If

$$\boldsymbol{\alpha} \equiv \begin{bmatrix} \alpha_{11} & \alpha_{12} \\ \alpha_{21} & \alpha_{22} \end{bmatrix} \quad \text{and} \quad \boldsymbol{\beta} \equiv \begin{bmatrix} \beta_{11} & \beta_{12} \\ \beta_{21} & \beta_{22} \end{bmatrix}$$

then (9.1.10) may be rewritten as

$$\boldsymbol{\alpha}\mathbf{w}(u(a)) + \boldsymbol{\beta}\mathbf{w}(u(b)) = \mathbf{0}$$

or in terms of partitioned matrices,†

$$[\boldsymbol{\alpha} \quad \boldsymbol{\beta}]\begin{bmatrix} \mathbf{w}(u(a)) \\ \mathbf{w}(u(b)) \end{bmatrix} = \mathbf{0} \tag{9.1.11}$$

We may thus represent the domain $D(T)$ as

$$D(T) = \left\{\mathbf{u} \in \mathcal{L}_2[a, b]\colon\ T\mathbf{u} \in \mathcal{L}_2[a, b];\ [\boldsymbol{\alpha} \quad \boldsymbol{\beta}]\begin{bmatrix} \mathbf{w}(u(a)) \\ \mathbf{w}(u(b)) \end{bmatrix} = \mathbf{0}\right\} \tag{9.1.12}$$

The specification of the operator **T** is now complete. Next we consider the adjoint operator $\mathbf{T}^*$. Since the formal operation $\mathbf{T}^*$ is given by (9.1.9), all that remains to identify $\mathbf{T}^*$ is to determine the boundary conditions that define $D(T^*)$.

The question arises as to whether the adjoint boundary conditions are unique. We had addressed this problem in Section 6.7 for some specific examples and observed that as long as the proper number of boundary conditions were attached to the operation $T$, the domain $D(T^*)$ is unique. The examples considered in Section 6.7 are simple and readily understood. Here, however, our attempt will be to be as general as possible in regard to the boundary conditions. Such a general treatment is greatly facilitated by the matrix formalism that follows. We adopt the usual practice of distinguishing a column vector, say $\mathbf{a} \equiv \begin{bmatrix} a_1 \\ a_2 \end{bmatrix}$, from its "row form" $[a_1 \quad a_2]$, which we denote by $\mathbf{a}^t$.

Since the right-hand side of (9.1.5) is a bilinear form, we rewrite it as

$$\langle T\mathbf{u}, \mathbf{v}\rangle - \langle \mathbf{u}, T^*\mathbf{v}\rangle = \Big[\mathbf{w}^t(v(x))\mathbf{P}(x)\mathbf{w}(u(x))\Big]_a^b \tag{9.1.13}$$

where

$$\mathbf{P}(x) \equiv \begin{bmatrix} p_1 - p_0' & p_0 \\ -p_0 & 0 \end{bmatrix}$$

†$\boldsymbol{\alpha}$ and $\boldsymbol{\beta}$ are $2 \times 2$ matrices, so $[\boldsymbol{\alpha} \quad \boldsymbol{\beta}]$ is the $2 \times 4$ matrix of coefficients in (9.1.10). The vector $\begin{bmatrix} \mathbf{w}(u(a)) \\ \mathbf{w}(u(b)) \end{bmatrix}$ is a column vector of four components.

The determinant of the matrix $\mathbf{P}$ is $p_0^2$, so that $\mathbf{P}$ is nonsingular in the interval of interest as long as $p_0(x)$ does not vanish there. Equation (9.1.13) may be further rewritten in terms of partitioned vectors and matrices as†

$$\langle T\mathbf{u}, \mathbf{v}\rangle - \langle \mathbf{u}, T^*\mathbf{v}\rangle = [\mathbf{w}^t(v(a)) \quad \mathbf{w}^t(v(b))]\begin{bmatrix} -\mathbf{P}(a) & \mathbf{0} \\ \mathbf{0} & \mathbf{P}(b) \end{bmatrix}\begin{bmatrix} \mathbf{w}(u(a)) \\ \mathbf{w}(u(b)) \end{bmatrix} \tag{9.1.14}$$

If $\mathbf{u} \in D(T)$, then (9.1.11) is true. This implies that the vector $\begin{bmatrix} \mathbf{w}(u(a)) \\ \mathbf{w}(u(b)) \end{bmatrix}$ belongs to the null space of the matrix operator $[\boldsymbol{\alpha} \quad \boldsymbol{\beta}]$, which premultiplies (column) vectors of $\mathcal{R}_4$ to produce vectors in $\mathcal{R}_2$ and postmultiplies (row) vectors of $\mathcal{R}_2$ to yield vectors in $\mathcal{R}_4$.

Consider first $[\boldsymbol{\alpha} \quad \boldsymbol{\beta}]$ as a premultiplying matrix operator, which acts on vectors in $\mathcal{R}_4$. Denoting the *null* space of this operator by $N$, we note in view of the matrix rank being 2 that the dimension of $N$ is $4 - 2$, or 2.

Next, $[\boldsymbol{\alpha} \quad \boldsymbol{\beta}]$ as a postmultiplying matrix operator transforms vectors of $\mathcal{R}_2$ into $\mathcal{R}_4$, so that its *range* space, denoted $R$ is in $\mathcal{R}_4$. Thus $R$ and $N$ are both subspaces of $\mathcal{R}_4$. Moreover, it can be shown that they are *orthogonal* subspaces, that is, $R \perp N$. Since $R$ also has dimension 2 (why?), we conclude further that $R^\perp = N$ and $N^\perp = R$. Now every vector in $R$ is uniquely associated with a vector in $\mathcal{R}_2$ through the postmultiplying operation of $[\boldsymbol{\alpha} \quad \boldsymbol{\beta}]$ (see Exercise 9.1.3). Thus the vector

$$[\mathbf{w}^t(v(a)) \quad \mathbf{w}^t(v(b))]\begin{bmatrix} -\mathbf{P}(a) & \mathbf{0} \\ \mathbf{0} & \mathbf{P}(b) \end{bmatrix}$$

which is clearly in $R$ by virtue of (9.1.14), may be written as $\mathbf{y}^t[\boldsymbol{\alpha} \quad \boldsymbol{\beta}]$, where $\mathbf{y} \in \mathcal{R}_2$ depends uniquely and linearly on $[\mathbf{w}^t(v(a)) \quad \mathbf{w}^t(v(b))]$. We will be concerned here only with the case for which $p_0$ does not vanish anywhere in the interval $[a, b]$, so that the matrix $\mathbf{P}(x)$ has an inverse everywhere. Thus

$$[\mathbf{w}^t(v(a)) \quad \mathbf{w}^t(v(b))] = \mathbf{y}^t[\boldsymbol{\alpha} \quad \boldsymbol{\beta}]\begin{bmatrix} -\mathbf{P}^{-1}(a) & \mathbf{0} \\ \mathbf{0} & \mathbf{P}^{-1}(b) \end{bmatrix} \tag{9.1.15}$$

---

†The vector $[\mathbf{w}^t(v(a)) \quad \mathbf{w}^t(v(b))]$ is the same as $[v(a) \quad v'(a) \quad v(b) \quad v'(b)]$ and

$$\begin{bmatrix} \mathbf{w}(u(a)) \\ \mathbf{w}(u(b)) \end{bmatrix} \text{ is the same as } \begin{bmatrix} u(a) \\ u'(a) \\ u(b) \\ u'(b) \end{bmatrix}$$

This mode of treatment is due to R. H. Cole, *Theory of Ordinary Differential Equations*, Appleton-Century-Crofts, New York, 1968.

Suppose now that we represent the adjoint boundary conditions by

$$[\boldsymbol{\alpha}^* \quad \boldsymbol{\beta}^*]\begin{bmatrix} \mathbf{w}(v(a)) \\ \mathbf{w}(v(b)) \end{bmatrix} = \mathbf{0} \tag{9.1.16}$$

where $[\boldsymbol{\alpha}^* \quad \boldsymbol{\beta}^*]$ is the associated matrix of coefficients.† Recognizing the transposed version of (9.1.16), we obtain from (9.1.15)

$$\mathbf{y}^t[\boldsymbol{\alpha} \quad \boldsymbol{\beta}]\begin{bmatrix} -\mathbf{P}^{-1}(a) & \mathbf{0} \\ \mathbf{0} & \mathbf{P}^{-1}(b) \end{bmatrix}\begin{bmatrix} \boldsymbol{\alpha}^{*t} \\ \boldsymbol{\beta}^{*t} \end{bmatrix} = \mathbf{0}$$

Since the foregoing equality must be true regardless of the choice of $[\mathbf{w}^t(v(a)) \quad \mathbf{w}^t(v(b))]$ [i.e., of $v(x)$] and hence $\mathbf{y}$, we conclude that

$$[\boldsymbol{\alpha} \quad \boldsymbol{\beta}]\begin{bmatrix} -\mathbf{P}^{-1}(a) & \mathbf{0} \\ \mathbf{0} & \mathbf{P}^{-1}(b) \end{bmatrix}\begin{bmatrix} \boldsymbol{\alpha}^{*t} \\ \boldsymbol{\beta}^{*t} \end{bmatrix} = 0 \tag{9.1.17}$$

or alternatively,

$$\boldsymbol{\alpha}\mathbf{P}^{-1}(a)\boldsymbol{\alpha}^{*t} = \boldsymbol{\beta}\mathbf{P}^{-1}(b)\boldsymbol{\beta}^{*t} \tag{9.1.18}$$

Thus (9.1.17) or (9.1.18) represents the relationship between the coefficients for the boundary conditions associated with $T$ and those for the adjoint boundary conditions. Equation (9.1.18) is equivalent to *four* equations in *eight* unknowns, so that the adjoint boundary conditions are *not* unique. (We had observed the same with the examples discussed in Section 6.7.) However, it is easily shown that the different sets of adjoint boundary conditions imply each other. For example, we may postmultiply (9.1.18) by a nonsingular $2 \times 2$ matrix, say $\boldsymbol{\gamma}$, to obtain a new matrix of coefficients $[\boldsymbol{\gamma}^t\boldsymbol{\alpha}^* \quad \boldsymbol{\gamma}^t\boldsymbol{\beta}^*]$ in place of $[\boldsymbol{\alpha}^* \quad \boldsymbol{\beta}^*]$ in the adjoint boundary conditions represented by (9.1.16). Notice as before that the domain $D(T^*)$ is not altered by such nonuniqueness of boundary conditions. Thus $\mathbf{T}^*$ is unique even if the adjoint boundary conditions are not.

It is not difficult to *find a* set of adjoint boundary conditions, although an entirely general treatment calls for some additional formalism. We present below such a formalism, which is more useful for the general coupled boundary conditions represented by (9.1.11). For relatively simple boundary conditions we advise a more direct approach, which is demonstrated through examples.

Consider again the boundary conditions (9.1.11). Recall that $[\boldsymbol{\alpha} \quad \boldsymbol{\beta}]$ has rank 2, so that by combining two appropriate columns of it a $2 \times 2$ nonsingular matrix can be found. In further discussion we will assume that the $2 \times 2$

†We regret the asterisk notation, which is in conflict with its use in denoting the complex conjugate of a number. However, its appropriateness in being associated with $T^*$ should hopefully be a compensating asset.

matrix formed by the $i$th and $j$th columns of $[\boldsymbol{\alpha} \quad \boldsymbol{\beta}]$ is nonsingular. Let $\boldsymbol{\gamma}_{ij}$ be a $2 \times 4$ matrix whose elements are all zero except for a unity in the first row and $i$th column and in the second row and $j$th column. Then the $2 \times 2$ matrix $[\boldsymbol{\alpha} \quad \boldsymbol{\beta}]\boldsymbol{\gamma}_{ij}^t$, abbreviated as $\boldsymbol{\delta}_{ij}$, is nonsingular. Suppose now that $k \neq i$, $l \neq j$; then the reader should prove that

$$\boldsymbol{\gamma}_{ij}^t\boldsymbol{\gamma}_{ij} + \boldsymbol{\gamma}_{kl}^t\boldsymbol{\gamma}_{kl} = \begin{bmatrix} 1 & 0 & 0 & 0 \\ 0 & 1 & 0 & 0 \\ 0 & 0 & 1 & 0 \\ 0 & 0 & 0 & 1 \end{bmatrix}$$

Thus the boundary condition (9.1.11) may be rewritten as

$$\boldsymbol{\delta}_{ij}\boldsymbol{\gamma}_{ij}\begin{bmatrix} \mathbf{w}(u(a)) \\ \mathbf{w}(u(b)) \end{bmatrix} = -\boldsymbol{\delta}_{kl}\boldsymbol{\gamma}_{kl}\begin{bmatrix} \mathbf{w}(u(a)) \\ \mathbf{w}(u(b)) \end{bmatrix}$$

Since $\boldsymbol{\delta}_{ij}$ is nonsingular, we have from the foregoing equation

$$\boldsymbol{\gamma}_{ij}\begin{bmatrix} \mathbf{w}(u(a)) \\ \mathbf{w}(u(b)) \end{bmatrix} = -\boldsymbol{\delta}_{ij}^{-1}\boldsymbol{\delta}_{kl}\boldsymbol{\gamma}_{kl}\begin{bmatrix} \mathbf{w}(u(a)) \\ \mathbf{w}(u(b)) \end{bmatrix} \tag{9.1.19}$$

It is important to recognize here that the two-dimensional vector $\boldsymbol{\gamma}_{kl}\begin{bmatrix} \mathbf{w}(u(a)) \\ \mathbf{w}(u(b)) \end{bmatrix}$ may be chosen *arbitrarily* in (9.1.19). Now from (9.1.14) we must have

$$[\mathbf{w}^t(v(a)) \quad \mathbf{w}^t(v(b))]\begin{bmatrix} -\mathbf{P}(a) & \mathbf{0} \\ \mathbf{0} & \mathbf{P}(b) \end{bmatrix}\{\boldsymbol{\gamma}_{ij}^t\boldsymbol{\gamma}_{ij} + \boldsymbol{\gamma}_{kl}^t\boldsymbol{\gamma}_{kl}\}\begin{bmatrix} \mathbf{w}(u(a)) \\ \mathbf{w}(u(b)) \end{bmatrix} = 0$$

Using (9.1.19), the foregoing equation leads to the adjoint boundary conditions

$$[\mathbf{w}^t(v(a)) \quad \mathbf{w}^t(v(b))]\begin{bmatrix} -\mathbf{P}(a) & \mathbf{0} \\ \mathbf{0} & \mathbf{P}(b) \end{bmatrix}\{-\boldsymbol{\gamma}_{ij}^t\boldsymbol{\delta}_{ij}^{-1}\boldsymbol{\delta}_{kl} + \boldsymbol{\gamma}_{kl}^t\} = \mathbf{0}$$

In view of (9.1.16) we must write

$$\begin{bmatrix} \boldsymbol{\alpha}^{*t} \\ \boldsymbol{\beta}^{*t} \end{bmatrix} = \begin{bmatrix} -\mathbf{P}(a) & \mathbf{0} \\ \mathbf{0} & \mathbf{P}(b) \end{bmatrix}\{-\boldsymbol{\gamma}_{ij}^t\boldsymbol{\delta}_{ij}^{-1}\boldsymbol{\delta}_{kl} + \boldsymbol{\gamma}_{kl}^t\} \tag{9.1.20}$$

It is left as an exercise to show that (9.1.20) is consistent with (9.1.17). This general formula is somewhat cumbersome, but frequently the adjoint boundary conditions are more easily obtained directly from equating the right-hand side of (9.1.14) to zero. The virtue of (9.1.20) is its existence even when $\mathbf{P}$ does not have an inverse at either $a$ or $b$.† We consider some examples below.

---

†See also Exercise 9.1.7 for an alternative approach to the adjoint boundary conditions.

*Example 1*

Consider the operator $\mathbf{T} \equiv \{T, D(T)\}$, where $T \equiv d^2/dx^2 - d/dx$ and

$$D(T) = \{\mathbf{u} \in \mathcal{L}_2[0, 1];\, T\mathbf{u} \in \mathcal{L}_2[0, 1]; \quad u(0) = 0, \quad u'(0) = 2u(1)\}$$

The boundary conditions for $T$, when put in the form of (9.1.11), yield

$$\boldsymbol{\alpha} \equiv \begin{bmatrix} 1 & 0 \\ 0 & 1 \end{bmatrix}, \qquad \boldsymbol{\beta} = \begin{bmatrix} 0 & 0 \\ -2 & 0 \end{bmatrix}$$

Notice that the rank of $[\boldsymbol{\alpha} \quad \boldsymbol{\beta}]$ is 2, as required. For the adjoint operator $\mathbf{T}^*$, the formal operation $T^*$ is obtained via (9.1.9) to be

$$T^* = \frac{d^2}{dx^2} + \frac{d}{dx}$$

Next we determine the adjoint boundary conditions using (9.1.20) first. The matrix $\mathbf{P}(x)$ is given by

$$\mathbf{P}(x) = \begin{bmatrix} 1 & 1 \\ -1 & 0 \end{bmatrix}$$

In applying formula (9.1.20), we must have $i = 1, j = 2$ (since the first two columns of $[\boldsymbol{\alpha} \quad \boldsymbol{\beta}]$ yield a nonsingular $2 \times 2$ matrix), $k = 3$ and $l = 4$. Clearly,

$$\boldsymbol{\delta}_{12} = \begin{bmatrix} 1 & 0 \\ 0 & 1 \end{bmatrix} \quad \text{so that} \quad \boldsymbol{\delta}_{12}^{-1} = \begin{bmatrix} 1 & 0 \\ 0 & 1 \end{bmatrix}, \qquad \boldsymbol{\delta}_{34} = \begin{bmatrix} 0 & 0 \\ -2 & 0 \end{bmatrix}$$

Thus

$$\begin{bmatrix} \boldsymbol{\alpha}^{*t} \\ \boldsymbol{\beta}^{*t} \end{bmatrix} = \begin{bmatrix} -1 & -1 & 0 & 0 \\ 1 & 0 & 0 & 0 \\ 0 & 0 & 1 & 1 \\ 0 & 0 & -1 & 0 \end{bmatrix} \left\{ - \begin{bmatrix} 1 & 0 \\ 0 & 1 \\ 0 & 0 \\ 0 & 0 \end{bmatrix} \begin{bmatrix} 1 & 0 \\ 0 & 1 \end{bmatrix} \begin{bmatrix} 0 & 0 \\ -2 & 0 \end{bmatrix} + \begin{bmatrix} 0 & 0 \\ 0 & 0 \\ 1 & 0 \\ 0 & 1 \end{bmatrix} \right\}$$

After some irksome manipulations, we obtain

$$\begin{bmatrix} \boldsymbol{\alpha}^{*t} \\ \boldsymbol{\beta}^{*t} \end{bmatrix} = \begin{bmatrix} -2 & 0 \\ 0 & 0 \\ 1 & 1 \\ -1 & 0 \end{bmatrix}$$

or

$$[\boldsymbol{\alpha}^* \quad \boldsymbol{\beta}^*] = \begin{bmatrix} -2 & 0 & 1 & -1 \\ 0 & 0 & 1 & 0 \end{bmatrix}$$

from which we obtain

$$2v(0) + v'(1) = 0, \qquad v(1) = 0$$

On the other hand, these boundary conditions are readily obtained from

$$0 = [(u'v - uv') + uv]_0^1 = u'(1)v(1) - u(1)v'(1) + u(1)v(1) - u'(0)v(0) + u(0)v'(0) - u(0)v(0)$$
$$= u'(1)v(1) + u(1)[-v'(1) + v(1) - 2v(0)]$$

where we have used the boundary conditions for $u$. Since $u'(1)$ and $u(1)$ can be chosen independently, we have

$$v(1) = 0, \qquad v'(1) + 2v(0) = 0$$

as before.

*Example 2*

For our second example, let $\mathbf{T} \equiv \{T, D(T)\}$, where $T \equiv x^2(d^2/dx^2) + x(d/dx) + 1$ and

$$D(T) = \{\mathbf{u} \in \mathcal{L}_2[0, 1]; T\mathbf{u} \in \mathcal{L}_2[0, 1]; u'(0) = 0, u'(1) + u(1) = 0\}$$

Now

$$T^* = x^2 \frac{d^2}{dx^2} + 3x \frac{d}{dx} + 2$$

$$\boldsymbol{\alpha} = \begin{bmatrix} 0 & 1 \\ 0 & 0 \end{bmatrix}, \qquad \boldsymbol{\beta} = \begin{bmatrix} 0 & 0 \\ 1 & 1 \end{bmatrix}, \qquad \mathbf{P}(x) = \begin{bmatrix} -x & x^2 \\ -x^2 & 0 \end{bmatrix}$$

Note that $\mathbf{P}(0) = \mathbf{0}$, so that it has no inverse and (9.1.18) is not defined. However, the adjoint boundary conditions can be obtained from (9.1.20) or more directly

$$0 = [x^2(u'v - v'u) + xuv]_0^1$$
$$= u'(1)v(1) - v'(1)u(1) + u(1)v(1)$$
$$= -v'(1)u(1)$$

from which the boundary conditions are $v'(1) = 0$. Note that in deriving the result above, we must have $v$ and $v'$ bounded at $x = 0$.

We will not pursue the operators of this section any further, since for the present we are interested only in self-adjoint operators. However, the present section is a useful prelude to the study of non-self-adjoint differential operators.

## EXERCISES

**9.1.1** Determine the appropriate substitute for (9.1.5) when $\mathbf{u}$ and $\mathbf{v}$ are complex-valued functions.

**9.1.2** If $p_0(x)$, $p_1(x)$, and $p_2(x)$ are complex-valued in $T$, find $T^*$.

**9.1.3** Consider the $2 \times 4$ matrix $[\boldsymbol{\alpha} \quad \boldsymbol{\beta}]$ defined in (9.1.10) and (9.1.11). Show that as a postmultiplying matrix operator its range space $R$ has dimension 2.

Show also that the inverse image in $\mathcal{R}_2$ of every vector in $R$ is unique, and that $R \perp N$.

**9.1.4** Show that the coefficient matrix of the adjoint boundary conditions as given by (9.1.20) satisfies the property (9.1.17) required of it.

**9.1.5** Review Exercise 6.7.1 and obtain the adjoint boundary conditions where the operator is non-self-adjoint.

**9.1.6** If $\mathbf{u} \equiv \{u(x)\}$, $\mathbf{v} = \{v(x)\}$ satisfy $T\mathbf{u} = \mathbf{0}$ and $T^*\mathbf{v} = \mathbf{0}$, respectively, show that

$$\frac{d}{dx}[\mathbf{w}^t(v(x))\mathbf{P}(x)\mathbf{w}(u(x))] = 0$$

and hence the equality (9.1.13).

**9.1.7** Let $[\boldsymbol{\alpha}_c \quad \boldsymbol{\beta}_c]$ be a matrix "complement" to $[\boldsymbol{\alpha} \quad \boldsymbol{\beta}]$ in the boundary condition (9.1.11) in the sense that the $4 \times 4$ matrix

$$\mathbf{H} = \begin{bmatrix} \boldsymbol{\alpha} & \boldsymbol{\beta} \\ \boldsymbol{\alpha}_c & \boldsymbol{\beta}_c \end{bmatrix}$$

is nonsingular ($\mathbf{H}$ exists obviously and is not unique). Show that (9.1.14) may be written as

$$\langle T\mathbf{u}, \mathbf{v}\rangle - \langle \mathbf{u}, T^*v\rangle = [\mathbf{w}^t(v(a)) \quad \mathbf{w}^t(v(b))]\mathbf{J}^t\mathbf{H}\begin{bmatrix} \mathbf{w}(u(a)) \\ \mathbf{w}(u(b)) \end{bmatrix} \tag{9.1.21}$$

by identifying the $4 \times 4$ matrix $\mathbf{J}$. Show further that the adjoint boundary conditions for $T^*$ may be found as

$$\begin{bmatrix} 0 & 0 & 1 & 0 \\ 0 & 0 & 0 & 1 \end{bmatrix}\mathbf{J}\begin{bmatrix} \mathbf{w}(v(a)) \\ \mathbf{w}(v(b)) \end{bmatrix} = \mathbf{0}$$

Thus as an alternative to (9.1.20) one may also take

$$[\boldsymbol{\alpha}^* \quad \boldsymbol{\beta}^*] = \begin{bmatrix} 0 & 0 & 1 & 0 \\ 0 & 0 & 0 & 1 \end{bmatrix}\mathbf{J}$$

Show also that a complement of $[\boldsymbol{\alpha}^* \quad \boldsymbol{\beta}^*]$ may be given by

$$[\boldsymbol{\alpha}_c^* \quad \boldsymbol{\beta}_c^*] = \begin{bmatrix} 1 & 0 & 0 & 0 \\ 0 & 1 & 0 & 0 \end{bmatrix}\mathbf{J}$$

so that

$$\mathbf{J} = \begin{bmatrix} \boldsymbol{\alpha}_c^* & \boldsymbol{\beta}_c^* \\ \boldsymbol{\alpha}^* & \boldsymbol{\beta}^* \end{bmatrix}$$

## 9.2 Self-Adjoint Differential Operators: Continuous Coefficients

For a differential operator $\mathbf{T} \equiv \{T, D(T)\}$ to be self-adjoint we must have $T = T^*$ and $D(T) = D(T^*)$. The first of these conditions can be met readily by referring to (9.1.6) or (9.1.9), which yields

$$2p_0' - p_1 = p_1, \qquad p_0'' - p_1' + p_2 = p_2$$

which are together equivalent to $p_0' = p_1$. We notice then that the operation (9.1.7) may be rewritten as

$$T \equiv -\frac{1}{r(x)}\frac{d}{dx}\left[p(x)\frac{d}{dx}\right] + q(x) \tag{9.2.1}$$

where we have put $p \equiv -p_0$ and $q(x) \equiv p_2(x)/r(x)$. The negative sign on $p$ is introduced for a future purpose but has no significance at the current stage. When $T = T^*$, we refer to it as a *formally self-adjoint operation.* Although the most general form of $T$ as given by (9.1.7) may not be formally self-adjoint, we show below that it is possible under very general conditions to rewrite $T$ as a formally self-adjoint expression as follows.

Let $\phi(x)$ be a function defined on $[a, b]$ such that it does not vanish anywhere. We rewrite (9.1.7) as

$$T = \frac{1}{r(x)\phi(x)}\left[p_0(x)\phi(x)\frac{d^2}{dx^2} + p_1(x)\phi(x)\frac{d}{dx} + p_2(x)\phi(x)\right]$$

For $T$ to be formally self-adjoint we must have $(p_0\phi)' = p_1\phi$ or if $p_0$ does not vanish anywhere in $[a, b]$,† we may write

$$(p_0\phi)' = \frac{p_1}{p_0}(p_0\phi)$$

which on integration produces

$$(p_0\phi) = \pm\exp\left[\int_a^x \frac{p_1(\xi)}{p_0(\xi)}\,d\xi\right]$$

where we have picked $\phi(a)$ to be such that $p_0(a)\phi(a) = \pm 1$. Thus

$$\phi(x) = \frac{1}{\pm p_0(x)}\exp\left[\int_a^x \frac{p_1(\xi)}{p_0(\xi)}\,d\xi\right] \tag{9.2.2}$$

Since $p_0(x)$ is presumed to be continuous and not to vanish anywhere in $[a, b]$, it must have a single sign throughout the interval. If $p_0(x)$ is positive, we select (9.2.2) with the positive sign, and if $p_0(x)$ is negative, we select (9.2.2) with the negative sign so that in either case $\phi(x)$ is *positive-valued.*

The operation $T$ becomes

$$T = \frac{1}{r(x)\phi(x)}\left[\frac{d}{dx}\left(p_0(x)\phi(x)\frac{d}{dx}\right)\right] + q(x) \tag{9.2.3}$$

---

†More generally, we may assume that $p_0(x)$ vanishes at one of the end points, say $a$. In this case, we may write (9.2.2) as

$$\phi(x) = \frac{1}{\pm p_0(x)}\exp\left[-\int_x^b \frac{p_1(\xi)}{p_0(\xi)}\,d\xi\right]$$

which permits the divergence of the integral above as long as $p_1/p_0$ is nonnegative and $\lim_{x\to a}\phi(x)$ exists. See Example 3 later in this section.

which is of the self-adjoint type (9.2.1). The inner product which must be used in defining the adjoint must, however, be modified further by inserting $r(x)\phi(x)$ in (9.1.8) in place of $r(x)$. Such a definition is permissible because $r(x)\phi(x)$ is a positive-valued function.

In this section we are concerned with operations of the type (9.2.1) for the case of *continuous* $p(x)$ in the interval $[a, b]$. Moreover, $p$ does not vanish anywhere except possibly at an end point. The matrix $\mathbf{P}(x)$ of the preceding section is now given by

$$\mathbf{P}(x) = \begin{bmatrix} 0 & -p(x) \\ p(x) & 0 \end{bmatrix} \tag{9.2.4}$$

which is *skew symmetric*. Equation (9.1.14) continues to hold with $T^*$ replaced by $T$.

$$\langle T\mathbf{u}, \mathbf{v}\rangle - \langle \mathbf{u}, T\mathbf{v}\rangle = [\mathbf{w}^t(v(a)) \quad \mathbf{w}^t(v(b))]\begin{bmatrix} -\mathbf{P}(a) & \mathbf{0} \\ \mathbf{0} & \mathbf{P}(b) \end{bmatrix}\begin{bmatrix} \mathbf{w}(u(a)) \\ \mathbf{w}(u(b)) \end{bmatrix} \tag{9.2.5}$$

What remains to obtain a self-adjoint operator is to find a domain $D(T)$ such that for $\mathbf{u}, \mathbf{v} \in D(T)$ the right-hand side of (9.2.5) vanishes. We recall the general boundary conditions (9.1.11) for $T$.

$$[\boldsymbol{\alpha} \quad \boldsymbol{\beta}]\begin{bmatrix} \mathbf{w}(u(a)) \\ \mathbf{w}(u(b)) \end{bmatrix} = 0 \tag{9.1.11}$$

Since we must have $D(T) = D(T^*)$ and in the notation of Section 9.1 the adjoint boundary conditions employed the coefficient matrix $[\alpha^* \quad \beta^*]$, we conclude that $\boldsymbol{\alpha} = \boldsymbol{\alpha}^*$, $\boldsymbol{\beta} = \boldsymbol{\beta}^*$. For $\mathbf{T}$ to be self-adjoint the right-hand side of (9.2.5) should vanish when the boundary conditions above are used. When matrix $\mathbf{P}(x)$ is nonsingular at both $a$ and $b$, (9.1.17) or (9.1.18) yield the conditions on the matrix $[\boldsymbol{\alpha} \quad \boldsymbol{\beta}]$ for $\mathbf{T}$ to be self-adjoint. Thus

$$[\boldsymbol{\alpha} \quad \boldsymbol{\beta}]\begin{bmatrix} -\mathbf{P}^{-1}(a) & \mathbf{0} \\ \mathbf{0} & \mathbf{P}^{-1}(b) \end{bmatrix} = \mathbf{0} \tag{9.2.6}$$

or alternatively,

$$\boldsymbol{\alpha}\mathbf{P}^{-1}(a)\boldsymbol{\alpha}^t = \boldsymbol{\beta}\mathbf{P}^{-1}(b)\boldsymbol{\beta}^t \tag{9.2.7}$$

When $\mathbf{P}(x)$ is singular at either $a$ or $b$, (9.1.20) may be used to obtain

$$\begin{bmatrix} \boldsymbol{\alpha}^t \\ \boldsymbol{\beta}^t \end{bmatrix} = \begin{bmatrix} -\mathbf{P}(a) & \mathbf{0} \\ \mathbf{0} & \mathbf{P}(b) \end{bmatrix}\{-\boldsymbol{\gamma}_{ij}^t\boldsymbol{\delta}_{ij}^{-1}\boldsymbol{\delta}_{kl} + \boldsymbol{\gamma}_{kl}^t\} \tag{9.2.8}$$

which, however, is not as suggestive of "symmetry" as either (9.2.6) or (9.2.7). Let us now consider some special cases of self-adjoint operators. We shall refer to the boundary conditions that lead to a self-adjoint operator as *self-adjoint boundary conditions*. Equation (9.2.8) represents the *necessary and sufficient* conditions for the boundary conditions to be self-adjoint whether or not $\mathbf{P}(a)$ or $\mathbf{P}(b)$ is singular.

***Definition 9.2.1*** The boundary conditions (9.1.11) are said to be *unmixed* if *neither* of the end points $a$ and $b$ appears in *both* boundary conditions.

The definition above essentially implies that one boundary condition involves only point $a$ and the other one involves only point $b$. Thus we may have $\alpha_{21} = \alpha_{22} = \beta_{11} = \beta_{12} = 0$ in the matrix $[\boldsymbol{\alpha} \quad \boldsymbol{\beta}]$. Next we prove the theorem below.

***Theorem 9.2.1*** Let $T$ be a formally self-adjoint operation as given by (9.2.1). Then unmixed boundary conditions are self-adjoint.

*Proof*: The proof is simple and depends on the fact that $\mathbf{P}(x)$ is nonsingular almost everywhere in $[a, b]$ and (9.2.7) holds since either side can be shown to vanish. ▮

Next we consider another type of boundary conditions.

***Definition 9.2.2*** The boundary conditions $u(a) = u(b)$, $u'(a) = u'(b)$ are said to be *periodic*.

It is left as an exercise to show that if neither $p(a)$ nor $p(b)$ vanishes, then periodic boundary conditions are self-adjoint if and only if $p(a) = p(b)$.

Let us consider some examples of self-adjoint differential operators.

*Example 1*

Let $\mathbf{T} \equiv \{T, D(T)\}$, where $T \equiv -d^2/dx^2$ and

$$D(T) = \{\mathbf{u} \in \mathfrak{L}_2[0, 1];\ T\mathbf{u} \in \mathfrak{L}_2[0, 1];\ u(0) + u(1) = 0,\ u'(0) + u'(1) = 0\}$$

$T$ is formally self-adjoint, since $T = T^*$. The boundary conditions are neither unmixed nor periodic (in fact, these are called *antiperiodic* boundary conditions), so that the self-adjointness of $\mathbf{T}$ does not follow immediately. Since

$$\mathbf{P}(x) = \begin{bmatrix} 0 & 1 \\ -1 & 0 \end{bmatrix}, \qquad \mathbf{P}^{-1}(x) = \begin{bmatrix} 0 & -1 \\ 1 & 0 \end{bmatrix}$$

Further,

$$\boldsymbol{\alpha} = \begin{bmatrix} 1 & 0 \\ 0 & 1 \end{bmatrix} = \boldsymbol{\beta}$$

so that (9.1.18) is trivially true. Thus $\mathbf{T}$ is self-adjoint.

*Example 2*

One encounters in the analysis of tubular chemical reactors the operation

$$T \equiv D\frac{d^2}{dx^2} - V\frac{d}{dx} - k, \qquad 0 < x < 1$$

with homogeneous boundary conditions $Du'(0) - Vu(0) = 0$, $u'(1) = 0$. This operation is not formally self-adjoint. We may, however, define in accord with (9.2.2) a positive function $\phi$ by

$$\phi(x) = \frac{1}{D}\exp\left(\int_0^x \frac{-V}{D}\,dx\right) = \frac{1}{D}e^{-Vx/D}$$

and following (9.2.3) rewrite $T$ as

$$T = \frac{D}{e^{-Vx/D}}\frac{d}{dx}\left(e^{-Vx/D}\frac{d}{dx}\right) - k$$

which is formally self-adjoint provided that we define the inner product as

$$\langle \mathbf{u}, \mathbf{v} \rangle = \int_0^1 e^{-Vx/D}u(x)v(x)\,dx$$

Since the boundary conditions are unmixed, from Theorem 9.2.1 the operator **T** is self-adjoint.

*Example 3*

Consider the operation

$$T \equiv x\frac{d^2}{dx^2} + 2\frac{d}{dx}, \qquad 0 < x < 1$$

The boundary conditions are $u'(0) = 0$, $u'(1) + au(1) = 0$.

To make $T$ formally self-adjoint, we must follow the procedure of Example 2. Note, however, that $p_0(x) \equiv x$ vanishes at $x = 0$, so that the footnote concerned with (9.2.2) becomes important. Thus

$$\phi(x) = \frac{1}{x}\exp\left(-\int_x^1 \frac{2}{\xi}\,d\xi\right) = \frac{1}{x}x^2 = x$$

so that $\phi(0)$ exists and is zero, although the integrand within the exponential argument is unbounded at $x = 0$. Thus $T$ is given by

$$T = \frac{1}{x}\frac{d}{dx}\left(x^2\frac{d}{dx}\right)$$

which is formally self-adjoint provided that the inner product is defined as

$$\langle \mathbf{u}, \mathbf{v} \rangle = \int_0^1 xu(x)v(x)\,dx$$

Since the boundary conditions are unmixed, we obtain a self-adjoint operator.

We suggest the following exercises for further examples.

**EXERCISES**

**9.2.1** Prove that periodic boundary conditions for the operation $T$ given by (9.2.1) are self-adjoint.

**9.2.2** Render the following operations formally self-adjoint and define the inner product in each case.

(a) $\dfrac{d^2}{dx^2} - 2x\dfrac{d}{dx}, \quad 0 < x < 1.$

(b) $(1 - x^2)\dfrac{d^2}{dx^2} - x\dfrac{d}{dx} + n^2, \quad -1 < x < 1.$

(c) $x(1 - x)\dfrac{d^2}{dx^2} - [q - (p + 1)x]\dfrac{d}{dx} - (p + n)n, \quad 0 < x < 1$ where $p$ and $q$ are real numbers with $q > 0, p - q > -1$ and $n$ is an integer.

**9.2.3** Identify the operations in Exercise 9.2.2 which will produce self-adjoint operators with periodic boundary conditions.

We have thus far defined a self-adjoint differential operator of the second order as comprising an operation $T$ given by (9.2.1) whose domain $D(T)$ is essentially characterized by the homogeneous boundary conditions (9.1.11) such that (9.2.8) is true. Since our eventual interest is to establish the grounds for the use of Theorem 7.4.3, we must show that $\mathbf{T}$ has a compact inverse. We shall take this up next.

### 9.2.1 Existence of a Compact Inverse

Consider the self-adjoint operator $\mathbf{T} = \{T, D(T)\}$, where $T$ is given by (9.2.1) and $D(T)$ is given by

$$D(T) = \{\mathbf{u} \in \mathcal{H}; T\mathbf{u} \in \mathcal{H}; \boldsymbol{\alpha}\mathbf{w}(u(a)) + \boldsymbol{\beta}\mathbf{w}(u(b)) = \mathbf{0}\}$$

in which $\mathcal{H}$ is the Hilbert space of functions square-integrable over the interval $[a, b]$ with $r(x)$ as a weight function. Further, self-adjointness implies the validity of (9.2.8). It is our objective now to find the inverse of this operator. Before we proceed to do this, some preliminary remarks are in order. First, if the inverse exists, we know that it is self-adjoint. Second, we will employ the method of variation of parameters to calculate the inverse for the general boundary conditions. Third, for the case of unmixed boundary conditions, we will show a very simple method for constructing the inverse.

Basically, the construction of the inverse is equivalent to the solution of the operator equation

$$\mathbf{T}\mathbf{u} = \mathbf{f}, \qquad \mathbf{f} \in \mathcal{H} \tag{9.2.9}$$

where we remind the reader that the use of the boldface symbol $\mathbf{T}$ (in accord with the notation elaborated in Section 6.7) implies that $\mathbf{u} \in D(T)$; that is, it satisfies the homogeneous boundary conditions. (Thus no special stipulation need be made to this effect.) Equation (9.2.9) implies that

$$-\frac{d}{dx}\left[p(x)\frac{du}{dx}\right] + q(x)r(x)u = r(x)f(x) \tag{9.2.10}$$

so that the vector $\mathbf{u}$ we are looking for is a function $u(x)$ which satisfies the differential equation (9.2.10) and the boundary conditions (9.1.11). The method of variation of parameters employs the two linearly independent solutions of the homogeneous equation $Tu = 0$ in constructing the inverse. The existence of the inverse is therefore contingent on the existence of the two solutions to the homogeneous equation

$$-\frac{d}{dx}\left(p\frac{dy}{dx}\right) + qry = 0 \tag{9.2.11}$$

Let us first observe that (9.2.11) may be rewritten as the following first-order system:

$$\frac{d}{dx}\begin{bmatrix} y \\ y' \end{bmatrix} = \begin{bmatrix} 0 & 1 \\ \frac{qr}{p} & -\frac{p'}{p} \end{bmatrix}\begin{bmatrix} y \\ y' \end{bmatrix} \tag{9.2.12}$$

When $y'$ is eliminated from (9.2.12), one obtains (9.2.11). The existence of a solution to (9.2.12) would imply the existence of a solution to (9.2.11). Consider the solution of (9.2.12) subject to the "initial" values $y(x_0) = c_1$, $y'(x_0) = c_2$, where $a \leq x_0 \leq b$ and $c_1$ and $c_2$ are arbitrarily fixed constants. In Section 3.7 we had occasion to discuss the existence of a solution to a system of ordinary *nonlinear* equations subject to prescribed initial values [see (3.7.19) and (3.7.20)]. The sufficient conditions† for the existence of a unique solution derived there are readily applied to the case of linear equations. However, the existence of solutions to linear equations of the type (9.2.12) holds under more general conditions, for the nature of which we refer the reader to other treatments.‡ To discuss this let us denote the matrix of coefficients in (9.2.12) by

$$\mathbf{A}(x) = \begin{bmatrix} 0 & 1 \\ \frac{qr}{p} & -\frac{p'}{p} \end{bmatrix}$$

We may be assured of a unique solution to the initial value problem just considered at $x_0$, even if $\mathbf{A}(x)$ is singular at $x_0$, by which is meant that at least one component of $\mathbf{A}$ becomes unbounded as $x \longrightarrow x_0$; $x_0$ is then referred

†This sufficiency condition refers to the fact that the right-hand side of (3.7.19) must satisfy the Lipschitz condition.

‡See *Theory of Ordinary Differential Equations*, by E. A. Coddington and N. Levinson, McGraw-Hill, New York, 1955.

to as the *point of singularity* or simply *singularity*. Of particular importance to applications of interest to us is what is referred to as *singularity of the first kind*, which is characterized by the behavior of $\mathbf{A}(x)$ near $x_0$. We write

$$\mathbf{A}(x) = (x - x_0)^{-1}\hat{\mathbf{A}}(x) \tag{9.2.13}$$

where $\hat{\mathbf{A}}(x)$ is *analytic* at $x = x_0$† and $\hat{\mathbf{A}}(x_0) \neq \mathbf{0}$. The matrix $\mathbf{A}(x)$ in (9.2.12) is singular at $x = x_0$ if either $p'/p$ or $qr/p$ or both are singular at $x = x_0$. Thus $(x - x_0)(p'/p)$ and $(x - x_0)(qr/p)$ must both have removable singularities at $x = x_0$ and be analytic at $x = x_0$ such that at least one of them does not vanish at $x = x_0$.

Within the context of a singularity of the first kind, it is possible to obtain a slight "improvement" on the requirement above (which we will find to be essential) by writing (9.2.11) as another first-order system by letting $z \equiv (x - x_0)y'$. We obtain

$$\frac{d}{dx}\begin{bmatrix} y \\ z \end{bmatrix} = \mathbf{B}(x)\begin{bmatrix} y \\ z \end{bmatrix} \tag{9.2.14}$$

where

$$\mathbf{B}(x) = \begin{bmatrix} 0 & (x - x_0)^{-1} \\ (x - x_0)\dfrac{qr}{p} & (x - x_0)^{-1} - \dfrac{p'}{p} \end{bmatrix}$$

Equation (9.2.14) is also obviously a first-order system with a coefficient matrix $\mathbf{B}(x)$ which is clearly different from $\mathbf{A}(x)$. Now if $x_0$ is a singularity of the first kind of the system (9.2.14), that is, $(x - x_0)\mathbf{B}(x)$ is analytic at $x_0$ without vanishing there, then

$$(x - x_0)\frac{p'}{p} \quad \text{and} \quad (x - x_0)^2\frac{qr}{p}$$

must be *analytic* at $x_0$. The first of these had already resulted from the system (9.2.13), but the second result now permits a "higher-order singularity" of the function $qr/p$ at $x_0$. In the literature, the singularity of the first kind is also referred to as the *regular singular point* of the operation $T$ given by (9.2.1).‡ On the other hand, if $p'/p$ and $qr/p$ are both analytic at $x = x_0$, $x_0$ is called an *ordinary point* of $T$.

---

†Analyticity of a function $f(x)$ at $x = x_0$ implies that it can be represented by $f(x) = \sum_{n=0}^{\infty} a_n(x - x_0)^n$, $a_n$'s being scalars, which converges in some neighborhood about $x_0$. When *every* coefficient of a matrix $\hat{\mathbf{A}}(x)$ is analytic at $x = x_0$, $\hat{\mathbf{A}}(x)$ is itself said to be analytic at $x = x_0$.

‡The concept of a regular singular point also holds for differential operations that are not formally self-adjoint. Thus for the operation (9.1.7), we require that $(x - x_0)(p_1/p_0)$ and $(x - x_0)^2(p_2/p_0)$ to be analytic if $x_0$ is to be a regular singular point.

Our discussion of ordinary and regular singular points of $T$ is motivated by the fact that the behavior of the solution at a point depends strongly on whether or not the point is ordinary or singular. Furthermore, the method of *finding* the solution is also governed by the nature of the point. Near an ordinary point, for example, the solution must be analytic, so that a series solution is possible. At a regular singular point a method due to Frobenius will lead to the solutions. A treatment of these methods is provided in Sections A.2 and A.3 of the Appendix. Furthermore, it is not necessary for us to use the actual solutions in the ensuing discussion.

Since we are guaranteed the existence of a unique solution to the initial value problem consisting of (9.2.12), $y(x_0) = c_1$ and $y'(x_0) = c_2$, we may generate the solution first with, say, $c_1 = 1$ and $c_2 = 0$. Let the solution be denoted $\begin{bmatrix} y_1 \\ y_1' \end{bmatrix}$ or, more succinctly, $\mathbf{w}(y_1)$. Similarly, we obtain the solution with $c_1 = 0$ and $c_2 = 1$ and represent it by $\mathbf{w}(y_2)$. Clearly,

$$\mathbf{w}(y_1(x_0)) = \begin{bmatrix} 1 \\ 0 \end{bmatrix}, \qquad \mathbf{w}(y_2(x_0)) = \begin{bmatrix} 0 \\ 1 \end{bmatrix}$$

are linearly independent. Further, $\mathbf{w}(y(x))$ depends *linearly* on $\mathbf{w}(y(x_0))$ so that the solution for the case $y(x_0) = c_1, y'(x_0) = c_2$ is given by $c_1\mathbf{w}(y_1) + c_2\mathbf{w}(y_2)$. The solutions $\mathbf{w}(y_1)$ and $\mathbf{w}(y_2)$ are linearly independent because

$$\mathbf{0} = c_1\mathbf{w}(y_1) + c_2\mathbf{w}(y_2) \rightarrow \mathbf{0} = \begin{bmatrix} c_1 \\ c_2 \end{bmatrix} \rightarrow c_1 = c_2 = 0$$

Indeed, $y_1$ and $y_2$ above are the linearly independent solutions of (9.2.11) to which we had referred earlier. Let us define

$$\mathbf{Y}(x) \equiv \begin{bmatrix} y_1(x) \\ y_2(x) \end{bmatrix}, \qquad \mathbf{W}(\mathbf{Y}(x)) \equiv \begin{bmatrix} y_1 & y_2 \\ y_1' & y_2' \end{bmatrix}$$

It may be recalled from Section 2.2 that $\mathbf{W}(\mathbf{Y}(x))$ is referred to as the *Wronskian matrix*. The nonvanishing of its determinant is a *sufficient* condition for its column vectors to be linearly independent. Now the column vectors of $\mathbf{W}(\mathbf{Y}(x))$ are $\mathbf{w}(y_1)$ and $\mathbf{w}(y_2)$, which are known to be linearly independent and it is of interest to determine whether the matrix is nonsingular. In the present context the Wronskian matrix, written more simply as $\mathbf{W}$, satisfies the differential equation

$$\frac{d\mathbf{W}}{dx} = \mathbf{A}\mathbf{W} \tag{9.2.15}$$

which is because $\mathbf{w}(y_1)$ and $\mathbf{w}(y_2)$ satisfy (9.2.12). For this reason $\mathbf{W}$ is called a *fundamental solution matrix* of the system (9.2.12).† It is certainly not unique since **W**, postmultiplied by an arbitrary $2 \times 2$ nonsingular matrix of constants, would also satisfy (9.2.15). If we let $W(x) \equiv \det \mathbf{W}(\mathbf{Y}(x))$, which is referred to as the *Wronskian*, it is readily shown that‡

$$\frac{dW}{dx} = W \operatorname{tr}(\mathbf{A}) \tag{9.2.16}$$

where $\operatorname{tr}(\mathbf{A})$ is the sum of the diagonal elements of the matrix **A**, which in this case is $-p'/p$. Solving (9.2.16) subject to its initial value at $x_0$ gives us

$$W(x) = W(x_0) \exp\left[\int_{x_0}^{x} \operatorname{tr}(\mathbf{A}(\xi))\, d\xi\right] \tag{9.2.17}$$

We already know that $W(x_0) \neq 0$ because $\mathbf{W}(\mathbf{Y}(x_0)) = \begin{bmatrix} 1 & 0 \\ 0 & 1 \end{bmatrix}$. Thus from (9.2.17) we conclude that $W(x) \neq 0$ for any $x \in [a, b]$, which of course is consistent with the linear independence of the vectors $\mathbf{w}(y_1)$ and $\mathbf{w}(y_2)$.

As observed earlier, we will not be concerned here with the actual determination of the fundamental matrix $\mathbf{W}(\mathbf{Y})$. Instead, we presume knowledge of this matrix to continue in our main quest, which is that of determining the inverse of the self-adjoint differential operator **T**. The *method of variation of parameters* is used to accomplish this.

The solution to the differential equation (9.2.10) is to be determined subject to the boundary conditions (9.1.11). The equivalent problem for the first-order system is given by

$$\frac{d}{dx}\begin{bmatrix} u \\ u' \end{bmatrix} = \begin{bmatrix} 0 & 1 \\ q\dfrac{r}{p} & -\dfrac{p'}{p} \end{bmatrix}\begin{bmatrix} u \\ u' \end{bmatrix} + \begin{bmatrix} 0 \\ -\dfrac{rf}{p} \end{bmatrix} \tag{9.2.18}$$

or, more succinctly,

$$\frac{d}{dx}\mathbf{w}(u) = \mathbf{A}(x)\mathbf{w}(u) + \mathbf{b}(x) \tag{9.2.19}$$

where $\mathbf{b}(x) \equiv \begin{bmatrix} 0 \\ -\dfrac{rf}{p} \end{bmatrix}$. When $u'(x)$ is eliminated from (9.2.18), one obtains

---

†If **K** is constant matrix, it is left for the student to verify that $\mathbf{W}(\mathbf{Y}(x))\mathbf{K} = \mathbf{W}(\mathbf{K}^t\mathbf{Y}(x))$.

‡Denoting the elements of **W** by $\{w_{ij}\}$ and the cofactor of $w_{ij}$ by $W_{ij}$, we have

$$\frac{dW}{dx} = \sum_j W_{ij}\frac{dw_{ij}}{dx} = \sum_j W_{ij}\sum_k a_{ik}w_{kj} = \sum_k a_{ik}\sum_j W_{ij}w_{kj} = \sum_k a_{ik}\delta_{ik}W$$
$$= W\sum_k a_{kk} = W \operatorname{tr}(\mathbf{A})$$

the inhomogeneous equation (9.2.10). In the method of variation of parameters one defines a transformation of variables by

$$\mathbf{w}(u) = \mathbf{W}\mathbf{C}(x), \qquad \mathbf{C}(x) \equiv \begin{bmatrix} c_1(x) \\ c_2(x) \end{bmatrix} \tag{9.2.20}$$

If $\mathbf{C}(x)$ were independent of $x$, it is clear that $\mathbf{WC}$ would be a solution of the homogeneous equation (9.2.12). Substitution of (9.2.20) into (9.2.19) yields

$$\frac{d}{dx}\mathbf{w}(u) = \frac{d\mathbf{W}}{dx}\mathbf{C} + \mathbf{W}\frac{d\mathbf{C}}{dx} = \mathbf{AWC} + \mathbf{W}\frac{d\mathbf{C}}{dx} = \mathbf{AWC} + \mathbf{b}$$

where we have used (9.2.15). Thus

$$\frac{d\mathbf{C}}{dx} = \mathbf{W}^{-1}\mathbf{b}$$

thanks to $\mathbf{W}$ being nonsingular. Integrating,† one obtains

$$\mathbf{C}(x) = \int_a^x \mathbf{W}^{-1}(\xi)\mathbf{b}(\xi)\, d\xi + \mathbf{k}$$

where $\mathbf{k} \equiv \begin{bmatrix} k_1 \\ k_2 \end{bmatrix} \in \mathcal{R}_2$ is unknown. Thus

$$\mathbf{w}(u) = \mathbf{W}(x)\int_a^x \mathbf{W}^{-1}(\xi)\mathbf{b}(\xi)\, d\xi + \mathbf{W}(x)\mathbf{k} \tag{9.2.21}$$

We may now invoke the boundary conditions (9.1.11) to get

$$[\boldsymbol{\alpha}\mathbf{W}(a) + \boldsymbol{\beta}\mathbf{W}(b)]\mathbf{k} = -\boldsymbol{\beta}\mathbf{W}(b)\int_a^b \mathbf{W}^{-1}(\xi)\mathbf{b}(\xi)\, d\xi \tag{9.2.22}$$

By letting

$$\mathbf{D}_0 \equiv [\boldsymbol{\alpha}\mathbf{W}(a) + \boldsymbol{\beta}\mathbf{W}(b)] \tag{9.2.23}$$

we note that the inverse of our operator $\mathbf{T}$ may exist if and only if $\mathbf{D}_0$ is nonsingular. Thus we must presume at this stage that $\mathbf{D}_0$ is nonsingular so that $\mathbf{T}^{-1}$ may exist. From (9.2.22) we obtain

$$\mathbf{k} = -\mathbf{D}_0^{-1}\boldsymbol{\beta}\mathbf{W}(b)\int_a^b \mathbf{W}^{-1}(\xi)\mathbf{b}(\xi)\, d\xi$$

Thus we have the solution to (9.2.19) as given by

$$\mathbf{w}(u) = \mathbf{W}(x)\int_a^x \mathbf{W}^{-1}(\xi)\mathbf{b}(\xi)\, d\xi - \mathbf{W}(x)\mathbf{D}_0^{-1}\boldsymbol{\beta}\mathbf{W}(b)\int_a^b \mathbf{W}^{-1}(\xi)\mathbf{b}(\xi)\, d\xi$$

which may be rewritten as

$$\mathbf{w}(u) = \int_a^b \mathbf{K}(x, \xi)\mathbf{b}(\xi)\, d\xi \tag{9.2.24}$$

---

†We have permitted ourselves the brevity of writing $\mathbf{W}(\mathbf{Y}(x))$ as $\mathbf{W}(x)$. The latter's notational insensitivity to $\mathbf{Y}$ should not be intolerable since the ultimate results are, in fact, independent of the choice of $\mathbf{Y}$ (see Exercise 9.2.9).

where

$$\mathbf{K}(x, \xi) = \begin{cases} \mathbf{W}(x)[\mathbf{I}_2 - \mathbf{D}_0^{-1}\boldsymbol{\beta}\mathbf{W}(b)]\mathbf{W}^{-1}(\xi), & a \leq \xi < x \\ -\mathbf{W}(x)\mathbf{D}_0^{-1}\boldsymbol{\beta}\mathbf{W}(b)\mathbf{W}^{-1}(\xi), & x < \xi \leq b \end{cases}$$

Here $\mathbf{I}_2$ is the $2 \times 2$ idem matrix $\begin{bmatrix} 1 & 0 \\ 0 & 1 \end{bmatrix}$. It is convenient to rewrite the foregoing expressions for the matrix $\mathbf{K}(x, \xi)$ as follows. We make the substitution $\mathbf{D}_0^{-1}\{\boldsymbol{\alpha}\mathbf{W}(a) + \boldsymbol{\beta}\mathbf{W}(b)\} = \mathbf{I}_2$ which follows from (9.2.23) to rewrite $\mathbf{K}(x, \xi)$ as

$$\mathbf{K}(x, \xi) = \begin{cases} \mathbf{W}(x)\mathbf{D}_0^{-1}\boldsymbol{\alpha}\mathbf{W}(a)\mathbf{W}^{-1}(\xi), & a \leq \xi < x \\ -\mathbf{W}(x)\mathbf{D}_0^{-1}\boldsymbol{\beta}\mathbf{W}(b)\mathbf{W}^{-1}(\xi), & x < \xi \leq b \end{cases} \tag{9.2.25}$$

Clearly, the continuity of the Wronskian matrix $\mathbf{W}(x)$ in $[a, b]$ implies the continuity of $\mathbf{K}(x, \xi)$ in the intervals $a \leq \xi < x$ and $x < \xi \leq b$. However, at $x = \xi$ there is a discontinuity because

$$\lim_{\xi \to x-0} \mathbf{K}(x, \xi) = \mathbf{W}(x)\mathbf{D}_0^{-1}\boldsymbol{\alpha}\mathbf{W}(a)\mathbf{W}^{-1}(x)$$

$$\lim_{\xi \to x+0} \mathbf{K}(x, \xi) = -\mathbf{W}(x)\mathbf{D}_0^{-1}\boldsymbol{\beta}\mathbf{W}(b)\mathbf{W}^{-1}(x)$$

Subtracting the two, we get

$$\begin{aligned} \lim_{\xi \to x-0} \mathbf{K}(x, \xi) - \lim_{\xi \to x+0} \mathbf{K}(x, \xi) &= \mathbf{W}(x)[\mathbf{D}_0^{-1}\{\boldsymbol{\alpha}\mathbf{W}(a) + \boldsymbol{\beta}\mathbf{W}(b)\}]\mathbf{W}^{-1}(x) \\ &= \mathbf{I}_2 \end{aligned} \tag{9.2.26}$$

Since in solving (9.2.10) we seek specifically the function $u$, let us observe that $u = \mathbf{e}_1^t\mathbf{w}(u)$, where $\mathbf{e}_1^t = [1 \quad 0]$. Further, we have

$$\mathbf{b}(\xi) = -\frac{r(\xi)}{p(\xi)} f(\xi)\mathbf{e}_2 \quad \text{where} \quad \mathbf{e}_2 \equiv \begin{bmatrix} 0 \\ 1 \end{bmatrix}$$

so that

$$u(x) = -\int_a^b \mathbf{e}_1^t\mathbf{K}(x, \xi)\mathbf{e}_2 \frac{r(\xi)f(\xi)}{p(\xi)}\, d\xi$$

It is convenient to define a function $G(x, \xi)$ called the *Green's function* by

$$u(x) = \int_a^b G(x, \xi)r(\xi)f(\xi)\, d\xi \tag{9.2.27}$$

so that

$$G(x, \xi) \equiv \frac{-\mathbf{e}_1^t\mathbf{K}(x, \xi)\mathbf{e}_2}{p(\xi)} \tag{9.2.28}$$

Through Eq. (9.2.27) we have in effect inverted the differential equation (9.2.10) [subject to the boundary condition (9.1.11)]. The inverse operator $\mathbf{T}^{-1}$ is then defined by

$$\mathbf{T}^{-1}\mathbf{f} \equiv \left\{\int_a^b G(x, \xi)r(\xi)f(\xi)\, d\xi\right\}, \qquad \mathbf{f} \in \mathcal{H} \tag{9.2.29}$$

We refer the reader at this stage to the discussion of compact operators in Section 5.3. We will show below that $G(x, \xi)$ is continuous in $[a, b] \times [a, b]$ so that the compactness of $\mathbf{T}^{-1}$ follows from Example 2 of Section 5.3.†

To show that $G(x, \xi)$ is continuous, we recognize from (9.2.28) that the continuity properties of $G(x, \xi)$ are inherited from those of $\mathbf{K}(x, \xi)$. Since $\mathbf{K}(x, \xi)$ has a discontinuity at $x = \xi$ given by (9.2.26), we write

$$\begin{aligned}\lim_{\xi \to x-0} G(x, \xi) - \lim_{\xi \to x+0} G(x, \xi) &= \frac{\mathbf{e}_1^t \mathbf{I}_2 \mathbf{e}_2}{p(x)} \\ &= \frac{\mathbf{e}_1^t \mathbf{e}_2}{p(x)} \\ &= 0\end{aligned}$$

Thus $G(x, \xi)$ is, in fact, continuous in $[a, b] \times [a, b]$ and $\mathbf{T}^{-1}$ *defined by (9.2.29) is therefore a compact operator*.‡ This is an extremely important conclusion.

In establishing the result above we have not used the fact that $\mathbf{T}$ is self-adjoint, that is, that the boundary conditions are self-adjoint. However, we have shown in Section 7.4 that the inverse of a self-adjoint operator is self-adjoint. It is left as an exercise for the student to show that the self-adjointness of $\mathbf{T}^{-1}$ implies that $G(x, \xi)$ is a symmetric function. From this viewpoint, that $G(x, \xi)$ as given by (9.2.28) is actually symmetric requires no special proof, but we will nevertheless establish this among other properties of the Green's function which we consider below. These properties are useful for its construction.

## Properties of the Green's function

The continuity of $G(x, \xi)$ in $[a, b] \times [a, b]$ has already been proved. Since the Wronskian matrix is differentiable. $G(x, \xi)$ is differentiable and we may write by differentiating (9.2.27),

$$u'(x) = \int_a^b \frac{\partial G}{\partial x}(x, \xi) r(\xi) f(\xi)\, d\xi \tag{9.2.30}$$

---

†In Section 5.3 we established that a continuous kernel gives rise to a compact operator using the normal inner product in $\mathfrak{L}_2[a, b]$. Compactness being dependent on the norm (and hence the inner product), the conclusions of Section 5.3 are applicable here only if the normal inner product in $\mathfrak{L}_2[a, b]$ and the weighted inner product (9.1.8) give rise to *equivalent metrics* (see Section 3.6, p. 93). Such equivalence exists, for example, when $r(x)$ is square integrable on $[a, b]$.

‡Even if $G(x, \xi)$ had been discontinuous at $x = \xi$, its continuity is *almost everywhere* in $[a, b] \times [a, b]$ so that it is square integrable in $[a, b] \times [a, b]$. This establishes that $\mathbf{T}^{-1}$ defined by (9.2.26) is a compact operator.

Since $u' = \mathbf{e}_2^t\mathbf{w}(u)$, (9.2.24) leads to

$$u'(x) = -\int_a^b \mathbf{e}_2^t\mathbf{K}(x, \xi)\mathbf{e}_2\frac{r(\xi)}{p(\xi)}f(\xi)\,d\xi \tag{9.2.31}$$

Equation (9.2.31) may be subtracted from (9.2.30) to get

$$0 = \int_a^b \left[\frac{\partial G}{\partial x}(x, \xi) + \frac{\mathbf{e}_2^t\mathbf{K}(x, \xi)\mathbf{e}_2}{p(\xi)}\right] r(\xi)f(\xi)\,d\xi$$

which must hold for every $\mathbf{f} \in \mathcal{H}$. Thus

$$\frac{\partial G}{\partial x} = -\frac{\mathbf{e}_2^t\mathbf{K}(x, \xi)\mathbf{e}_2}{p(\xi)} \tag{9.2.32}$$

Equation (9.2.32) is also obtainable directly from differentiating (9.2.28) and is left as an exercise. Again the continuity of $\mathbf{K}(x, \xi)$ in $a \leq \xi < x$ and $x < \xi \leq b$ implies the continuity of $\partial G/\partial x$ there. However, at $x = \xi$ we obtain

$$\begin{aligned}\lim_{\xi \to x-0} \frac{\partial G}{\partial x}(x, \xi) - \lim_{\xi \to x+0} \frac{\partial G}{\partial x}(x, \xi) &= -\frac{e_2^t I_2 e_2}{p(x)} \\ &= -\frac{e_2^t e_2}{p(x)} \\ &= -\frac{1}{p(x)}\end{aligned} \tag{9.2.33}$$

Thus the derivative takes a jump at $x = \xi$ of magnitude $1/p(x)$.

Finally, in order to show the symmetry of $G(x, \xi)$ for self-adjoint boundary conditions let us observe that

$$G(x, \xi) = \begin{cases} \dfrac{-\mathbf{e}_1^t\mathbf{W}(x)\mathbf{D}_0^{-1}\boldsymbol{\alpha}\mathbf{W}(a)\mathbf{W}^{-1}(\xi)\mathbf{e}_2}{p(\xi)}, & a \leq \xi \leq x \\ \dfrac{\mathbf{e}_1^t\mathbf{W}(x)\mathbf{D}_0^{-1}\boldsymbol{\beta}\mathbf{W}(b)\mathbf{W}^{-1}(\xi)\mathbf{e}_2}{p(\xi)}, & x \leq \xi \leq b \end{cases}$$

We will simplify the foregoing expression as follows. First we recall (9.2.16) and rewrite it as

$$\frac{d}{dx}[pW] = 0$$

which obtains because $\mathrm{tr}(\mathbf{A}) = -p'/p$. Thus $pW$ is a constant on the interval $[a, b]$. Next we recognize that

$$\mathbf{e}_1^t\mathbf{W}(x) = \mathbf{e}_1^t\mathbf{W}(\mathbf{Y}(x)) = \mathbf{Y}^t(x)$$

which was defined preparatory to (9.2.15). By further letting

$$\mathbf{Z}(x) = \frac{1}{p(x)}\mathbf{W}^{-1}(x)\mathbf{e}_2 \tag{9.2.34}$$

we may write the Green's function as

$$G(x, \xi) = \begin{cases} -\mathbf{Y}^t(x)\mathbf{D}_0^{-1}\boldsymbol{\alpha}\mathbf{W}(a)\mathbf{Z}(\xi), & a \le \xi \le x \\ \mathbf{Y}^t(x)\mathbf{D}_0^{-1}\boldsymbol{\beta}\mathbf{W}(b)\mathbf{Z}(\xi), & x \le \xi \le b \end{cases} \tag{9.2.35}$$

Now we will show that an interesting relationship exists between $\mathbf{Y}(x)$ and $\mathbf{Z}(x)$. We refer to Exercise 9.1.6 to observe that

$$\mathbf{W}^t(x)\mathbf{P}(x)\mathbf{W}(x) = \mathbf{J}pW \tag{9.2.36}$$

where $\mathbf{J}$ is the skew-symmetric matrix $\begin{bmatrix} 0 & 1 \\ -1 & 0 \end{bmatrix}$. Equation (9.2.36) yields

$$\mathbf{W}^{-1}(x) = \frac{\mathbf{J}^t\mathbf{W}^t(x)\mathbf{P}(x)}{pW} \tag{9.2.37}$$

where we have used the property $\mathbf{J}^{-1} = \mathbf{J}^t$. Further, it is not difficult to show that

$$\mathbf{J}^t\mathbf{W}^t(x) = \mathbf{J}^t\mathbf{W}^t(\mathbf{Y}(x)) = \mathbf{W}^t(\mathbf{J}^t\mathbf{Y}(x))$$

so that

$$\begin{aligned} \mathbf{W}^{-1}(x)\mathbf{e}_2 &= \frac{\mathbf{W}^t(\mathbf{J}^t\mathbf{Y}(x))\mathbf{P}(x)\mathbf{e}_2}{pW} \\ &= \frac{\mathbf{W}^t(\mathbf{J}^t\mathbf{Y}(x))\mathbf{e}_1 p(x)}{pW} \\ &= \frac{\mathbf{J}^t\mathbf{Y}(x)p(x)}{pW} \end{aligned}$$

which in the light of (9.2.34) brings us to the interesting conclusion that

$$\mathbf{Z}(x) = \frac{1}{pW}\mathbf{J}^t\mathbf{Y}(x) \tag{9.2.38}$$

so that the Green's function becomes

$$G(x, \xi) = \begin{cases} \dfrac{-\mathbf{Y}^t(x)\mathbf{D}_0^{-1}\boldsymbol{\alpha}\mathbf{W}(\mathbf{Y}(a))\mathbf{J}^t\mathbf{Y}(\xi)}{pW(\mathbf{Y})}, & a \le \xi \le x \\ \dfrac{\mathbf{Y}^t(x)\mathbf{D}_0^{-1}\boldsymbol{\beta}\mathbf{W}(\mathbf{Y}(b))\mathbf{J}^t\mathbf{Y}(\xi)}{pW(\mathbf{Y})}, & x \le \xi \le b \end{cases} \tag{9.2.39}$$

Equation (9.2.39) is also the prescription for *constructing* the Green's function. As an ultimate reminder of the dependence of $\mathbf{W}$ on $\mathbf{Y}$ we have reinserted the latter in the former's argument. However, that the Green's function is independent of the choice of $\mathbf{Y}(x)$ is brought forth through Exercise 9.2.9.

We shall now confirm that the Green's function $G(x, \xi)$ is symmetric when the boundary conditions are self-adjoint. We will assume that $\mathbf{P}^{-1}(x)$ exists at both end points so that the relation (9.2.7) holds in view of the self-adjoint boundary conditions. For specificity we assume that $\xi < x$. Then

$$G(x, \xi) - G(\xi, x) = \frac{-1}{pW}[\mathbf{Y}^t(x)\mathbf{D}_0^{-1}\boldsymbol{\alpha}\mathbf{W}(a)\mathbf{J}^t\mathbf{Y}(\xi) + \mathbf{Y}^t(\xi)\mathbf{D}_0^{-1}\boldsymbol{\beta}\mathbf{W}(b)\mathbf{J}^t\mathbf{Y}(x)]$$

$$= \frac{-1}{pW}\mathbf{Y}^t(x)[\mathbf{D}_0^{-1}\boldsymbol{\alpha}\mathbf{W}(a)\mathbf{J}^t + \mathbf{J}\mathbf{W}^t(b)\boldsymbol{\beta}^t(\mathbf{D}_0^{-1})^t]\mathbf{Y}(\xi)$$

From (9.2.36) and the properties $\mathbf{J}\mathbf{J}^t = \mathbf{I}_2$, $\mathbf{J}^2 = -\mathbf{I}_2$, we have

$$\frac{1}{pW}\mathbf{W}(a)\mathbf{J}^t = \mathbf{P}^{-1}(a)\mathbf{W}^t(a)^{-1}, \qquad \frac{1}{pW}\mathbf{J}\mathbf{W}^t(b) = -\mathbf{W}^{-1}(b)\mathbf{P}^{-1}(b)$$

In what follows we will make use of the above and the definition of $\mathbf{D}_0$ as given by (9.2.23). Thus

$$\begin{aligned} G(x, \xi) - G(\xi, x) &= -\mathbf{Y}^t(x)[\mathbf{D}_0^{-1}\boldsymbol{\alpha}\mathbf{P}^{-1}(a)\mathbf{W}^t(a)^{-1} \\ &\qquad - \mathbf{W}^{-1}(b)\mathbf{P}^{-1}(b)\boldsymbol{\beta}^t(\mathbf{D}_0^{-1})^t]\mathbf{Y}(\xi) \\ &= -\mathbf{Y}^t(x)[\mathbf{D}_0^{-1}\boldsymbol{\alpha}\mathbf{P}^{-1}(a)\{\boldsymbol{\alpha}^t(\mathbf{D}_0^t)^{-1} + \mathbf{W}^t(a)^{-1}\mathbf{W}^t(b)\boldsymbol{\beta}^t(\mathbf{D}_0^t)^{-1}\} \\ &\qquad - \{\mathbf{D}_0^{-1}\boldsymbol{\alpha}\mathbf{W}(a)\mathbf{W}(b)^{-1} + \mathbf{D}_0^{-1}\boldsymbol{\beta}\}\mathbf{P}^{-1}(b)\boldsymbol{\beta}^t(\mathbf{D}_0^{-1})^t]\mathbf{Y}(\xi) \\ &= -\mathbf{Y}^t(x)[\mathbf{D}_0^{-1}\{\boldsymbol{\alpha}\mathbf{P}^{-1}(a)\boldsymbol{\alpha}^t - \boldsymbol{\beta}\mathbf{P}^{-1}(b)\boldsymbol{\beta}^t\}(\mathbf{D}_0)^{-1} \\ &\qquad + \mathbf{D}_0^{-1}\boldsymbol{\alpha}\{\mathbf{P}^{-1}(a)\mathbf{W}^t(a)^{-1}\mathbf{W}^t(b) - \mathbf{W}(a)\mathbf{W}(b)^{-1}\mathbf{P}^{-1}(b)\} \\ &\qquad \boldsymbol{\beta}^t(\mathbf{D}_0^t)^{-1}]\mathbf{Y}(\xi) \\ &= 0 \end{aligned}$$

which follows from the fact that the term in the first set of braces vanishes by virtue of (9.2.7), while that in the second set of braces vanishes because of the application of (9.2.36) at $x = a$ and $x = b$. The expected symmetry property of $G(x, \xi)$ is therefore confirmed. We leave the student to establish other properties of the Green's function through the exercises.

The most important result of this section is that the operator $\mathbf{T}^{-1}: \mathcal{H} \to D(T)$ defined by (9.2.29) is self-adjoint and compact. Thus spectral theorem 7.4.3 becomes applicable to the unbounded operator $\mathbf{T}$. In establishing this result we have assumed that $\mathbf{T}$ has an inverse. If $\mathbf{T}$ does not have an inverse but $(\mathbf{T} - \lambda_0\mathbf{I})^{-1}$ exists for some real number $\lambda_0$, we could have proceeded entirely on analogous lines to identify the Green's function representing $(\mathbf{T} - \lambda_0\mathbf{I})^{-1}$ to again reach the conclusion just made about $\mathbf{T}$, that is, spectral theorem 7.4.3 implies that the operator $\mathbf{T}$ has a discrete set of eigenvalues $\{\lambda_j\}$ and an orthonormal basis of eigenvectors $\{\mathbf{z}_j\}$ in $\mathcal{H}$ such that $|\lambda_j| \to \infty$. We will consider the determination of the eigenvalues and eigenvectors in the next section. Before concluding this section, we will consider some examples of construction of the Green's function. The construction becomes particularly simple in the case of unmixed boundary conditions. Since such boundary conditions occur frequently, let us inquire into this simplification. It is easily shown that the Green's function $G(x, \xi)$ must satisfy the boundary conditions (see Exercise 9.2.8) of the operator $\mathbf{T}$. In the unmixed case, the boundary condition at $x = a$ must be satisfied by $G(x, \xi)$ in the interval

$x < \xi \leq b$, while that at $x = b$ must be satisfied by $G(x, \xi)$ in the interval $a \leq \xi < x$. In view of the symmetry of the Green's function, one need then only employ the linearly independent solutions, $y_1(x)$ and $y_2(x)$, of $Ty(x) = 0$ as follows.

We let

$$g_i(x) = \mathbf{k}_i^t \mathbf{Y}(x), \qquad i = 1, 2$$

where $\mathbf{k}_i^t \equiv [k_{1,i} \quad k_{2,i}]$ and require that $g_1(x)$ and $g_2(x)$ satisfy the boundary conditions at $x = a$ and $x = b$, respectively. Then we may write

$$G(x, \xi) = \begin{cases} g_1(\xi)g_2(x), & a \leq \xi \leq x \\ g_1(x)g_2(\xi), & x \leq \xi \leq b \end{cases} \tag{9.2.40}$$

which automatically satisfies the requirements of symmetry and continuity at $x = \xi$. The condition (9.2.33), which represents the jump in the derivative at $x = \xi$, subjects $g_1$ and $g_2$ to the constraint

$$[g_1(x)g_2'(x) - g_1'(x)g_2(x)] = \frac{-1}{p(x)}$$

or alternatively,

$$1 = p[g_1' g_2 - g_1' g_2] \tag{9.2.41}$$

the right-hand side being the constant $pW$ when $W$ is calculated for the functions $g_1(x)$ and $g_2(x)$. Since $g_1$ and $g_2$ are required to satisfy only one boundary condition each, the vectors $\mathbf{k}_i^t$ are known only up to a multiplicative constant, which can be evaluated through (9.2.41). We consider some examples below.

*Example 1*

Let $T = -d^2/dx^2$, $D(T) = \{\mathbf{u} \in \mathfrak{L}_2[0, 1];\ T\mathbf{u} \in \mathfrak{L}_2[0, 1];\ u(0) = u(1) = 0\}$. Clearly, the operator $\mathbf{T}$ is self-adjoint. Since the boundary conditions are unmixed, the Green's function is constructed readily. First we determine the independent solutions of $Ty = 0$, which yields $d^2y/dx^2 = 0$, so that $y_1(x) = 1$ and $y_2(x) = x$.

$$g_1(x) = k_{1,1} + k_{2,1}x, \qquad g_1(0) = 0 \longrightarrow g_1(x) = k_{2,1}x$$
$$g_2(x) = k_{1,2} + k_{2,2}x, \qquad g_2(1) = 0 \longrightarrow g_2(x) = k_{1,2}(1 - x)$$

Since $p(x) \equiv 1$, from (9.2.41),

$$k_{2,1}(1 - x) + k_{2,1}x = k_{2,1} = 1 \quad \text{or} \quad k_{2,1} = 1$$

Thus

$$G(x, \xi) = \begin{cases} \xi(1 - x), & 0 \leq \xi \leq x \\ x(1 - \xi), & x \leq \xi \leq 1 \end{cases}$$

*Example 2*

For the second example, consider $T \equiv -d^2/dx^2$ and

$$D(T) = \{\mathbf{u} \in \mathcal{L}_2[0, 1]; T\mathbf{u} \in \mathcal{L}_2[0, 1]; u(0) + u(1) = 0, u'(0) + u'(1) = 0\}$$

We have seen how the foregoing antiperiodic boundary conditions make **T** self-adjoint. Here the boundary conditions are mixed. To construct the Green's function, observe that $p(x) \equiv 1$.

$$\boldsymbol{\alpha} = \begin{bmatrix} 1 & 0 \\ 0 & 1 \end{bmatrix}, \qquad \boldsymbol{\beta} = \begin{bmatrix} 1 & 0 \\ 0 & 1 \end{bmatrix}, \qquad \mathbf{Y}(x) = \begin{bmatrix} 1 \\ x \end{bmatrix}$$

$$\mathbf{W}(\mathbf{Y}(x)) = \begin{bmatrix} 1 & x \\ 0 & 1 \end{bmatrix}, \qquad \mathbf{D}_0 = \begin{bmatrix} 1 & 0 \\ 0 & 1 \end{bmatrix}\begin{bmatrix} 1 & 0 \\ 0 & 1 \end{bmatrix} + \begin{bmatrix} 1 & 0 \\ 0 & 1 \end{bmatrix}\begin{bmatrix} 1 & 1 \\ 0 & 1 \end{bmatrix}$$

$$= \begin{bmatrix} 2 & 1 \\ 0 & 2 \end{bmatrix}$$

$$W(\mathbf{Y}) = 1$$

From (9.2.39) we have for $\xi \le x$,

$$G(x, \xi) = -[1 \quad x]\begin{bmatrix} \frac{1}{2} & -\frac{1}{4} \\ 0 & \frac{1}{2} \end{bmatrix}\begin{bmatrix} 1 & 0 \\ 0 & 1 \end{bmatrix}\begin{bmatrix} 1 & 0 \\ 0 & 1 \end{bmatrix}\begin{bmatrix} 0 & -1 \\ 1 & 0 \end{bmatrix}\begin{bmatrix} 1 \\ \xi \end{bmatrix}$$

$$= -[1 \quad x]\begin{bmatrix} -\frac{1}{4} & -\frac{1}{2} \\ \frac{1}{2} & 0 \end{bmatrix}\begin{bmatrix} 1 \\ \xi \end{bmatrix} = -\tfrac{1}{4} + \tfrac{1}{2}(x - \xi)$$

Thus

$$G(x, \xi) = \begin{cases} \frac{1}{4} - \frac{1}{2}(x - \xi), & 0 \le \xi \le x \\ \frac{1}{4} - \frac{1}{2}(\xi - x), & x \le \xi \le 1 \end{cases}$$

*Example 3*

$$\text{Let } \mathcal{H} \equiv \left\{u\colon \int_0^1 xu^2(x)\,dx < \infty, \langle u, v\rangle \equiv \int_0^1 xu(x)v(x)\,dx\right\}.$$

$$T = -\frac{1}{x}\frac{d}{dx}\left\{x\frac{d}{dx}\right\}, \qquad D(T) = \{\mathbf{u} \in \mathcal{H}; T\mathbf{u} \in \mathcal{H}; \lim_{x\to 0} xu'(x) = 0\dagger,$$
$$u(1) = 0\}$$

The boundary conditions are unmixed. The independent solutions of $Ty = 0$ are given by $y_1 = \ln x$, $y_2 = 1$,

$$g_1(x) = k_{1,1}\ln x + k_{2,1}, \qquad \lim_{x\to 0} xg_1'(x) = k_{1,1} = 0 \to g_1(x) = k_{2,1}$$

$$g_2(x) = k_{1,2}\ln x + k_{2,2}, \qquad g_2(1) = 0, \qquad k_{2,2} = 0 \to g_2(x) = \ln x$$

---

†The domain $D(T)$ is unaltered if instead we demand boundedness of $u$ at $x = 0$. This condition typically arises at a singular point.

where we have let $k_{1,2} = 1$. From (9.2.41)

$$x\left[-k_{2,1}\frac{1}{x}\right] = 1 \quad \text{or} \quad k_{2,1} = -1$$

Thus

$$G(x, \xi) = \begin{cases} \ln \dfrac{1}{x}, & 0 \leq \xi \leq x \\ \ln \dfrac{1}{\xi}, & x \leq \xi \leq 1 \end{cases}$$

Suppose that we try to use (9.2.39) to obtain $G(x, \xi)$. Then

$$y(x) = \begin{bmatrix} \ln x \\ 1 \end{bmatrix}$$

$$\mathbf{W}(\mathbf{Y}(x)) = \begin{bmatrix} \ln x & 1 \\ \dfrac{1}{x} & 0 \end{bmatrix}, \qquad W(\mathbf{Y}) = -\frac{1}{x}, \qquad pW = -1$$

To calculate $\boldsymbol{\alpha}\mathbf{W}(\mathbf{Y}(0))$ we must take

$$\lim_{x \to 0} \begin{bmatrix} 0 & x \\ 0 & 0 \end{bmatrix} \begin{bmatrix} \ln x & 1 \\ \dfrac{1}{x} & 0 \end{bmatrix} = \begin{bmatrix} 1 & 0 \\ 0 & 0 \end{bmatrix}$$

On the other hand,

$$\boldsymbol{\beta}\mathbf{W}(\mathbf{Y}(1)) = \begin{bmatrix} 0 & 0 \\ 1 & 0 \end{bmatrix} \begin{bmatrix} 0 & 1 \\ 1 & 0 \end{bmatrix} = \begin{bmatrix} 0 & 0 \\ 0 & 1 \end{bmatrix}$$

$$\mathbf{D}_0 = \begin{bmatrix} 1 & 0 \\ 0 & 0 \end{bmatrix} + \begin{bmatrix} 0 & 0 \\ 0 & 1 \end{bmatrix} = \begin{bmatrix} 1 & 0 \\ 0 & 1 \end{bmatrix}$$

and for $\xi \leq x$ we have

$$\begin{aligned} G(x, \xi) &= -[\ln x \quad 1] \begin{bmatrix} 1 & 0 \\ 0 & 1 \end{bmatrix} \begin{bmatrix} 1 & 0 \\ 0 & 0 \end{bmatrix} \begin{bmatrix} 0 & -1 \\ 1 & 0 \end{bmatrix} \begin{bmatrix} \ln \xi \\ 1 \end{bmatrix} \Big/ (-1) \\ &= -\ln x = \ln \frac{1}{x} \end{aligned}$$

which is the same as before.

*Example 4*

For the fourth example we take $T = -d^2/dx^2$ and

$$D(T) = \{\mathbf{u} \in \mathfrak{L}_2[0, 1];\, T\mathbf{u} \in \mathfrak{L}_2[0, 1];\, u'(0) = 0,\, u'(1) = 0\}$$

Here we find that $Ty = 0$ with $\mathbf{y} \in D(T)$ (or $\mathbf{Ty} = \mathbf{0}$) yields a nonzero solution, so that the inverse of $\mathbf{T}$ does not exist. Alternatively, if $\mathbf{D}_0$ is computed, it will be found to be singular. In this case in order to show that $(\mathbf{T} - \lambda_0\mathbf{I})$

has a compact inverse, we must compute $(T - \lambda_0 I)\mathbf{y} = 0$. Thus

$$\frac{d^2y}{dx^2} + \lambda_0 y = 0, \qquad y_1(x) = \sin\sqrt{\lambda_0}\, x, \quad y_2(x) = \cos\sqrt{\lambda_0}\, x$$

and

$$g_1(x) = k_{1,1}\sin\sqrt{\lambda_0}\, x + k_{2,1}\cos\sqrt{\lambda_0}\, x,\ g_1'(0) = 0$$
$$\rightarrow k_{1,1} = 0 \rightarrow g_1(x) = \cos\sqrt{\lambda_0}\, x$$

$$g_2(x) = k_{1,2}\sin\sqrt{\lambda_0}\, x + k_{2,2}\cos\sqrt{\lambda_0}\, x,\ g_2'(1) = 0 \rightarrow k_{1,2}$$
$$= k_{2,2}\tan\sqrt{\lambda_0}$$

$$g_2(x) = k_{2,2}(\sin\sqrt{\lambda_0}\, x\tan\sqrt{\lambda_0} + \cos\sqrt{\lambda_0}\, x) = \frac{k_{2,2}}{\cos\sqrt{\lambda_0}}\cos\sqrt{\lambda_0}\,(1-x)$$

Using (9.2.41), we have

$$1 = \left[\frac{\sqrt{\lambda_0}\sin\sqrt{\lambda_0}\, x\cos\sqrt{\lambda_0}\,(1-x)}{\cos\sqrt{\lambda_0}} + \frac{\sqrt{\lambda_0}\cos\sqrt{\lambda_0}\, x\sin\sqrt{\lambda_0}\,(1-x)}{\cos\sqrt{\lambda_0}}\right]k_{2,2}$$
$$= \frac{\sqrt{\lambda_0}}{\cos\sqrt{\lambda_0}}\sin\sqrt{\lambda_0}\, k_{2,2} = \sqrt{\lambda_0}\tan\sqrt{\lambda_0}\, k_{2,2}$$

Thus

$$g_1(x) = \cos\sqrt{\lambda_0}\, x, \qquad g_2(x) = \frac{\cos\sqrt{\lambda_0}\,(1-x)}{\sqrt{\lambda_0}\sin\sqrt{\lambda_0}}$$

and

$$G(x,\xi) = \begin{cases} \dfrac{\cos\sqrt{\lambda_0}\,\xi\cos\sqrt{\lambda_0}\,(1-x)}{\sqrt{\lambda_0}\sin\sqrt{\lambda_0}}, & 0 \le \xi \le x \\[2ex] \dfrac{\cos\sqrt{\lambda_0}\, x\cos\sqrt{\lambda_0}\,(1-\xi)}{\sqrt{\lambda_0}\sin\sqrt{\lambda_0}}, & x \le \xi \le 1 \end{cases}$$

which clearly exists as long as $\sqrt{\lambda_0}\sin\sqrt{\lambda_0} \neq 0$. Note particularly that it does not exist when $\lambda_0 = 0$.

In all of the examples considered above, it was possible to solve for the homogeneous equations readily. However, in more general situations such analytical solutions are not so readily identified. In the interest of avoiding an elaborate discussion of other methods which are only peripheral to the central issue here, we defer them to the Appendix. We will also be faced with this problem in Section 9.2.2, where our concern is first to find the spectrum of the self-adjoint operators under discussion here.

Although the main objective of this section was to establish the applicability of spectral theorem 7.4.3, there are other results of importance to us. In obtaining the Green's function we have in effect solved the equation

$$\mathbf{Tu} = \mathbf{f}, \qquad \mathbf{f} \in \mathcal{H} \tag{9.2.9}$$

Thus the spectral inversion of **T** as discussed in Section 7.4 for solving the foregoing equation is quite unnecessary for the operators of this section. In considering problems of the type (9.2.9), there arises, however, an important variation, which is that of solving

$$T\mathbf{v} = \mathbf{f}, \qquad \mathbf{f} \in \mathcal{H} \tag{9.2.42}$$

where $\mathbf{v}$ satisfies the *inhomogeneous* boundary conditions†

$$\boldsymbol{\alpha}\mathbf{w}(\mathbf{v}(a)) + \boldsymbol{\beta}\mathbf{w}(\mathbf{v}(b)) = \boldsymbol{\gamma} \tag{9.2.43}$$

where $\boldsymbol{\gamma}$ is a constant vector. It is therefore clear that $\mathbf{v} \notin D(T)$, which is why we have used the operation $T$ in (9.2.42) instead of **T**. The solution to the problem (9.2.42) subject to (9.2.43) can be readily obtained in terms of the Green's function as follows.

Referring back to (9.2.21) and (9.2.22), if $v$ were to replace $u$ in the former, then the right-hand side of the latter should include additively the vector $\boldsymbol{\gamma}$ that appears in (9.2.43). Thus $\mathbf{w}(v)$ may be obtained by modifying the right-hand side of (9.2.24) to read

$$\mathbf{w}(v) = \int_a^b \mathbf{K}(x, \xi)\mathbf{b}(\xi)\, d\xi + \mathbf{W}(x)D_0^{-1}\boldsymbol{\gamma} \tag{9.2.44}$$

Clearly, then, we must have

$$v(x) = \int_a^b G(x, \xi)r(\xi)f(\xi)\, d\xi + \mathbf{e}_1^t\mathbf{W}(x)\mathbf{D}_0^{-1}\boldsymbol{\gamma} \tag{9.2.45}$$

which is equivalent to writing

$$v(x) = u(x) + \mathbf{Y}^t(x)\mathbf{D}_0^{-1}\boldsymbol{\gamma} \tag{9.2.46}$$

where $\mathbf{u} \equiv \{u(x)\}$ satisfies (9.2.9). We have thus obtained the solution to (9.2.42) subject to the inhomogeneous boundary conditions (9.2.43).

## EXERCISES

**9.2.4** Show that if **A** in (9.2.12) has a singularity at $x = x_0$, it cannot satisfy the Lipschitz condition there.

**9.2.5** In transforming (9.2.10) into a first-order system, identify the matrix coefficients if the component $z$ in (9.2.14) had been defined by $z \equiv (x - x_0)^2 y'$. What is the order of the singularity of this matrix at $x_0$?

**9.2.6** Show that $\mathbf{T}^{-1}$ as defined by (9.2.29) is self-adjoint with respect to the inner product (9.1.8) if and only if $G(x, \xi)$ is symmetric, that is, $G(x, \xi) = G(\xi, x)$.

**9.2.7** Establish the result (9.2.31) by direct differentiation of (9.2.27).

---

†Note that the matrices $\boldsymbol{\alpha}$ and $\boldsymbol{\beta}$ are the same as those occurring in $D(T)$, the domain of $T$.

**9.2.8** Show that the Green's function $G(x, \xi)$ in (9.2.39) satisfies the boundary conditions (9.1.11).

**9.2.9** Let $\mathbf{J} \equiv \begin{bmatrix} 1 & 0 \\ -1 & 0 \end{bmatrix}$. Show that for any $2 \times 2$ constant matrix $\mathbf{K}$, $\mathbf{K}^t\mathbf{J}^t\mathbf{K} = \mathbf{J}^t \det \mathbf{K}$. Use this result to show that the Green's function as given by (9.2.39) remains the same when $\mathbf{Y}$ is replaced by $\mathbf{KY}$, that is, that the Green's function is independent of the particular fundamental matrix used in expressing it.

**9.2.10** Reconcile the form (9.2.40) for the Green's function where $g_i(x) = \mathbf{k}_i^t\mathbf{Y}(x)$, $i = 1, 2$, with that given by (9.2.39) by finding relationships between the matrices appearing in (9.2.39) and the vectors $\mathbf{k}_1$ and $\mathbf{k}_2$.

**9.2.11** Construct the inverse operator of $\mathbf{T} = \{T, D(T)\}$, where

$$T \equiv -\frac{1}{x^2}\frac{d}{dx}\left(x^2\frac{d}{dx}\right), \qquad 0 < x < 1$$

with $D(T) \subset \mathcal{H}$, where

$$\mathcal{H} = \left\{\mathbf{u}: \int_0^1 x^2u^2(x)\,dx < \infty,\ \langle \mathbf{u}, \mathbf{v}\rangle = \int_0^1 x^2u(x)v(x)\,dx\right\}$$

$$D(T) = \{\mathbf{u} \in \mathcal{H};\ T\mathbf{u} \in \mathcal{H};\ \lim_{x\to 0} x^2u'(x) = 0,\ u'(1) + \beta u(1) = 0\}$$

**9.2.12** Find the Green's function for the operator $\mathbf{T} = \{T, D(T)\}$, where

$$T = -\frac{D}{e^{-Vx/D}}\frac{d}{dx}\left(e^{-Vx/D}\frac{d}{dx}\right) + k, \qquad 0 < x < 1$$

with $D(T) \subset \mathcal{H}$, where

$$\mathcal{H} = \left\{\mathbf{u}: \int_0^1 e^{-Vx/D}u^2(x)\,dx < \infty,\ \langle \mathbf{u}, \mathbf{v}\rangle = \int_0^1 e^{-Vx/D}u(x)v(x)\,dx\right\}$$

$$D(T) = \{\mathbf{u} \in \mathcal{H};\ T\mathbf{u} \in \mathcal{H};\ -Du'(0) + Vu(0) = 0,\ u'(1) = 0\}$$

Express the steady-state concentration $c(x)$ of a reactant undergoing first-order irreversible transformation isothermally in an axially dispersed tubular reactor in terms of the Green's function above. The reactor equation is given by

$$D\frac{d^2c}{dx^2} - V\frac{dc}{dx} - kc = 0, \qquad 0 < x < 1$$

subject to the boundary conditions

$$-D\frac{dc}{dx}(0) + Vc(0) = Vc_f, \qquad \frac{dc}{dx}(1) = 0$$

**9.2.13** Let $\mathbf{T} = \{T, D(T)\}$, where $T = -d^2/dx^2$ and

$$D(T) = \{\mathbf{u} \in \mathcal{L}_2[0, 1];\ T\mathbf{u} \in \mathcal{L}_2[0, 1];\ u(0) = u(1),\ u'(0) = u'(1)\}$$

Does the Green's function exist for $\mathbf{T}$?

**9.2.14** In one-dimensional steady-state heat conduction such as in a slab between $x = 0$ and $x = 1$, which is heated internally by a source distribution $f(x)$ and whose faces $x = 0$ and $x = 1$ are maintained at temperatures $v_0$ and $v_1$, respectively, we obtain the equation for the temperature distribution $v(x)$:

$$k\frac{d^2v}{dx^2} + f(x) = 0, \qquad v(0) = v_0, \quad v(1) = v_1$$

where $k$ is the thermal conductivity of the slab. Obtain a formula for $v(x)$ in terms of an appropriate Green's function.

### 9.2.2 Evaluation of Spectra

Having established in Section 9.2.1 that the self-adjoint operator $\mathbf{T}$ considered therein has a compact self-adjoint inverse [or that for some real $\lambda_0$, $(\mathbf{T} - \lambda_0\mathbf{I})$ has a compact inverse], the implication of Theorem 7.4.3 provided for the existence of a countable set of eigenvalues $\{\lambda_j\}$ of $T$ and an orthonormal basis $\{\mathbf{z}_j\}$ in $\mathcal{H}$. Our aim in this section is to demonstrate how the eigenvalues and eigenvectors may be computed. Thus we must address the eigenvalue problem

$$\mathbf{T}\mathbf{z} = \lambda\mathbf{z} \tag{9.2.47}$$

by which is meant $T\mathbf{z} = \lambda\mathbf{z}$ and $\mathbf{z} \in D(T)$. The strategy of solution is identical to that in Section 9.2.1. We will express $z(x, \lambda)$ as a linear combination of the linearly independent solutions $u_1(x, \lambda)$ and $u_2(x, \lambda)$ of the homogeneous differential equation

$$-\frac{1}{r(x)}\frac{d}{dx}\left[p(x)\frac{du}{dx}\right] + q(x)u = \lambda u \tag{9.2.48}$$

Again to preserve the generality of our discussion we will defer the actual solution of (9.2.48) to a later stage of this section. Letting

$$z(x, \lambda) = \mathbf{U}^t(x, \lambda)\mathbf{k}, \qquad \mathbf{U}(x, \lambda) \equiv \begin{bmatrix} u_1(x, \lambda) \\ u_2(x, \lambda) \end{bmatrix}, \qquad \mathbf{k} \equiv \begin{bmatrix} k_1 \\ k_2 \end{bmatrix} \tag{9.2.49}$$

where $k_1$ and $k_2$ are constants, the imposition of boundary conditions (9.1.11) [since we are considering self-adjoint operators (9.1.11) as subject to the constraints (9.2.7) or (9.2.8)] yields

$$\mathbf{D}_\lambda\mathbf{k} = \mathbf{0} \tag{9.2.50}$$

where we have let

$$\mathbf{D}_\lambda \equiv \boldsymbol{\alpha}\mathbf{W}(\mathbf{U}(a, \lambda)) + \boldsymbol{\beta}\mathbf{W}(\mathbf{U}(b, \lambda)) \tag{9.2.51}$$

We seek nontrivial solutions of (9.2.50), the condition for which is

$$D_\lambda = 0 \tag{9.2.52}$$

where $D_\lambda$ is the determinant of $\mathbf{D}_\lambda$. Equation (9.2.52) is called the *characteristic equation* of $\mathbf{T}$, which must be solved to obtain the eigenvalues.† Clearly,

---

†When the boundary conditions are relatively simple, it may be more convenient to substitute the solution (9.2.49) into the boundary conditions than using (9.2.52)

the foregoing is a nonlinear algebraic equation in $\lambda$ whose solution normally requires numerical methods, but in some simple situations the eigenvalues are completely identifiable. Indeed, consistent with the result of spectral theorem 7.4.3, (9.2.52) will yield a denumerably infinite number of solutions all of which are the eigenvalues $\{\lambda_j\}$ of the operator $\mathbf{T}$. If $\lambda = 0$ is an eigenvalue, then clearly $\mathbf{T}$ does not have an inverse and the nontrivial solution of $\mathbf{Tz} = \mathbf{0}$ will yield the eigenvector corresponding to the zero eigenvalue. In subsequent discussions we will let

$$\mathbf{z}_j \equiv \{z(x, \lambda_j)\}, \qquad j = 1, 2, \ldots$$

be the eigenvectors that form an orthonormal basis in $\mathcal{H}$. To compute $\mathbf{z}_j$, we recognize that $z(x, \lambda_j) = \mathbf{U}^t(x, \lambda_j)\mathbf{k}_j$, where $\mathbf{k}_j$ is given by

$$\mathbf{D}_{\lambda_j}\mathbf{k}_j = \mathbf{0} \tag{9.2.53}$$

which has a maximum of two linearly independent solutions. When there is only one solution of (9.2.53), there is only one eigenvector $\mathbf{z}_j$ and $\mathbf{k}_j$ is uniquely obtained by the requirement that $||\mathbf{z}_j|| = 1$. Here $N_{\lambda_j}$ has dimension 1.

The two linearly independent solutions of (9.2.53) occur when $\mathbf{D}_{\lambda_j} = \mathbf{0}$, in which case $N_{\lambda_j}$ has dimension 2 and there are two linearly independent eigenvectors $\mathbf{z}_j^{(1)}$ and $\mathbf{z}_j^{(2)}$, which corresponds to the eigenvalue $\lambda_j$. Thus we write

$$z^{(i)}(x, \lambda_j) = \mathbf{U}^t(x, \lambda_j)\mathbf{k}_j^{(i)}, \qquad i = 1, 2 \tag{9.2.54}$$

where $\mathbf{k}_j^{(1)}$ and $\mathbf{k}_j^{(2)}$ are the linearly independent solutions of (9.2.53). The actual choice can be made unique by requiring that

$$||\mathbf{z}_j^{(1)}|| = 1, \qquad ||\mathbf{z}_j^{(2)}|| = 1, \qquad \langle \mathbf{z}_j^{(1)}, \mathbf{z}_j^{(2)} \rangle = 0$$

the last of which is to preserve the orthogonality between the different eigenvectors. This orthogonality is accomplished by Gram–Schmidt orthogonalization. We leave the reader to verify that if $k_{2,j}^{(1)} = 0$ in $\mathbf{k}_j^{(1)}$, and $k_{1,j}^{(1)}$ normalizes $\mathbf{z}_j^{(1)}$, then

$$k_{1,j}^{(2)} = -[k_{1,j}^{(1)}]^2 \langle \mathbf{u}_{2,j}, \mathbf{u}_{1,j} \rangle k_{2,j}^{(2)} \tag{9.2.55}$$

where $\mathbf{u}_{i,j} \equiv \{u_i(x, \lambda_j)\}$, $i = 1, 2$, and the normalization condition for $\mathbf{z}_j^{(2)}$ together specify $k_{2,j}^{(2)}$. Thus we have $\mathbf{z}_j^{(1)} = \{k_{1,j}^{(1)} u_1(x, \lambda_j)\}$ and $\mathbf{z}_j^{(2)} = \{k_{1,j}^{(2)} u_1(x, \lambda_j) + k_{2,j}^{(2)} u_2(x, \lambda_j)\}$.

If $\mathbf{P}_j$ is the self-adjoint projection of $\mathcal{H}$ onto $N_{\lambda_j}$, then $\mathbf{T}$ may be represented by

$$\mathbf{T} = \sum_{j=1}^{\infty} \lambda_j \mathbf{P}_j$$

which can only operate on vectors in $D(\mathbf{T})$. The compact inverse of $\mathbf{T}$, on the other hand, given by

$$\mathbf{T}^{-1} = \sum_{j=1}^{\infty} \lambda_j^{-1} \mathbf{P}_j$$

is defined for all elements of $\mathcal{H}$. If $\mathbf{f} \in \mathcal{H}$, then we may write

$$\begin{aligned}\mathbf{T}^{-1}\mathbf{f} &= \sum_{j=1}^{\infty} \lambda_j^{-1}\mathbf{P}_j\mathbf{f} \\ &= \sum_{j=1}^{\infty} \lambda_j^{-1}\langle \mathbf{f}, \mathbf{z}_j\rangle \mathbf{z}_j \\ &= \sum_{j=1}^{\infty} \langle \mathbf{f}, \lambda_j^{-1}\mathbf{z}_j\rangle \mathbf{z}_j\end{aligned}$$

We now refer the reader to Section 6.3, relating to tensor products of inner product spaces. Specifically, (6.3.6) may be used to express $\mathbf{T}^{-1}\mathbf{f}$ obtained above as

$$\begin{aligned}\mathbf{T}^{-1}\mathbf{f} &= \sum_{j=1}^{\infty} \langle \mathbf{f}, \lambda_j^{-1}(\mathbf{z}_j \otimes \mathbf{z}_j)\rangle \\ &= \langle \mathbf{f}, \sum_{j=1}^{\infty} \lambda_j^{-1}(\mathbf{z}_j \otimes \mathbf{z}_j)\rangle, \qquad \mathbf{f} \in \mathcal{H}\end{aligned} \tag{9.2.56}$$

This second relation arises from the fact that the equality in the first is in the sense of convergence with respect to the norm in $\mathcal{H}$. It is interesting to compare the above with (9.2.29), which may be rewritten as

$$\mathbf{T}^{-1}\mathbf{f} = \langle \mathbf{f}, \mathbf{G}\rangle \tag{9.2.57}$$

where $\mathbf{G} \in \mathcal{H} \otimes \mathcal{H}$ presents an alternative viewpoint of the Green's function. Subtracting (9.2.57) from (9.2.56), we obtain

$$0 = \langle \mathbf{f}, \sum_{j=1}^{\infty} \lambda_j^{-1}(\mathbf{z}_j \otimes \mathbf{z}_j) - \mathbf{G}\rangle \qquad \forall \;\; \mathbf{f} \in \mathcal{H}$$

which leads to the result

$$\mathbf{G} = \sum_{j=1}^{\infty} \lambda_j^{-1}(\mathbf{z}_j \otimes \mathbf{z}_j) \tag{9.2.58}$$

Equation (9.2.58) is called *Mercer's expansion* and strikingly reflects the symmetric nature of the Green's function.

Quite frequently the eigenvalues of $\mathbf{T}$ are of the same sign; that is, the operator is nonpositive or nonnegative definite. It is possible to find conditions under which the operator $\mathbf{T}$ is, for example, positive definite. We will restrict considerations to unmixed boundary conditions represented by

$$u(a) + \alpha u'(a) = 0, \qquad u(b) + \beta u'(b) = 0$$

We examine the quantity $\langle \mathbf{Tu}, \mathbf{u}\rangle$, which is given by

$$\begin{aligned}\langle \mathbf{Tu}, \mathbf{u}\rangle &= \int_a^b u(x)\left\{-\frac{d}{dx}\left[p(x)\frac{du}{dx}\right] + q(x)r(x)u\right\} dx \\ &= [-p(x)u(x)u'(x)]_a^b + \int_a^b [p(x)u'^2(x) + q(x)r(x)u^2(x)]\, dx\end{aligned}$$

If we assume that $p$ and $q$ are nonnegative in $[a, b]$ and $(a, b)$, respectively, then the nonnegativity of $\mathbf{T}$ depends on the sign of the expression

$$-p(b)u'(b)u(b) + p(a)u'(a)u(a)$$

Clearly, if $p(a)$ and $p(b)$ are positive, $\mathbf{T}$ is *nonnegative* only if

$$\alpha \leq 0, \qquad \beta \geq 0 \tag{9.2.59}$$

Of course, since either or both of $\alpha$ and $\beta$ may be infinity, the sign is required only on the finite coefficient. This is as much as one can say about the eigenvalues of $\mathbf{T}$ without solving the characteristic equation (9.2.52). Indeed, the solution of the characteristic equation is possible only after the linearly independent solutions of (9.2.48) have been obtained. We consider some examples below in some of which the eigenvalues are obtained analytically and in others numerical evaluation is required.

*Example 1*

Consider the operator in Example 1 in Section 9.2.1. The solutions $u_1(x)$ and $u_2(x)$ of

$$Tu = -\frac{d^2u}{dx^2} = \lambda u$$

are given by $u_1(x) = \sin\sqrt{\lambda}\,x$ and $u_2(x) = \cos\sqrt{\lambda}\,x$, and we let $z(x, \lambda) = k_1 \sin\sqrt{\lambda}\,x + k_2 \cos\sqrt{\lambda}\,x$. Since the boundary conditions $z(0, \lambda) = z(1, \lambda) = 0$ are rather simple, it is possible to apply the boundary conditions directly.

$$z(0, \lambda) = 0 \longrightarrow k_2 = 0$$

$$z(1, \lambda) = 0 \longrightarrow \sin\sqrt{\lambda} = 0$$

since $z(x, \lambda) \not\equiv 0 \longrightarrow k_1 \neq 0$. The characteristic equation is $\sin\sqrt{\lambda} = 0$, which gives $\sqrt{\lambda} = \pm j\pi, j = 1, 2, \ldots$. Denoting the $j$th eigenvalue by $\lambda_j$, we obtain

$$\lambda_j = j^2\pi^2, \qquad z_j(x) \equiv z(x, \lambda_j) = k_{1,j} \sin j\pi x, \qquad j = 1, 2, \ldots$$

The constant $k_{1,j}$ is to be obtained using $||\mathbf{z}_j|| = 1$. Thus

$$||\mathbf{z}_j|| = k_{1j}^2 \int_0^1 \sin^2 j\pi x \, dx = \tfrac{1}{2}k_{1j}^2 = 1$$

so that $k_{1,j} = \sqrt{2}$ and $\mathbf{z}_j = \{\sqrt{2} \sin j\pi x\}$. We have thus obtained the entire spectrum analytically. Notice how the eigenvalues $\{j^2\pi^2\}$ are all positive, which is consistent with the fact that (9.2.59) holds for $\mathbf{T}$ because $\alpha = \beta = 0$.

Finally, Mercer's expansion (9.2.58) must imply that the Green's function evaluated in Section 9.2.1 satisfies the relationship

$$G(x, \xi) = \sum_{j=1}^{\infty} \frac{2}{j^2\pi^2} \sin j\pi x \sin j\pi\xi = \begin{cases} \xi(1-x), & 0 \le \xi \le x \\ x(1-\xi), & x \le \xi \le 1 \end{cases}$$

whose convergence is with respect to the $\mathfrak{L}_2$-norm on $\mathfrak{L}_2[a, b] \otimes \mathfrak{L}_2[a, b]$.†

*Example 2*

Next we examine Example 2 of Section 9.2.1. Again, as in Example 1, we have $u_1(x, \lambda) = \sin\sqrt{\lambda}\,x$, $u_2(x, \lambda) = \cos\sqrt{\lambda}\,x$. Since the boundary conditions are mixed, we will prefer the route of (9.2.52). Thus

$$\mathbf{W}(\mathbf{U}(x)) = \begin{bmatrix} \sin\sqrt{\lambda}\,x & \cos\sqrt{\lambda}\,x \\ \sqrt{\lambda}\cos\sqrt{\lambda}\,x & -\sqrt{\lambda}\sin\sqrt{\lambda}\,x \end{bmatrix}$$

$$\mathbf{D}_\lambda = \begin{bmatrix} 1 & 0 \\ 0 & 1 \end{bmatrix}\begin{bmatrix} 0 & 1 \\ \sqrt{\lambda} & 0 \end{bmatrix} + \begin{bmatrix} 1 & 0 \\ 0 & 1 \end{bmatrix}\begin{bmatrix} \sin\sqrt{\lambda} & \cos\sqrt{\lambda} \\ \sqrt{\lambda}\cos\sqrt{\lambda} & -\sqrt{\lambda}\sin\sqrt{\lambda} \end{bmatrix}$$

$$= \begin{bmatrix} \sin\sqrt{\lambda} & 1 + \cos\sqrt{\lambda} \\ \sqrt{\lambda}\,(1 + \cos\sqrt{\lambda}) & -\sqrt{\lambda}\sin\sqrt{\lambda} \end{bmatrix}$$

Thus the characteristic equation is given by

$$D_\lambda = -\sqrt{\lambda}\sin^2\sqrt{\lambda} - \sqrt{\lambda}\,(1 + \cos\sqrt{\lambda})^2 = 0$$

The roots of the equation above are 0 and those of $\cos\sqrt{\lambda} = -1$. To ascertain if 0 is an eigenvalue, we must determine whether the corresponding eigenvector is nonzero. (In this case it is not necessary since we already know that the Green's function has been constructed!). Since $\lambda = 0$ is not an eigenvalue, we have the eigenvalues of $\mathbf{T}$ given by

$$\lambda_j = (2j - 1)^2\pi^2, \qquad j = 1, 2, \ldots$$

all positive, which could have been anticipated because

$$\langle \mathbf{Tu}, \mathbf{u} \rangle = -u'(1)u(1) + u'(0)u(0) + \int_0^1 u'^2(x)\,dx$$

$$= \int_0^1 u'^2(x)\,dx > 0$$

To calculate the eigenvector $\mathbf{z}_j$ we must solve $\mathbf{D}_{\lambda_j}\mathbf{k}_j = \mathbf{0}$. On substituting the eigenvalue $\lambda_j$, we clearly have $\mathbf{D}_{\lambda_j} = \mathbf{0}$ for every $j$. Thus there are two linearly

---

†Because of the continuity of $G(x, \xi)$ in $[a, b] \times [a, b]$, there is also *uniform pointwise convergence* in this case. For example, if $x = \xi = \frac{1}{2}$, we obtain the expansion

$$\sum_{j=1}^{\infty} \frac{1}{(2j-1)^2} = \frac{\pi^2}{8}$$

which is well known.

independent eigenvectors $\mathbf{z}_j^{(1)}$ and $\mathbf{z}_j^{(2)}$ given by

$$\mathbf{z}_j^{(1)} = \{k_{1,j} \sin (2j - 1)\pi x\}, \qquad \mathbf{z}_j^{(2)} = \{k_{2,j} \cos (2j - 1)\pi x\}$$

where

$$k_{1,j}^2 \int_0^1 \sin^2 (2j - 1)\pi x \, dx = 1, \qquad k_{2,j}^2 \int_0^1 \cos^2 (2j - 1)\pi x \, dx = 1$$

from which it is readily seen that $k_{1,j} = k_{2,j} = \sqrt{2}$. We have thus evaluated the spectrum of **T**. Note here that the non-zero eigenvalues are repeated.

*Example 3*

Our third example has the same operator as in Example 3 of Section 9.2.1. Thus $T\mathbf{u} = \lambda\mathbf{u}$ gives

$$-\frac{1}{x}\frac{d}{dx}\left(x\frac{du}{dx}\right) = \lambda u$$

which may also be written as

$$x^2\frac{d^2u}{dx^2} + x\frac{du}{dx} + \lambda^2 x^2 u = 0$$

The solution of this equation has been considered in Section A.3 of the Appendix. The linearly independent solutions of the differential equation above are given by $u_1(x, \lambda) = J_0(\sqrt{\lambda}\, x)$ and $u_2(x, \lambda) = Y_0(\lambda x)$, so that

$$z(x, \lambda) = k_1 J_0(\sqrt{\lambda}\, x) + k_2 Y_0(\sqrt{\lambda}\, x)$$

Since $J_0'(0) = 0$ and $Y_0(\sqrt{\lambda}\, x)$ has a logarithmic singularity at $x = 0$, $\lim_{x\to 0} xY_0'(\sqrt{\lambda}\, x) \neq 0$ and the application of the boundary condition at $x = 0$ yields $k_2 = 0$.† Thus $z(x, \lambda) = k_1 J_0(\sqrt{\lambda}\, x)$. The boundary condition at $x = 1$ yields the characteristic equation

$$J_0(\sqrt{\lambda}) = 0$$

Since $J_0$ is a function oscillating between positive and negative values with $\lim_{\lambda\to\infty} J_0(\sqrt{\lambda}) = 0$, the roots of the equation above are infinitely many. Furthermore, $J_0(-\sqrt{\lambda}) = J_0(\sqrt{\lambda})$, so that if $\sqrt{\lambda}$ is a root, $-\sqrt{\lambda}$ is also a root. Since the eigenvalue is $\lambda$, one need bother only about the positive roots of $J_0$. The roots of $J_0$ are tabulated in standard texts,‡ although it is frequently more convenient from a computational standpoint to use numerical methods based on quasi-linearization. Denoting the eigenvalues by $\{\lambda_j\}$, we have

$$\mathbf{z}_j = \{k_{1,j} J_0(\sqrt{\lambda_j}\, x)\}$$

---

† $Y_0$ is eliminated also on the basis of boundedness at $x = 0$.

‡See, for example, *Conduction of Heat in Solids*, by H. S. Carslaw and J. C. Jaeger, Oxford University Press, London, 1959,

whose normalization yields

$$k_{1,j}^{-2} = \int_0^1 x J_0^2(\sqrt{\lambda_j}\, x)\, dx$$

We may now draw from the treasury of properties of the Bessel functions (see Section A.4.1 of the Appendix) to obtain the integral above as

$$k_{1,j}^{-2} = \tfrac{1}{2} J_0'^2(\sqrt{\lambda_j})$$

Thus we obtain

$$k_{1,j} = \sqrt{2}\,[J_0'(\sqrt{\lambda_j})]^{-1}$$

This is as far as we can go without numerical computation. Notice how Mercer's expansion for this example gives

$$2 \sum_{j=1}^{\infty} \frac{J_0(\sqrt{\lambda_j}\, x) J_0(\sqrt{\lambda_j}\, \xi)}{\lambda_j J_0'^2(\sqrt{\lambda_j})} = \begin{cases} \ln \dfrac{1}{x}, & 0 \le \xi < x \\ \ln \dfrac{1}{\xi}, & x \le \xi \le 1 \end{cases}$$

*Example 4*

The example here frequently arises in the solution of boundary value problems connected with heat conduction or mass diffusion in spheres. Let

$$\mathcal{H} = \left\{\mathbf{u}\colon \int_0^1 x^2 u^2(x)\, dx < \infty;\ \langle \mathbf{u}, \mathbf{v} \rangle = \int_0^1 x^2 u(x) v(x)\, dx\right\}$$

Defined

$$\mathbf{T} = \{T, D(T)\} \quad \text{where} \quad T \equiv -\frac{1}{x_2}\frac{d}{dx}\left(x^2 \frac{d}{dx}\right)$$

and

$$D(T) = \{\mathbf{u} \in \mathcal{H};\ T\mathbf{u} \in \mathcal{H};\ u'(1) = -\beta u(1),\ u \text{ bounded at } x = 0 \text{ or } \lim_{x \to 0} x^2 u'(x) = 0\}$$

That $\mathbf{T}$ is self-adjoint is readily perceived. The eigenvalue problem $\mathbf{Tu} = \lambda\mathbf{u}$ yields

$$\frac{1}{x^2}\frac{d}{dx}\left(x^2 \frac{du}{dx}\right) + \lambda u = 0$$

By letting $xu = v$, it is easy to see that the differential equation above yields

$$\frac{d^2 v}{dx^2} + \lambda v = 0$$

whose linearly independent solutions are clearly $\sin\sqrt{\lambda}\, x$ and $\cos\sqrt{\lambda}\, x$. Thus we obtain $u_1(x, \lambda) = (1/x)\sin\sqrt{\lambda}\, x$ and $u_2(x, \lambda) = (1/x)\cos\sqrt{\lambda}\, x$. The boundedness condition at $x = 0$ eliminates $u_2$, so that

$$z(x, \lambda) = \frac{k_1}{x}\sin\sqrt{\lambda}\, x$$

The characteristic equation for the eigenvalues is obtained from the boundary condition at $x = 1$, which yields

$$\sin\sqrt{\lambda} - \sqrt{\lambda}\cos\sqrt{\lambda} = \beta\sin\sqrt{\lambda}$$

or

$$\tan\sqrt{\lambda} = \frac{\sqrt{\lambda}}{1-\beta}$$

which has infinitely many roots yielding positive values $\{\lambda_j\}$. The infinite nature of the number of eigenvalues is readily perceived by plotting $\tan\sqrt{\lambda}$ and $\sqrt{\lambda}/(1-\beta)$ versus $\sqrt{\lambda}$ and noting their intersections. The actual values are readily obtained on a digital computer. When $\beta = 1$, we obtain $\lambda_j = (2j-1)^2(\pi^2/4)$, $j = 1, 2, \ldots$ ; when $\beta > 1$, $\lambda_j > (2j-1)^2(\pi^2/4)$; and $\beta < 1 \rightarrow \lambda_j < (2j-1)^2(\pi^2/4)$. Eventually, however, for large enough $j$, regardless of the value of $\beta$ we have $\lambda_j \sim (2j-1)^2(\pi^2/4)$. The eigenvector

$$\mathbf{z}_j = \left\{\frac{k_{1,j}}{x}\sin\sqrt{\lambda_j}\,x\right\}$$

is normalized by letting

$$1 = \|\mathbf{z}_j\|^2 = k_{1,j}^2\int_0^1 \sin^2\sqrt{\lambda_j}\,x\,dx = \tfrac{1}{2}k_{1,j}^2\left(1 - \frac{\sin\sqrt{\lambda_j}\cos\sqrt{\lambda_j}}{\sqrt{\lambda_j}}\right)$$

$$= \tfrac{1}{2}k_{1,j}^2\left(1 - \frac{\cos^2\sqrt{\lambda_j}}{1-\beta}\right), \qquad \beta \neq 1$$

or

$$k_{1,j} = \sqrt{\frac{2(1-\beta)}{\sin^2\sqrt{\lambda_j} - \beta}}, \quad \beta \neq 1; \qquad k_{1,j} = \sqrt{2}, \quad \beta = 1$$

It is left for the student to go through Exercise 9.2.11, which asks for the inverse of **T** in this example. As in the previous examples, Mercer's expansion can be used to express the Green's function representing $\mathbf{T}^{-1}$ in terms of the eigenfunctions of **T**.

### Normalization of the Eigenvector

In the examples considered above, the eigenvectors were normalized by directly evaluating their norms. Here we show a method that does not require this procedure. Suppose that we consider the self-adjoint operator $\mathbf{T} = \{T, D(T)\}$, where $T$ is the operation (9.2.1) and whose $j$th eigenvalue is $\lambda_j$ with the corresponding eigenvector $\mathbf{z}_j \equiv \{z(x, \lambda_j)\}$. We further let $\mathbf{z}(\lambda) \equiv \{z(x, \lambda)\}$, which satisfies

$$Tz(x, \lambda) = \lambda z(x, \lambda)$$

and *one* of the homogeneous boundary conditions in $D(T)$. Then clearly

$\lim_{\lambda \to \lambda_j} \mathbf{z}(\lambda) = \mathbf{z}_j$. Now from (9.2.5) we have

$$\langle T\mathbf{z}(\lambda), \mathbf{z}_j \rangle - \langle \mathbf{z}(\lambda), \mathbf{T}\mathbf{z}_j \rangle = [\mathbf{w}^t(z_j(a)) \quad \mathbf{w}^t(z_j(b))] \begin{bmatrix} \mathbf{P}(a) & \mathbf{0} \\ \mathbf{0} & -\mathbf{P}(b) \end{bmatrix} \begin{bmatrix} \mathbf{w}(z(a, \lambda)) \\ \mathbf{w}(z(b, \lambda)) \end{bmatrix} \tag{9.2.60}$$

where $z(x, \lambda_j)$ has been abbreviated as $z_j(x)$. The left-hand side of the equality above is also given by $(\lambda - \lambda_j)\langle \mathbf{z}(\lambda), \mathbf{z}_j \rangle$, differentiating which with respect to $\lambda$ yields

$$\frac{d}{d\lambda}[(\lambda - \lambda_j)\langle \mathbf{z}(\lambda), \mathbf{z}_j \rangle] = \langle \mathbf{z}(\lambda), \mathbf{z}_j \rangle + (\lambda - \lambda_j)\langle \mathbf{z}_\lambda(\lambda), \mathbf{z}_j \rangle$$

where the subscript $\lambda$ is used to denote differentiation with respect to $\lambda$. Differentiating also the right-hand side of (9.2.60) and setting $\lambda = \lambda_j$, we obtain with the recognition that $\|\mathbf{z}_j\| = 1$,

$$1 = [\mathbf{w}^t(z_j(a)) \quad \mathbf{w}^t(z_j(b))] \begin{bmatrix} \mathbf{P}(a) & \mathbf{0} \\ \mathbf{0} & -\mathbf{P}(b) \end{bmatrix} \begin{bmatrix} \mathbf{w}(\mathbf{z}_\lambda(a, \lambda_j)) \\ \mathbf{w}(\mathbf{z}_\lambda(b, \lambda_j)) \end{bmatrix} \tag{9.2.61}$$

Equation (9.2.61) may be viewed as the normalization condition for the undetermined constant appearing in $\mathbf{z}_j$. From (9.2.49),

$$z(x, \lambda_j) = \mathbf{U}^t(x, \lambda_j)\mathbf{k}_j, \qquad z(x, \lambda) = \mathbf{U}^t(x, \lambda)\mathbf{k}(\lambda) \tag{9.2.62}$$

where $\mathbf{k}_j$ satisfies (9.2.53) and $\mathbf{k}(\lambda)$ is obtained by letting $z(x, \lambda)$ satisfy one of the homogeneous boundary conditions. Clearly, $\mathbf{k}_j = \mathbf{k}(\lambda_j)$. Furthermore, we readily see that

$$z_\lambda(x, \lambda_j) = \mathbf{U}^t(x, \lambda_j)\mathbf{k}_\lambda(\lambda_j) + \mathbf{U}^t_\lambda(x, \lambda_j)\mathbf{k}_j \tag{9.2.63}$$

It is now possible to substitute (9.2.62) and (9.2.63) into (9.2.61) and produce a formula $\mathbf{k}_j$ in order that $\|\mathbf{z}_j\| = 1$. The formula is uncouth, but its advantage is that no integration is involved. The entire procedure is not difficult in simple examples. Consider Example 3 of Section 9.2.2. Let us do the normalization by the method just presented. Accounting for the boundary condition at $x = 0$, we have

$$z(x, \lambda) = k_1 J_0(\sqrt{\lambda}\, x)$$

Notice in particular that $k_1$ does not depend on $\lambda$ in this example. Clearly, $z_\lambda(x, \lambda) = (k_1 x/2\sqrt{\lambda})J_0'(\sqrt{\lambda}\, x)$. Using primes for differentiating with respect to $x$, (9.2.61) is equivalent to

$$\begin{aligned} 1 &= [x\{z_j'(x)z_\lambda(x, \lambda_j) - z_j(x)z_\lambda'(x, \lambda_j)\}]_0^1 \\ &= [z_j'(1)z_\lambda(1, \lambda_j) - z_j(1)z_\lambda'(1, \lambda_j)] \\ &= \frac{k_{1,j}^2}{2}\left[\sqrt{\lambda_j}\, J_0'(\sqrt{\lambda_j})\frac{1}{\sqrt{\lambda_j}}J_0'(\sqrt{\lambda_j})\right] \\ &= \frac{k_{1,j}^2}{2}(J_0'\sqrt{\lambda_j})^2 \end{aligned}$$

so that $k_{1,j} = \sqrt{2}[J_0'(\sqrt{\lambda_j})]^{-1}$, which is the same result as that obtained by integration. In some situations direct integration may be difficult and the procedure described above is preferable.

## EXERCISES

**9.2.15** Compute the eigenvalues and eigenvectors of the operators below. In each case obtain the normalized eigenvectors by direct integration and compare with the method outlined above. Where the compact inverse exists, find the Green's function and identify Mercer's expansion.

(a) $T = \dfrac{d^2}{dx^2}$, $D(T) = \{\mathbf{u} \in \mathcal{L}_2[0, 1]; T\mathbf{u} \in \mathcal{L}_2[0, 1]; u'(0) = u'(1) = 0\}$.

(b) $T = \dfrac{d^2}{dx^2}$, $D(T) = \{\mathbf{u} \in \mathcal{L}_2[0, 1]; T\mathbf{u} \in \mathcal{L}_2[0, 1]; u(0) = u'(1) = 0\}$.

(c) Operator of Exercise 9.2.11.

(d) Operator of Exercise 9.2.12.

(e) $\mathcal{H} = \left\{\mathbf{u}: \int_a^b xu^2(x)\,dx < \infty; \langle \mathbf{u}, \mathbf{v}\rangle = \int_a^b xu(x)v(x)\,dx\right\}$

$$T = -\frac{1}{x}\frac{d}{dx}\left(x\frac{d}{dx}\right), \quad D(T) = \{\mathbf{u} \in \mathcal{H}; T\mathbf{u} \in \mathcal{H}; u(a) = u(b) = 0\}.$$

(f) $\mathcal{H} = \left\{\mathbf{u}: \int_a^b x^2u^2(x)\,dx < \infty; \langle \mathbf{u}, \mathbf{v}\rangle = \int_a^b x^2u(x)v(x)\,dx\right\}$

$$T = -\frac{1}{x^2}\frac{d}{dx}\left(x^2\frac{d}{dx}\right), \quad D(T) = \{\mathbf{u} \in \mathcal{H}; T\mathbf{u} \in \mathcal{H}; u'(a) = u'(b) = 0\}.$$

**9.2.16** In some cases end points $a$ and $b$ may *both* be regular singular points of the differential expression. Here no boundary conditions other than boundedness at the end points need be specified. To understand how eigenvalues may be computed in such situations, consider $\mathcal{H} = \mathcal{L}_2[-1, 1]$ and $\mathbf{T} = \{T, D(T)\}$, where

$$T = -\frac{d}{dx}\left[(x^2 - 1)\frac{d}{dx}\right]$$

$$D(T) = \{\mathbf{u} \in \mathcal{H}; T\mathbf{u} \in \mathcal{H}; u(\pm 1) < \infty\}$$

Apply the method of Frobenius (see Appendix A.3) to the eigenvalue problem and show that the boundedness conditions in $D(T)$ yield the eigenvalues $\lambda_j = j(j+1), j = 0, 1, 2, \ldots$. Note that the eigenvectors of $\mathbf{T}$, when normalized, are the so-called *Legendre polynomials*.

The student should take note of the interesting situation in Exercise 9.2.16. The Legendre polynomials arise as a special case of polynomial eigenfunctions of the more general *hypergeometric differential operators*. Other examples are *Tchebycheff polynomials* in the interval $[-1, 1]$, *Jacobi* polynomials in the interval $[0, 1]$, and so on, which form orthonormal bases in Hilbert spaces with different inner products. There are also examples of polynomials that arise as orthonormal bases in Hilbert spaces on *infinite* intervals. Thus the *Laguerre polynomials* belong to $\mathcal{L}_2\{(0, \infty): e^{-x}\}$, Hermite polynomials form an orthonormal basis in $\mathfrak{L}_2\{-\infty, \infty): e^{-x^2}\}$, and so on. For an excellent treatment of such polynomials, the reader is referred to Courant and Hilbert.† The operators of this chapter involved no infinite intervals. Indeed, we had pointed out that self-adjoint Sturm-Liouville differential operators on finite intervals would possess only discrete spectra. The infinite-interval cases are more complicated and the *possibility* of continuous spectra exists; however, that the existence of continuous spectra in such cases is not necessary follows from the fact that the hypergeometric differential operators, which give rise to Laguerre polynomials, Hermite polynomials, and so on, are defined on infinite domains and yet have only discrete spectra. More light on this issue is outside the scope of this book. The interested reader is referred to Titchmarsh.‡

### 9.2.3 Boundary Value Problems

We are now in a position to show how we may solve many boundary value problems that comprise the self-adjoint differential operators of this section. The method of solution is as elaborated in Section 7.5. Thus, once the self-adjoint operator is recognized and its spectrum identified, the solution of the problem is written immediately in terms of the self-adjoint projections. We will consider some examples of interest to engineers before which some general remarks are in order.

Suppose that $\mathbf{T} = \{T, D(T)\}$ is a self-adjoint operator, where $T$ is given by (9.2.1) with $D(T)$ as in (9.1.12) such that the self-adjointness constraint on the boundary condition either in the form of (9.2.7) or (9.2.8) is satisfied. We are interested in the solution of the boundary value problem in $\mathbf{v}(t) = \{v(x, t)\}$, which satisfies

$$-\frac{d}{dt}\mathbf{v}(t) = T\mathbf{v}(t) - \mathbf{f}(t), \qquad \mathbf{v}(0) = \mathbf{v}_0 \tag{9.2.64}$$

Further, $\mathbf{v}(t)$ satisfies the boundary conditions given by

$$\boldsymbol{\alpha}\mathbf{w}(v(a, t)) + \boldsymbol{\beta}\mathbf{w}(v(b, t)) = \boldsymbol{\gamma}(t) \tag{9.2.65}$$

where

$$\boldsymbol{\gamma}(t) \equiv \begin{bmatrix} \gamma_1(t) \\ \gamma_2(t) \end{bmatrix} \in \mathfrak{R}_2 \qquad \forall \ t$$

Clearly, $\mathbf{v}(t)$ does not belong to the domain of $T$. In the concluding part of Section 7.5 we had discussed a method of solving (9.2.64), provided that $\mathbf{v}(t)$

---

†*Methods of Mathematical Physics*, Vol. 1, by R. Courant and D. Hilbert, Interscience, 1953.

‡See, for example, *Eigenfunction Expansions* by E. C. Titchmarsh, Clarendon Press, Oxford, Part I, 2nd ed. 1962; Part II, 1958.

can be expressed as $\mathbf{u}(t) + \mathbf{g}(t)$ such that $\mathbf{u}(t) \in D(T)$ and $T\mathbf{g}(t) \in \mathcal{H}$. We had encountered conditions of the type (9.2.65) [see, e.g., (9.2.43)] in dealing with equations of the type (9.2.9). The transformation of variables (9.2.46) used there can also be employed here (see Exercise 9.2.18), but we will take a more general approach based on the solution (7.5.15) of Section 7.5. The solution (7.5.15) would seem to depend on the choice of $\mathbf{g}(t)$; we will show below that it actually does not. Since $\mathbf{v}(t) = \mathbf{u}(t) + \mathbf{g}(t)$ and $\mathbf{u}(t) \in D(T)$, we must have $\mathbf{g}(t)$ also satisfy (9.2.65). Now all vectors $\mathbf{g}(t)$ satisfying (9.2.65) and $T\mathbf{g}(t) \in \mathcal{H}$ [which include the unknown vector $\mathbf{v}(t)$] have an interesting common property. To realize this, it is necessary to return to (9.2.5), which we reproduce here for the pair of vectors $\mathbf{g}(t)$ and $\mathbf{z}$, with $\mathbf{z} \in D(T)$.

$$\langle T\mathbf{g}(t), \mathbf{z}\rangle - \langle \mathbf{g}(t), \mathbf{Tz}\rangle = [\mathbf{w}^t(z(a)) \quad \mathbf{w}^t(z(b))]\begin{bmatrix} -\mathbf{P}(a) & \mathbf{0} \\ \mathbf{0} & \mathbf{P}(b) \end{bmatrix}\begin{bmatrix} \mathbf{w}(g(a,t)) \\ \mathbf{w}(g(b,t)) \end{bmatrix} \tag{9.2.66}$$

Based on the same arguments that led to (9.1.15), we must have a vector $\mathbf{y}^t(\mathbf{z})$, depending *linearly* on $\mathbf{z}$, such that

$$[\mathbf{w}^t(z(a)) \quad \mathbf{w}^t(z(b))]\begin{bmatrix} -\mathbf{P}(a) & \mathbf{0} \\ \mathbf{0} & \mathbf{P}(b) \end{bmatrix} = \mathbf{y}^t(\mathbf{z})[\boldsymbol{\alpha} \quad \boldsymbol{\beta}]$$

which, together with the fact that $\mathbf{g}(t)$ satisfies (9.2.65), converts (9.2.66) into

$$\langle T\mathbf{g}(t), \mathbf{z}\rangle - \langle \mathbf{g}(t), \mathbf{Tz}\rangle = \mathbf{y}^t(\mathbf{z})\boldsymbol{\gamma}(t) \tag{9.2.67}$$

In order to identify $\mathbf{y}^t(\mathbf{z})$ for the most general situation, we must return to the development in Section 9.1, which led to the formula (9.1.20). Thus we had defined there a nonsingular $2 \times 2$ matrix $\boldsymbol{\delta}_{ij}$, and a $4 \times 2$ matrix $\boldsymbol{\gamma}^t_{ij}$, in terms of which it is possible to show that (see Exercise 9.2.17)

$$\mathbf{y}^t(\mathbf{z}) = \boldsymbol{\delta}_{ij}^{-1}[\mathbf{w}^t(z(a)) \quad \mathbf{w}^t(z(b))]\begin{bmatrix} -\mathbf{P}(a) & \mathbf{0} \\ \mathbf{0} & \mathbf{P}(b) \end{bmatrix}\boldsymbol{\gamma}^t_{ij}$$

so that $\mathbf{y}^t(\mathbf{z})$ is identified. Clearly, $\mathbf{y}^t(\mathbf{z})$ is independent of $\mathbf{g}(t)$, so that (9.2.67) is also independent of $\mathbf{g}(t)$. Moreover, $\mathbf{v}(t)$ also satisfies (9.2.67). The value of the formula above lies in its generality, but for relatively simple examples we suggest that the right-hand side of (9.2.67) be *directly* obtained by incorporating the inhomogeneous boundary condition (9.2.65) for $\mathbf{g}(t)$ into the formula (9.2.5). The examples presented in this section will establish the methodology more clearly. For the present we assume that $\mathbf{y}^t(\mathbf{z})$ is known. For the $j$th eigenvector $\mathbf{z}_j$ we may write from (9.2.67)

$$\langle T\mathbf{g}(t) - \lambda_j\mathbf{g}(t), \mathbf{z}_j\rangle = \mathbf{y}^t(\mathbf{z}_j)\boldsymbol{\gamma}(t)$$

from which

$$\mathbf{P}_j\{T\mathbf{g}(t) - \lambda_j\mathbf{g}(t)\} = \mathbf{y}^t(\mathbf{z}_j)\boldsymbol{\gamma}(t)\mathbf{z}_j$$

The solution (7.5.15) thus leads to the solution in its most general form for the boundary value problems of this section as

$$\mathbf{v}(t) = \sum_{j=1}^{\infty} e^{-\lambda_j t}\left[\mathbf{P}_j\mathbf{v}_0 + \int_0^t e^{\lambda_j t'}\{\mathbf{P}_j\mathbf{f}(t') - \mathbf{y}^t(\mathbf{z}_j)\boldsymbol{\gamma}(t')\mathbf{z}_j\}\, dt'\right] \tag{9.2.68}$$

We now present a more convenient solution of (9.2.64) subject to (9.2.65) for the case where $\mathbf{f}$ and $\boldsymbol{\gamma}$ do not depend on $t$. For this case we let

$$T\tilde{\mathbf{v}} = \mathbf{f}$$

$$\boldsymbol{\alpha}\mathbf{w}(\tilde{v}(a)) + \boldsymbol{\beta}\mathbf{w}(\tilde{v}(b)) = \boldsymbol{\gamma}$$

whose solution has been discussed in Section 9.2.1 [see (9.2.42) and (9.2.43) and the discussion following them]. By letting

$$u(x, t) = v(x, t) - \tilde{v}(x)$$

it is clear that $\mathbf{u}(t) \in D(T)$ and (9.2.64) yields

$$-\frac{d}{dt}\mathbf{u}(t) = \mathbf{T}\mathbf{u}(t), \qquad \mathbf{u}(0) = \mathbf{v}_0 - \tilde{\mathbf{v}}$$

whose solution is given by

$$\mathbf{v}(t) - \tilde{\mathbf{v}} = \mathbf{u}(t) = \sum_{j=1}^{\infty} e^{-\lambda_j t}\langle \mathbf{v}_0 - \tilde{\mathbf{v}}, \mathbf{z}_j\rangle \mathbf{z}_j \tag{9.2.69}$$

Note that directly using the solution (9.2.68) by substituting for $\mathbf{f}$ and $\boldsymbol{\gamma}$ is less efficient because it is equivalent to using (9.2.69) with $\tilde{\mathbf{v}}$ computed from its series expansion in terms of $\{\mathbf{z}_j\}$.

We now consider some examples.

*Example 1. Heat Conduction in a One-dimensional Slab*

We consider an unsteady-state heat conduction problem connected with a one-dimensional slab of solid material of thickness $l$ which has a volumetric heat source distributed in space and time, represented by a function $F(x', t')$, where $x'$ is the single spatial coordinate and $t'$ is time. The material has conductivity $k$, density $\rho$, and specific heat $C_p$; $\alpha \equiv k/\rho C_p$ represents the thermal diffusivity. The outer surface of the slab exchanges heat with the surroundings, which vary in temperature with time. The heat transfer coefficients at $x' = 0$ and $x' = l$ are $h_0$ and $h_l$, respectively. The environment temperatures at $x' = 0$ and $x' = l$ are $T_0(t')$ and $T_l(t')$, respectively. The equations governing the transfer of heat between the slab and its surroundings are readily written down with some background of transport phenomena.†

†See, for example, *Transport Phenomena*, by R. B. Bird, W. E. Stewart, and E. N. Lightfoot, Wiley, New York, 1960.

The differential equation for the transport of energy in the slab is given by

$$\frac{\partial T}{\partial t'} = \alpha \frac{\partial^2 T}{\partial x'^2} + \frac{1}{\rho C_p} F(x', t'), \qquad 0 < x' < l, \quad t' > 0 \tag{9.2.70}$$

The boundary conditions are

$$x' = 0, \qquad k\frac{\partial T}{\partial x'} = h_0[T - T_0(t')], \quad t' > 0 \tag{9.2.71}$$

$$x' = l, \qquad -k\frac{\partial T}{\partial x'} = h_l[T - T_l(t')], \quad t' > 0 \tag{9.2.72}$$

The initial temperature of the slab must be specified and it is stated as

$$T(x', 0) = T_i(x'), \qquad 0 \leq x' \leq l \tag{9.2.73}$$

It is convenient to work with nondimensional variables defined by

$$x \equiv \frac{x'}{l}, \qquad t \equiv \frac{\alpha t'}{l^2}, \qquad v(x, t) \equiv \frac{T - T_0(0)}{T_i(0) - T_0(0)}, \qquad \beta_0 \equiv \frac{h_0 l}{k}, \qquad \beta_1 \equiv \frac{h_l l}{k}$$

$$f(x, t) \equiv \frac{l^2 F(x', t')}{k[T_i(0) - T_0(0)]}, \qquad v_0(x) = \frac{T_i(x') - T_0(0)}{T_i(0) - T_0(0)}$$

$$\gamma_1(t) = \frac{T_0(t') - T_0(0)}{T_i(0) - T_0(0)}\beta_0, \qquad \gamma_2(t) = \frac{T_l(t') - T_0(0)}{T_i(0) - T_0(0)}\beta_1$$

In terms of the dimensionless variables, the differential equation (9.2.70), boundary conditions (9.2.71) and (9.2.72), and the initial condition (9.2.73) become, respectively,

$$-\frac{\partial v}{\partial t} = -\frac{\partial^2 v}{\partial x^2} - f(x, t), \qquad 0 < x < 1, \quad t > 0 \tag{9.2.74}$$

$$x = 0, \qquad \frac{\partial v}{\partial x} = \beta_0 v + \gamma_1(t), \quad \beta_0 > 0 \tag{9.2.75}$$

$$x = 1, \qquad -\frac{\partial v}{\partial x} = \beta_1 v + \gamma_2(t), \quad \beta_1 > 0 \tag{9.2.76}$$

$$v(x, 0) = v_0(x) \tag{9.2.77}$$

We recognize the operation $-d^2/dx^2$ as being of the formally self-adjoint type. The boundary conditions are unmixed but inhomogeneous. We have thus a situation which corresponds exactly to that of (9.2.64) and (9.2.65). More precisely, we have the Hilbert space $\mathcal{H} \equiv \mathcal{L}_2[0, 1]$ and the operator $\mathbf{T} = \{T, D(T)\}$, where

$$T \equiv -\frac{d^2}{dx^2}, \qquad D(T) = \{\mathbf{f} \in \mathcal{H};\, T\mathbf{f} \in \mathcal{H};\, f'(0) = \beta_0 f(0), f'(1) = -\beta_1 f(1)\}$$

Notice particularly how the domain $D(T)$ has been identified by dropping the inhomogeneous terms in (9.2.75) and (9.2.76). Because the boundary conditions are unmixed, $\mathbf{T}$ is clearly a self-adjoint operator. The solution to the boundary value problem is contained in (9.2.68) and all that remains is to identify the eigenvalues and eigenvectors of **T**. Using the methods of Section 9.2.2, we obtain the characteristic equation

$$\tan\sqrt{\lambda} = \frac{(\beta_1 + \beta_0)\sqrt{\lambda}}{\lambda - \beta_0\beta_1}$$

whose solution for the eigenvalues $\{\lambda_j\}$ is readily implemented on a digital computer. The eigenvalues are all positive since

$$\langle \mathbf{Tf}, \mathbf{f}\rangle = \beta_1 f(1)^2 + \beta_0 f(0)^2 + \int_0^1 f'^2(x)\, dx$$

is positive. The normalized eigenvector $\mathbf{z}_j$ is given by

$$z(x, \lambda_j) = k_{1,j}\left(\sin\sqrt{\lambda_j}\, x + \frac{\sqrt{\lambda_j}}{\beta_0}\cos\sqrt{\lambda_j}\, x\right)$$

where

$$k_{1,j}^{-2} = \frac{1}{2}\left(1 + \frac{\lambda_j}{\beta_0^2}\right) + \frac{(\beta_1 + \beta_0)\sqrt{\lambda_j}[(\lambda_j/\beta_0)(\beta_1 + \beta_0) - (1/2\sqrt{\lambda_j}) \times (1 - \lambda_j/\beta_0^2)(\lambda_j - \beta_0\beta_1)]}{[(\beta_1 + \beta_0)\sqrt{\lambda_j} + (\lambda_j - \beta_1\beta_0)]^2}$$

Other forms are also possible for $k_{1,j}$. Thus the solution can be obtained by substituting the eigenvalues and eigenvectors into (9.2.68). It remains to identify the term $\mathbf{y}^t(\mathbf{z}_j)\boldsymbol{\gamma}(t)$ on the right-hand side of (9.2.68). This is readily done here because for any function $\mathbf{g}(t)$ satisfying the inhomogeneous boundary conditions (9.2.75) and (9.2.76), we obtain

$$\langle T\mathbf{g}(t), \mathbf{z}_j\rangle - \langle \mathbf{g}(t), \mathbf{T}\mathbf{z}_j\rangle = z_j(1)\gamma_1(t) + z_j(0)\gamma_2(t) \tag{9.2.78}$$

Equation (9.2.78) identifies $\mathbf{y}^t(\mathbf{z}_j)$ as the row vector $[z_j(1) \quad z_j(0)]$. The solution to the boundary value problem is therefore completely known through (9.2.68).

If we consider the case of time-independent heat source $F(x')$ and constant temperatures $T_0$ and $T_l$ at $x' = 0$ and $x' = l$, then the solution to the problem may be written as (9.2.69), in which $\tilde{\mathbf{v}}$ is the "steady-state" temperature distribution

$$-\frac{d^2\tilde{\mathbf{v}}}{dx^2} = \mathbf{f}$$

$$-\tilde{v}'(0) + \beta_0\tilde{v}(0) = \gamma_1$$
$$\tilde{v}'(1) + \beta_1\tilde{v}(1) = \gamma_2$$

where $\gamma_1$ and $\gamma_2$ are constants. Using the methods of Section 9.2.1, the solution $\tilde{v}(x)$, as given by (9.2.46), is

$$\tilde{v}(x) = \int_0^1 G(x, \xi)f(\xi)\, d\xi + \frac{1}{\beta_0 + \beta_1 + \beta_0\beta_1} \times [\{1 + \beta_1(1 - x)\}\gamma_1 + (1 + \beta_0 x)\gamma_2]$$

the details of which are left as an exercise. The Green's function, again via the methods of Section 9.2.1, is found to be

$$G(x, \xi) = \begin{cases} g_1(\xi)g_2(x), & 0 \leq \xi \leq x \\ g_1(x)g_2(\xi), & x \leq \xi \leq 1 \end{cases}$$

with

$$g_1(x) \equiv \frac{(1 + \beta_1)(1 + \beta_0 x)}{\beta_0 + \beta_1 + \beta_0\beta_1}, \qquad g_2(x) = 1 - \frac{\beta_1}{1 + \beta_1} x$$

When $\tilde{v}(x)$ is substituted into (9.2.69), the solution to the "unsteady-state" heat conduction problem is completely determined.

The student is advised to study this example carefully since it is covered in some detail and lays out the basic strategy for the solution of boundary value problems. We will reiterate this strategy before passing on to the next example. Nondimensionalizing the variables is a useful procedure because it frees one from the deluge of physical properties that normally frequent physical equations and replaces them with fewer parameters. Further, the procedure sometimes directly leads to homogeneous boundary conditions in the dimensionless variable; indeed, other reasons for nondimensionalization are well known and require no special articulation here. In the examples to follow we will directly deal with equations in their dimensionless form.

Once the dimensionless equations are at hand, the first step is to identify the Hilbert space of interest. Frequently, this is obvious. In that the inner product may require some adjustments at times, this procedure may call for a certain amount of ingenuity. The identification of the self-adjoint operator consisting of the formal operation and its domain follows next. In the examples of this section, this is accomplished by eliminating the inhomogeneous terms in the boundary conditions. By using the methods of Section 9.2.2, the evaluation of the spectrum of the operator is completed. Once the eigenvalues and normalized eigenvectors are known, the solution is known from the methods of Section 7.5.

*Example 2. The Graetz Problem*†

This problem arises in the transfer of energy or mass in fully developed laminar flow through a tube. In terms of dimensionless variables with $t$ as the axial coordinate and $x$ as the transverse (or radial) coordinate, the differential equation for the dimensionless temperature or concentration $v(x, t)$

†See, for example, *Heat and Mass Transfer*, by E. R. G. Eckert and R. F. Drake, McGraw-Hill, New York, 1959.

may be written as

$$(1 - x^2)\frac{\partial v}{\partial t} - \frac{1}{x}\frac{\partial}{\partial x}\left(x\frac{\partial v}{\partial x}\right) = F(x, t), \qquad t > 0, \quad 0 < x < 1 \quad (9.2.79)$$

Equation (9.2.79) neglects axial molecular transport, an assumption that is valid for "high-Peclet-number flows." The boundary conditions arise from the entrance condition

$$v(x, 0) = v_0(x) \qquad (9.2.80)$$

and at the wall

$$v(1, t) = \phi(t), \qquad t > 0 \qquad (9.2.81)$$

which corresponds to specifying the wall temperature distribution. (A different problem arises when the flux is specified at the wall.) The other condition is one of axisymmetry at the center $x = 0$, which implies that

$$\frac{\partial v}{\partial x}(0, t) = 0, \qquad t > 0 \qquad (9.2.82)$$

although it will turn out that a simple boundedness condition at $x = 0$ will be adequate since axisymmetry is already implied by exclusion of terms corresponding to the tangential coordinate in the differential equation.

To solve the boundary value problem above, we rewrite (9.2.79) as

$$\frac{\partial v}{\partial t} = -\frac{1}{x(1 - x^2)}\frac{\partial}{\partial x}\left(x\frac{\partial v}{\partial x}\right) - f(x, t) \qquad (9.2.83)$$

where $f(x, t) \equiv [1/(1 - x^2)]F(x, t)$. We note that the operation $T$ given by

$$T \equiv -\frac{1}{x(1 - x^2)}\frac{d}{dx}\left(x\frac{d}{dx}\right)$$

is formally self-adjoint, provided that we define the Hilbert space of interest as

$$\mathcal{H} = \left\{\mathbf{f}\colon \int_0^1 x(1 - x^2)f^2(x)\,dx < \infty;\ \langle\mathbf{f}, \mathbf{g}\rangle \equiv \int_0^1 x(1 - x^2)f(x)g(x)\,dx\right\}$$

Next we define the operator $\mathbf{T} = \{T, D(T)\}$, where

$$D(T) = \{\mathbf{w} \in \mathcal{H};\ T\mathbf{w} \in \mathcal{H};\ w(1) = 0;\ w'(0) = 0 \text{ or } w \text{ bounded at } x = 0\}$$

Obviously, $\mathbf{T}$ is self-adjoint with eigenvalues $\{\lambda_j\}$ and eigenvectors $\{\mathbf{z}_j\}$ which must be determined. The boundary value problem is precisely of the form (9.2.64) subject to (9.2.65). The solution is therefore given by (9.2.68) in terms of $\{\lambda_j\}$ and $\{\mathbf{z}_j\}$. In this case, the term $\mathbf{y}^t(\mathbf{z}_j)\boldsymbol{\gamma}(t)$ on the right-hand side of (9.2.68) is readily found to be $z_j'(1)\phi(t)$. Alternatively, one could introduce a new variable $u(x, t) = v(x, t) - \phi(t)$, which will now satisfy homogeneous boundary conditions. This would require $\phi(t)$ to be differentiable with respect to $t$. To obtain the spectrum of the operator $\mathbf{T}$, we set $\mathbf{Tz} = \lambda\mathbf{z}$, which yields

$$\frac{d^2z}{dx^2} + \frac{1}{x}\frac{dz}{dx} + \lambda(1 - x^2)z = 0 \tag{9.2.84}$$

This differential equation can be solved by the method of Frobenius. However, this equation has been investigated thoroughly and its solution is expressed in terms of what are known as Kummer's functions.† These functions are discussed briefly in Section A.3 of the Appendix. Here we simply express the solution in terms of symbols. Thus we may write

$$z(x, \lambda) = ke^{-\sqrt{\lambda}x^2/2}\, {}_1F_1[\tfrac{1}{2} - \tfrac{1}{4}\sqrt{\lambda}\,; 1; \sqrt{\lambda}\, x^2] \tag{9.2.85}$$

which satisfies the boundedness condition at $x = 0$, while a second linearly independent solution has been eliminated because it is unbounded at $x = 0$. The quantity ${}_1F_1[\ \ ;\ \ ;\ \ ]$ is called Kummer's function and is elaborated on in Section A.3 of the Appendix. The characteristic equation for the eigenvalues is obtained by the homogeneous version of boundary condition (9.2.81). Thus

$${}_1F_1[\tfrac{1}{2} - \tfrac{1}{4}\sqrt{\lambda}\,; 1; \sqrt{\lambda}\,] = 0 \tag{9.2.86}$$

That the eigenvalues are all positive follows from the fact that

$$\langle \mathbf{Tw}, \mathbf{w} \rangle = \int_0^1 xw'^2(x)\, dx > 0$$

The characteristic equation (9.2.86) has been investigated for the eigenvalues $\{\lambda_j\}$.‡ The eigenvector $\mathbf{z}_j \equiv \{z(x, \lambda_j)\}$ is given by

$$z(x, \lambda_j) = k_j e^{-\sqrt{\lambda_j}x^2/2}\, {}_1F_1[\tfrac{1}{2} - \tfrac{1}{4}\sqrt{\lambda_j}; 1; \sqrt{\lambda_j}\, x^2]$$

where $k_j$ is evaluated by imposing that $\|\mathbf{z}_j\| = 1$. In this instance, the method of Section 9.2.2 becomes a handy tool, yielding

$$1 = \frac{\partial z}{\partial x}(x, \lambda_j)\big|_{x=1} \frac{\partial z}{\partial \lambda}(1, \lambda)\big|_{\lambda=\lambda_j}$$

which provides the required equation for $k_j$. We have thus obtained all the information required to implement the solution (9.2.68) to the Graetz problem.

*Example 3. Multicomponent Diffusion*

In Section 8.5 we dealt with multicomponent diffusion. It was established there that the matrix of diffusion coefficients **D** was a self-adjoint operator on the Hilbert space comprising $\mathcal{R}_n$ and the inner product (8.5.13). The strategy of solving multicomponent diffusion problems was to extend the solution of

---

†See, for example, *Confluent Hypergeometric Functions* by L. J. Slater, Cambridge University Press, London, 1960.

‡See, for example, W. Nusselt, *Z. Ver. Dtsch. Eng.*, **54**, 1154, 1910. See R. Siegel, E. M. Sparrow and T. M. Hallman, *Appl. Sci. Res.*, **A7**, 386, 1958, for eigenvalues when (9.2.81) is replaced by a value for the derivative.

a corresponding binary diffusion problem, represented abstractly by (8.5.10), as shown in (8.5.11). Equation (8.5.11) was further interpreted in the light of the spectral representation (8.5.14) of **D**. Thus the abstract solution to the multicomponent diffusion problem was contained in (8.5.16). The purpose of this example is simply to obtain, in the case of one-dimensional diffusion, an expression for the solution of the binary diffusion problem. The one-dimensional binary diffusion problem can, of course, be any of the boundary value problems in this subsection.

Suppose that an $n$-component mixture undergoes diffusion between two parallel planes $x = 0$ and $x = 1$. The concentration of the different components in the mixture are described by the vector

$$\mathbf{c}(x, t) = \begin{bmatrix} c_1(x, t) \\ c_2(x, t) \\ \cdot \\ \cdot \\ \cdot \\ c_n(x, t) \end{bmatrix} \in \mathcal{R}_n \qquad \forall \quad t, x$$

We may also represent the multicomponent concentration vector as an element of the $n$-fold direct sum space $\mathcal{H} \equiv \mathcal{L}_2[0, 1] \oplus \ldots \oplus \mathcal{L}_2[0, 1]$. Thus we may write the concentration vector as $\mathbf{c}(t) = \{\mathbf{c}(x, t)\}$, where the "components" are vectors in $\mathcal{R}_n$ for each fixed $x$ and $t$. Alternatively, we may write

$$\mathbf{c}(t) = \begin{bmatrix} \mathbf{c}_1(t) \\ \mathbf{c}_2(t) \\ \cdot \\ \cdot \\ \cdot \\ \mathbf{c}_n(t) \end{bmatrix} \qquad \mathbf{c}_i(t) \in \mathcal{L}_2[0, 1] \qquad \forall \quad t; \quad i = 1, 2, \ldots, n$$

The diffusion equation is given by

$$\mathbf{D}\frac{\partial^2}{\partial x^2}\mathbf{c}(x, t) = \frac{\partial}{\partial t}\mathbf{c}(x, t), \qquad 0 < x < 1, \quad t > 0 \tag{9.2.87}$$

The boundary conditions at $x = 0$ and $x = 1$ are assumed to be

$$\frac{\partial}{\partial x}\mathbf{c}(0, t) = \boldsymbol{\beta}_0[\mathbf{c}(0, t) - \boldsymbol{\gamma}_1(t)] \tag{9.2.88}$$

$$-\frac{\partial}{\partial x}\mathbf{c}(1, t) = \boldsymbol{\beta}_1[\mathbf{c}(1, t) - \boldsymbol{\gamma}_2(t)] \tag{9.2.89}$$

where $\boldsymbol{\gamma}_1(t), \boldsymbol{\gamma}_2(t) \in \mathcal{R}_n \ \forall\ t$, and $\boldsymbol{\beta}_0$ and $\boldsymbol{\beta}_1$ are constant $(n \times n)$ matrices. Notice that this boundary value problem is the multicomponent mass transfer version of Example 1 of this section. The boundary conditions also introduce coupling between the species. For the method of Section 8.5 to be applicable, it is essential to assume that $\boldsymbol{\beta}_0$ and $\boldsymbol{\beta}_1$ may be expressed as functions of the diffusion coefficient matrix **D**.† Thus

$$\boldsymbol{\beta}_0 = \boldsymbol{\phi}_0(\mathbf{D}), \qquad \boldsymbol{\beta}_1 = \boldsymbol{\phi}_1(\mathbf{D}) \tag{9.2.90}$$

where $\phi_0(\cdot)$ and $\phi_1(\cdot)$ are scalars if their arguments are scalars. The initial condition is represented by

$$\mathbf{c}(x, 0) = \mathbf{c}_0(x) \tag{9.2.91}$$

Again as a vector in $\mathcal{H}$, the initial concentrations are represented by $\mathbf{c}_0 = \{\mathbf{c}_0(x)\}$. The methodology of Section 8.5 requires that we solve the problem

$$\begin{aligned} &\mu \frac{\partial^2 v}{\partial x^2} = \frac{\partial v}{\partial t}, \qquad 0 < x < 1, \quad t > 0 \\ &\frac{\partial}{\partial x} v(0, t) = \phi_0(\mu)[v(0, t) - \gamma_1(t)] \\ &-\frac{\partial}{\partial x} v(1, t) = \phi_1(\mu)[v(1, t) - \gamma_2(t)] \\ &v(x, 0) = v_0(x) \end{aligned} \tag{9.2.92}$$

where $\mu$ is a real parameter. The operator of interest may be represented as $\mathbf{T}_\mu = \{T_\mu, D(T_\mu)\}$ with

$$T_\mu = -\mu \frac{d^2}{dx^2}, \qquad D(T_\mu) = \{\mathbf{u} \in \mathcal{L}_2[0, 1]; T_\mu \mathbf{u} \in \mathcal{L}_2[0, 1]; u'(0) = \phi_0(\mu)u(0), \; -u'(1) = \phi_1(\mu)u(1)\}$$

Obviously, $\mathbf{T}_\mu$ is self-adjoint and positive definite for each $\mu$, so that it has positive eigenvalues $\{\lambda_j(\mu)\}$ and corresponding eigenvectors $\{\mathbf{z}_j(\mu)\}$. Note in particular that the eigenvalues and the eigenvectors depend on the parameter $\mu$. Denoting the $j$th self-adjoint projection by $\mathbf{P}_j(\mu)$, the solution to the boundary value problem (9.2.92) may be written as

$$\mathbf{v}(t, \mu) = \sum_{j=1}^{\infty} e^{-\lambda_j(\mu)t} \Big[ \mathbf{P}_j(\mu)\mathbf{v}_0 + \int_0^t e^{\lambda_j(\mu)t'} \{z_j(0, \mu)\phi_0(\mu)\gamma_1(t') + z_j(1, \mu)\phi_1(\mu)\gamma_2(t')\}\mathbf{z}_j(\mu)\, dt' \Big]$$

We let the spectral representation of **D** be given by

$$\mathbf{D} = \sum_{k=1}^{n} \mu_k \boldsymbol{\pi}_k \tag{9.2.94}$$

where $\mu_k$ is the $k$th eigenvalue of **D** and $\boldsymbol{\pi}_k$ the corresponding self-adjoint

---

†This assumption is tenable only when the diffusion coefficients in the domain and in the ambient bear a constant ratio to each other.

projection. Following the methods of Section 8.5, the solution to the multicomponent diffusion problem is written as

$$\mathbf{c}(t) = \sum_{k=1}^{n}\sum_{j=1}^{\infty} e^{-\lambda_j(\mu_k)t}\Big[\boldsymbol{\pi}_k\{\mathbf{P}_j(\mu_k)\mathbf{c}_0\}$$

$$+ \int_0^t e^{\lambda_j(\mu_k)t'}\{z_j(0, \mu_k)\phi_0(\mu_k)\boldsymbol{\pi}_k\boldsymbol{\gamma}_1(t') + z_j(1, \mu_k)\phi_1(\mu_k)\boldsymbol{\pi}_k\boldsymbol{\gamma}_2(t')\}\mathbf{z}_j(\mu_k)\, dt'\Big]$$

In interpreting the solution above, one must recognize that for each $j$ and $k$, $\mathbf{P}_j(\mu_k)\mathbf{c}_0$ is a vector in $\mathcal{R}_n$ obtained by applying $\mathbf{P}_j(\mu_k)$ to each of the components of $\mathbf{c}_0$, which are vectors in $\mathcal{L}_2[0, 1]$. The problem is thus solved, its numerical evaluation depending on the calculation of first the eigenvalues $\mu_1, \mu_2, \ldots, \mu_k$ of $\mathbf{D}$ and its corresponding eigenvectors, next computing, for each of the $\mu_k$'s, the eigenvalues $\{\lambda_j(\mu_k)\}$ of the operator $\mathbf{T}_{\mu_k}$ and the corresponding eigenvectors $\{\mathbf{z}_j(\mu_k)\}$.

For a somewhat different but equivalent approach to the problem, the student is referred to Exercise 9.2.20. Exercises 9.2.21 and 9.2.22, which combine diffusion with chemical reaction, are also of interest in this regard.

## EXERCISES

**9.2.17** From the development in Section 9.1, show that the vector $\mathbf{y}^t(\mathbf{z})$ in (9.2.67) is given by

$$\mathbf{y}^t(z) = \boldsymbol{\delta}_{ij}^{-1}[\mathbf{w}^t(z(a)) \quad \mathbf{w}^t(z(b))]\begin{bmatrix} -\mathbf{P}(a) & \mathbf{0} \\ \mathbf{0} & \mathbf{P}(b) \end{bmatrix}\boldsymbol{\gamma}_{ij}^t$$

where $\boldsymbol{\delta}_{ij} = [\boldsymbol{\alpha} \quad \boldsymbol{\beta}]\boldsymbol{\gamma}_{ij}^t$; $\boldsymbol{\gamma}_{ij}^t$ is as defined in Section 9.1.

**9.2.18** Consider the boundary-initial value problem given by (9.2.64) and (9.2.65). Let $\lambda_0$ be a real number for which the matrix $\mathbf{D}_{\lambda_0}$, defined by (9.2.51), is nonsingular. (How do you know that such a $\lambda_0$ exists?) Show that the transformed variable $u(x, t) = v(x, t) - \mathbf{U}^t(x, \lambda_0)\mathbf{D}_{\lambda_0}^{-1}\boldsymbol{\gamma}(t)$ satisfies the homogeneous boundary conditions, that is, $\mathbf{u}(t) \in D(T)$.

**9.2.19** Extend the analysis of Example 1 of this section to unsteady-state heat transfer in concentric cylinders and spheres; that is, solve the boundary initial value problem

$$-\frac{\partial v}{\partial t} = -\frac{1}{x^m}\frac{\partial}{\partial x}\left(x^m\frac{\partial v}{\partial t}\right) - f(x, t), \qquad a < x < b, \quad t > 0$$

$$x = a, \qquad \frac{\partial v}{\partial x} = \beta_0 v + \gamma_1(t)$$

$$x = b, \qquad -\frac{\partial v}{\partial x} = \beta_1 v + \gamma_2(t)$$

$$v(x, 0) = v_0(x).$$

**9.2.20** Consider again the multicomponent diffusion system of Example 3 in this section. Let $\mathcal{H}$ be the $n$-fold direct sum space $\mathcal{L}_2[0, 1] \oplus \ldots \oplus \mathcal{L}_2[0, 1]$, in which a vector $\mathbf{u}$ is represented by $\mathbf{u} = \{\mathbf{u}(x)\}$ or by

$$\mathbf{u} \equiv \begin{bmatrix} \mathbf{u}_1 \\ \mathbf{u}_2 \\ \cdot \\ \cdot \\ \cdot \\ \mathbf{u}_n \end{bmatrix}, \qquad \mathbf{u}_i \in \mathcal{L}_2[0, 1], \quad i = 1, 2, \ldots, n$$

The inner product on $\mathcal{H}$ is defined by

$$\langle \mathbf{u}, \mathbf{v} \rangle = \sum_{i=1}^{n} (\mathbf{u}_i, \mathbf{v}_i)$$

where $(\cdot, \cdot)$ is the inner product on $\mathcal{L}_2[0, 1]$. Define the differential operator $\mathbf{T} = \{T, D(T)\}$, where

$$T = -\mathbf{D}\frac{d^2}{dx^2}$$

$$D(T) = \{\mathbf{u} \in \mathcal{H}; T\mathbf{u} \in \mathcal{H}; u'(0) = \beta_0 u(0), -u'(1) = \beta_1 u(1)\}$$

with $\beta_0$ and $\beta_1$ as stipulated in (9.2.90).

(a) Show that $\mathbf{T}$ is self-adjoint in $\mathcal{H}$.

(b) To obtain the spectral representation of $\mathbf{T}$, solve the eigenvalue problem $\mathbf{Tz} = \lambda\mathbf{z}$ by a standard matrix diagonalization procedure or otherwise through to its characteristic equation.

(c) Show that the solution to the boundary-initial value problem (9.2.87)–(9.2.91) is the same as (9.2.95).

**9.2.21** Consider the first-order reaction mixture of Section 8.5 diffusing under unsteady-state conditions thorough a porous catalyst medium with $\alpha$ as the catalytic surface per unit volume. Assume that the diffusion is one-dimensional, occurring between parallel planes between $x = 0$ and $x = 1$ (the problem of radial diffusion in cylinders and spheres may be treated with only minor changes). Further, $\mathbf{D}$ may be taken to be *diagonal* and the reaction rate constant matrix $\hat{\mathbf{K}}$ may be recalled from Section 8.5. Unsteady-state diffusion is then described by

$$\mathbf{D}\frac{\partial^2}{\partial x^2}\mathbf{c}(x, t) + \alpha\hat{\mathbf{K}}\mathbf{c}(x, t) = \frac{\partial}{\partial t}\mathbf{c}(x, t), \qquad 0 < x < 1, \quad t > 0$$

where $\mathbf{c}(x, t)$ is as defined in Example 3 of this section. The boundary conditions at $x = 0$ and $x = 1$ may be taken as (9.2.88) and (9.2.89), while the initial condition is given by (9.2.91). Let $\mathcal{H}$ be the $n$-fold direct sum space $\mathcal{L}_2[0, 1] \oplus \ldots \oplus \mathcal{L}_2[0, 1]$. Denoting the vectors in $\mathcal{H}$ as in Exercise 9.2.19, define the inner product as

$$\langle \mathbf{u}, \mathbf{v} \rangle = \sum_{i=1}^{n} \frac{1}{a_i}(\mathbf{u}_i, \mathbf{v}_i)$$

where $(\cdot, \cdot)$ is the inner product in $\mathcal{L}_2[0, 1]$ and $a_1, a_2, \ldots, a_n$ are the equilibrium mole fractions of Section 8.5.

Define the differential operator $\mathbf{T} = \{T, D(T)\}$ by $T = -\mathbf{D}(d^2/dx^2) - \alpha\hat{\mathbf{K}}$ and $D(T)$ as in Exercise 9.2.17.

(a) Show that $\mathbf{T}$ is self-adjoint and positive definite.

(b) Suggest a method of numerically solving the eigenvalue problem.

(c) Present the solution to the boundary-initial value problem in terms of the spectral resolution of $\mathbf{T}$.

**9.2.22** Suppose that the reaction mixture of Exercise 9.2.2 is contained between two catalytic surfaces at $x = \pm 1$. Pure diffusion occurs in the bulk, while reaction occurs at the surface. Set up and solve the unsteady-state problem using an operator formalism.

### 9.2.4 Convergence of Eigenfunction Expansions in $\mathcal{L}_2\{[a, b] : r(x)\}$

We have seen from the spectral theorem in Section 7.4 that the eigenvectors $\{\mathbf{z}_j\}$ of a self-adjoint operator $\mathbf{T}$ defined on $\mathcal{H}$ are an orthonormal basis in $\mathcal{H}$. The expansion of an arbitrary element $\mathbf{f} \in \mathcal{H}$ in terms of $\{\mathbf{z}_j\}$ converges to $\mathbf{f}$ with respect to the metric generated by the inner product in $\mathcal{H}$. In this section many boundary value problems involving self-adjoint differential operators have been solved as expansions in terms of eigenvectors. The Hilbert space of interest is $\mathcal{H} \equiv \mathcal{L}_2\{[a, b]: r(x)\}$ and it is important, especially from the point of view of applications, to ask if the eigenvector expansion converges at specific values of $x \in [a, b]$ to the value of the function (which is expanded) at $x$. We address this problem by first proving the following theorem.

***Theorem 9.2.2*** Let $\mathbf{T}$ be the general self-adjoint differential operator of this section with $\mathbf{T}$ given by (9.2.1) and domain $D(T)$. Further, let $\{\lambda_j\}$ and $\{\mathbf{z}_j\}$ be the eigenvalues and eigenvectors of $\mathbf{T}$, respectively. Then the linear manifold spanned by $\{\mathbf{z}_j\}$ is dense in $D(T)$ with respect to the norm $\|\cdot\|_\infty$.

*Proof*: We already know from the spectral theorem that $L(\mathbf{z}_1, \mathbf{z}_2, \ldots)$ is dense in $\mathcal{H}$ [and hence in $D(T)$] with respect to $\|\cdot\|_2$). Let $\mathbf{T}^{-1}$ exist and $G(x, \xi)$ be the Green's function. Then $G$ is continuous on $[a, b] \times [a, b]$ and denote $M \equiv \sup_{x,\xi\in[a,b]} G(x, \xi)\sqrt{r(\xi)}$. For any $\mathbf{u} \in \mathcal{H}$,

$$\left|\int_a^b r(\xi)G(x, \xi)u(\xi)\, d\xi\right| \le \int_a^b M\sqrt{r(\xi)}u(\xi)\, d\xi$$
$$\le M(b - a)^{1/2}\|u\|_2 \qquad (9.2.96)$$

the last following from the Schwarz inequality. Suppose that $m$ and $n$ are integers such that $m > n$. Now

$$\mathbf{T}^{-1}\sum_{j=n}^{m}\langle\mathbf{u}, \mathbf{z}_j\rangle\mathbf{z}_j = \sum_{j=n}^{m}\frac{1}{\lambda_j}\langle\mathbf{u}, \mathbf{z}_j\rangle\mathbf{z}_j$$

From (9.2.95)

$$\left|\sum_{j=n}^{m}\frac{1}{\lambda_j}\langle\mathbf{u},\mathbf{z}_j\rangle\mathbf{z}_j(x)\right| \leq M(b-a)^{1/2}\left\{\sum_{j=n}^{m}|\langle\mathbf{u},\mathbf{z}_j\rangle|^2\right\}^{1/2}$$

The Bessel inequality then implies that $\sum_{j=1}^{\infty}(1/\lambda_j)\langle\mathbf{u},\mathbf{z}_j\rangle z_j(x)$ converges uniformly since the right-hand side of the last inequality can be made arbitrarily small independently of $x$. However, from the spectral theorem,

$$\mathbf{T}^{-1}\mathbf{u} = \sum_{j=1}^{\infty}\langle\mathbf{T}^{-1}\mathbf{u},\mathbf{z}_j\rangle\mathbf{z}_j = \sum_{j=1}^{\infty}\frac{1}{\lambda_j}\langle\mathbf{u},\mathbf{z}_j\rangle\mathbf{z}_j$$

Since $\mathbf{T}^{-1}\mathbf{u} \in D(T)$, it is also continuous in $[a, b]$, so that $\sum_{j=1}^{\infty}(1/\lambda_j)\langle\mathbf{u}, \mathbf{z}_j\rangle\mathbf{z}_j$ converges uniformly to $\mathbf{T}^{-1}\mathbf{u}$ for every $\mathbf{u} \in \mathcal{H}$. To prove that $L(\mathbf{z}_1, \mathbf{z}_2, \ldots)$ is dense in $D(T)$ with respect to $\|\cdot\|_\infty$, let $\mathbf{f} \in D(T)$ and $\mathbf{u} = \mathbf{Tf}$. Then $\mathbf{f} = \mathbf{T}^{-1}\mathbf{u}$. We have just shown that $\sum_{j=1}^{\infty}(1/\lambda_j)\langle\mathbf{u},\mathbf{z}_j\rangle\mathbf{z}_j$ converges uniformly to $\mathbf{T}^{-1}\mathbf{u}$, which is $\mathbf{f}$. Because

$$\sum_{j=1}^{\infty}\frac{1}{\lambda_j}\langle\mathbf{u},\mathbf{z}_j\rangle\mathbf{z}_j = \sum_{j=1}^{\infty}\langle\mathbf{T}^{-1}\mathbf{u},\mathbf{z}_j\rangle\mathbf{z}_j = \sum_{j=1}^{\infty}\langle\mathbf{f},\mathbf{z}_j\rangle\mathbf{z}_j$$

we have shown that for any $\mathbf{f} \in D(T)$, its eigenfunction expansion *converges uniformly* to $\mathbf{f}$, which is the same as $L(\mathbf{z}_1, \mathbf{z}_2, \ldots)$ being dense in $D(T)$ with respect to $\|\cdot\|_\infty$. ▮

The conclusion of Theorem 9.2.2 is important because it implies that once we grant the existence of a unique solution $\mathbf{u}(t)$† to initial-boundary value problems of Section 9.2.3 in which $\mathbf{u}(t) \in D(T)$ for each $t$, its eigenfunction expansion *converges uniformly* with respect to $x$ at all times.

In Section 9.2.3 we also dealt with initial-boundary value problems in which the solution $\mathbf{v}(t)$ did not belong to $D(T)$ because of inhomogeneities appearing in the boundary conditions.‡ In this case Theorem 9.2.2 is inapplicable. We state here without proof a more general theorem.§

---

†The problem of existence of a unique solution to initial-boundary problems is a different issue which is not covered here. For such issues, see *Lectures on Partial Differential Equations*, by I. G. Petrovsky, Interscience, New York, 1954.

‡This situation may appear paradoxical because $\mathbf{v}(t)$, which satisfies *inhomogenous* boundary conditions, is obtained as an expansion in terms of eigenfunctions satisfying *homogeneous* boundary conditions. The paradox would disappear when it is recognized that the limit of any series as $x$ approaches a boundary point cannot be evaluated by taking the limit inside the summation. A sufficient condition for interchanging the limit and the sum is that the series converge uniformly in the closed interval.

§See *Methods of Mathematical Physics*, by R. Courant and D. Hilbert, Vol. 1, Interscience, New York, 1953, p. 427.

***Theorem 9.2.3*** Let **f** be absolutely continuous on $[a, b]$. Then the eigenfunction expansion (referred to in Theorem 9.2.2) of **f** converges uniformly in all subdomains free of points of discontinuity. At the points of discontinuity the series converges to the arithmetic mean of the left- and right-hand limits.

The foregoing theorem implies that the eigenfunction expansion of a function will converge uniformly in intervals in which the function is continuous even if the function does not satisfy the boundary conditions. Thus the spectral expansion (9.2.68) of $\mathbf{v}(t)$, the solution to the boundary value problem (to which we had referred just before stating Theorem 9.2.3), will converge uniformly in intervals *not* containing the boundary points, provided that the inhomogeneities [such as $\mathbf{f}(t)$, $\boldsymbol{\gamma}(t)$, etc.] satisfy suitable constraints under which the existence and uniqueness proofs for $\mathbf{v}(t)$ are available (see footnote on previous page). In intervals containing the boundary point the expansion can be shown to satisfy the inhomogeneous boundary condition only in the limiting sense as the boundary point is approached.

Considerable caution is advised on differentiating spectral solutions for computing the derivatives of the solution. Unless the formally differentiated series is uniformly convergent, the derivative of the solutions cannot be computed by differentiating the series (see Exercise 9.2.23).

## EXERCISES

**9.2.23** Solve

$$\frac{\partial^2 v}{\partial x^2} = \frac{\partial v}{\partial t}, \qquad 0 < x < 1, \quad t > 0$$

$$v(x, 0) = 0$$

$$v(x, 0) = 0$$

$$v(0, t) = 1$$

$$v(1, t) = 0$$

first by using (9.2.68) as the solution, and second by using (9.2.69). Suppose that $\partial v(0, t)/\partial x$ is required. Which of the solutions above can be differentiated?

**9.2.24** Show *rigorously* that

$$u(x, t) = 2 \sum_{j=1}^{\infty} \int_0^1 u_0(\xi) \sin j\pi\xi \, d\xi \sin j\pi x \, e^{-j^2\pi^2 t}$$

satisfies

$$\frac{\partial^2 u}{\partial x^2} = \frac{\partial u}{\partial t}$$

$$u(x, 0) = u_0(x)$$

$$u(0, t) = u(1, t) = 0$$

## 9.3 Self-Adjoint Differential Operators: Discontinuous Coefficients

In this section we focus again on the formally self-adjoint operation $T$ as given by (9.2.1):

$$T \equiv -\frac{1}{r(x)}\frac{d}{dx}\left[p(x)\frac{d}{dx}\right] + q(x) \tag{9.3.1}$$

where we now entertain discontinuities in the function $p(x)$ in the interval $[a, b]$. There may be a finite number of points at which $p(x)$ may be discontinuous, but for the purposes of our development we need only one such point of discontinuity. Accordingly, we assume that $p(x)$ is discontinuous at some point $x_1 \in (a, b)$; that is, if $p(x_1-)$ and $p(x_1+)$ are the left- and right-hand limits, respectively, we have $p(x_1-) \neq p(x_1+)$. As before, $p(x)$ is a positive function and in particular $p(x_1-)$ and $p(x_1+)$ are both positive.

The Hilbert space of interest is again $\mathcal{H} \equiv \mathcal{L}_2\{[a, b]: r(x)\}$, in which the inner product is (9.1.8). In view of the discontinuity of $p(x)$ at $x = x_1$, the formula (9.2.5) must be replaced by

$$\begin{aligned}\langle T\mathbf{u}, \mathbf{v}\rangle - \langle \mathbf{u}, T\mathbf{v}\rangle &= [\mathbf{w}^t(v(a)) \quad \mathbf{w}^t(v(b))]\begin{bmatrix} -\mathbf{P}(a) & \mathbf{0} \\ \mathbf{0} & \mathbf{P}(b)\end{bmatrix}\begin{bmatrix}\mathbf{w}(u(a)) \\ \mathbf{w}(u(b))\end{bmatrix} \\ &\quad - [\mathbf{w}^t(v(x_1-)) \quad \mathbf{w}^t(v(x_1+)]\begin{bmatrix} -\mathbf{P}(x_1-) & \mathbf{0} \\ \mathbf{0} & \mathbf{P}(x_1+)\end{bmatrix}\begin{bmatrix}\mathbf{w}(u(x_1-)) \\ \mathbf{w}(u(x_1+))\end{bmatrix}\end{aligned} \tag{9.3.2}$$

where $\mathbf{u}, \mathbf{v} \in \mathcal{H}$ and $T\mathbf{u}, T\mathbf{v} \in \mathcal{H}$. Our objective is to formulate domains $D(T)$ such that $\mathbf{T} = \{T, D(T)\}$ is self-adjoint. If the boundary conditions at $a$ and $b$ are described by (9.1.11) such that (9.2.7) or, more generally, (9.2.8) is true, then the first term on the right-hand side of (9.3.2) vanishes. In order for the second term on the right-hand side of (9.3.2) to vanish, contraints must be imposed on $\mathbf{u}$ and $\mathbf{v}$ at $x = x_1$; for a self-adjoint operator the constraints on $\mathbf{u}$ and $\mathbf{v}$ must be the same. Since the differential operation is of second order, two linearly independent constraints must be found. The most general representation of such constraints, referred to as *discontinuity* conditions, is accomplished by

$$[\boldsymbol{\alpha}_1^- \quad \boldsymbol{\alpha}_1^+]\begin{bmatrix}\mathbf{w}(u(x_1-)) \\ \mathbf{w}(u(x_1+))\end{bmatrix} = 0 \tag{9.3.3}$$

where $\boldsymbol{\alpha}_1^-$ and $\boldsymbol{\alpha}_1^+$ are $2 \times 2$ matrices and $[\boldsymbol{\alpha}_1^- \quad \boldsymbol{\alpha}_1^+]$ is a $2 \times 4$ matrix of rank 2. Equation (9.3.3) is identical to the formulation of the boundary conditions (9.1.11). Obviously, the arguments for self-adjointness leading to (9.2.7) apply also to (9.3.3), and we must therefore conclude that

$$\boldsymbol{\alpha}_1^+\mathbf{P}^{-1}(x_1+)\boldsymbol{\alpha}_1^{+t} = \boldsymbol{\alpha}_1^-\mathbf{P}^{-1}(x_1^-)\boldsymbol{\alpha}_1^{-t} \tag{9.3.4}$$

which is *necessary and sufficient* for the second term on the right-hand side of (9.3.2) to vanish. To summarize, the differential operator $\mathbf{T} = \{T, D(T)\}$, where $T$ is given by (9.3.1) and $D(T)$ is defined as

$$D(T)=\left\{\mathbf{u}\in\mathcal{H};\ T\mathbf{u}\in\mathcal{H};\ [\boldsymbol{\alpha}\quad\boldsymbol{\beta}]\begin{bmatrix}\mathbf{w}(u(a))\\ \mathbf{w}(u(b))\end{bmatrix}=\mathbf{0},\ [\boldsymbol{\alpha}_1^-\quad\boldsymbol{\alpha}_1^+]\begin{bmatrix}\mathbf{w}(u(x_1-))\\ \mathbf{w}(u(x_1+))\end{bmatrix}=\mathbf{0}\right\}$$

is self-adjoint *if and only if* (9.2.7) [or (9.2.8)] *and* (9.3.4) are both valid. Before we consider some special domains of interest to applications, it is useful to indicate a modification in the foregoing development when a slight variation is introduced in the inner product in $\mathcal{H}$. Suppose that we define the inner product as

$$\langle\mathbf{u},\mathbf{v}\rangle=\int_a^{x_1} r(x)u(x)v(x)\,dx+\delta_1\int_{x_1}^{b} r(x)u(x)v(x)\,dx \tag{9.3.5}$$

where $\delta_1$ is a positive number. In terms of this inner product, (9.3.2) becomes

$$\langle T\mathbf{u},\mathbf{v}\rangle-\langle\mathbf{u},T\mathbf{v}\rangle=[\mathbf{w}^t(v(a))\quad\mathbf{w}^t(v(b))]\begin{bmatrix}-\mathbf{P}(a) & \mathbf{0}\\ \mathbf{0} & \delta_1\mathbf{P}(b)\end{bmatrix}\begin{bmatrix}\mathbf{w}(u(a))\\ \mathbf{w}(u(b))\end{bmatrix}$$

$$-[\mathbf{w}^t(v(x_1-))\quad\mathbf{w}^t(v(x_1+))]\begin{bmatrix}-\mathbf{P}(x_1-) & \mathbf{0}\\ \mathbf{0} & \delta_1\mathbf{P}(x_1+)\end{bmatrix}\begin{bmatrix}\mathbf{w}(u(x_1-))\\ \mathbf{w}(u(x_1+))\end{bmatrix} \tag{9.3.6}$$

The self-adjointness criteria (9.2.7) and (9.3.4) will then modify to

$$\boldsymbol{\alpha}\mathbf{P}^{-1}(a)\boldsymbol{\alpha}^t=\boldsymbol{\beta}\delta_1^{-1}\mathbf{P}^{-1}(b)\boldsymbol{\beta}^t \tag{9.3.7}$$

and

$$\boldsymbol{\alpha}_1^+\delta_1^{-1}\mathbf{P}^{-1}(x_1+)\boldsymbol{\alpha}_1^{+t}=\boldsymbol{\alpha}_1^-\mathbf{P}^{-1}(x_1-)\boldsymbol{\alpha}_1^{-t} \tag{9.3.8}$$

respectively. We now prove a couple of theorems that are very useful for applications.

***Theorem 9.3.1*** Let $\mathcal{H}\equiv\mathcal{L}_2\{[a,b]\colon r(x)\}$ with the regular inner product (9.1.8), and $T$ be the formally self-adjoint operation (9.3.1) with $p(x)$ discontinuous at $x_1$. Further, let $D(T)$ be the subspace of elements $\mathbf{u}\in\mathcal{H}$ such that $T\mathbf{u}\in\mathcal{H}$ satisfy self-adjoint boundary conditions at $a$ and $b$, and at $x=x_1$ satisfy discontinuity conditions

(i) $p(x_1-)u'(x_1-)=p(x_1+)u'(x_1+)$.

(ii) $-u(x_1-)+u(x_1+)=\gamma p(x_1-)u'(x_1-),\quad \gamma\geq 0$.

Then $\mathbf{T}=\{T, D(T)\}$ is self-adjoint.

*Proof*: The proof consists in establishing (9.3.4). Clearly,

$$\boldsymbol{\alpha}_1^+=\begin{bmatrix}0 & -p(x_1+)\\ 1 & 0\end{bmatrix},\qquad \boldsymbol{\alpha}_1^-=\begin{bmatrix}0 & p(x_1-)\\ -1 & -\gamma p(x_1-)\end{bmatrix},$$

$$\mathbf{P}^{-1}(x)=\begin{bmatrix}0 & p^{-1}(x)\\ -p^{-1}(x) & 0\end{bmatrix}$$

$$\boldsymbol{\alpha}_1^+\mathbf{P}^{-1}(x_1+)\boldsymbol{\alpha}_1^{+t}=\begin{bmatrix}0 & -p(x_1+)\\ 1 & 0\end{bmatrix}\begin{bmatrix}0 & p^{-1}(x_1t)\\ -p^{-1}(x_1+) & 0\end{bmatrix}\begin{bmatrix}0 & 1\\ -p(x_1+) & 0\end{bmatrix}$$

$$=\begin{bmatrix}0 & 1\\ -1 & 0\end{bmatrix}$$

$$\boldsymbol{\alpha}_1^- \mathbf{P}^{-1}(x_1-)\boldsymbol{\alpha}_1^{-t} = \begin{bmatrix} 0 & p(x_1-) \\ -1 & -\gamma p(x_1-) \end{bmatrix} \begin{bmatrix} 0 & p^{-1}(x_1-) \\ -p^{-1}(x_1-) & 0 \end{bmatrix}$$

$$\begin{bmatrix} 0 & -1 \\ p(x_1-) & -\gamma p(x_1-) \end{bmatrix} = \begin{bmatrix} 0 & 1 \\ -1 & 0 \end{bmatrix}$$

which establishes (9.3.4) and hence the self-adjointness of **T**. ∎

For positive values of $\gamma$, the domain $D(T)$ above consists of functions $u(x)$ that are *discontinuous* at $x_1$. If $\gamma = 0$, then $u(x)$ is continuous at $x = x_1$. These domains are very important in applications to heat transfer problems in heterogeneous media, where interfaces may exist with interfacial resistance but with no heat capacity. We will consider specific examples later.

We prove next a second theorem which is also important to applications.

***Theorem 9.3.2*** Let $\mathcal{H} \equiv \mathcal{L}_2\{[a, b]: r(x)\}$ with inner product (9.3.5), where $\delta_1 > 0$ is to be determined, and $T$ be the formally self-adjoint operation (9.3.1) with $p(x)$ discontinuous at $x_1$. Further, let $D(T)$ be the subspace of elements $\mathbf{u} \in \mathcal{H}$ such that $T\mathbf{u} \in \mathcal{H}$ satisfy *unmixed* boundary conditions at $a$ and $b$, and at $x = x_1$:

(i) $p(x_1-)u'(x_1-) = p(x_1+)u'(x_1+)$.

(ii) $-u(x_1-) + \kappa u(x_1+) = \gamma p(x_1-)u'(x_1-), \quad \gamma \geq 0$.

Then $\mathbf{T} = \{T, D(T)\}$ is self-adjoint if $\delta_1 = \kappa$.

*Proof*: We must establish (9.3.7) and (9.3.8). Since the boundary conditions are unmixed, (9.3.7) is satisfied regardless of the value of $\delta_1$. The condition under which (9.3.8) is valid must be determined. Clearly,

$$\boldsymbol{\alpha}_1^+ = \begin{bmatrix} 0 & -p(x_1+) \\ \kappa & 0 \end{bmatrix}, \qquad \boldsymbol{\alpha}_1^- = \begin{bmatrix} 0 & p(x_1-) \\ -1 & -\gamma p(x_1-) \end{bmatrix}$$

$$\boldsymbol{\alpha}_1^+ \delta_1^{-1} \mathbf{P}^{-1}(x_1+)\boldsymbol{\alpha}_1^{+t} = \delta_1^{-1} \begin{bmatrix} 0 & -p(x_1+) \\ \kappa & 0 \end{bmatrix} \begin{bmatrix} 0 & p^{-1}(x_1+) \\ -p^{-1}(x_1+) & 0 \end{bmatrix}$$

$$\begin{bmatrix} 0 & \kappa \\ -p(x_1+) & 0 \end{bmatrix} = \frac{\kappa}{\delta_1} \begin{bmatrix} 0 & 1 \\ -1 & 0 \end{bmatrix}$$

$$\boldsymbol{\alpha}_1^- \mathbf{P}^{-1}(x_1-)\boldsymbol{\alpha}_1^{-t} = \begin{bmatrix} 0 & p(x_1-) \\ -1 & -\gamma p(x_1-) \end{bmatrix} \begin{bmatrix} 0 & p^{-1}(x_1-) \\ -p(x_1-) & 0 \end{bmatrix}$$

$$\begin{bmatrix} 0 & -1 \\ p(x_1-) & -\gamma p(x_1-) \end{bmatrix} = \begin{bmatrix} 0 & 1 \\ -1 & 0 \end{bmatrix}$$

If $\delta_1 = \kappa$, (9.3.8) holds, so that **T** is self-adjoint. ∎

In this case notice that even if $\gamma = 0$, the domain $D(T)$ consists of functions that are discontinuous at $x = x_1$. The domain in Theorem 9.3.2 is important in applications to mass diffusion problems in heterogeneous media where the interface may provide resistance to mass transfer ($\gamma > 0$) across it. The extension of the foregoing development to $n$ points $\{x_1, x_2, \ldots, x_n\}$ at which $p(x)$ is discontinuous is straightforward and is left as an exercise (see Exercise (9.3.1).

We present some examples below.

*Example 1*

Consider one-dimensional heat conduction in a two-layered medium with the first layer between $x = 0$ and $x = x_1$ of thermal conductivity $k_1$ and the second of thermal conductivity $k_2$ between $x = x_1$ and $x = 1$. At the interface $x = x_1$, the physical conditions are such that the temperature and energy flux be continuous. If the surfaces at $x = 0$ and $x = 1$ are maintained at zero temperature, the unsteady-state heat conduction equation may be written for the entire two-layered medium as

$$-\frac{1}{c(x)}\frac{\partial}{\partial x}\left[k(x)\frac{\partial \Theta}{\partial x}\right] = -\frac{\partial \Theta}{\partial t}, \qquad 0 < x < 1, \quad t > 0$$

where $\Theta$ is the temperature and $k(x)$ is the thermal conductivity of the medium with

$$k(x) = \begin{cases} k_1, & 0 < x < x_1 \\ k_2, & x_1 < x < 1 \end{cases}$$

and $c(x)$, the heat capacity per unit volume, is given by

$$c(x) = \begin{cases} c_1, & 0 < x < x_1 \\ c_2, & x_1 < x < 1 \end{cases}$$

The boundary conditions are

$$\Theta(0, t) = \Theta(1, t) = 0$$

At the interface we require that

$$\Theta(x_1-, t) = \Theta(x_1+, t)$$
$$k_1\frac{\partial \Theta}{\partial x}(x_1-, t) = k_2\frac{\partial \Theta}{\partial x}(x_1+, t)$$

The Hilbert space of interest is clearly $\mathcal{H} \equiv \mathcal{L}_2\{[0, 1]: c(x)\}$, and if we define

$$T = -\frac{1}{c(x)}\frac{d}{dx}\left[k(x)\frac{d}{dx}\right]$$

$$D(T) = \{\mathbf{u} \in \mathcal{H};\ T\mathbf{u} \in \mathcal{H};\ u(0) = u(1) = 0,\ u(x_1-) = u(x_1+),\ k(x_1-)u'(x_1-) = k(x_1+)u'(x_1+)\}$$

From Theorem 9.3.1, the operator $\mathbf{T} = \{T, D(T)\}$ is self-adjoint and one writes the entire heat conduction problem as

$$\mathbf{T}\boldsymbol{\Theta}(t) = -\frac{d}{dt}\boldsymbol{\Theta}(t)$$

It is important to recognize in the formalism here that one describe the problem of conduction in the *two* media by a *single* equation.

Similarly, one may also formulate one-dimensional (radial) heat conduction problems in cylinders and spheres of multiplayered media. For the two-layered medium with material 1 inside material 2, the heat equation is

$$-\frac{1}{x^m c(x)}\frac{\partial}{\partial x}\left[x^m k(x)\frac{\partial \Theta}{\partial x}\right] = -\frac{\partial \Theta}{\partial t}, \qquad 0 < x < 1, \quad t > 0$$

where $k(x)$ and $c(x)$ are as defined before. For cylinders $m = 1$ and for spheres $m = 2$. Suppose now that at the interface at $x = x_1$ there exists a heat transfer resistance because of a third material or a thin gap (where no energy can be stored). Then the physical equations for the temperature $\Theta$ at $x_1$ are

$$\begin{aligned} -k_1\frac{\partial \Theta}{\partial x}(x_1-, t) &= -k_2\frac{\partial \Theta}{\partial x}(x_1+, t) \\ &= h[\Theta(x_1-, t) - \Theta(x_1+, t)] \end{aligned}$$

where $h$ is called a *heat transfer coefficient*. Let the outer surface $x = 1$ be maintained at zero temperature. At the center $x = 0$, the temperature must be bounded. The foregoing requirements lead to the Hilbert space $\mathcal{H} \equiv \mathcal{L}_2\{[0, 1]: x^m c(x)\}$ and the operator $\mathbf{T} = \{T, D(T)\}$, where

$$T = -\frac{1}{x^m c(x)}\frac{d}{dx}\left[x^m k(x)\frac{d}{dx}\right],$$

$$D(T) = \{\mathbf{u} \in \mathcal{H};\ T\mathbf{u} \in \mathcal{H};\ u(0+) < \infty, u(1) = 0,$$
$$k(x_1-)u'(x_1-) = k'(x_1+)u'(x_1+) = h[u(x_1+) - u(x_1-)]\}$$

Again from Theorem 9.3.2, the operator $\mathbf{T}$ is self-adjoint. We have thus seen that heat conduction problems in multilayered media lead to equations featuring self-adjoint operators which have just been identified. We shall take up the solutions of these problems at a later stage.

*Example 2*

In mass diffusion problems in multilayered media, the concentrations referred to different media will not, in general, be continuous at the interface. If there is no interfacial resistance to mass transfer, local equilibrium may be assumed. Thus interface conditions corresponding to those in Theorem 9.3.2 will arise. As in Example 1 above, mass diffusion problems will lead to self-adjoint operators, provided that the end boundary conditions are suitably specified.

The following exercises are strongly recommended.

**EXERCISES**

**9.3.1** Consider the operation $T$ as given by (9.3.1) with $p(x)$ discontinuous at $n$ points $x_1, x_2, \ldots, x_n \in (a, b)$ such that $x_{i-1} < x_i$. Identify the Hilbert space $\mathcal{H}$ and the inner product to be used if the conditions at $x_i$ to be imposed on $\mathbf{u} \in \mathcal{H}$ are

$$p(x_i-)u'(x_i-) = p(x_i+)u'(x_i+)$$

$$\gamma_i p(x_i-)u'(x_i-) = [u(x_i+) - \delta_i u(x_i-)]$$

in order that a self-adjoint operator **T** may be obtained with suitable boundary conditions.

**9.3.2** Let $T$ be as in (9.3.1) with $p(x)$ discontinuous at $x_1 \in (a, b)$. The domain of $T$ is such that the end boundary conditions are unmixed. With $\mathcal{H} \equiv \mathcal{L}_2\{[a, b]: r(x)\}$ and the regular inner product (9.1.8), determine if the operator is self-adjoint in each of the following cases.

(a) $$u(x_1-) = u(x_1+)$$
$$p(x_1-)u'(x_1-) = p(x_1+)u'(x_1+) - \gamma u(x_1).$$

(b) $$u(x_1-) = u(x_1+)$$
$$p(x_1-)u'(x_1-) = \gamma p(x_1+)u'(x_1+).$$

(c) $$u(x_1-) = \kappa u(x_1+)$$
$$p(x_1-)u'(x_1-) = \gamma p(x_1+)u'(x_1+)$$

(d) $$u'(x_1-) = u'(x_1+)$$
$$p(x_1-)u(x_1-) = p(x_1+)u(x_1+).$$

(e) $$u'(x_1-) = \kappa u'(x_1+)$$
$$p(x_1-)u(x_1-) = p(x_1+)u(x_1+).$$

In instances where **T** is not self-adjoint, determine an inner product in $\mathcal{H}$ such that **T** becomes self-adjoint.

**9.3.3**† A fluid flows through a two-section tube, each section being of uniform cross section; at the joint the cross section changes abruptly from its "upstream" value of $A_1$ to its downstream value of $A_2$. The flow velocity in each section is uniform, with $v_1$ in section 1 and $v_2$ in section 2. Since the fluid is incompressible, $v_1 A_1 = v_2 A_2 \equiv Q$. A solute is dispersed in this fluid and it is found convenient to describe the dispersion by an "axial dispersion model" leading to the differential equation

$$\frac{1}{A(x)}\left\{\frac{\partial}{\partial x}\left[D(x)A(x)\frac{\partial c}{\partial x}\right] - Q\frac{\partial c}{\partial x}\right\} = \frac{\partial c}{\partial t}, \qquad 0 < x < 1, \quad t > 0$$

$$\text{where } D(x) = \begin{Bmatrix} D_1, & 0 < x < x_1, & A_1 \\ D_2, & x_1 < x < 1, & A_2 \end{Bmatrix} = A(x)$$

---

†See D. Ramkrishna and N. R. Amundson, *Chem. Eng. Sci.*, *29*, 1457, 1974.

The boundary conditions at $x = 0$ and $x = 1$ are

$$x = 0, \quad -D_1 A_1 \frac{\partial c}{\partial x} + Qc = Qc_f; \qquad x = 1, \qquad \frac{\partial c}{\partial x} = 0$$

At $x = x_1$ it is suggested that the concentration be continuous and

$$D_1 A_1 \frac{\partial}{\partial x} c(x_1 -, t) = D_2 A_2 \frac{\partial}{\partial x} c(x_1 +, t)$$

Define:
(a) A suitable, formally self-adjoint operation $T$.
(b) The Hilbert space of interest and the inner product.
(c) The domain $D(T)$ so that **T** is self-adjoint.
(d) Write the boundary value problem as an operator equation.
(*Hint:* Solve Exercise 9.3.2 before attempting this problem.)

### 9.3.1 Existence of a Compact Inverse

The objective of this section (similar to that of Section 9.2.1) is to show that the differential operators of this section have a compact self-adjoint inverse making spectral theorem 7.4.3 applicable. However, we will be considerably more brief here in that we will take the approach of simply presenting the results in the form of a constructive recipe for the Green's function representing the inverse operator. Thus we will spare ourselves lengthy proofs and tortuous derivations justifying the steps of construction. That the inverse operator is self-adjoint follows from the self-adjointness of the differential operator. Also, certain properties of the Green's function will be taken for granted. If $G(x, \xi)$ is the Green's function, it will be presumed that

1. $G(x, \xi) = G(\xi, x)$; that is, the function is symmetric. This property was established in detail for the operators of Section 9.2.
2. For fixed $\xi$ (or $x$), $G(x, \xi)$ is a function of $x$ (or $\xi$) which is in the domain of $T$, that is, it satisfies the homogeneous boundary conditions and conditions (9.3.3) at the point of discontinuity. A compact way of stating this fact is $\mathbf{G} \equiv \{G(x, \xi)\} \in D(T) \otimes D(T)$. (Note that **G** is not viewed here as an operator.)
3. $G(x, \xi)$ satisfies the jump condition (9.2.33) in its partial derivative, which may be restated as

$$\left[ p(x) \frac{\partial G}{\partial x}(x, \xi) \right]_{\xi = x-}^{\xi = x+} = 1, \qquad x, \xi \in [a, x_1) \tag{9.3.9}$$

This formula also holds for $x, \xi \in [x_1, b]$ when we are dealing with inner product (9.1.8), but for (9.3.5) the jump condition must read

$$\left[ \delta_1 p(x) \frac{\partial G}{\partial x}(x, \xi) \right]_{\xi = x-}^{\xi = x+} = 1, \qquad x, \xi \in [x_1, b] \tag{9.3.10}$$

The foregoing factors will guide us in the construction of the Green's function and hence the inverse of the operator **T**. The inversion of **Tu** = **f** may be represented by

$$\mathbf{u} = \langle \mathbf{G}, \mathbf{f} \rangle \tag{9.3.11}$$

where the reader is reminded that because **G** is in the tensor product space $\mathcal{H} \otimes \mathcal{H}$, the inner product on the right-hand side of (9.3.11) is a vector in $\mathcal{H}$, or more precisely, in $D(T)$.

### Construction of the Green's function

We will consider only the case of unmixed boundary conditions. This is generally the case in those applications of interest to us. The form of the Green's function $G(x, \xi)$ will depend on the locations of $x$ and $\xi$ in $[a, b]$. The interval $[a, b]$ may be viewed as the union of $[a, x_1)$ and $[x_1, b]$; clearly, $x$ and $\xi$ may be distributed between these two intervals or may both lie in either of them. If $x$ and $\xi$ lie in the same interval, their relative locations ($x > \xi$ or $x < \xi$) dictate the form. The symmetry property 1 makes matters easier because one need consider only one of the two possibilities. To be clear, let us enumerate the various possibilities with due regard to symmetry.

(i) $\xi \in [a, x_1)$, $x \in [a, x_1)$.
(ii) $\xi \in [a, x_1)$, $x \in [x_1, b]$.
(iii) $\xi \in [x_1, b]$, $x \in [x_1, b]$.

In each of cases (i) and (iii) two further subcases arise depending on whether $x < \xi$ or $x > \xi$, making a total of four cases among them. Case (ii) gives rise to another possibility by interchanging $x$ and $\xi$. Thus there are *six* cases in *all*. The symmetry property 1 makes it essential only to specify the form of the Green's function for cases (i), (ii), and (iii), the others being obtainable by interchanging $x$ and $\xi$. In case (i) we will assume that $\xi < x$, and in case (ii), $\xi > x$, allowing $x$ to be near $x_1$ in either case.

The construction strategy is as follows. We first solve

$$Ty(x) = 0, \qquad x \in (a, x_1)$$

for its linearly independent solutions $y_1^-(x)$ and $y_2^-(x)$, and

$$Ty(x) = 0, \qquad x \in (x_1, b)$$

to obtain its linearly independent solutions $y_1^+(x)$ and $y_2^+(x)$.† Next we set

$$\mathbf{Y}^{\pm}(x) \equiv \begin{bmatrix} y_1^{\pm}(x) \\ y_2^{\pm}(x) \end{bmatrix}$$

†In some situations there is no distinction between the solutions of $Ty = 0$ for $x \in [a, x_1]$ and those of $Ty = 0$ for $x \in (x_1, b]$. Thus $\mathbf{Y}^-(x)$ and $\mathbf{Y}^+(x)$ are the same in these cases.

and form *four* linear combinations given by

$$g_i^{\pm}(x) = \mathbf{Y}^{\pm t}(x)\mathbf{k}_i^{\pm}, \qquad \mathbf{k}_i^{\pm} \equiv \begin{bmatrix} k_{1,i}^{\pm} \\ k_{2,i}^{\pm} \end{bmatrix}, \quad i = 1, 2 \tag{9.3.12}$$

in which the four column vectors $\mathbf{k}_1^-, \mathbf{k}_2^-, \mathbf{k}_1^+, \mathbf{k}_2^+$, belonging to $\mathcal{R}_2$, contain *eight* constants to be determined. We let $g_1^-(x)$ satisfy the boundary conditions at $x = a$, and $g_2^+(x)$ satisfy the boundary condition at $x = b$. Thus $\mathbf{k}_1^-$ and $\mathbf{k}_2^+$ contain only one constant each, making a total of *six* constants.

We will associate the function $g_2^-(x)$ with case (i) in order to represent the Green's function for this case as

$$G(x, \xi) = g_1^-(\xi)g_2^-(x), \qquad \xi < x \tag{9.3.13}$$

The symmetry property 1 requires that if $x < \xi$ in case (i), we simply interchange $x$ and $\xi$ in (9.3.13). Similarly, we associate $g_1^+(x)$ with case (iii) so that $G(x, \xi)$ may be represented by

$$G(x, \xi) = g_1^+(x)g_2^+(\xi), \qquad x < \xi \tag{9.3.14}$$

Again symmetry can be obtained by interchanging $x$ and $\xi$ in (9.3.14) for $\xi < x$. Finally, for case (ii), $G(x, \xi)$ has the form

$$G(x, \xi) = g_1^-(\xi)g_2^+(x) \tag{9.3.15}$$

As before, by interchanging $x$ and $\xi$ in (9.3.15) we account for the case in which $x \in [a, x_1)$ and $\xi \in [x_1, b]$.

The Green's function is completely determined when the six constants in the vectors $\{\mathbf{k}_i^{\pm}\}_{i=1}^2$ are known. Equation (9.3.13) suggests that the unknown constant in the linear combination for $g_1^-$ can be "absorbed" by the two constants in $g_2^-$. Similarly, from (9.3.14) it is clear that the unknown constant in $g_2^+$ may be absorbed by the two constants in $g_1^+$. Equation (9.3.15) suggests that the two unknown constants in $g_1^-$ and $g_2^+$ may be multiplied to get a *single* constant. We thus conclude that there are only *five constants* to be determined in all, and hence *five equations* must be identified. Since we have fully exploited the symmetry property. Properties 2 and 3 of the Green's function are the only source of these equations.

Let us fix $\xi \in [a, x_1]$. Then $G(x, \xi)$ must be in $D(T)$. Since the boundary conditions have already been accounted for, we must turn to the conditions (9.3.3) at $x = x_1$. As $G(x, \xi)$ is given by (9.3.13) for $\xi < x \leq x_1$ and by (9.3.15) for $x \geq x_1$, Property 2 applied to (9.3.3) must imply that

$$\boldsymbol{\alpha}_1^-\mathbf{w}(g_1^-(x_1)) + \boldsymbol{\alpha}_1^+\mathbf{w}(g_2^+(x_1)) = \mathbf{0}$$

or alternatively,

$$\boldsymbol{\alpha}_1^-\mathbf{W}(\mathbf{Y}^-(x_1))\mathbf{k}_2^- + \boldsymbol{\alpha}_1^+\mathbf{W}(\mathbf{Y}^+(x_1))\mathbf{k}_2^+ = \mathbf{0} \tag{9.3.16}$$

In an entirely analogous manner, by fixing $\xi \in [x_1, b]$ it is possible to show that

$$\boldsymbol{\alpha}_1^-\mathbf{W}(\mathbf{Y}^-(x_1))\mathbf{k}_1^- + \boldsymbol{\alpha}_1^+\mathbf{W}(\mathbf{Y}^+(x_1))\mathbf{k}_1^+ = \mathbf{0} \tag{9.3.17}$$

Equations (9.3.16) and (9.3.17) may be rewritten as

$$[\boldsymbol{\alpha}_1^- \mathbf{W}(\mathbf{Y}^-(x_1)) \quad \boldsymbol{\alpha}_1^+ \mathbf{W}(\mathbf{Y}^+(x_1))]\begin{bmatrix} \mathbf{k}_1^- & \mathbf{k}_2^- \\ \mathbf{k}_1^+ & \mathbf{k}_2^+ \end{bmatrix} = \mathbf{0} \tag{9.3.18}$$

which represents *four* equations in all. The fifth equation that remains clearly arises from Property 3 of the Green's function, that is, (9.3.9), which in the present context implies that

$$p[g_1^{-\prime} g_2^- - g_1^- g_2^{-\prime}] = 1 \tag{9.3.19}$$

This equation is clearly applied only to the interval $[a, x_1)$. A second equation results by repeating (9.3.19) for $g_1^+$ and $g_2^+$ but is not linearly independent. When the inner product (9.3.5) is involved, (9.3.10) implies that

$$\delta_1 p[g_1^{+\prime} g_2^+ - g_1^+ g_2^{+\prime}] = 1 \tag{9.3.20}$$

Again, however, (9.3.19) and (9.3.20) are not linearly independent and only one of them is needed.

We shall present some examples that will clarify further the method just outlined. Once the Green's function is constructed, (9.3.11) can be used to show that $\mathbf{Tu} = \mathbf{f}$; in other words, the Green's function, viewed as an integral operator, represents the inverse of $\mathbf{T}$. It is not difficult to see that $G(x, \xi)$ is a Hilbert–Schmidt kernel, so that $\mathbf{T}^{-1}$ is compact and self-adjoint and thus spectral theorem 7.4.3 becomes applicable. Thus $\mathbf{T}$ has a countably infinite number of eigenvalues $\{\lambda_j\}$ (with $\lambda_j \longrightarrow \infty$) and corresponding eigenvectors $\{\mathbf{z}_j\}$ which form an orthonormal basis in $\mathcal{H}$.

*Example 1*

We will construct the Green's function for Example 1 considered earlier in the section, that is, $\mathcal{H} = \mathcal{L}_2[a, b]$, $T = \{T, D(T)\}$ with

$$T = -\frac{1}{c(x)}\frac{d}{dx}\left[k(x)\frac{d}{dx}\right], \qquad D(T) = \{\mathbf{u} \in \mathcal{H};\ T\mathbf{u} \in \mathcal{H};\ u(0) = u(1) = 0,$$

$$u(x_1-) = u(x_1+),\ k(x_1-)u'(x_1-) = k(x_1+)u'(x_1+)\}$$

We proceed in steps. First, $Ty = 0$ yields $y(x) = 1$ and $y_2(x) = x$ as the independent solutions in both subintervals $(0, x_1)$ and $(x_1, 1)$. Since $g_1^-(0) = 0$ and $g_2^+(1) = 0$, we must have

$$g_1^-(x) = k_{1,1}^- x, \qquad g_2^+(x) = k_{1,2}^+(1 - x)$$

Further,

$$g_2^-(x) = k_{1,2}^- + k_{2,2}^- x, \qquad g_1^+(x) = k_{1,1}^+ + k_{2,1}^+ x$$

Application of (9.3.18) gives

$$g_1^-(x_1) = g_1^+(x_1), \qquad k_1 g_1^{-\prime}(x_1) = k_2 g_1^{+\prime}(x_1)$$

$$g_2^-(x_1) = g_2^+(x_1), \qquad k_1 g_2^{-\prime}(x_1) = k_2 g_2^{+\prime}(x_1)$$

These equations lead to

$$g_2^-(x) = k_{1,2}^+\left[1 - x_1 + \frac{k_2}{k_1}(x_1 - x)\right]$$

$$g_1^+(x) = k_{1,1}^-\left[x_1 - \frac{k_1}{k_2}(x_1 - x)\right]$$

Clearly, the different forms for the Green's function in (9.3.13) to (9.3.15) contain only the product $k_{1,1}^- k_{1,2}^+$. To obtain this we use (9.3.19), which yields

$$k_{1,1}^- k_{1,2}^+ = \frac{1}{k_1(1 - x_1) + k_2 x_1}$$

We have now the complete Green's function as

$$G(x, \xi) = \begin{cases} \dfrac{\xi[1 - x_1 + (k_2/k_1)(x_1 - x)]}{[k_1(1 - x_1) + k_2 x_1]}, & \xi, x \in [a_1 x_1), \quad \xi < x \\[2ex] \dfrac{[x_1 - (k_1/k_2)(x_1 - x)](1 - \xi)}{k_1(1 - x_1) + k_2 x_1}, & x, \xi \in [x_1, b], \quad x < \xi \\[2ex] \dfrac{\xi(1 - x)}{k_1(1 - x_1) + k_2 x_1}, \quad \xi \in [a, x_1), & x \in [x_1, b] \end{cases}$$

In obtaining the Green's function, we have in effect accomplished the steady-state solution of the heat transfer problem consisting of a volumetric heat source in the two-layered medium whose boundaries are maintained at zero temperature.

*Example 2*

For our second example, let us consider radial diffusion in a coaxial two-layered cylinder (where the diffusion coefficients are $D_1$ and $D_2$ in the inner and outer layers, respectively) with zero concentration at the outer surface, an equilibrium relationship at the interface implying a concentration discontinuity at $x = x_1$ of the type $u(x_1-) = \kappa u(x_1+)$, and continuity of fluxes. The Hilbert space is $\mathcal{H} \equiv \mathcal{L}_2\{[0, 1]: x\}$, with the inner product (9.3.5) in which $\delta_1 = \kappa$ (see Theorem 9.3.2). The operator is $T = \{T, D(T)\}$, where

$$T = -\frac{1}{x}\frac{d}{dx}\left[xD(x)\frac{d}{dx}\right],$$

$$D(T) = \{\mathbf{u} \in \mathcal{H};\ T\mathbf{u} \in \mathcal{H};\ u(0+) < \infty,\ u(1) = 0$$

$$u(x_1-) = \kappa u(x_1+),\ D_1 u'(x_1-) = D_2 u'(x_1+)\}$$

The function $D(x)$ has the value $D_1$ in $[0, x_1)$ and $D_2$ in $(x_1, 1]$. To construct the Green's function, we first observe that $Ty = 0$ produces $y_1(x) = 1$ and $y_2(x) = \ln x$. Note that, as in Example 1, these independent solutions are the same for both $(0, x_1)$ and $(x_1, 1)$. Thereafter the recipe is readily followed to obtain

$$g_1^-(0) < \infty \rightarrow g_1^-(x) = k_{1,1}^-$$
$$g_2^+(1) = 0 \rightarrow g_2^+(x) = k_{1,2}^+ \ln x$$
$$\left.\begin{aligned} g_1^-(x_1) &= \kappa g_1^+(x_1) \\ D_1 g_1^{-\prime}(x_1) &= D_2 g_1^{+\prime}(x_1) \end{aligned}\right\} \rightarrow g_1^-(x) = \frac{1}{\kappa} k_{1,1}^-$$
$$\left.\begin{aligned} g_2^-(x_1) &= \kappa g_2^+(x_1) \\ D_1 g_2^{-\prime}(x_1) &= D_2 g_2^{+\prime}(x_1) \end{aligned}\right\} \rightarrow g_2^-(x) = k_{1,2}^+\left(\kappa \ln x_1 + \frac{D_2}{D_1} \ln \frac{x}{x_1}\right)$$

Equation (9.3.19) gives

$$D_1[g_1^{-\prime} g_2^- - g_1^- g_2^{-\prime}] = 1 \rightarrow k_{1,1}^- k_{1,2}^+ = -\frac{1}{D_2}$$

The Green's function $G(x, \xi)$ is thus given by

$$G(x, \xi) = \begin{cases} -\dfrac{1}{D_2}\left(\kappa \ln x_1 + \dfrac{D_2}{D_1} \ln \dfrac{x}{x_1}\right), & 0 \le \xi < x < x_1 \\ -\dfrac{1}{D_2 \kappa} \ln \xi, & x_1 \le x < \xi \le 1 \\ -\dfrac{1}{D_2} \ln x, & 0 \le \xi < x_1 \le x \le 1 \end{cases}$$

In writing the solution to the equation $\mathbf{Tu} = \mathbf{f}$, we must invoke (9.3.11) to obtain

$$u(x) = \int_0^{x_1} \xi G(x, \xi) f(\xi)\, d\xi + \kappa \int_{x_1}^1 \xi G(x, \xi) f(\xi)\, d\xi$$

which may be further expanded as

$$u(x) = \begin{cases} -\displaystyle\int_0^x \frac{\xi}{D_2}\left(\kappa \ln x_1 + \frac{D_2}{D_1} \ln \frac{x}{x_1}\right) f(\xi)\, d\xi \\ -\displaystyle\int_x^{x_1} \frac{\xi}{D_2}\left(\kappa \ln x_1 + \frac{D_2}{D_1} \ln \frac{\xi}{x_1}\right) f(\xi)\, d\xi - \kappa \int_{x_1}^1 \frac{\xi}{D_2} \ln \xi f(\xi)\, d\xi, \\ \qquad\qquad 0 \le x < x_1 \\ -\displaystyle\int_0^{x_1} \frac{\xi}{D_2} \ln x f(\xi)\, d\xi - \int_{x_1}^x \frac{\xi}{D_2} \ln x f(\xi)\, d\xi - \int_x^1 \frac{\xi}{D_2} \ln \xi f(\xi)\, d\xi, \\ \qquad\qquad x_1 < x \le 1 \end{cases}$$

The strategy of construction above can be readily extended to the case of a finite number of discontinuities of $p(x)$ (see Exercise 9.3.6).

## EXERCISES

**9.3.4** Construct the Green's function for the operators in Exercise 9.3.2.

**9.3.5** Construct the Green's function for the operator in Exercise 9.3.3 and obtain the *steady-state* concentration of the solute in the tube.

**9.3.6** Consider the operator $T = \{T, D(T)\}$ with

$$T = -\frac{1}{r(x)}\frac{d}{dx}\left[p(x)\frac{d}{dx}\right] + q(x)$$

where $p(x)$ has $n$ points $\{x_i\}_{i=1}^n$ of discontinuity in $[a, b]$. Let $\mathcal{H} \equiv \mathcal{L}_2\{[a, b]: r(x)\}$ and

$$D(T) = \{\mathbf{u} \in \mathcal{H};\ T\mathbf{u} \in \mathcal{H};\ u(a) + \alpha u'(a) = 0,\ u(b) + \beta u'(b) = 0,$$
$$\boldsymbol{\alpha}_i^- \mathbf{w}(u(x_i-)) + \boldsymbol{\alpha}_i^+ \mathbf{w}(u(x_i+)) = \mathbf{0},\ i = 1, 2, \ldots, n\}$$

such that

$$\boldsymbol{\alpha}_i^+ \mathbf{P}^{-1}(x_i+)\boldsymbol{\alpha}_i^{+t} = \boldsymbol{\alpha}_i^- \mathbf{P}^{-1}(x_i-)\boldsymbol{\alpha}_i^{-t}$$

holds for each $i$, thus rendering $T$ self-adjoint.

Suppose that we want to construct the Green's function for **T**. Denote by $I_i$ the interval $[x_{i-1}, x_i)$, $i = 1, 2, \ldots, n + 1$, where $x_0 = a$, $x_{n+1} = b$. Let

$$\mathbf{Y}_i(x) = \begin{bmatrix} y_{1,i}(x) \\ y_{2,i}(x) \end{bmatrix}$$

where $y_{1,i}(x)$ and $y_{2,i}(x)$ are the linearly independent solutions $Ty(x) = 0$, $x \in I_i$. Define functions $g_1^{(i)}(x)$ and $g_2^{(i)}(x)$ by

$$g_r^{(i)}(x) = \mathbf{Y}_i^t(x)\mathbf{k}_r^{(i)}, \qquad \mathbf{k}_r^{(i)} = \begin{bmatrix} k_{1,r}^{(i)} \\ k_{2,r}^{(i)} \end{bmatrix}, \qquad r = 1, 2$$

Then the Green's function $G(x, \xi)$ is expressed by

$$G(x, \xi) = \begin{cases} g_1^{(i)}(\xi)g_2^{(i)}(x), & \xi, x \in I_i,\quad \xi < x \\ g_1^{(i)}(\xi)g_2^{(j)}(x), & \xi \in I_i,\quad x \in I_j,\quad i < j \end{cases}$$

Show that

(a) $[1 \quad \alpha]\mathbf{W}(\mathbf{Y}_i(a))\mathbf{k}_1^{(1)} = 0$, $[1 \quad \beta]\mathbf{W}(\mathbf{Y}_{n+1}(b))\mathbf{k}_2^{(n+1)} = 0$.

(b) $[\boldsymbol{\alpha}_i^- \mathbf{W}(\mathbf{Y}_i(x_i)) \quad \boldsymbol{\alpha}_i^+ \mathbf{W}(\mathbf{Y}_{i+1}(x_i))]\begin{bmatrix} \mathbf{k}_1^{(i)} & \mathbf{k}_2^{(i)} \\ \mathbf{k}_1^{(i+1)} & \mathbf{k}_2^{(i+1)} \end{bmatrix} = \mathbf{0}$.

Express the discontinuity condition in the partial derivative of the Green's function.

**9.3.7** Solve the boundary value problem

$$T\mathbf{v} = \mathbf{f}$$

$$\left.\begin{aligned} \boldsymbol{\alpha}\mathbf{w}(v(a)) + \boldsymbol{\beta}\mathbf{w}(v(b)) &= \boldsymbol{\gamma} \\ \boldsymbol{\alpha}^-\mathbf{w}(v(x_1-)) + \boldsymbol{\alpha}^+\mathbf{w}(v(x_1+)) &= \boldsymbol{\gamma}_1 \end{aligned}\right\} \quad \boldsymbol{\gamma}, \boldsymbol{\gamma}_1 \ni \mathcal{R}_2$$

where $T$ is the general self-adjoint operator of Section 9.3 for the case of a single discontinuity of $p(x)$ at $x_1$. [*Hint:* Let $\mathbf{Y}^-(x)$ and $\mathbf{Y}^+(x)$ be as defined above (9.3.12) and

$$\mathbf{D}_0 \equiv \begin{bmatrix} \boldsymbol{\alpha}\mathbf{W}(\mathbf{Y}^-(a)) & \boldsymbol{\beta}\mathbf{W}(\mathbf{Y}^+(b)) \\ \boldsymbol{\alpha}^-\mathbf{W}(\mathbf{Y}^-(x_1-)) & \boldsymbol{\alpha}^+\mathbf{W}(\mathbf{Y}^+(x_1+)) \end{bmatrix}$$

Use the transformation $v(x) = u(x) + [\mathbf{Y}^{-t}(x) \quad \mathbf{Y}^{+t}(x)]\mathbf{D}_0^{-1}\begin{bmatrix}\boldsymbol{\gamma}\\ \boldsymbol{\gamma}_1\end{bmatrix}$ and show that $\mathbf{u} \in D(T)$ and $\mathbf{Tu} = \mathbf{f}$.] Note that although we have used the general boundary conditions here, insofar as the construction of the Green's function discussed in this section requires unmixed boundary condition, this constraint must be realized.

### 9.3.2 Evaluation of Spectra

It is in the order of things next to inquire into the determination of the eigenvalues and eigenvectors of the self-adjoint operator $\mathbf{T}$ of this section. Thus we must solve $\mathbf{Tz} = \lambda\mathbf{z}$. We seek to represent the eigenvector as a linear combination of the linearly independent solutions of the differential equation $Tu(x) = \lambda u(x)$. Naturally, the solution of this equation would depend on the interval for which it is considered. We let $u_1^-(x, \lambda)$, $u_2^-(x, \lambda)$ be the linearly independent solutions for the interval $(a, x_1)$, and $u_1^+(x, \lambda)$, $u_2^+(x, \lambda)$ be the linearly independent solutions for the interval $(x_1, b)$. Further, we let

$$\mathbf{U}^{\pm}(x, \lambda) = \begin{bmatrix} u_1^{\pm}(x, \lambda) \\ u_2^{\pm}(x, \lambda) \end{bmatrix}$$

and represent the eigenvector as

$$z(x, \lambda) = \begin{cases} \mathbf{U}^{-t}(x, \lambda)\mathbf{k}^-, & x \in [a, x_1) \\ \mathbf{U}^{+t}(x, \lambda)\mathbf{k}^+, & x \in (x_1, b] \end{cases} \tag{9.3.21}$$

where $\mathbf{k}^-$ and $\mathbf{k}^+$ contain four constants to be determined so that the boundary conditions (9.1.11) and constraints (9.3.3) at $x_1$ are satisfied. Boundary conditions (9.1.11) imply that

$$\boldsymbol{\alpha}\mathbf{W}(\mathbf{U}^-(a, \lambda))\mathbf{k}^- + \boldsymbol{\beta}\mathbf{W}(\mathbf{U}^+(b, \lambda))\mathbf{k}^+ = \mathbf{0} \tag{9.3.22}$$

From (9.3.3) one obtains

$$\boldsymbol{\alpha}_1^-\mathbf{W}(\mathbf{U}^-(x_1, \lambda))\mathbf{k}^- + \boldsymbol{\alpha}_1^+\mathbf{W}(\mathbf{U}^+(x_1, \lambda))\mathbf{k}^+ = \mathbf{0} \tag{9.3.23}$$

Equations (9.3.22) and (9.3.23) may be more compactly written as

$$\mathbf{D}_\lambda\mathbf{k} = \mathbf{0} \tag{9.3.24}$$

where

$$\mathbf{D}_\lambda \equiv \begin{bmatrix} \boldsymbol{\alpha}\mathbf{W}(\mathbf{U}^-(a, \lambda)) & \boldsymbol{\beta}\mathbf{W}(\mathbf{U}^+(b, \lambda)) \\ \boldsymbol{\alpha}_1^-\mathbf{W}(\mathbf{U}^-(x_1, \lambda)) & \boldsymbol{\alpha}_1^+\mathbf{W}(\mathbf{U}^+(x_1, \lambda)) \end{bmatrix}, \qquad \mathbf{k} \equiv \begin{bmatrix} \mathbf{k}^- \\ \mathbf{k}^+ \end{bmatrix}$$

For $\mathbf{z} \neq \mathbf{0}$, we must have $\mathbf{k} \neq \mathbf{0}$ so that the *characteristic equation* for the eigenvalues is given by

$$D_\lambda = 0 \tag{9.3.25}$$

where $D_\lambda$ is the determinant of $\mathbf{D}_\lambda$. Equation (9.3.25) is a nonlinear algebraic equation and is in general solvable only numerically. Theorem 7.4.3 assures us an infinite number of eigenvalues $\{\lambda_j\}$; the eigenvector $\mathbf{z}_j$ corresponding to $\lambda_j$ is obtained by solving

$$\mathbf{D}_{\lambda_j}\mathbf{k}_j = \mathbf{0} \tag{9.3.26}$$

Since $\mathbf{D}_{\lambda_j}$ is a $4 \times 4$ matrix, the maximum number of linearly independent solutions of (9.3.26), and hence the dimension of $N_{\lambda_j}$, is *four*. If $\mathbf{D}_{\lambda_j}$ has rank $r_j$ $(0 \leq r_j \leq 3)$, the dimension of $N_{\lambda_j}$ is $(4 - r_j)$. When there are more than one linearly independent eigenvectors, they may be orthonormalized by the Gram–Schmidt procedure, which has been encountered before. Thus it is possible to find an orthonormal basis $\{\mathbf{z}_j^{(1)} \ldots \mathbf{z}_j^{(4-r_j)}\}$ for the eigenspace $N_{\lambda_j}$. The self-adjoint projection $\mathbf{P}_j$ of $\mathcal{H}$ onto $N_{\lambda_j}$ is defined by

$$\mathbf{P}_j\mathbf{u} \equiv \sum_{i=1}^{4-r_j} \langle \mathbf{u}, \mathbf{z}_j^{(i)} \rangle \mathbf{z}_j^{(i)}, \qquad \mathbf{u} \in \mathcal{H} \tag{9.3.27}$$

where $\|\mathbf{z}_j^{(i)}\| = 1$, $i = 1, 2, \ldots, 4 - r_j$. We have thus shown how to determine the spectral representation of $\mathbf{T}$. From this point onward, the manner in which the foregoing spectral resolution is employed to solve equations involving $\mathbf{T}$ is identical to that in Section 9.2. For example, Mercer's expansion (9.2.58) for the Green's function is valid. If positive definiteness is important, it is established by examining $\langle \mathbf{Tu}, \mathbf{u} \rangle$ for abitrary $\mathbf{u} \in D(T)$. Thus unmixed boundary conditions such as $u(a) + \alpha u'(a) = 0$, $u(b) + \beta u'(b) = 0$ must be subject to (9.2.59) in addition to which constraint will appear on the coefficients connected with equations concerning behavior at $x_1$. Suppose that we consider the operator which is addressed in Theorem 9.3.1. Then $\langle \mathbf{Tu}, \mathbf{u} \rangle$, excluding all other terms which we assume to be nonnegative, will yield

$$\frac{1}{\gamma}[u(x_1+) - u(x_1-)]^2$$

from which we conclude that $\gamma$ must be *positive* in order that $\mathbf{T}$ is positive definite.

As a demonstration of the evaluation of spectra, let us consider Examples 1 and 2 in Section 9.3.1.

*Example 1*

Consider the operator of Example 1 of Section 9.3.1. We obtain the characteristic equation for the eigenvalues as follows. In the interval $(0, x_1)$, $Tu(x) = \lambda u(x)$ is

$$-\frac{d^2u^-}{dx^2} = \frac{\lambda}{\alpha_1}u^-, \qquad \alpha_1 \equiv \frac{k_1}{c_1}$$

whose linearly independent solutions are $u_1^-(x, \lambda) = \sin \sqrt{\lambda/\alpha_1}\, x$ and $u_2^-(x, \lambda) = \cos \sqrt{\lambda/\alpha_1}\, x$. Similarly, in $(x_1, 1)$ we have

$$-\frac{d^2u^+}{dx^2} = \frac{\lambda}{\alpha_2}u^+, \qquad \alpha_2 \equiv \frac{k_2}{c_2}$$

with solutions

$$u_1^+(x, \lambda) = \sin\sqrt{\frac{\lambda}{\alpha_2}}(1 - x), \qquad u_2^+(x, \lambda) = \cos\sqrt{\frac{\lambda}{\alpha_2}}(1 - x)$$

[We pick this form because of the boundary condition $u(1) = 0$.] Thus

$$\mathbf{W}(\mathbf{U}^-(x, \lambda)) = \begin{bmatrix} \sin\sqrt{\frac{\lambda}{\alpha_1}}x & \cos\sqrt{\frac{\lambda}{\alpha_1}}x \\ \sqrt{\frac{\lambda}{\alpha_1}}\cos\sqrt{\frac{\lambda}{\alpha_1}}x & -\sqrt{\frac{\lambda}{\alpha_1}}\sin\sqrt{\frac{\lambda}{\alpha_1}}x \end{bmatrix}$$

$$\mathbf{W}(\mathbf{U}^+(x, \lambda)) = \begin{bmatrix} \sin\sqrt{\frac{\lambda}{\alpha_2}}(1 - x) & \cos\sqrt{\frac{\lambda}{\alpha_2}}(1 - x) \\ -\sqrt{\frac{\lambda}{\alpha_2}}\cos\sqrt{\frac{\lambda}{\alpha_2}}(1 - x) & \sqrt{\frac{\lambda}{\alpha_2}}\sin\sqrt{\frac{\lambda}{\alpha_2}}(1 - x) \end{bmatrix}$$

Now

$$\boldsymbol{\alpha} = \begin{bmatrix} 1 & 0 \\ 0 & 0 \end{bmatrix}, \quad \boldsymbol{\beta} = \begin{bmatrix} 0 & 0 \\ 0 & 1 \end{bmatrix}, \quad \boldsymbol{\alpha}_1^- = \begin{bmatrix} 0 & k_1 \\ -1 & 0 \end{bmatrix}, \quad \boldsymbol{\alpha}_1^+ = \begin{bmatrix} 0 & -k_2 \\ 1 & 0 \end{bmatrix}$$

so that

$$\mathbf{D}_\lambda = \left[\begin{matrix} 0 & 1 \\ 0 & 0 \\ k_1\sqrt{\frac{\lambda}{\alpha_1}}\cos\sqrt{\frac{\lambda}{\alpha_1}}x_1 & -k_1\sqrt{\frac{\lambda}{\alpha_1}}\sin\sqrt{\frac{\lambda}{\alpha_1}}x_1 \\ -\sin\sqrt{\frac{\lambda}{\alpha_1}}x_1 & -\cos\sqrt{\frac{\lambda}{\alpha_1}}x_1 \end{matrix}\right.$$

$$\left.\begin{matrix} 0 & 0 \\ 0 & 1 \\ k_2\sqrt{\frac{\lambda}{\alpha_2}}\cos\sqrt{\frac{\lambda}{\alpha_2}}(1 - x_1) & -k_2\sqrt{\frac{\lambda}{\alpha_2}}\sin\sqrt{\frac{\lambda}{\alpha_2}}(1 - x_1) \\ \sin\sqrt{\frac{\lambda}{\alpha_2}}(1 - x_1) & \cos\sqrt{\frac{\lambda}{\alpha_2}}(1 - x_1) \end{matrix}\right]$$

The characteristic equation becomes

$$D_\lambda = k_1\sqrt{\frac{\lambda}{\alpha_1}}\sin\sqrt{\frac{\lambda}{\alpha_2}}(1 - x_1)\cos\sqrt{\frac{\lambda}{\alpha_1}}x_1 + k_2\sqrt{\frac{\lambda}{\alpha_2}}\sin\sqrt{\frac{\lambda}{\alpha_1}}\cos\sqrt{\frac{\lambda}{\alpha_2}}(1 - x_1)$$
$$= 0$$

This characteristic equation could have been more easily obtained by recognizing that for $x \in [0, x_1)$, $z^-(x, \lambda) = k_1^- \sin\sqrt{\lambda/\alpha_1}\, x$, and for $x \in (x_1, 1)$, $z^+(x, \lambda) = k_1^+ \sin\sqrt{\lambda/\alpha_2}(1 - x)$ between which the boundary conditions are met. On applying the conditions $z^-(x_1, \lambda) = z^+(x_1, \lambda)$ and $k_1 z^{+\prime}(x_1, \lambda)$

$= k_2 z^{-\prime}(x, \lambda)$, the same characteristic equation is obtained. The general recipe is more useful in more complicated problems; $\lambda = 0$ is not an eigenvalue because $z(x, 0) \equiv 0$. The other eigenvalues must be obtained from the simplified equation

$$-\frac{k_1}{k_2}\sqrt{\frac{\alpha_2}{\alpha_1}} = \frac{\tan\sqrt{\lambda/\alpha_1}\,x_1}{\tan\sqrt{\lambda/\alpha_2}(1 - x_1)}$$

This is as far as we can go without numerical computation. When $\{\lambda_j\}$ are known, the equation $\mathbf{D}_{\lambda_j}\mathbf{k}_j = 0$ yields $\mathbf{k}_j$. In this case it turns out that there is only one $z_j$ for each $\lambda_j$. Moreover, it is easy to verify that

$$z_j(x) = \begin{cases} k_{1,j}^{-} \sin\sqrt{\dfrac{\lambda_j}{\alpha_1}}\,x, & x \in [0, x_1) \\ k_{i,j}^{-}\dfrac{\sin\sqrt{\lambda_j/\alpha_1}\,x_1}{\sin\sqrt{\lambda_j/\alpha_2}(1 - x_1)} \sin\sqrt{\dfrac{\lambda_j}{\alpha_2}}(1 - x), & x \in (x_1, 1] \end{cases}$$

The constant $k_{1,j}^{-}$ is fixed by normalization of $z_j$, which yields

$$1 = (k_{1,j}^{-})^2\left[c_1\int_0^{x_1} \sin^2\sqrt{\frac{\lambda_j}{\alpha_1}}\,x\,dx + c_2\frac{\sin^2\sqrt{\lambda_j/\alpha_1}\,x_1}{\sin^2\sqrt{\lambda_j/\alpha_2}(1 - x_1)} \cdot \int_{x_1}^{1} \sin^2\sqrt{\frac{\lambda_j}{\alpha_2}}(1 - x)\,dx\right]$$

where $c_1 \equiv k_1/\alpha_1$ and $c_2 \equiv k_2/\alpha_2$. An analytical expression for $k_{1,j}^{-}$ is easily obtained as

$$k_{1,j}^{-} = \left\{\frac{c_1}{2}\left(x_1 - \frac{1}{2}\sqrt{\frac{\alpha_1}{\lambda_j}}\sin^2\sqrt{\frac{\lambda_j}{\alpha_1}}\,x_1\right) + \frac{c_2}{2}\frac{\sin^2\sqrt{\lambda_j/\alpha_1}\,x_1}{\sin\sqrt{\lambda_j/\alpha_2}(1 - x_1)} \cdot \left[1 - x_1 - \frac{1}{2}\sqrt{\frac{\lambda_j}{\alpha_2}}(1 - x_1)\right]\right\}^{-1/2}$$

The spectral analysis of $\mathbf{T}$ is now complete except for the numerical evaluation of eigenvalues.

*Example 2*

We consider Example 2 of Section 9.3.1, in which the operator $T$ arose from the description of radial diffusion in a cylinder. Thus we let $\mathcal{H} = \mathcal{L}_2\{[0, 1]: x\}$ with inner product (9.3.5):

$$T \equiv -\frac{1}{x}\frac{d}{dx}\left[xD(x)\frac{d}{dx}\right], D(T) = \{\mathbf{u} \in \mathcal{H};\ T\mathbf{u} \in \mathcal{H};\ u(0) < \infty, u(1) = 0;$$
$$u(x_1-) = \kappa u(x_1+);\ D(x_1-)u'(x_1-) = D(x_1+)u'(x_1+)\}$$

Note that

$$D(x) = \begin{cases} D_1, & x \in [0, x_1) \\ D_2, & x \in (x_2, 1] \end{cases}$$

The equation $Tu(x) = \lambda u(x)$ leads to

$$-\frac{1}{x}\frac{d}{dx}\left(x\frac{du}{dx}\right) = \frac{\lambda}{D_i}u \qquad \begin{cases} i = 1 & \text{if } x \in (0, x_1) \\ i = 2 & \text{if } x \in (x, 1) \end{cases}$$

Solving, one obtains $u_1^-(x, \lambda) = J_0(\sqrt{\lambda/D_1}\, x)$, $u_2^-(x, \lambda) = Y_0(\sqrt{\lambda/D_1}\, x)$, $u_1^+(x, \lambda) = J_0(\sqrt{\lambda/D_2}\, x)$, $u_2^+(x, \lambda) = Y_0(\sqrt{\lambda/D_2}\, x)$. Instead of following the general formulation of the characteristic equation, we use the boundary condition at $x = 0$ to write $z^-(x, \lambda) = k_1^- J_0(\sqrt{\lambda/D_1}\, x)$, since $Y_0$ is unbounded at the origin. By letting $z^+(x, \lambda) = k_1^+ J_0(\sqrt{\lambda/D_2}\, x) + k_2^+ Y_0(\sqrt{\lambda/D_2}\, x)$, the boundary condition at $y = 1$ and the conditions at $x = x_1$ yield the characteristic equation $\Delta_\lambda = 0$, where $\Delta_\lambda$ is the determinant of

$$\mathbf{\Delta}_\lambda \equiv \begin{bmatrix} 0 & J_0\left(\sqrt{\frac{\lambda}{D_2}}\right) & Y_0\left(\sqrt{\frac{\lambda}{D_2}}\right) \\ D_1\sqrt{\frac{\lambda}{D_1}}J_0'\left(\sqrt{\frac{\lambda}{D_1}}x_1\right) & -D_2\sqrt{\frac{\lambda}{D_2}}J_0'\left(\sqrt{\frac{\lambda}{D_2}}x_1\right) & -D_2\sqrt{\frac{\lambda}{D_2}}Y_0'\left(\sqrt{\frac{\lambda}{D_2}}x_1\right) \\ J_0\left(\sqrt{\frac{\lambda}{D_1}}x_1\right) & -\kappa J_0\left(\sqrt{\frac{\lambda}{D_2}}x_1\right) & -\kappa Y_0\left(\sqrt{\frac{\lambda}{D_2}}x_1\right) \end{bmatrix}$$

Further processing of the determinant above is left to the student. The eigenvectors are determined by solving the equation

$$\mathbf{\Delta}_{\lambda_j}\begin{bmatrix} k_{1,j}^- \\ k_{1,j}^+ \\ k_{2,j}^+ \end{bmatrix} = 0$$

To normalize the eigenvector, the inner product (9.3.5) must be used with $\delta_1 = \kappa$.

## EXERCISES

**9.3.8** Obtain the characteristic equation for the eigenvalues of the operators in Exercise 9.3.2. In each case find conditions pertaining to the signs of the parameters in the discontinuity conditions under which the operator becomes positive or nonnegative definite.

**9.3.9** Find the characteristic equation for the operator of Exercise 9.3.3.

**9.3.10** Obtain the characteristic equation for the general operator **T** with self-adjoint boundary condition and discontinuity conditions as in Exercise 9.3.6.

**9.3.11** A method of normalization of the eigenvector $\mathbf{z}_j$ was given for the operators of Section 9.2 which does not involve direct integration of the eigenfunction [see (9.2.61)]. Assuming that $N_{\lambda_j}$ has dimension 1 for the operator **T** in Section 9.3.3, obtain a normalization condition for eigenvector $\mathbf{z}_j$ similar to (9.2.61). Apply the formula to Example 1 of this section and verify the expression for the constant $k_{1,j}^-$ derived therein.

### 9.3.3 Boundary Value Problems

Again, the methods of Section 7.5 are very valuable in the solution of boundary value problems in which the operators of this section are encountered. The method of solution proceeds precisely as in Section 9.2.3, except for the details in regard to the spectral analysis of the operator. A second element of detail pertains to the dependent variable $\mathbf{v}(t)$ not being in the domain $D(T)$ because of inhomogeneities appearing in the boundary *and* discontinuity conditions. More specifically, we may address boundary-initial value problems of the type

$$-\frac{d}{dt}\mathbf{v}(t) = T\mathbf{v}(t) - \mathbf{f}(t), \qquad \mathbf{v}(0) = \mathbf{v}_0 \tag{9.3.28}$$

where $T$ is the formal operation (9.3.1) with $p(x)$ discontinuous at $x_1 \in (a, b)$. In addition, $\mathbf{v}(t)$ satisfies the inhomogeneous boundary conditions

$$\boldsymbol{\alpha}\mathbf{w}(v(a, t)) + \boldsymbol{\beta}\mathbf{w}(v(b, t)) = \boldsymbol{\gamma}(t) \tag{9.3.29}$$

*and* inhomogeneous discontinuity conditions given by

$$\boldsymbol{\alpha}^-\mathbf{w}(v(x_1-, t)) + \boldsymbol{\alpha}^+\mathbf{w}(v(x_1+, t)) = \boldsymbol{\gamma}_1(t) \tag{9.3.30}$$

In (9.3.29) and (9.3.30), $\boldsymbol{\gamma}(t)$ and $\boldsymbol{\gamma}_1(t)$ are known vectors in $\mathcal{R}_2$ for each $t$. The matrices $\boldsymbol{\alpha}$, $\boldsymbol{\beta}$, $\boldsymbol{\alpha}^-$, and $\boldsymbol{\alpha}^+$ satisfy suitable constraints for the operator $\mathbf{T}$, defined by dropping the inhomogeneities in the boundary and discontinuity conditions, to be self-adjoint. [For example, recall conditions such as (9.2.7) and (9.3.4) when the regular inner product (9.1.8) is used, or conditions (9.3.7) and (9.3.8) when inner product (9.3.5) is used.] We will assume further that the operator $\mathbf{T}$ is nonnegative and that its spectral information is available (i.e, the eigenvalues $\{\lambda_j\}$ and the self-adjoint projections $\{\mathbf{P}_j\}$ are known).

The solution of the boundary-initial value problem is in fact given by (7.5.15) if we can identify the term $\mathbf{P}_j(T\mathbf{g}(t) - \lambda_j\mathbf{g}(t))$, where $\mathbf{g}(t)$ is *any* function satisfying (9.3.29) and (9.3.30). In the discussion in Section 9.2.3 of the operators of Section 9.2, we had shown that the foregoing term is independent of $\mathbf{g}(t)$. A similar result can be established here also.

From (9.3.2), we may write for any $\mathbf{z} \in D(T)$,

$$\langle T\mathbf{g}(t), \mathbf{z}\rangle - \langle \mathbf{g}(t), T\mathbf{z}\rangle = [\mathbf{w}^t(z(a)) \quad \mathbf{w}^t(z(b))]\begin{bmatrix} -\mathbf{P}(a) & 0 \\ 0 & \mathbf{P}(b) \end{bmatrix}\begin{bmatrix} \mathbf{w}(g(a, t)) \\ \mathbf{w}(g(b, t)) \end{bmatrix}$$

$$- [\mathbf{w}^t(z(x_1-)) \quad \mathbf{w}^t(z(x_1+))]\begin{bmatrix} -\mathbf{P}(x_1-) & 0 \\ 0 & \mathbf{P}(x_1+) \end{bmatrix}\begin{bmatrix} \mathbf{w}(g(x_1-, t)) \\ \mathbf{w}(g(x_1+, t)) \end{bmatrix}$$

Using arguments precisely analogous to those which led to (9.1.15), we had identified a unique vector $\mathbf{y}^t(z) \in \mathcal{R}_2$, depending linearly on $\mathbf{z}$, in Section 9.2.1 such that the first term on the right-hand side of the equation above is

$\mathbf{y}^t(z)\boldsymbol{\gamma}(t)$ [see (9.2.67)]. The argument is applicable also to the second term on the right-hand side, which may be represented by $-\mathbf{y}_1^t(\mathbf{z})\boldsymbol{\gamma}_1(t)$, where $\mathbf{y}_1^t(\mathbf{z}) \in \mathcal{R}_2$ also depends linearly on $\mathbf{z}$, and $\boldsymbol{\gamma}_1(t)$ originates from (9.3.30). The identification of $\mathbf{y}_1^t(\mathbf{z})$ is again based on the same arguments as those for $\mathbf{y}^t(z)$ and the student is referred to Exercise 9.3.12. In many applications, however, $\mathbf{y}^t(z)$ is obtained more directly, as will be evident from an example to be treated presently.

From the foregoing arguments we are able to write

$$P_j[T\mathbf{g}(t) - \lambda_j \mathbf{z}_j] = [\mathbf{y}^t(z_j)\boldsymbol{\gamma}(t) - \mathbf{y}_1^t(\mathbf{z}_j)\boldsymbol{\gamma}_1(t)]\mathbf{z}_j \tag{9.3.31}$$

Thus the solution to the boundary value problem becomes

$$\mathbf{v}(t) = \sum_{j=1}^{\infty} e^{-\lambda_j t}[\mathbf{P}_j\mathbf{v}_0 + \int_0^t e^{\lambda_j t'}\{\mathbf{P}_j\mathbf{f}(t') - \mathbf{y}^t(\mathbf{z}_j)\boldsymbol{\gamma}(t')\mathbf{z}_j + \mathbf{y}_1^t(\mathbf{z}_j)\boldsymbol{\gamma}_1(t')\mathbf{z}_j\}\, dt'] \tag{9.3.32}$$

The simplification that obtains when $\mathbf{f}$, $\boldsymbol{\gamma}$, and $\boldsymbol{\gamma}_1$ are independent of time, which was discussed in Section 9.2.3, is also applicable here. Thus solution (9.2.69) is also applicable here, in the evaluation of which the student is advised to refer to Exercise 9.3.7.

Let us consider a sample application.

*Example of Transient Conduction in Composite Slab*

The physical situation here is essentially that of Example 1 of Section 9.3. Thus we have a two-layered slab of unit thickness along the direction of heat conduction, located between $x = 0$ and $x = 1$, with properties stated as before. At the interface $x = x_1$, we will assume that a thin surface source exists generating heat uniformly but varying with time. We also allow a volumetric heat source term in the medium which may vary with $x$ and $t$. At $x = 0$ and $x = 1$, we specify the temperature as a function of time. The relevant equations are

$$-\frac{1}{c(x)}\frac{\partial}{\partial x}\left[k(x)\frac{\partial \Theta}{\partial x}\right] - f(x, t) = \frac{\partial \Theta}{\partial t}, \qquad 0 < x < 1, \quad t > 0 \tag{9.3.33}$$

$$\Theta(0, t) = \gamma_1(t) \tag{9.3.34}$$

$$\Theta(1, t) = \gamma_2(t) \tag{9.3.35}$$

$$\Theta(x_1-, t) = \Theta(x_1+, t) \tag{9.3.36}$$

$$-k_1\frac{\partial \Theta}{\partial x}(x_1-, t) = -k_2\frac{\partial \Theta}{\partial x}(x_1+, t) + f_1(t) \tag{9.3.37}$$

$$\Theta(x, 0) = \Theta_0(x) \tag{9.3.38}$$

where $f(x, t)$ and $f_1(t)$ arise from the volumetric and surface heat sources, respectively. The foregoing boundary-initial value problem suggests the Hilbert space $\mathcal{H} \equiv \mathcal{L}_2\{[0, 1]: c(x)\}$ and the operator **T** of Example 1 of Section 9.3.1. The vector $\boldsymbol{\Theta}(t) = \{\Theta(x, t)\}$ satisfies

$$-\frac{d}{dt}\boldsymbol{\Theta}(t) = T\boldsymbol{\Theta}(t) - \mathbf{f}(t), \qquad \boldsymbol{\Theta}(0) = \boldsymbol{\Theta}_0 \tag{9.3.39}$$

which comes from (9.3.33) and (9.3.40). Because $\boldsymbol{\Theta}(t)$ does not belong to $D(T)$, we do not use the boldface symbol for $T$. Since (9.3.39) is identical to that of (9.3.28), the solution is given by (9.3.32) with $\Theta$ in place of $v$, except for the identity of the terms $\mathbf{y}^t(\mathbf{z}_j)\boldsymbol{\gamma}(t)$ and $\mathbf{y}_1^t(\mathbf{z}_j)\boldsymbol{\gamma}_1(t)$. Let us see how to determine these terms directly in the present case. If $\mathbf{g}(t)$ satisfies (9.3.34)–(9.3.37), it is readily seen that

$$\begin{aligned}\langle \mathbf{Tg}(t), \mathbf{z}_j\rangle - \lambda_j\langle \mathbf{g}(t), \mathbf{z}_j\rangle &= k_2[g(1, t)z_1'(1)] - k_1[g(0, t)z_j'(0)] \\ &\quad + k_1\Big[g(x_1-, t)z_j'(x_1-) - \frac{\partial}{\partial x}g(x_1, t)z_j(x_1-)\Big] \\ &\quad - k_2\Big[g(x_1+, t)z_j'(x_1+) - \frac{\partial}{\partial x}g(x_1+, t)z_j(x_1+)\Big] \\ &= k_2\gamma_2(t)z_j'(1) - k_1\gamma_1(t)z_j'(0) + z_j(x_1)f_1(t)\end{aligned}$$

The right-hand side consists of known quantities, provided that the spectral information on **T** has been determined. In Section 9.3.2 we had determined the characteristic equation for **T** and the expression for the eigenvector $\mathbf{z}_j$. On substituting into the solution, one obtains

$$\begin{aligned}\boldsymbol{\Theta}(t) = \sum_{j=1}^{\infty} e^{-\lambda_j t}\Big[&\mathbf{P}_j\Theta_0 \\ &+ \int_0^t e^{\lambda_j t'}\Big\{\mathbf{P}_j\mathbf{f}(t') + k_2\gamma_2(t')k_{1,j}^{-}\sqrt{\frac{\lambda_j}{\alpha_2}}\frac{\sin\sqrt{\lambda_j/\alpha_1}\,x_1}{\sin\sqrt{\lambda_j/\alpha_2}(1 - x_1)}\mathbf{z}_j \\ &+ k_1\gamma_1(t')k_{1,j}^{-}\sqrt{\frac{\lambda_j}{\alpha_1}}\mathbf{z}_j - k_{1,j}^{-}\sin\sqrt{\frac{\lambda_j}{\alpha_1}}\,x_1 f_1(t')\mathbf{z}_j\Big\}\,dt'\Big]\end{aligned} \tag{9.3.40}$$

Several other applications are suggested through the exercises that appear below. Obviously, the solutions to boundary value problems can be expressed only in terms of the eigenvalues and eigenvectors. Insofar as numerical methods are required for the computation of the spectra, the spectral solutions obtained here are not numerically complete. Hence the exercises must also be interpreted in this light.

Although most exercises involve boundary value problems of the type (9.3.28), in Section 7.5 we had shown how other equations may be solved just as well. In this connection the student is referred particularly to Exercise 9.3.17.

**EXERCISES**

**9.3.12** Let $\boldsymbol{\varepsilon}_{ij}$ be a nonsingular $2 \times 2$ matrix obtained from the $i$th and $j$th columns of the $2 \times 4$ matrix $[\boldsymbol{\alpha}^- \quad \boldsymbol{\alpha}^+]$ and $\boldsymbol{\gamma}_{ij}$ be the $2 \times 4$ matrix defined in Section 9.1. Then establish that $\mathbf{y}_1^t(\mathbf{z})$ appearing in (9.3.31) is given by

$$\mathbf{y}_1^t(\mathbf{z}) = \boldsymbol{\varepsilon}_{ij}^{-1}[\mathbf{w}^t(z(x_1-)) \quad \mathbf{w}^t(z(x_1+))]\begin{bmatrix} -\mathbf{P}(x_1-) & \mathbf{0} \\ \mathbf{0} & \mathbf{P}(x_1+) \end{bmatrix}\boldsymbol{\gamma}_{ij}^t$$

**9.3.13** Solve the boundary-initial value problem (9.3.33)–(9.3.38) for the case where $\mathbf{f}, f_1, \gamma_1$, and $\gamma_2$ are constants via the strategy of Section 9.2.3 leading to the solution (9.2.69).

**9.3.14** A vertical container has two immiscible liquids with the heavier liquid between $x = 0$ and $x = x_1$ and the lighter liquid between $x = x_1$ and $x = 1$. Set up and solve the problem of a solute diffusing in the system assuming equilibrium at the interface. No solute escapes from the free surface and the two liquids are completely quiescent (no free convection or interfacial effects). Assume a linear equilibrium relationship.

**9.3.15** Solve the transient reactor problem of Exercise 9.3.3 for an arbitrary initial concentration.

**9.3.16** Consider the "conjugated Graetz problem," which features two concentric tubes; the inner tube of radius $x_1$ has a fluid in fully developed flow with a velocity profile $q_1(x)$ and the outer tube of radius unity has a second fluid flowing concurrently with velocity $q_2(x)$. The two fluids exchange energy through the intervening wall, whose thickness, but not thermal resistance, may be neglected. The outer fluid loses energy through the tube surface to the surroundings at a constant temperature. Conduction is negligible in the direction of the flow.

The foregoing situation leads to the energy differential equation

$$-\frac{1}{r(x)}\frac{\partial}{\partial x}\left[xk(x)\frac{\partial v}{\partial x}\right] - f(x) = -\frac{\partial v}{\partial t}$$

where

$$r(x) = \begin{cases} xc_1q_1(x), & x \in [0, x_1) \\ xc_2q_2(x), & x \in (x_1, 1] \end{cases}$$

$t$ represents the axial coordinate and $f(x)$ arises from the viscous dissipation term.

The exchange of energy between the two fluids is described by

$$-k(x_1-)\frac{\partial v}{\partial x}(x_1-, t) = -k(x_1+)\frac{\partial v}{\partial x}(x_1+, t)$$

$$= h[v(x_1-, t) - v(x_1+, t)]$$

The entering fluid temperatures are described by a single function $v_0(x)$, $0 < x < 1$. Obtain a spectral solution for this boundary value problem.

**9.3.17** Consider the forced longitudinal vibration of a vertical prismatic bar with its upper end fixed at $x = 0$. The bar consists of $(n + 1)$ materials of different moduli of elasticity joined end to end, the joint occurring at $x_1, x_2, \ldots, x_n$; each section is of uniform cross section but varying from one section to another. The $i$th section has cross section $A_i$, modulus $E_i$, and specific weight $\gamma_i$. The bottom end, located at $x = 1$, is free. Denoting the displacement from equilibrium by $v(x, t)$, the vibration of the bar is described by

$$-\frac{g}{A(x)}\frac{\partial}{\partial x}\left[\frac{E(x)A(x)}{\gamma(x)}\frac{\partial v}{\partial x}(x, t)\right] - f(x, t)$$

$$= -\frac{\partial^2 v}{\partial t^2}(x, t), \quad t > 0; \qquad x \in I_i, \quad i = 1, 2, \ldots, n + 1$$

where $I_i \equiv (x_{i-1}, x_i)$, $E(x)$, $A(x)$, and $\gamma(x)$ are piecewise constant with values $E_i$, $A_i$, and $\gamma_i$, respectively, in $I_i$; $g$ is the acceleration due to gravity. $F(x, t)$ is the distributed applied force. (Note that $x_0 = 0$, $x_{n+1} = 1$.) At $x_i$ we must have

$$v(x_i-, t) = v(x_i+, t), \qquad i = 1, 2, \ldots, n$$

$$\frac{E_i A_i}{\gamma_i}\frac{\partial}{\partial x}v(x_i-, t) = \frac{E_{i+1}A_{i+1}}{\gamma_{i+1}}\frac{\partial}{\partial x}v(x_i+, t)$$

The boundary conditions are

$$v(0, t) = 0, \qquad \frac{\partial}{\partial x}v(1, t) = 0$$

The initial conditions are

$$v(x, 0) = v_0(x), \qquad \frac{\partial v}{\partial t}(x, 0) = v_1(x)$$

Cast the problem in the form

$$-\frac{d^2}{dt^2}\mathbf{v}(t) = \mathbf{T}\mathbf{v}(t) - \mathbf{f}(t), \qquad \mathbf{v}(0) = \mathbf{v}_0, \qquad \frac{d}{dt}\mathbf{v}(0) = \mathbf{v}_1$$

by identifying $\mathbf{T}$ as a self-adjoint operator and obtain the solution.

## 9.4 Self-Adjoint Differential Operators: "Weighted" Boundaries and Discontinuities

This section deals with another important class of differential operators. The focal differential expression is again (9.3.1), but one or more of the boundary points $a$ and $b$, and/or the points of discontinuity of $p(x)$, may have special "weights." By "weight" of a point here, we imply that in one-dimensional intervals such as $[a, b]$, the point in question has a concentrated (positive) "length" measure, unlike ordinary points. Thus if $a$ is a weighted point, it has a certain length associated with it; similarly with $b$ and points $x_1, x_2, \ldots$ of discontinuity of $p(x)$. When a weighted point is involved, a typical vector $\mathbf{u}$ among those of interest to us must not only feature $\{\mathbf{u}(x)\}$, $x \in [a, b]$, but also make a special inclusion of the value at the weighted point. Note particularly that the significance of the weighted point is that it contributes to the norm $||\mathbf{u}||$, unlike any other (nonweighted) single point of the interval $[a, b]$. Clearly, such vectors have been encountered before as originating from direct sums of linear spaces. If, for example, the end points $a$ and $b$ are both weighted, we would be dealing with the space $\mathcal{L}_2\{[a, b]: r(x)\} \oplus \mathcal{R}_2$, in which the elements of primary interest to us would be

$$\mathbf{u} \equiv \begin{bmatrix} \{u(x)\} \\ \begin{bmatrix} u(a) \\ u(b) \end{bmatrix} \end{bmatrix}$$

in which $\{u(x)\} \in \mathcal{L}_2\{[a, b]: r(x)\}$ and $\begin{bmatrix} u(a) \\ u(b) \end{bmatrix} \in \mathcal{R}_2$. If $a$ and $b$ are not weighted points but instead there are $n$ points $x_1, x_2, \ldots, x_n$ of discontinuity of $p(x)$ which are weighted, the space concerned would be $\mathcal{L}_2\{[a, b]: r(x)\} \oplus \mathcal{R}_n$ with particular interest in vectors $\mathbf{u}$ expressed as

$$\mathbf{u} \equiv \begin{bmatrix} \{u(x)\} \\ \begin{bmatrix} u(x_1) \\ u(x_2) \\ \cdot \\ \cdot \\ \cdot \\ u(x_n) \end{bmatrix} \end{bmatrix}$$

and so on. Before we proceed to formulate the operators of this section it will be convenient to establish a notation. If any expression, depending on $x \in [a, b]$, is enclosed within square brackets carrying a point in the interval as a *superscript*, the notation represents the limit of the expression as $x$ approaches the superscript. A *subscript* is used to denote the limit of the *negative* of the expression contained as $x$ tends to the subscript. If *both* a subscript and a superscript are present, it represents the usual notation of the limit of the enclosed expression as $x$ approaches the superscript *minus* the limit as $x$ approaches the subscript. Summarizing, we have

$$-\lim_{x \to a} [\cdot] \equiv [\cdot]_a, \qquad \lim_{x \to b} [\cdot] \equiv [\cdot]^b$$

$$\lim_{x \to x_1+} [\cdot] - \lim_{x \to x_1-} [\cdot] \equiv [\cdot]_{x_1-}^{x_1+}$$

and so on.

The formal operation of interest here is a *matrix differential expression* $L$ of the type

$$L \equiv \begin{bmatrix} -\dfrac{1}{r(x)}\dfrac{d}{dx}\left[p(x)\dfrac{d}{dx}\right] + q(x) & 0 & 0 \\ \alpha_a\left[-p(x)\dfrac{d}{dx}\right]^a & 0 & 0 \\ \alpha_b\left[-p(x)\dfrac{d}{dx}\right]_b & 0 & 0 \end{bmatrix} \tag{9.4.1}$$

where both $a$ and $b$ are weighted boundary points, and $\alpha_a$ and $\alpha_b$ are *positive* constants. Since the differential expression in the upper left-hand corner has been denoted by $T$ in (9.3.1), we may use this to simplify our notation somewhat. If only *one* of the end points is weighted, (9.4.1) must be replaced by

$$\begin{bmatrix} T & 0 \\ \alpha_a\left[-p(x)\dfrac{d}{dx}\right]^a & 0 \end{bmatrix} \quad \text{or} \quad \begin{bmatrix} T & 0 \\ \alpha_b\left[-p(x)\dfrac{d}{dx}\right]_b & 0 \end{bmatrix} \tag{9.4.2}$$

according as whether $a$ or $b$ is weighted, respectively. If $p(x)$ has discontinuities at interior points $x_1, x_2, \ldots, x_n$ in $(a, b)$ all of which are weighted, then

$$L \equiv \begin{bmatrix} T & 0 & \ldots & 0 \\ \alpha_1\left[-p(x)\dfrac{d}{dx}\right]_{x_1-}^{x_1+} & 0 & \ldots & 0 \\ \alpha_2\left[-p(x)\dfrac{d}{dx}\right]_{x_2-}^{x_2+} & 0 & \ldots & 0 \\ \vdots & \vdots & & \vdots \\ \alpha_n\left[-p(x)\dfrac{d}{dx}\right]_{x_n-}^{x_n+} & 0 & \ldots & 0 \end{bmatrix}, \qquad \alpha_1, \alpha_2, \ldots, \alpha_n > 0 \tag{9.4.3}$$

In all of the operations (9.4.1) to (9.4.3) the Hilbert space concerned is obtained by direct-summing $\mathcal{L}_2\{[a, b]: r(x)\}$ with finite-dimensional space $\mathcal{R}_2$ in the case of (9.4.1), $\mathcal{R}_1$ (or simply $\mathcal{R}$) for either operation in (9.4.2), and $\mathcal{R}_n$ for (9.4.3). Notice that the elements below $T$ in the first column in each case are linear (unbounded) functionals on elements of $\mathcal{L}_2\{[a, b]: r(x)\}$.

The question arises as to whether more general operations may be considered in which the zero elements may be replaced by other operations. Within the scope of this section, we will retain the elements appearing to the right of $T$ in the first row as zero. However, the other zero elements, which form a square null matrix of order equal to the number of weights, may be replaced by a self-adjoint matrix. We shall consider this case also.

Let us see how the formal operations considered above lead to self-adjoint operators. We begin with operation (9.4.1). The Hilbert space here is $\mathcal{H}$

$\equiv \mathcal{L}_2\{[a, b]: r(x)\} \oplus \mathcal{R}_2$, comprising element $\mathbf{u}$ denoted by

$$\mathbf{u} \equiv \begin{bmatrix} \{u(x)\} \\ \begin{bmatrix} u_1 \\ u_2 \end{bmatrix} \end{bmatrix}$$

The inner product on $\mathcal{H}$ is represented by

$$\langle \mathbf{u}, \mathbf{v} \rangle = \int_a^b r(x)u(x)v(x)\,dx + r_1u_1v_1 + r_2u_2v_2 \tag{9.4.4}$$

where $r_1$ and $r_2$ are positive constants at our disposal. For suitable elements $\mathbf{u}$ we may write

$$\begin{bmatrix} T & 0 & 0 \\ \alpha_a\left[-p(x)\dfrac{d}{dx}\right]^a & 0 & 0 \\ \alpha_b\left[-p(x)\dfrac{d}{dx}\right]_b & 0 & 0 \end{bmatrix} \begin{bmatrix} u(x) \\ u_1 \\ u_2 \end{bmatrix} = \begin{bmatrix} Tu(x) \\ -p(a)u'(a)\alpha_a \\ +p(b)u'(b)\alpha_b \end{bmatrix}$$

In terms of the inner product (9.4.4) it is readily shown that

$$\begin{aligned} \langle L\mathbf{u}, \mathbf{v} \rangle - \langle \mathbf{u}, L\mathbf{v} \rangle &= p(a)u'(a)[v(a) - \alpha_a r_1 v_1] - p(b)u'(b)[v(b) - \alpha_b r_2 v_2] \\ &\quad - p(a)v'(a)[u(a) - \alpha_a r_1 u_1) + p(b)v'(b)[u(b) - \alpha_b r_2 u_2] \end{aligned} \tag{9.4.5}$$

which vanishes if and only if $u(a) = \alpha_a r_1 u_1$, $u(b) = \alpha_b r_2 u_2$ and $v(a) = \alpha_a r_1 v_1$, $v(b) = \alpha_b r_2 v_2$. In (9.4.5) we have assumed that $p(x)$ is continuous in $[a, b]$. If we define the domain

$$D(L) = \{\mathbf{u} \in \mathcal{H};\, L\mathbf{u} \in \mathcal{H};\, u(a) = \alpha_a r_1 u_1,\ u(b) = \alpha_b r_2 u_2\} \tag{9.4.6}$$

then $\mathbf{L} = \{L, D(L)\}$ is a self-adjoint operator. Frequently, the relationship between $u(a)$ and $u_1$, and that between $u(b)$ and $u_2$, arises from the physics, as examples considered subsequently will show. If it is required, for example, that $u(a) = u_1$ and $u(b) = u_2$, then we must choose $r_1 = 1/\alpha_a$ and $r_2 = 1/\alpha_b$.

If only one of the boundary points is weighted, say $a$, then $L$ must be given by the expression on the left-hand side of (9.4.2) and we must define $\mathcal{H} = \mathcal{L}_2\{[a, b]: r(x)\} \oplus \mathcal{R}$ with $\mathbf{u} \in \mathcal{H}$ given by

$$\mathbf{u} \equiv \begin{bmatrix} \{u(x)\} \\ u_1 \end{bmatrix}$$

If the inner product is chosen as

$$\langle \mathbf{u}, \mathbf{v} \rangle = \int_a^b r(x)u(x)v(x)\,dx + r_1u_1v_1, \qquad r_1 > 0 \tag{9.4.7}$$

then

$$\begin{aligned} \langle L\mathbf{u}, \mathbf{v} \rangle - \langle \mathbf{u}, L\mathbf{v} \rangle &= p(a)u'(a)[v(a) - \alpha_a r_1 v_1] - p(b)[u'(b)v(b) \\ &\quad - u(b)v'(b)] - p(a)v'(a)[u(a) - \alpha_a r_1 u_1] \end{aligned} \tag{9.4.8}$$

Again $p(x)$ has been assumed to be continuous in $[a, b]$. Defining

$$D(L) = \{\mathbf{u} \in \mathcal{H};\, L\mathbf{u} \in \mathcal{H};\, u(a) = \alpha_a r_1 u_1,\ -u'(b) = \beta u(b)\} \tag{9.4.9}$$

where the unmixed boundary condition at $x = b$, together with $u(a) = \alpha_a r_1 u_1$, makes the right-hand side of (9.4.8) vanish. Notice particularly that a boundary condition is required at an unweighted end point. Thus $\mathbf{L} = \{L, D(L)\}$ is a self-adjoint operator.

As another variation, consider $L$ as in (9.4.1), with $p(x)$ discontinuous at unweighted point $x_1 \in (a, b)$. The Hilbert space is again $\mathcal{H} \equiv \mathcal{L}_2\{[a, b]: r(x)\} \oplus \mathcal{R}_2$ with inner product as in (9.4.4). Then if $D(L)$ in (9.4.6) is augmented to read

$$D(L) = \{\mathbf{u} \in \mathcal{H};\, L\mathbf{u} \in \mathcal{H}$$

$$u(a) = \alpha_a r_1 u_1,\ u(b) = \alpha_b r_2 v_2;\, \boldsymbol{\alpha}_1^- \mathbf{w}(u(x_1-)) + \boldsymbol{\alpha}_1^+ \mathbf{w}(u(x_1+)) = \mathbf{0}\}$$

$\mathbf{L} = \{L, D(L)\}$ is self-adjoint, provided that $\boldsymbol{\alpha}^-$ and $\boldsymbol{\alpha}^+$ satisfy the relationship

$$r_1^{-1} \boldsymbol{\alpha}_1^- \mathbf{P}^{-1}(x_1-) \boldsymbol{\alpha}^{-t} = r_2^{-1} \boldsymbol{\alpha}_1^+ \mathbf{P}^{-1}(x_1+) \boldsymbol{\alpha}^{+t} \tag{9.4.10}$$

which follows from arguments identical to those that led to (9.3.8).

Suppose now that we consider the operation $L$ in (9.4.3). The Hilbert space is $\mathcal{H} \equiv \mathcal{L}_2\{[a, b]: r(x)\} \oplus \mathcal{R}_n$ with elements

$$\mathbf{u} \equiv \begin{bmatrix} \{u(x)\} \\ u_1 \\ u_2 \\ \cdot \\ \cdot \\ \cdot \\ u_n \end{bmatrix}$$

and inner product

$$\langle \mathbf{u}, \mathbf{v} \rangle = \int_a^b r(x) u(x) v(x)\, dx + \sum_{i=1}^n r_i u_i v_i \tag{9.4.11}$$

If $\mathbf{u}$ and $\mathbf{v}$ satisfy unmixed boundary conditions at $a$ and $b$, then

$$\begin{aligned} \langle L\mathbf{u}, \mathbf{v} \rangle - \langle \mathbf{u}, L\mathbf{v} \rangle = \sum_{i=1}^n \{ & p(x_i+) u'(x_i+)[v(x_i+) - \alpha_i r_i v_i] \\ & - p(x_i+) v'(x_i+)[u(x_i+) - \alpha_i r_i u_i] \\ & - p(x_i-) u'(x_i-)[v(x_i-) - \alpha_i r_i v_i] \\ & + p(x_i-) v'(x_i-)[u(x_i-) - \alpha_i r_i u_i]\} \end{aligned} \tag{9.4.12}$$

The right-hand side of (9.4.12) will vanish if we pick $u(x_i+) = u(x_i-) = \alpha_i r_i u_i$ and an identical relationship for $v$. Thus

$$D(L) = \{\mathbf{u} \in \mathcal{H};\, T\mathbf{u} \in \mathcal{H};\, u(a) - \alpha u'(a) = 0,\ u(b) + \beta u'(b) = 0$$

$$u(x_i+) = u(x_i-) = \alpha_i r_i u_i,\ i = 1, 2, \ldots, n\}$$

yields a self-adjoint operator $\mathbf{L} = \{L, D(L)\}$.

We have now covered a variety of self-adjoint operators. They can be neatly summarized into the following theorem with the added generalization concerning the zero elements in $L$ to which reference was made earlier.

***Theorem 9.4.1*** Let $T$ be the differential expression

$$-\frac{1}{r(x)}\frac{d}{dx}\left[p(x)\frac{d}{dx}\right] + q(x)$$

with $p(x)$ discontinuous at $x_1, x_2, \ldots, x_m$ consisting of disjoint subsets $x'_1, x'_2, \ldots, x'_{m-n}$ which are not weighted, and $x''_1, x''_2, \ldots, x''_n$ which are weighted.

Let $\mathcal{H} = \mathcal{L}_2\{[a, b]: r(x)\} \oplus \mathcal{R}_n$ with inner product (9.4.11) and

$$L \equiv \begin{bmatrix} T & 0 & \cdots & 0 \\ -\alpha_1\left[p(x)\frac{d}{dx}\right]_{x_1''-}^{x''+} & k_{11} & \cdots & k_{1m} \\ & \vdots & & \vdots \\ -\alpha_n\left[p(x)\frac{d}{dx}\right]_{x_n''-}^{x_n''+} & k_{m1} & \cdots & k_{mm} \end{bmatrix}$$

with

$$\begin{aligned} D(L) = \{\mathbf{u} \in \mathcal{H}; L\mathbf{u} \in \mathcal{H}; \boldsymbol{\alpha}\mathbf{w}(u(a)) + \boldsymbol{\beta}\mathbf{w}(u(b)) = \mathbf{0}; \\ \boldsymbol{\alpha}_i^-\mathbf{w}(u(x_i'-)) + \boldsymbol{\alpha}_i^+\mathbf{w}(u(x_i'+)) = \mathbf{0}, \quad i = 1, 2, \ldots, n - m; \\ u(x_j''-) = u(x_j''+) = \alpha_j r_j u_j, \quad j = 1, 2, \ldots, n\} \end{aligned}$$

Then $\mathbf{L} = \{L, D(L)\}$ is self-adjoint if and only if

(i) $\boldsymbol{\alpha}\mathbf{P}^{-1}(a)\boldsymbol{\alpha}^t = \boldsymbol{\beta}\mathbf{P}^{-1}(b)\boldsymbol{\beta}^t$.

(ii) $\boldsymbol{\alpha}_i^-\mathbf{P}^{-1}(x_i-)\boldsymbol{\alpha}_1^{-t} = \boldsymbol{\alpha}_i^+\mathbf{P}^{-1}(x_i+)\boldsymbol{\alpha}_i^{+t}, \quad i = 1, 2, \ldots, n.$

(iii) $r_i k_{ij} = r_j k_{ji}$.

*Proof*: The proof is merely collecting results already established earlier. Thus (i) comes from Section 9.2, (ii) from Section 9.3 (see Theorem 9.3.1), and (iii) from Section 8.1 [Eq. (8.1.4)]. ▮

Let us consider some examples of applications in which the operators considered above are encountered.

*Example 1. Unsteady Binary Diffusion Between Two Finite Reservoirs*

Consider two reservoirs of volumes $V_0$ and $V_1$ connected by a capillary of length $l$ and cross section $A$, each containing a binary gaseous mixture at any instant. The reservoirs are perfectly mixed at all times. The entire (closed) assembly is maintained at constant temperature so that the total pressure must also remain constant. Diffusion occurs between the two reservoirs through the capillary, which can be described in terms of the concentration of one of the species, denoted $c(x', t')$, where $x'$ is the space coordinate and $t'$ is time. Defining dimensionless variables

$$x \equiv \frac{x'}{l}, \qquad t \equiv \frac{Dt'}{l^2}, \qquad u(x, t) = \frac{c}{c_T}$$

and parameters $\alpha_0 \equiv Al/V_0$, $\alpha_1 \equiv Al/V_1$, where $D$ is the diffusion coefficient and $c_T$ is the total average concentration which remains constant, the relevant equations are

$$-\frac{\partial^2 u}{\partial x^2} = -\frac{\partial u}{\partial t}, \qquad 0 < x < 1, \quad t > 0 \tag{9.4.13}$$

$$-\alpha_0 \frac{\partial u}{\partial x}(0+, t) = -\frac{du_1}{dt}, \qquad t > 0 \tag{9.4.14}$$

$$\alpha_1 \frac{\partial u}{\partial x}(1-, t) = -\frac{du_2}{dt}, \qquad t > 0 \tag{9.4.15}$$

where $u_1$ and $u_2$ are dimensionless reservoir concentrations. Now we expect the concentration in the capillary at $x = 0$ (which is connected to reservoir of volume $V_0$) to be the same as that in the reservoir. Similarly, the concentration at $x = 1$ within the capillary must be equal to that in the reservoir of volume $V_1$, which is $u_2$. Thus we must have

$$u(0+, t) = u_1, \qquad u(1-, t) = u_2 \tag{9.4.16}$$

Equations (9.4.13)–(9.4.15) suggest the formal operation

$$L \equiv \begin{bmatrix} -\dfrac{d^2}{dx^2} & 0 & 0 \\ \alpha_0\left[-\dfrac{d}{dx}\right]^{x=0} & 0 & 0 \\ \alpha_1\left[-\dfrac{d}{dx}\right]_{x=1} & 0 & 0 \end{bmatrix} \tag{9.4.17}$$

which is of the type (9.4.1) with $p(x) \equiv 1$; $x = 0$ and $x = 1$ are weighted points. The Hilbert space is $\mathcal{H} \equiv \mathcal{L}_2[0, 1] \oplus \mathcal{R}_2$ and in view of (9.4.16), we must have the inner product

$$\langle \mathbf{u}, \mathbf{v} \rangle = \int_0^1 u(x)v(x)\,dx + \frac{1}{\alpha_0}u_1 v_1 + \frac{1}{\alpha_1}u_2 v_2 \tag{9.4.18}$$

which is the same as (9.4.4) with $r_1 = 1/\alpha_0$ and $r_2 = 1/\alpha_1$. The domain $D(L)$ should be given by

$$D(L) = \{\mathbf{u} \in \mathcal{H};\, T\mathbf{u} \in \mathcal{H};\, u(0+) = u_1, u(1-) = u_2\} \tag{9.4.19}$$

which yields the self-adjoint operator $\mathbf{L} = \{L, D(L)\}$. The entire boundary initial value problem (9.4.13)–(9.4.15) may be stated succinctly as

$$-\frac{d}{dt}\mathbf{u}(t) = \mathbf{T}\mathbf{u}(t) \tag{9.4.20}$$

In specifying the initial condition $\mathbf{u}(0)$, notice in particular that one must specify $\{u(x, 0)\}$, $u_1(0)$, and $u_2(0)$. We shall take up the solution of this problem at a subsequent stage. Consult Exercise 9.4.1 for an extension of this example involving nonzero terms along the diagonal elements in (9.4.17) in the second and third rows.

*Example 2. Radial Conduction Through Composite Cylinder*

Consider unsteady radial conduction through concentric cylinders comprising materials of different thermal conductivities and volumetric heat capacities. The innermost cylinder of radius $R_1$ has conductivity $k_1$ and heat capacity $c_1$. There are two successive outer layers, the inner one with properties $k_2$ and $c_2$ located between $R_1$ and $R_2$ and the outermost with properties $k_3$ and $c_3$ stretching between $R_2$ and $R_3$. The radial coordinate $x'$ and time $t'$ are transformed to dimensionless coordinates $x = x'/R_3$ and $t = k_1 t'/c_1 R_3^2$, respectively. The outermost surface exchanges energy with the surroundings at temperature $T_f$. The dimensionless temperature $v(x, t) \equiv (T - T_f)/T_f$ must satisfy the differential equation

$$-\frac{1}{r(x)}\frac{\partial}{\partial x}\left[p(x)\frac{\partial v}{\partial x}\right] - f(x, t) = -\frac{\partial v}{\partial t} \tag{9.4.21}$$

where

$$p(x) = \begin{Bmatrix} x, & 0 < x < x_1, & x \\ \dfrac{xk_2}{k_1}, & x_1 < x < x_2, & \dfrac{xc_2}{c_1} \\ \dfrac{xk_3}{k_1}, & x_2 < x < 1, & \dfrac{xc_3}{c_1} \end{Bmatrix} = r(x)$$

Equation (9.4.21) makes provision for volumetric heating. In addition, we must have boundary conditions

$$v(0+, t) < \infty, \qquad -\frac{\partial v}{\partial x}(1, t) = \beta v(1, t), \quad \beta > 0 \tag{9.4.22}$$

and discontinuity (interface) conditions

$$v(x_i-, t) = v(x_i+, t), \qquad p(x_i-)\frac{\partial v}{\partial x}(x_i-, t) = p(x_i+)\frac{\partial v}{\partial x}(x_i+, t),$$
$$i = 1, 2$$

The foregoing problem is clearly an example of those encountered in Section

9.3. However, let us consider a slight variation here. It is left for the student to show that

$$p(x_1-)\frac{\partial v}{\partial x}(x_1-,t) - p(x_2+)\frac{\partial v}{\partial x}(x_2+,t) - \int_{x_1}^{x_2} r(x)f(x,t)\,dx = -\frac{\partial}{\partial t}\int_{x_1}^{x_2} r(x)v(x,t)\,dx \tag{9.4.23}$$

Now suppose that the thermal conductivity of the intermediate layer $k_2$ is considerably larger than $k_1$ or $k_3$; then one might make the approximation

$$v(x_1-,t) \simeq v(x,t) \simeq v(x_2+,t), \qquad x_1 < x < x_2$$

so that (9.4.23) becomes

$$\alpha_1\left[-p(x)\frac{\partial}{\partial x}v(x,t)\right]_{x_1^-}^{x_2^+} - f_1(t) = -\frac{d}{dt}v_1(t) \tag{9.4.24}$$

where we have set $\alpha_1^{-1} \equiv \frac{1}{2}(c_2/c_1)(x_2^2 - x_1^2)$, $f_1(t) \equiv \alpha_1 \int_{x_1}^{x_2} r(x)f(x,t)\,dx$, and $v_1(t) \equiv v(x,t)$, $x_1 < x < x_2$.

Equations (9.4.21), (9.4.22), and (9.4.24) lead to an operator of this section with a slight difference. The points $x_1$ and $x_2$ at which $p(x)$ is discontinuous are *weighted together*. This variation brings about no essential change in the formalism. The Hilbert space in this case is defined as $\mathcal{H} \equiv \mathcal{L}_2\{[0, x_1] \cup [x_2, 1]: r(x)\} \oplus \mathcal{R}$ with the inner product

$$\langle \mathbf{u}, \mathbf{v}\rangle = \int_0^{x_1} r(x)u(x)v(x)\,dx + \int_{x_2}^{1} r(x)u(x)v(x)\,dx + r_1u_1v_1 \tag{9.4.25}$$

The constant $r_1$ is readily found. Clearly,

$$L \equiv \begin{bmatrix} T & 0 \\ \alpha_1\left[-p(x)\frac{d}{dx}\right]_{x_1^-}^{x_2^+} & 0 \end{bmatrix}$$

with

$$D(L) = \{\mathbf{u} \in \mathcal{H};\, L\mathbf{u} \in \mathcal{H};\, u(0+) < \infty,$$
$$-u'(1) = \beta u(1),\, u_1 = u(x_1-) = u(x_2+)\}$$

will render $\mathbf{L} = \{L, D(L)\}$ self-adjoint if $r_1 = 1/\alpha_1$. The entire boundary-initial value problem may now be written as

$$-\frac{d}{dt}\mathbf{v}(t) = \mathbf{T}\mathbf{v}(t) - \mathbf{f}(t) \tag{9.4.26}$$

where

$$\mathbf{f}(t) = \begin{bmatrix} \{f(x,t)\} \\ f_1(t) \end{bmatrix}$$

**EXERCISES**

**9.4.1** Let the reservoirs in Example 1 be coated with a catalyst that allows a virtually irreversible first-order transformation of the species [whose concentration was considered in (9.4.13)]. The rate constant $k$ may be associated with a dimensionless parameter $\delta \equiv kl^2/D$. Identify the operator **L** and show that it is self-adjoint.

**9.4.2** Establish (9.4.23) in Example 2.

**9.4.3** Generalize Theorem 9.4.1 for the case when the inner product in $\mathfrak{L}_2\{[a, b]: r(x)\}$ is replaced by

$$\langle \mathbf{u}, \mathbf{v} \rangle = \sum_{i=0}^{n} \delta_i \int_{x'_i}^{x'_{i+1}} r(x)u(x)v(x)\, dx$$

where $x'_0 = a$, $x'_{n+1} = b$.

**9.4.4**† A liquid is contained within a closed isothermal system with a pure gas above it at some pressure initially. The vapor pressure of the liquid may be considered small relative to the total pressure $p$. Gas diffusion drops the pressure in the system and it is desired to calculate this change in pressure as a function of time. The gas may be assumed to be ideal and the solubility may be related to the pressure by a linear relationship of the form $c = (1/K)p$, where $K$ is a constant. The volume in the gas phase remains essentially constant at $V$. Neglecting free convection effects, formulate the transient diffusion problem assuming equilibrium at the interface. Show that the problem can be described in terms of an operator **T** with $T$ of the type (9.4.2) and $D(T)$ of the type (9.4.9).

**9.4.5** Formulate the problem of heat transfer to the sphere in Section 2.12 in terms of an operator in this section by identifying the Hilbert space, the formal operation, and its domain.

**9.4.6**‡ A first-order chemical reaction is carried out in a tubular reactor that discharges into a stirred reactor. The entire assembly is held at a constant temperature. Reaction occurs in the tube as well as the stirred vessel. The axial dispersion model applies in the tube so that the differential equation in $c$, the concentration of reactant, is given by

$$D\frac{\partial^2 c}{\partial x^2} - v\frac{\partial c}{\partial x} - kc = \frac{\partial c}{\partial t}, \qquad 0 < x < l, \quad t > 0$$

The boundary condition at the inlet to the tube is

$$-D\frac{\partial c}{\partial x}(0+, t) + vc(0+, t) = vc_f$$

where $c_f$ is the feed to the tubular section. The boundary condition at $x = l$

†See D. Ramkrishna, *Advances in Transport Processes*, Vol. III, edited by A. S. Mujumdar and R. A. Mashelkar, Wiley Eastern, New Delhi, 1983.

‡See D. Ramkrishna and N. R. Amundson, *Chem. Eng. Sci.*, *29*, 1353, 1974.

is obtained by a "reactant balance" for the stirred reactor. If $c_1$ is the concentration in the stirred vessel with a constant holdup volume $V$, one gets

$$vAc(1-0, t) - DA\frac{\partial c}{\partial x}(1-0, t) - Vkc_1 - qc_1 = V\frac{dc_1}{dt}$$

The initial condition is given by $c(x, 0) = c_0(x)$ in the tube and $c_1(0) = c_{1,0}$ in the stirred pot; $q = vA$ is the flow out of the pot. Cast this problem in terms of a suitable self-adjoint operator from this section.

**9.4.7** Extend the formulation of Exercise 9.4.6 for the case of $m$ stirred pots of volumes $V_1, V_2, \ldots, V_m$ located at $x_1, x_2, \ldots, x_m$ along the tubular reactor. The boundary condition at $x = 0$ is the same as that in Exercise 9.4.6. At $x = l$ use the boundary condition

$$\frac{\partial c}{\partial x}(l-0, t) = 0$$

Identify the self-adjoint operator featured in this problem.

**9.4.8** Extend the formulation of Exercise 9.4.7 for the case of the first-order reaction system of Section 8.2, consisting of $n$ reacting species. Identify the self-adjoint operator involved.

### 9.4.1 Existence of a Compact Inverse

This section (similar to Sections 9.2.1 and 9.3.1) is to show that the operators of Section 9.4 have a discrete spectrum. Interestingly, the task here is virtually one of collating the results in Section 9.2.1 and 9.3.1. We will therefore not be very exhaustive in our demonstration but select only two principal cases.

Consider, for example, the operator $\mathbf{L} = \{L, D(L)\}$, where $L$ is given by (9.4.1) and $D(L)$ by (9.4.6). In constructing the inverse of $\mathbf{L}$, we seek to solve†

$$\mathbf{Lu} = \mathbf{f}, \qquad \mathbf{f} \equiv \begin{bmatrix} \{f(x)\} \\ \begin{bmatrix} f_1 \\ f_2 \end{bmatrix} \end{bmatrix} \in \mathcal{H} \equiv \mathcal{L}_2\{[a, b]: r(x)\} \oplus \mathcal{R}_2 \tag{9.4.27}$$

which implies that

$$Tu(x) = f(x) \tag{9.4.28}$$

$$-\alpha_a p(a)u'(a) = f_1 \tag{9.4.29}$$

$$\alpha_b p(b)u'(b) = f_2 \tag{9.4.30}$$

The solution of (9.4.28)–(9.4.30) has already been encountered in Section 9.2.1. We can solve for $u(x)$ in terms of the Green's function of the operator $\mathbf{T} = \{T, D(T)\}$, where $D(T) = \{\mathbf{g}, T\mathbf{g} \in \mathcal{L}_2\{[a, b]: r(x)\};\ g'(a) = g'(b) = 0\}$

---

†Note that if $q(x) \equiv 0$ in $T$, $\mathbf{L}$ has no inverse. In this case one must establish that for some $\lambda_0$, $(\mathbf{L} - \lambda_0\mathbf{I})$ has an inverse (see Exercise 9.4.9).

[see Eq. (9.2.46)]. Denoting the Green's function for **T** by $G_T(x, \xi)$, we may write

$$u(x) = \int_a^b r(\xi)G_T(x, \xi)f(\xi)\,d\xi + \mathbf{Y}^t(x)\mathbf{D}_0^{-1}\begin{bmatrix} f_1 \\ f_2 \end{bmatrix} \tag{9.4.31}$$

where $\mathbf{Y}^t(x)$ is the row vector containing the linearly independent solutions, $y_1(x)$ and $y_2(x)$, of $Ty(x) = 0$. The matrix $\mathbf{D}_0$ in the present context is given by

$$\mathbf{D}_0 = \begin{bmatrix} 0 & -\alpha_a p(a) \\ 0 & 0 \end{bmatrix}\begin{bmatrix} y_1(a) & y_2(a) \\ y_1'(a) & y_2'(a) \end{bmatrix} + \begin{bmatrix} 0 & 0 \\ 0 & \alpha_b p(b) \end{bmatrix}\begin{bmatrix} y_1(b) & y_2(b) \\ y_1'(b) & y_2'(b) \end{bmatrix}$$

$$= \begin{bmatrix} -\alpha_a p(a)y_1'(a) & -\alpha_a p(a)y_2'(a) \\ \alpha_b p(b)y_1'(b) & \alpha_b p(b)y_2'(b) \end{bmatrix} \tag{9.4.32}$$

Since $D(L)$ requires that $u(a) = \alpha_a r_1 u_1$, $u(b) = \alpha_b r_2 u_2$, we obtain

$$u_1 = \frac{1}{\alpha_a r_1}\left\{\int_a^b r(\xi)G_T(a, \xi)f(\xi)\,d\xi + \mathbf{Y}^t(a)\mathbf{D}_0^{-1}\begin{bmatrix} f_1 \\ f_2 \end{bmatrix}\right\} \tag{9.4.33}$$

$$u_2 = \frac{1}{\alpha_b r_2}\left\{\int_a^b r(\xi)G_T(b, \xi)f(\xi)d\xi + \mathbf{Y}^t(b)\mathbf{D}_0^{-1}\begin{bmatrix} f_1 \\ f_2 \end{bmatrix}\right\} \tag{9.4.34}$$

Equations (9.4.31), (9.4.33), and (9.4.34) may be combined to represent the solution **u** to (9.4.27) as

$$\mathbf{u} \equiv \begin{bmatrix} \{u(x)\} \\ \begin{bmatrix} u_1 \\ u_2 \end{bmatrix} \end{bmatrix} = \begin{bmatrix} \mathbf{T}^{-1} & \mathbf{0} \\ \dfrac{1}{\alpha_a r_1}[\mathbf{T}^{-1}]^a & \dfrac{1}{\alpha_a r_1}\mathbf{Y}^t(a)\mathbf{D}_0^{-1} \\ \dfrac{1}{\alpha_b r_2}[\mathbf{T}^{-1}]^b & \dfrac{1}{\alpha_b r_2}\mathbf{Y}^t(b)\mathbf{D}_0^{-1} \end{bmatrix}\begin{bmatrix} \{f(x)\} \\ \begin{bmatrix} f_1 \\ f_2 \end{bmatrix} \end{bmatrix} \tag{9.4.35}$$

Thus $\mathbf{L}^{-1}$ is represented on the right-hand side of (9.4.35) by a partitioned matrix in which the elements in the second column are $2 \times 2$ matrices operating on the vector $\left[\begin{smallmatrix} f_1 \\ f_2 \end{smallmatrix}\right]$. That $\mathbf{L}^{-1}$ is compact follows readily from the fact that $\mathbf{T}^{-1}$ is compact. The details are left as an exercise.

As another example, consider the operator **L**, with $L$ as in (9.4.3) and $D(L)$, which appears below (9.4.12). If we consider only one weighted discontinuity of $p(x)$ at $x_1$, then $\mathbf{Lu} = \mathbf{f}$ will yield

$$Tu(x) = f(x)$$

$$u(a) - \alpha u'(a) = 0$$

$$u(b) + \beta u'(b) = 0$$

$$\alpha_1[-p(x_1+)u'(x_1+) + p(x_1-)u'(x_1-)] = f_1$$

$$u(x_1-) = u(x_1+)$$

The solution of the foregoing problem is considered in even greater generality in Exercise 9.3.7. Proceeding exactly as in the preceding case, one arrives at an operator $\mathbf{L}^{-1}$ that is readily perceived as being compact. The details are extremely cumbersome, however.

Thus the operators of this section also have discrete eigenvalues $\{\lambda_j\}$ and corresponding eigenvectors $\{z_j\}$ which form an orthonormal basis in $\mathcal{H}$.

**EXERCISES**

**9.4.9** Let $q(x) \equiv 0$ in the expression (9.3.1) for $T$. Consider operation $L$ as given by (9.4.1) with $D(L)$ by (9.4.6). Although $\mathbf{L}$ has no inverse, show that spectral theorem 7.4.3 nevertheless applies to $\mathbf{L}$.

**9.4.10** Construct the Green's function for the operators in
(a) Example 1 of Section 9.4.
(b) Example 2 of Section 9.4.
(c) Exercise 9.4.6 and hence calculate the concentration of the reactant at the outlet under steady-state conditions.

### 9.4.2 Evaluation of Spectra

In view of the general treatments in Sections 9.2.2 and 9.3.2 for evaluating the spectra of the respective operators of Sections 9.2 and 9.3, a similar discussion becomes quite unnecessary for the operators of this section. This is mainly because the eigenvalue problem $\mathbf{Lz} = \lambda\mathbf{z}$ leads to a virtually identical system of equations as those encountered in Sections 9.2.2 and 9.3.2. It will therefore suffice to consider some examples directly for the purposes of demonstration.

*Example 1*

We will evaluate the spectrum of the operator in the binary diffusion problem encountered as Example 1 in Section 9.4. Thus $\mathbf{L}$ is given by the operation $L$ in (9.4.17) and $D(L)$ as given by (9.4.19). To solve the eigenvalue problem $\mathbf{Lz} = \lambda\mathbf{z}$, we have

$$-\frac{d^2}{dx^2}z(x) = \lambda z(x)$$

$$-\alpha_0 z'(0) = \lambda z_1 = \lambda z(0)$$

$$\alpha_1 z'(1) = \lambda z_2 = \lambda z(1)$$

Solving the differential equation, $z(x, \lambda) = k_1 \sin\sqrt{\lambda}\, x + k_2 \cos\sqrt{\lambda}\, x$, which on substituting into the equations at $x = 0$, $x = 1$,

$$-\alpha_0\sqrt{\lambda}\, k_1 - \lambda k_2 = 0$$

$$(\alpha_1\sqrt{\lambda}\cos\sqrt{\lambda} - \lambda\sin\sqrt{\lambda})k_1 + (-\alpha_1\sqrt{\lambda}\sin\sqrt{\lambda} - \lambda\cos\sqrt{\lambda})k_2 = 0$$

the characteristic equation becomes

$$\begin{vmatrix} -\alpha_0\sqrt{\lambda} & -\lambda \\ \alpha_1\sqrt{\lambda}\cos\sqrt{\lambda} - \lambda\sin\sqrt{\lambda} & -\alpha_1\sqrt{\lambda}\sin\sqrt{\lambda} - \lambda\cos\sqrt{\lambda} \end{vmatrix} = 0$$

Indeed, $\lambda = 0$ is an eigenvalue with corresponding eigenvector $z_0$, given by

$$\mathbf{z}_0 = k_0\begin{bmatrix} \{1\} \\ \begin{bmatrix} 1 \\ 1 \end{bmatrix} \end{bmatrix}, \qquad k_0 = \left(1 + \frac{1}{\alpha_0} + \frac{1}{\alpha_1}\right)^{-1/2}$$

so that $\|\mathbf{z}_0\| = 1$. The other eigenvalues are obtained by simplifying the characteristic equation to

$$\tan\sqrt{\lambda} = \sqrt{\lambda}\,\frac{\alpha_0 + \alpha_1}{\lambda - \alpha_0\alpha_1}$$

and solving it numerically. That the eigenvalues are all nonnegative follows from the fact that

$$\langle \mathbf{Lu}, \mathbf{u}\rangle = \int_0^1 -u(x)\frac{d}{dx}\left[p(x)\frac{du}{dx}\right]dx + \frac{1}{\alpha_0}\{-\alpha_0 u_1 u'(0)\} + \frac{1}{\alpha_1}\{\alpha_1 u_2 u'(1)\}$$

$$= \int_0^1 p(x)\left(\frac{du}{dx}\right)^2 dx \geq 0$$

The $j$th eigenvector $\mathbf{z}_j$ corresponding to the $j$th eigenvalue $\lambda_j$ is given by

$$\mathbf{z}_j = k_{1,j}\begin{bmatrix} \left\{\sin\sqrt{\lambda_j}x - \dfrac{\alpha_0}{\sqrt{\lambda_j}}\cos\sqrt{\lambda_j}\,x\right\} \\ -\dfrac{\alpha_0}{\sqrt{\lambda_j}} \\ \sin\sqrt{\lambda_j} - \dfrac{\alpha_0}{\sqrt{\lambda_j}}\cos\sqrt{\lambda_j} \end{bmatrix}$$

where

$$1 = k_{1,j}^2\left[\int_0^1\left(\sin\sqrt{\lambda_j}x - \frac{\alpha_0}{\sqrt{\lambda_j}}\cos\sqrt{\lambda_j}x\right)^2 dx + \frac{1}{\alpha_0}\frac{\alpha_0^2}{\lambda_j} + \frac{1}{\alpha_1}\left(\sin\sqrt{\lambda_j} - \frac{\alpha_0}{\sqrt{\lambda_j}}\cos\sqrt{\lambda_j}\right)^2\right]$$

$$= k_{1,j}^2\left[\frac{1}{2}\left(1 + \frac{\alpha_0^2}{\lambda_j} - \frac{\alpha_0}{\sqrt{\lambda_j}} + \frac{2\alpha_0}{\lambda_j}\right) + \frac{\sin 2\sqrt{\lambda_j}}{4\sqrt{\lambda_j}}\left(\frac{\alpha_0^2}{\lambda_j} - 1\right) + \frac{\alpha_0\cos 2\sqrt{\lambda_j}}{2\sqrt{\lambda_j}}\right]$$

in order that $\|\mathbf{z}_j\| = 1$.

*Example 2*

Let us determine the characteristic equation of the operator **L** in Example 2, considered earlier in this section, on radial conduction through a composite cylinder. $\mathbf{Lz} = \lambda \mathbf{z}$ yields

$$-\frac{1}{x}\frac{d}{dx}\left(x\frac{dz}{dx}\right) = \lambda z \qquad 0 < x < x_1$$

$$-\frac{1}{x}\frac{d}{dx}\left(x\frac{dz}{dx}\right) = \frac{\lambda}{\gamma_1} z \qquad x_2 < x < 1$$

$$\alpha_1[-\kappa z'(x_2+) + z'(x_1-)] = \lambda z_1 = \lambda z(x_1)$$

$$z(0+) < \infty$$

$$z(x_1) = z(x_2)$$

$$-z'(1) = \beta z(1)$$

where $\kappa \equiv k_3/k_1$ and $\gamma_1 = \kappa(c_3/c_1)$. The solutions of the differential equations are given by

$$z(x, \lambda) = \begin{cases} k_1^{(1)} J_0(\sqrt{\lambda}\, x), & 0 < x < x_1 \\ k_1^{(2)} J_0(\sqrt{\lambda/\gamma_1}\, x) + k_2^{(2)} Y_0(\sqrt{\lambda/\gamma_1}\, x), & x_2 < x < 1 \end{cases}$$

where we have accounted for the boundedness of $z(0+, \lambda)$. Substituting the foregoing into the remaining equations, we obtain

$$\alpha_1[k_1^{(1)}\sqrt{\lambda}\, J_0'(\sqrt{\lambda}\, x_1) - \kappa\sqrt{\lambda/\gamma_1}\,\{k_1^{(2)} J_0'(\sqrt{\lambda/\gamma_1}\, x_2) + k_2^{(2)} Y_0'(\sqrt{\lambda/\gamma_1}\, x_2)\}] = \lambda k_1^{(1)} J_0(\sqrt{\lambda}\, x_1)$$

$$k_1^{(1)} J_0(\sqrt{\lambda}\, x_1) = k_1^{(2)} J_0(\sqrt{\lambda/\gamma_1}\, x_2) + k_2^{(2)} Y_0(\sqrt{\lambda/\gamma_1}\, x_2)$$

$$-\sqrt{\lambda/\gamma_1}\,[k_1^{(2)} J_0'\sqrt{\lambda/\gamma_1}) + k_2^{(2)} Y_0'(\sqrt{\lambda/\gamma_1})] = \beta[k_1^{(2)} J_0(\sqrt{\lambda/\gamma_1}) + k_2^{(2)} Y_0(\sqrt{\lambda/\gamma_1})]$$

which leads to the characteristic equation

$$\begin{vmatrix} \alpha_1\sqrt{\lambda}J_0'(\sqrt{\lambda}x_1) - \lambda J_0(\sqrt{\lambda}x_1) & -\kappa\sqrt{\lambda/\gamma_1}J_0'(\sqrt{\lambda/\gamma_1}x_2) & -\kappa\sqrt{\lambda/\gamma_1}\,Y_0'(\sqrt{\lambda/\gamma_1}x_2) \\ J_0(\sqrt{\lambda}x_1) & -J_0(\sqrt{\lambda/\gamma_1}x_2) & -Y_0(\sqrt{\lambda/\gamma_1}x_2) \\ 0 & \sqrt{\lambda/\gamma_1}J_0'(\sqrt{\lambda/\gamma_1}) + \beta J_0(\sqrt{\lambda/\gamma_1}) & \sqrt{\lambda/\gamma_1}\,Y_0'(\sqrt{\lambda/\gamma_1}) + \beta Y_0(\sqrt{\lambda/\gamma_1}) \end{vmatrix} = 0$$

Here $\lambda = 0$ is not an eigenvalue because the boundary condition at $x = 1$ permits only the trivial solution in the entire interval. In a manner analogous to that for Example 1, it is possible to show that the eigenvalues are all positive. These eigenvalues are to be determined numerically by solving the nonlinear algebraic equation that accrues from the determinantal equation. When the eigenvalues are determined, the eigenvectors are readily solved for. The normalization condition $\|\mathbf{z}_j\| = 1$ identifies the constants $k_{1,j}^{(1)}$, $k_{1,j}^{(2)}$, and $k_{2,j}^{(2)}$ uniquely. The spectral information on the operator **L** would then be complete.

**EXERCISE**

**9.4.11** Identify the characteristic equation for the operators in
(a) Exercise 9.4.1.
(b) Exercise 9.4.4.
(c) Exercise 9.4.5.
(d) Exercise 9.4.6.

### 9.4.3 Boundary Value Problems

Once the boundary value problem is represented in terms of a self-adjoint operator, the mode of exploitation of the methods of Section 7.5 to solving equations involving the operators of Sections 9.2 and 9.3 carries over just as well to the operators of this section. A general discussion is therefore quite redundant.

Examples 1 and 2 of Section 9.4 are solved readily in terms of the eigenvalues and eigenvectors of $L$. The solution is determined by (9.2.68) in accordance with the methods of Section 7.5. The student is advised to solve as many of the following exercises as possible.

**EXERCISES**

**9.4.12** Obtain the spectral solutions for the boundary initial value problems in
(a) Exercise 9.4.1.
(b) Exercise 9.4.4.
(c) Transient version of Exercise 9.4.5.
(d) Exercise 9.4.6.

**9.4.13** Reconsider Exercise 9.3.17, which concerns longitudinal vibrations of a prismatic bar. Retain all conditions of the problem but consider concentrated weight $W_i$ at the $i$th joint, where $i$ varies from $1, 2, \ldots, n$. A force $f_i(t)$ is continually applied at the $i$th joint and the local vibration is described by

$$\frac{g}{w_i}\left[E_iA_i\frac{\partial}{\partial x}v(x_i-, t) - E_{i+1}A_{i+1}\frac{\partial}{\partial x}v(x_i+, t)\right] + f_i(t) = -\frac{d^2v_i}{dt^2}$$

where $v_i$ is the displacement of joint $i$. The initial conditions for the $i$th joint are stated as

$$v_i(0) = v_{i,0}, \qquad v_i'(0) = v_{i,1}$$

Formulate the boundary-initial value problem in terms of a self-adjoint operation of Section 9.4 and obtain a spectral solution.

## 9.5 General Even-Order Differential Operators

We have so far dealt with second-order differential operators. Since the applications of interest have been in real Hilbert spaces, odd-order differential expressions are ruled out as formally self-adjoint differential operations (which is why first-order expressions did not enter our discussion). In this section we begin by establishing the fact just mentioned and consider the general even-order differential operator. Much of the spadework for this has already been accomplished by the manner in which we dealt with the second-order differential expressions. Thus several results established for second-order operators readily apply as well for the general even-order operators.

Consider the differential expression

$$T \equiv a_0(x)\frac{d^n}{dx^n} + a_1(x)\frac{d^{n-1}}{dx^{n-1}} + \ldots + a_{n-1}(x)\frac{d}{dx} + a_n(x) \tag{9.5.1}$$

where, for the present, we will assume that $a_{n-k}(x)$ is $k$ times differentiable for $x$ in some interval, say $[a, b]$. The operation (9.5.1) will apply to a suitable subspace of $\mathfrak{L}_2[a, b]$.

In defining the adjoint operation $T^*$, we must let $\langle T\mathbf{u}, \mathbf{v}\rangle - \langle \mathbf{u}, T^*\mathbf{v}\rangle$ depend entirely on the boundary values of the functions $u(x)$ and $v(x)$, and their derivatives of order up to $(n - 1)$. In preparation let us denote $d^ku/dx^k$ by $u^{(k)}$ and observe that

$$\begin{aligned} vu^{(k)} &= [vu^{(k-1)}]' - v'u^{(k-1)} \\ &= [vu^{(k-1)} - v'u^{(k-2)}]' + v''u^{(k-2)} \\ &= [vu^{(k-1)} - v'u^{(k-2)} + v''u^{(k-3)}]' - v^{(3)}u^{(k-3)} \\ &= [\sum_{j=0}^{k-1} (-1)^{k-1-j}u^{(j)}v^{(k-1-j)}]' + (-1)^k v^{(k)}u \end{aligned}$$

The foregoing result may be used to write

$$\begin{aligned} vTu = v\sum_{k=0}^{n} a_{n-k}u^{(k)} &= [\sum_{k=1}^{n}\sum_{j=0}^{k-1} (-1^{k-1-j}u^{(j)}(a_{n-k}v)^{(k-1-j)}]' \\ &\quad + u\sum_{k=0}^{n}(-1)^k(a_{n-k}v)^{(k)} \end{aligned} \tag{9.5.2}$$

which prompts the definition

$$T^*v = \sum_{k=0}^{n} (-1)^k(a_{n-k}v)^{(k)}$$

for the formal adjoint operation. Using Leibnitz's formula for higher-order derivatives of the product of two functions, we may write

$$T^* = \sum_{k=0}^{n} \left\{\sum_{r=0}^{n-k} \binom{k+r}{r}(-1)^{k+r}a_{n-k-r}^{(r)}\right\}\frac{d^k}{dx^k} \tag{9.5.3}$$

If we define the Wronskian vector $\mathbf{w}(u(x))$ as

$$\mathbf{w}(u(x)) = \begin{bmatrix} u(x) \\ u'(x) \\ \cdot \\ \cdot \\ \cdot \\ u^{(n-1)}_{(x)} \end{bmatrix}$$

then a few algebraic manipulations will lead to the formula

$$\langle T\mathbf{u}, \mathbf{v}\rangle - \langle \mathbf{u}, T^*\mathbf{v}\rangle = \Big[\mathbf{w}^t(v(x))\mathbf{P}(x)\mathbf{w}(u(x))\Big]_a^b \tag{9.5.4}$$

where the matrix $\mathbf{P}$ has elements $p_{ij}$, defined by†

$$p_{ij} = \begin{cases} \sum_{h=i}^{n-j+1} (-1)^h \binom{h-i}{i-1} a_{n-h-j+1}^{(h-i)}, & 1 \leq i \leq n-j+1 \\ 0 \qquad\qquad , & n-j+1 < i \leq n \end{cases} \tag{9.5.5}$$

Note that (9.5.4) is a generalization of (9.1.13).

It is now possible to obtain an elegant criterion for $T$ to be formally self-adjoint, that is, $T = T^*$. Equation (9.5.2) implies that

$$vTu - uT^*v = [\mathbf{w}^t(v(x))\mathbf{P}(x)\mathbf{w}(u(x))]' \tag{9.5.6}$$

Precisely analogous arguments must lead to a matrix $\mathbf{Q}(x)$ such that

$$\begin{aligned} uT^*v - vT^{**}u &= [\mathbf{w}^t(u(x))\mathbf{Q}(x)\mathbf{w}(v(x))]' \\ &= [\mathbf{w}^t(v(x))\mathbf{Q}^t(x)\mathbf{w}(u(x))]' \end{aligned} \tag{9.5.7}$$

where $T^{**}$ is the formal adjoint of the operation $T^*$. Adding (9.5.6) and (9.5.7), one gets

$$v(T - T^{**})u = [\mathbf{w}^t(v(x))\{\mathbf{P}(x) + \mathbf{Q}^t(x)\}\mathbf{w}(u(x))]'$$

Now the expression in brackets on the right-hand side of the foregoing expression is bilinear in $u$ and $v$ and may be expressed as

$$\mathbf{w}^t(v)(\mathbf{P} + \mathbf{Q}^t)\mathbf{w}(u) = r_1 b + r_2 v' + \ldots + r_n v^{(n-1)}$$

where $r_1, r_2, \ldots, r_n$ depend linearly on the choice of $u$. Thus we may write

$$v(T - T^{**})u = [r_1 v + r_2 v' + \ldots + r_n v^{(n-1)}]'$$

The expression on the left-hand side of the identity above involves no derivatives of $v$, which implies that $r_1 = r_2 = \ldots = r_n = 0$. Thus we obtain

$$(T - T^{**})u = 0$$

†See *Theory of Ordinary Differential Equations*, by R. H. Cole, Appleton-Century-Crofts, New York, 1968, p. 63.

for *all* admissible functions $u$ so that $T = T^{**}$. Also, we have

$$\mathbf{w}^t(v)(\mathbf{P} + \mathbf{Q}^t)\mathbf{w}(u) = 0$$

for every choice of $u$ and $v$; we must therefore have

$$\mathbf{P} + \mathbf{Q}^t = \mathbf{0} \quad \text{or} \quad \mathbf{P} = -\mathbf{Q}^t$$

Thus for a *formally self-adjoint operation*, we arrive at the important relationship

$$\mathbf{P} = -\mathbf{P}^t \tag{9.5.8}$$

that is, $\mathbf{P}$ must be a *skew-symmetric* matrix. Since the determinant of $\mathbf{P}$ cannot be identically equal to zero, we conclude that *$n$ must be even* for a self-adjoint expression.†

An elegant representation of the general even-order self-adjoint differential operation [corresponding to the form (9.2.1) for the second-order expression] is as follows.‡

Let $p_0, p_1, \ldots, p_n$ be functions such that $p_i$ has $(n - i)$ continuous derivatives (although those conditions may be relaxed further to piecewise properties such as in Section 9.3). Then letting $D \equiv d/dx$, the general $2n$th-order formally self-adjoint expression may be written as

$$T = p_n D^0 - D[p_{n-1}D - D\{p_{n-2}D^2 - \ldots - D(p_1 D^{n-1} - Dp_0 D^n)\ldots\}] \tag{9.5.9}$$

For an even simpler representation let

$$\begin{aligned} D^{[k]} &= D^k, & k &= 0, 1, 2, \ldots, n-1 \\ D^{[n]} &= p_0 D^n & & \\ D^{[n+k]} &= p_k D^{n-k} - DD^{[n+k-1]}, & k &= 1, 2, \ldots, n \end{aligned} \tag{9.5.10}$$

The operation $T$ may then be represented by the simple form

$$T = D^{[2n]} \tag{9.5.11}$$

Denoting $D^{[k]}u \equiv u^{[k]}$, we may represent (9.5.6) by the alternative form

$$vTu - uTv = \sum_{k=1}^{n} \{u^{[k-1]}v^{[2n-k]} - u^{[2n-k]}v^{[k-1]}\}' \tag{9.5.12}$$

---

†For $\mathbf{P} = -\mathbf{P}^t$, $P = -(1)^n P$, where $P$ is the determinant of $\mathbf{P}$. If $n$ is odd, then $P \equiv 0$, which makes $\mathbf{P}$ singular. Thus $n$ must be even.

‡See, for example, *Theory of Linear Operators in Hilbert Space*, Vol. 2, by N. Akhizer and M. Glazman, translated from the Russian by Merlynd Nestell, Frederick Ungar, New York, 1961.

From (9.5.6) with $T = T^*$, we obtain

$$\langle T\mathbf{u}, \mathbf{v}\rangle - \langle \mathbf{u}, T\mathbf{v}\rangle = [\mathbf{w}^t(v(a)) \quad \mathbf{w}^t(v(b))]\begin{bmatrix} -\mathbf{P}(a) & \mathbf{0} \\ \mathbf{0} & \mathbf{P}(b) \end{bmatrix}\begin{bmatrix} \mathbf{w}(u(a)) \\ \mathbf{w}(u(b)) \end{bmatrix} \tag{9.5.13}$$

which is the same as (9.1.14). Alternatively, from (9.5.12) we have the form†

$$\langle T\mathbf{u}, \mathbf{v}\rangle - \langle \mathbf{u}, T\mathbf{v}\rangle = \sum_{k=1}^{n} \Big[ u^{[k-1]}v^{[2n-k]} - u^{[2n-k]}v^{[k-1]} \Big]_a^b \tag{9.5.14}$$

The advantage of the form (9.5.13) is that several results established for the second-order self-adjoint expression carry over virtually without change for the even-order operator. We will therefore retain this form, although the merit of (9.5.14) lies in its automatically reflecting the formal self-adjointness of the expression $T$. Finally we admit also the differential expressions

$$T \equiv \frac{1}{r(x)} \sum_{k=0}^{2n} a_{2n-k} \frac{d^k}{dx^k} \quad \text{or} \quad T = \frac{1}{r(x)} D^{[2n]}, \qquad r(x) > 0 \tag{9.5.15}$$

the first of which has $T^*$ given by the right-hand side of (9.5.3) divided by $r(x)$, while the second is formally self-adjoint, provided that the inner product is given by (9.1.8). Again (9.5.4) holds for the first operation while (9.5.14) holds for the second with the inner product given by (9.1.8).

### 9.5.1 Self-Adjoint Operators

As long as $a_0(x), a_1(x), \ldots, a_{2n-k}(x)$ satisfy the maximum differentiability constraints, that is, $a_{2n-k}$ is $k$ times continuously differentiable) and the condition of skew symmetry (9.5.8), the situation here is identical to that of Section 9.2. Thus define the domain

$$D(T) = \{\mathbf{u} \in \mathcal{H}; T\mathbf{u} \in \mathcal{H}; \boldsymbol{\alpha}\mathbf{w}(u(a)) + \boldsymbol{\beta}\mathbf{w}(u(b)) = \mathbf{0}\} \tag{9.5.16}$$

where $\mathcal{H} \equiv \mathcal{L}_2\{[a, b]: r(x)\}$ and $\boldsymbol{\alpha}$ and $\boldsymbol{\beta}$ are $2n \times 2n$ matrices such that the partitioned matrix $[\boldsymbol{\alpha} \quad \boldsymbol{\beta}]$ has rank $2n$. Following exactly the same proof as in Section 9.2, we conclude that $\mathbf{T} = \{T, D(T)\}$ is self-adjoint if and only if

$$\boldsymbol{\alpha}\mathbf{P}^{-1}(a)\boldsymbol{\alpha}^t = \boldsymbol{\beta}\mathbf{P}^{-1}(b)\boldsymbol{\beta}^t$$

That $\mathbf{T}$ has a compact inverse follows from the fact that the development of the Green's function in Section 9.2.1 applies here with only minor changes.

---

†This expression is the counterpart of the formula

$$\langle T\mathbf{u}, \mathbf{v}\rangle - \langle \mathbf{u}, T\mathbf{v}\rangle = \Big[ p(x)\{u(x)v'(x) - u'(x)v(x)\} \Big]_a^b$$

where $T$ is given by (9.2.1) and the inner product by (9.1.8).

The formula (9.2.35) for the Green's function may be recalled

$$G(x, \xi) = \begin{cases} -\mathbf{Y}^t(x)\mathbf{D}_0^{-1}\boldsymbol{\alpha}\mathbf{W}(a)\mathbf{Z}(\xi), & a \leq \xi \leq x \\ \mathbf{Y}^t(x)\mathbf{D}_0^{-1}\boldsymbol{\beta}\mathbf{W}(b)\mathbf{Z}(\xi), & x \leq \xi \leq b \end{cases} \tag{9.5.17}$$

which holds here provided that $\mathbf{Y}^t(x) = [y_1(x), y_2(x), \ldots, y_{2n}(x)]$ is the vector of the $2n$ linearly independent solutions of the homogeneous differential equation $Ty(x) = 0$, where $T$ is given by (9.5.15). The Wronskian matrix $\mathbf{W}(x)$ is of order $2n$ and is given by

$$\mathbf{W}(x) \equiv \mathbf{W}(\mathbf{Y}(x)) \equiv \begin{bmatrix} y_1 & y_2 & \cdots & y_{2n} \\ y_1' & y_2' & \cdots & y_{2n}' \\ \cdot & & & \\ \cdot & & & \\ \cdot & & & \\ y_1^{(2n-1)} & y_2^{(2n-1)} & \cdots & y_{2n}^{(2n-1)} \end{bmatrix} \tag{9.5.18}$$

In (9.5.17) $\mathbf{D}_0$ retains its form (9.2.13) with the change implied by (9.5.18). Following (9.2.34), the vector $\mathbf{Z}(x)$ is given by

$$\mathbf{Z}(x) = -\frac{1}{a_0(x)}\mathbf{W}^{-1}(\mathbf{Y}(x))\mathbf{e}_2$$

where $\mathbf{e}_2^t = [0, 1, 0, \ldots, 0]$ is the transpose of $\mathbf{e}_2 \in \mathcal{R}_{2n}$.

The Green's function above and its first $(2n - 2)$ derivatives are continuous while the $(2n - 1)$st derivative has an upward jump of $1/a_0(x)$ at $x = \xi$. The important implication, however, is that $\mathbf{T}^{-1}$ is compact so that spectral theorem 7.4.3 becomes applicable. Thus $\mathbf{T}$ has a discrete spectrum and a self-adjoint spectral representation.

The evaluation of the spectrum proceeds exactly as in Section 9.2.2. Thus (9.5.52) represents the characteristic equation, that is, $D_\lambda = 0$, where $D_\lambda$ is the determinant of the matrix $\mathbf{D}_\lambda$, given by (9.2.51) provided that $\mathbf{W}(\mathbf{U}(x, \lambda))$ is the $(2n \times 2n)$ Wronskian matrix of the vector $\mathbf{U}(x, \lambda)$ of the $2n$ linearly independent solutions of $Tu(x) = \lambda u(x)$. The $j$th eigenvector $\mathbf{z}_j$ is obtained by solving (9.2.53) and writing $z(x, \lambda_j) = \mathbf{U}^t(x, \lambda_j)\mathbf{k}_j$. The solution of boundary value problems follows exactly the prescription of Section of 9.2.3.

When the coefficients in $T$ are piecewise smooth, self-adjoint operators obtain with domains $D(T)$ as defined in Section 9.3. As an example, let us consider the general fourth-order formally self-adjoint differential expression as determined by (9.5.9). Clearly, this is given by

$$T = p_2(x) - \frac{d}{dx}\left[p_1(x)\frac{d}{dx}\right] + \frac{d^2}{dx^2}\left[p_0(x)\frac{d^2}{dx^2}\right] \tag{9.5.19}$$

The coefficients, whose piecewise smoothness properties are important, are $p_1(x)$ and $p_0(x)$. While it is possible that $p_1(x)$ and $p_0(x)$ may be discontinuous at different points, there is no loss of generality in assuming that they occur at the same points. If only one discontinuity exists at $x = x_1$ the relationship (9.5.14) becomes

$$\langle T\mathbf{u}, \mathbf{v}\rangle - \langle \mathbf{u}, T\mathbf{v}\rangle = \Big[ p_1(x)\{u(x)v'(x) - u'(x)v(x)\} \Big]_{x_1+}^{x_1-} + \Big[ p_0(x)\{u'(x)v''(x) - u''(x)v'(x)\} \Big]_{x_1+}^{x_1-} - \Big[ u'(x)\{p_0(x)v''(x)\}' - v'(x)\{p_0(x)u''(x)\}' \Big]_{x_1+}^{x_1-} \quad (9.5.20)$$

where we have assumed that the boundary conditions at $a$ and $b$ are such that their contributions to the right-hand side of (9.5.20) vanish. Following Section 9.3, the first term on the right-hand side of (9.5.20) may be allowed to vanish by requiring $u$ (and $v$), for example, to satisfy $u(x_1-) = u(x_1+)$, $p_1(x_1-)u'(x_1-) = p_1(x_1+)u'(x_1+)$ (see Theorem 9.3.1). The second term can vanish by requiring that

$$\frac{p_0(x_1-)}{p_1(x_1-)}u''(x_1-) = \frac{p_0(x_1+)}{p_1(x_1+)}u''(x_1+)$$

The third term may be made to vanish by imposing that

$$\frac{1}{p_1(x_1-)}\{p_0(x_1-)u''(x_1-)\}' = \frac{1}{p_1(x_1+)}\{p_0(x_1+)u''(x_1+)\}'$$

We have thus obtained four independent discontinuity conditions which, together with four homogeneous boundary conditions, make up the domain $D(T)$. There are many such domains that are possible, but all of them can be most elegantly brought together by the discontinuity conditions

$$\boldsymbol{\alpha}^{-}\mathbf{w}(u(x_1-)) + \boldsymbol{\alpha}^{+}\mathbf{w}(u(x_1+)) = \mathbf{0} \quad (9.5.21)$$

which is the same as (9.3.3) except for the fact that now $\boldsymbol{\alpha}^{-}$ and $\boldsymbol{\alpha}^{+}$ are $2n \times 2n$ matrices such that the matrix $[\boldsymbol{\alpha}^{-} \quad \boldsymbol{\alpha}^{+}]$ has rank $2n$ and

$$\boldsymbol{\alpha}^{-}\mathbf{P}^{-1}(x_1-)\boldsymbol{\alpha}^{-t} = \boldsymbol{\alpha}^{+}\mathbf{P}^{-1}(x_1+)\boldsymbol{\alpha}^{+t} \quad (9.5.22)$$

Indeed, (9.5.22) is the same as (9.3.4). It is left as an exercise to show that the domain above does satisfy the constraint (9.5.22) (see Exercise 9.5.1). A sample application follows.

### Lateral Vibrations of a Prismatic Bar

Consider the forced lateral vibrations of a beam of unit length which has a "built-in" left end at $x = 0$ and a "free" right end at $x = 1$. The bar has two sections of different materials of different properties, cross sections, and moments of inertia, the first from $x = 0$ to $x = x_1$ with associated quantities subscripted by 1, and the second from $x = x_1$ to $x = 1$ with quantities subscripted by 2. The displacement from equilibrium, denoted $u(x, t)$, satisfies the differential equation

$$\frac{q}{\gamma A}\frac{\partial^2}{\partial x^2}\left[EI\frac{\partial^2}{\partial x^2}u(x, t)\right] = -\frac{\partial^2 u}{\partial t^2} + f(x, t) \tag{9.5.23}$$

where $E$ is the modulus of elasticity, $I$ is the moment of inertia, $\gamma$ is the specific weight, and $A$ is the cross-sectional area; $f(x, t)$ is the applied force distribution and $g$ is the acceleration due to gravity. Clearly, the differential expression on the left-hand side of (9.5.23) is a formally self-adjoint operation with $p_0(x) \equiv EI$ which is piecewise constant, $p_1(x) \equiv 0$, and $r(x) = \gamma A$. The boundary conditions at the built-in end are

$$u(0, t) = 0, \qquad \frac{\partial u}{\partial x}(0, t) = 0$$

while at the free end

$$\frac{\partial^2 u}{\partial x^2}(1, t) = 0, \qquad \frac{\partial^3 u}{\partial x^3}(1, t) = 0$$

Other physical considerations require that the deflection $u$ and its first derivative $\partial u/\partial x$ must be continuous in $[0, 1]$ at all times; further $EI(\partial^2 u/\partial x^2)$ and $\partial/\partial x(EI(\partial^2 u/\partial x^2))$ must also be continuous.† In particular, the foregoing continuity properties must hold at $x = x_1$. Thus

$$u(x_1-, t) = u(x_1+, t)$$

$$\frac{\partial u}{\partial x}(x_1-, t) = \frac{\partial u}{\partial x}(x_1+, t)$$

$$E_1 I_1\frac{\partial^2 u}{\partial x^2}(x_1-, t) = E_2 I_2\frac{\partial^2 u}{\partial x^2}(x_1+, t)$$

$$\frac{\partial}{\partial x}\left[E_1 I_1\frac{\partial^2 u}{\partial x^2}(x_1-, t)\right] = \frac{\partial}{\partial x}\left[E_2 I_2\frac{\partial^2 u}{\partial x^2}(x_1+, t)\right]$$

Initial conditions must be specified for $u$ and $\partial u/\partial t$ as

$$u(x, 0) = u_0(x), \qquad \frac{\partial u}{\partial t}(x, 0) = u_1(x)$$

To solve the above boundary-initial value problem, we define the Hilbert space $\mathcal{H} = \mathcal{L}_2\{[0, 1]: \gamma A\}$ with the inner product

$$\langle \mathbf{u}, \mathbf{v}\rangle = \frac{\gamma_1 A_1}{g}\int_0^{x_1} u(x)v(x)\,dx + \frac{\gamma_2 A_2}{g}\int_{x_1}^1 u(x)v(x)\,dx$$

†See, for example, *Vibration Problems in Engineering*, 3rd ed., by S. Timoshenko and D. H. Young, D. Van Nostrand, Princeton, N.J., 1959, for the physical interpretation of these quantities.

and the operator $T = \{T, D(T)\}$, where

$$T \equiv \frac{g}{\gamma A}\frac{d^2}{dx^2}\left(EI\frac{d^2}{dx^2}\right)$$

and

$$D(T) = \{\mathbf{u} \in \mathcal{H};\ T\mathbf{u} \in \mathcal{H};\ u(x_1-) = u(x_1+),\ u'(x_1-) = u'(x_1+),$$
$$E_1I_1u''(x_1-) = E_2I_2u''(x_1+),\ E_1I_1u'''(x_1-) = E_1I_2u'''(x_1+);$$
$$u(0) = u'(0) = 0,\ u''(1) = u'''(1) = 0\}$$

It is readily seen by substitution of the conditions in $D(T)$ that the expression for $\langle T\mathbf{u}, \mathbf{v}\rangle - \langle \mathbf{u}, T\mathbf{v}\rangle$ vanishes so that $\mathbf{T}$ is self-adjoint. The boundary-initial value problem is written as

$$-\frac{d^2}{dt^2}\mathbf{u}(t) = \mathbf{T}\mathbf{u}(t) - \mathbf{f}(t), \qquad \mathbf{u}(0) = \mathbf{u}_0, \qquad \frac{d}{dt}\mathbf{u}(0) = \mathbf{u}_1 \tag{9.5.24}$$

This problem is readily solved by the methods of Section 7.5 (see also Exercise 7.5.3).

The eigenvalues and eigenvectors are obtained by solving $Tu(x) = \lambda u(x)$, which yields four linearly independent solutions and may be represented in the notation of Section 9.3.2 as

$$U^+(x, \lambda) = \begin{bmatrix} \sin\sqrt{\mu^+}\,x \\ \cos\sqrt{\mu^+}\,x \\ \sinh\sqrt{\mu^+}\,x \\ \cosh\sqrt{\mu^+}\,x \end{bmatrix}, \qquad U^-(x, \lambda) = \begin{bmatrix} \sin\sqrt{\mu^-}\,x \\ \cos\sqrt{\mu^-}\,x \\ \sinh\sqrt{\mu^-}\,x \\ \cosh\sqrt{\mu^-}\,x \end{bmatrix} \tag{9.5.25}$$

where $\mu^- = (\gamma_1A_1/gE_1I_1)\lambda$ and $\mu^+ = (\gamma_2A_2/gE_2I_2)\lambda$. From the domain $D(T)$ it is easy to identify the matrices $\boldsymbol{\alpha}$, $\boldsymbol{\beta}$, $\boldsymbol{\alpha}^-$, and $\boldsymbol{\alpha}^+$ so that the boundary conditions are represented by (9.1.11) and the discontinuity conditions by (9.5.21). The characteristic equation is then given by (9.3.25), that is,

$$D_\lambda = 0$$

where $D_\lambda$ is the determinant of the matrix $\mathbf{D}_\lambda$ to be computed from (9.3.24) using (9.5.24). The steps that remain follow the prescription of Section 9.3.2 for the computation of the eigenvalues $\{\lambda_j\}$ and the normalized eigenvectors $\{z_j\}$.

Finally, we consider operators corresponding to those in Section 9.4. These arise when the boundary points and/or the points of discontinuity of the coefficients of the differential equation have weights (see Section 9.4). Consider $T$ as given by the *second* form in (9.5.5). Let us assume that the coefficients appearing in $T$ are smooth in $(a, b)$ except at $x = x_1$, which we consider to be weighted. Let the boundary conditions at $a$ and $b$ be unmixed. Consider the differential expression

$$L \equiv \begin{bmatrix} \frac{1}{r(x)} D^{[2n]} & 0 & \dots & 0 \\ \alpha_1 \Big[ D^{[2n-1]} \Big]_{x_1-}^{x_1+} & 0 & \dots & 0 \\ \alpha_2 \Big[ D^{[2n-2]} \Big]_{x_1-}^{x_1+} & 0 & \dots & 0 \\ \vdots & & & \\ \alpha_n \Big[ D^{[n]} \Big]_{x_1-}^{x_1+} & 0 & \dots & 0 \end{bmatrix} \tag{9.5.26}$$

acting on elements of $\mathcal{H} = \mathcal{L}_2\{[a, b]: r(x)\} \oplus \mathcal{R}_n$. An element $u \in \mathcal{H}$ is denoted by

$$u \equiv \begin{bmatrix} \{u(x)\} \\ \begin{bmatrix} u_1 \\ \cdot \\ \cdot \\ \cdot \\ u_n \end{bmatrix} \end{bmatrix}$$

and the inner product on $\mathcal{H}$ is defined by

$$\langle u, v \rangle = \int_a^b r(x)u(x)v(x)\, dx + \sum_{k=1}^{n} r_k u_k v_k \tag{9.5.27}$$

where $r_1, r_2, \dots, r_n$ are positive numbers which are to be chosen suitably. Notice particularly that the generalization here (of the operators of Section 9.4) leads to a direct sum of $\mathcal{L}_2\{[a, b]: r(x)\}$ with $\mathcal{R}_n$ for each weighted point.† From (9.5.14) one may write

$$\begin{aligned} \langle L\mathbf{u}, \mathbf{v} \rangle - \langle \mathbf{u}, L\mathbf{v} \rangle = \sum_{k=1}^{n} \{ & v^{[2n-k]}(x_1-)[u^{[k-1]}(x_1-) - \alpha_k r_k u_k] \\ & - v^{[2n-k]}(x_1+)[u^{[k-1]}(x_1+) - \alpha_k r_k u_k] \\ & - u^{[2n-k]}(x_1-)[v^{[k-1]}(x_1-) - \alpha_k r_k v_k] \\ & + u^{[2n-k]}(x_1+)[v^{[k-1]}(x_1+) - \alpha_k r_k v_k]\} \end{aligned} \tag{9.5.28}$$

where the contributions from the (unmixed) boundary conditions to the right-hand side have been allowed to vanish. It should be obvious that a rich choice of domains $D(L)$ is possible for the right-hand side of (9.5.28) to vanish. An example is

$$\begin{aligned} D(L) = \{\mathbf{u} \in \mathcal{H};\ L\mathbf{u} \in \mathcal{H};\ & u^{[2n-k]}(x_1-) = u^{[2n-k]}(x_1+), \\ & u^{[k-1]}(x_1-) = u^{[k-1]}(x_1+),\ k = 1, 2, \dots, n\} \end{aligned}$$

†In Section 9.4 each weighted point involved a direct sum with $\mathcal{R}$, as it should, since for second-order operators, $n = 1$.

which will make $\mathbf{L} = \{L, D(L)\}$ self-adjoint, provided that we take $r_k = \alpha_k^{-1}$ for each $k$. Indeed, several other domains are possible.

The spectral analysis of $\mathbf{L}$ proceeds in much the same way as the other operators with which we have dealt. The solution of boundary value problems comprising $\mathbf{L}$ is also accomplished by the methods of Section 7.5. The student is advised to attempt Exercise 9.5.2 in this regard.

**EXERCISES**

**9.5.1** Let $T$ be given by (9.5.19) with $p_0(x)$ and $p_1(x)$ discontinuous at $x_1 \in (a, b)$. Cast the discontinuity conditions $u(x_1-) = u(x_1+)$, $p_1(x_1-)u'(x_1-) = p_1(x_1+)u'(x_1+)$,

$$\frac{p_0(x_1-)}{p_1(x_1-)}u''(x_1-) = \frac{p_0(x_1+)}{p_1(x_1+)}u''(x_1+)$$

and

$$\frac{1}{p_1(x_1-)}\{p_0(x_1-)u''(x_1-)\}' = \frac{1}{p_1(x_1+)}\{p_0(x_1+)u''(x_1+)\}'$$

in the form (9.5.21) and show that (9.5.22) is satisfied.

**9.5.2**† Consider the example of the lateral vibrations of a prismatic bar discussed in this section. Suppose that at $x = x_1$, a concentrated weight $W_1$ exists so that the last of the discontinuity conditions must be replaced by the differential equation

$$\frac{-g}{W_1}\frac{\partial}{\partial x}\left[EI\frac{\partial^2 u}{\partial x^2}\right]_{x_1^-}^{x_1^+} = -\frac{\partial^2 u}{\partial t^2}(x_1, t) + f_1(t)$$

where $f_1(t)$ is the applied force at $x_1$. The foregoing equation must be considered together with (9.5.23). Retaining the same boundary and the remaining discontinuity conditions as in the example, formulate the boundary-initial value problem (how would you state the initial conditions?) in terms of a suitable self-adjoint operator. Show how you would perform the spectral analysis and obtain a spectral solution to the problem.

## 9.6 Concluding Remarks

What has been covered in this chapter is much of what we had set out to do in this book. We have obtained solutions to several boundary-initial value problems ranging widely in complexity in terms of the spectral decomposition of the self-adjoint differential operators involved. We emphasize here that although the boundary-initial value problems featured *partial* differential

†See D. Ramkrishna and N. R. Amundson, *J. Appl. Mech.*, *41*, 1106, 1974.

equations, the operators were themselves *ordinary* differential.† The partial differential nature of the equations solved arise from the appearance of derivatives with respect to the second independent variable $t$, which may be referred to as an *evolutionary* parameter. This is because the system described by the solution vector [$\mathbf{u}(t)$ or $\mathbf{v}(t)$] "evolves" continuously along the path of variation of $t$. The equation [in $\mathbf{u}(t)$ or $\mathbf{v}(t)$ such as (9.2.64)] is itself referred to as *evolutionary equation*.

Section 9.2 has dealt with self-adjoint differential operators with continuous coefficients whose applications lie in problems connected with the behavior of single-phase materials with physical properties that do not vary discontinuously with position. Thus transport processes in homogeneous media, vibration problems of elastic bodies of continuously varying shapes and moduli of deformation, and so on, are some of the common beneficiaries of the techniques of Section 9.2. Because of being restricted to ordinary differential operators, one is confined to solution of one-dimensional transport problems, vibration of one-dimensional bodies, and so on.

Sections 9.3 and 9.4 deal with situations where discontinuous coefficients appear in the operations. In physical problems such discontinuities arise because of abrupt changes in the material media across "interfaces." Generally, interfaces have no capacity and applications in this context have been the subject matter of Section 9.3. Under other circumstances it is possible to encounter "thin" intermediate "sandwich" media whose capacitance (for the transported quantity such as mass, energy, or momentum) cannot be neglected. The resulting formulations are then different (from those for interfaces without capacitance in Section 9.3) and are covered in Section 9.4.

The self-adjoint formulations in Sections 9.3 to 9.5 clearly establish that a class of self-adjoint problems can only be recognized as such by the proper choice of Hilbert space and inner product, formal operation, and its associated domain. Without such a manipulation the operator appears non-self-adjoint and a delinquent misfit for the mold of self-adjoint theory.

Our demonstrations of spectral analyses of self-adjoint operators have been limited to cases where the equation $Tu(x, \lambda) = \lambda u(x, \lambda)$ can be solved analytically or identified as well-categorized special functions (such as Bessel's, Kummer's, etc.). In other situations approximation methods are required. The Rayleigh–Ritz method deserves special mention, although there indeed are several others available.

There are a number of very interesting facts about the eigenvalues and eigenvectors of the operators encountered in this chapter to which we have had no opportunity to refer. For example, results are available on the asymptotic distribution of eigenvalues, that is, the nature of the large (in

†The subject of partial differential operators is taken up in Chapter 10.

absolute value) eigenvalues. There also exist oscillation theorems concerning the nature of the eigenfunctions that have useful applications. Some references in regard to this are suggested in the Further Reading.

We have restricted considerations to operators that have only discrete spectra. When the independent variable is allowed to take on infinite domains (instead of the compact interval $[a, b]$), the eigenvalues can become *continuously* spaced. Thus a continuous spectrum may be encountered in such cases. Obviously, this situation arises when the differential operator does not have a compact inverse. Although we have not dealt with examples of continuous spectra, they are indeed quite important and have numerous applications.

Section 9.1 had initiated discussion on differential operations that were not formally self-adjoint. Such operations of course lead to non-self-adjoint differential operators. Further, we have seen that for a self-adjoint operator we must not only have a formally self-adjoint differential expression but also have suitable boundary conditions. The analysis of non-self-adjoint differential operators is considerably more difficult. In Chapter 11 an introductory treatment of non-self-adjoint operators has been included.

The boundary-initial value problems of this chapter are also amenable to solution by the method of Laplace transforms. The solutions thus obtained would then be completely equivalent to those developed here, for in the method of Laplace transforms, the inversion of the transform (of the solution) would entail the calculation of the residues at the poles (which will coincide with the eigenvalues). The residue at a pole will coincide with the component of the solution in the eigenspace corresponding to the eigenvalue or pole. The entire equivalence is contained in the elegantly condensed version of the representation of the identity by

$$\mathbf{I} = \frac{1}{2\pi i}\oint_c [\mathbf{T} - \lambda\mathbf{I}]^{-1}\, d\lambda$$

to which reference was made in Section 7.7. The operator $(\mathbf{T} - \lambda\mathbf{I})^{-1}$ is of course identified by the Green's function developed in Section 9.2.1. Recognizing that the curve $c$ must enclose all the singularities of $(\mathbf{T} - \lambda\mathbf{I})^{-1}$ (i.e., the eigenvalues) the foregoing representation of the identity becomes

$$\mathbf{I} = \sum_{j=1}^{\infty} \frac{1}{(\nu_j - 1)!}\frac{d^{\nu_j - 1}}{d\lambda^{\nu_j - 1}}[(\lambda - \lambda_j)^{\nu_j}(\mathbf{T} - \lambda\mathbf{I})^{-1}]_{\lambda=\lambda_j}$$

which is a result of the residue theorem applied to a pole $\lambda = \lambda_j$ of order $\nu_j$. Note that $\nu_j$ becomes the dimension of $N(\mathbf{T} - \lambda_j\mathbf{I})$ because $\mathbf{T}$ is self-adjoint and

$$\mathbf{P}_j = \frac{1}{(\nu_j - 1)!}\frac{d^{\nu_j - 1}}{d\lambda^{\nu_j - 1}}[(\lambda - \lambda_j)^{\nu_j}\{\mathbf{T} - \lambda\mathbf{I}\}^{-1}]_{\lambda=\lambda_j}$$

where $\mathbf{P}_j$ is the self-adjoint projection corresponding to eigenvalue $\lambda_j$. This reasoning may be continued to show exactly how solutions to the boundary-initial value problems (of this chapter) obtained by the method of Laplace transforms are equivalent to the spectral solutions presented here, but we feel that the fact is transparent enough to terminate the current development. The important thing to note, however, is that the theory of self-adjoint operators has produced a structure whose beauty is matched by the ease with which it yields to computation. This structure lies subtly submerged in the methodology of the Laplace transform.

## FURTHER READING

For the determination of eigenvalues of Sturm–Liouville operators by approximate methods, several books are available. In particular, we refer the reader to

*Approximate Solution of Operator Equations*, by M. A, Kraisnoselskii, G. M. Vainikko, P. O. Zabreibo, Ya. B. Rutitskii, and V. Ya Stetsenko, Wolters-Noordhoff, Groningen, The Netherlands, 1972.

In regard to asymptotic distribution of eigenvalues, oscillation, and comparison theorems, the reader should consult

*Methods of Mathematical Physics*, Vol. 1, by Courant and D. Hilbert, Interscience, New York, 1953, and

*Theory of Ordinary Differential Equations*, by E. A. Coddington and N. Levinson, McGraw-Hill, New York, 1955.

For an extensive treatment of differential operators with continuous spectra, we refer to

*Eigenfunction Expansions Associated with Second Order Differential Equations*, by E. C. Titchmarsh, Clarendon Press, Oxford, Part I, 2nd ed., 1962; Part II, 1958.

# Partial Differential Operators 10

## 10.0 Foreword

Since the initial-boundary value problems of Chapter 9 arose from partial differential equations the title of this chapter is apt to be confusing. It must be noted that although the equations of Chapter 9 were *partial* differential, the operators of interest possessed operations that were *ordinary* differential expressions. In this chapter our focus is on operators whose operations are themselves partial differential in nature. Thus evolution equations governing the initial-boundary value problem will be partial differential equations in *more* than *two* variables, unlike those of Chapter 9. On the other hand, steady-state problems will themselves be partial differential equations. The student is advised to be familiar with Chapters 7 and 9 before embarking on this chapter.

## 10.1 Introduction

Our study of partial differential operators is necessarily very limited in scope. There have been enormous developments in the theory of partial differential equations in the last three or four decades in the mathematical literature. These developments are generally concerned with the existence and properties of solutions. Actual solutions frequently require the aid of approximate methods even when the operators are self-adjoint, Thus although spectral solutions exist for self-adjoint problems, the determination of eigenvalues and eigenvectors is not often possible without suitable approximate methods. Such approximate procedures being outside the scope of this book, we focus only on those cases where spectral solutions are obtained analytically in the same sense as they were in Chapter 9. In this connection the method of *separation* of *variables* is central to the calculation of eigenvalues and eigenvectors. In presenting the method of separation of variables, however, we adopt an abstract formulation due to Friedman† which is both elegant and useful. In this formulation the partial differential operator is expressed as the sum of tensor products of operators on different Hilbert spaces.‡

## 10.2 Partial Differential Expressions of Second Order

A linear partial differential expression in $n$ variables of order 2 may be represented by

$$Tu \equiv -\left(\sum_{i=1}^{n}\sum_{j=1}^{n} a_{ij}\frac{\partial^2 u}{\partial x_i\,\partial x_j} + \sum_{i=1}^{n} b_i\frac{\partial u}{\partial x_i} - cu\right) \tag{10.2.1}$$

where $\{a_{ij}\}_{i,j=1}^{n}$, $\{b_i\}_{i=1}^{n}$, and $c$ are functions of $\mathbf{x} \equiv (x_1, x_3, \ldots, x_n)$ in some compact region $\Omega$ in $\mathcal{R}_n$. The classification of partial differential equations is based on the eigenvalues of the $n \times n$ matrix of real coefficients $\{a_{ij}\}$. The matrix above may be assumed to be symmetric without any loss of generality, that is, $a_{ij} = a_{ji}$, so that all the $n$ eigenvalues are real-valued and dependent on $\mathbf{x}$. Suppose that $n_+$ and $n_-$ are the numbers of positive and negative eigenvalues, respectively. Partial differential expressions are classified as *elliptic*, *hyperbolic*, and *parabolic* according as $n_- = n$ or $n_+ = n$ (i.e., all eigenvalues are of the same sign), $n_\pm > 0$ and $n_+ + n_- = n$ (i.e., there are positive *and* negative eigenvalues, but zero is not an eigenvalue) and $n_+ + n_- < n$ (i.e., zero is an eigenvalue), respectively. Insofar as the eigenvalues depend on position, the category to which a partial differential expression belongs may vary from one region to another in $\mathcal{R}_n$. For reasons to be stated later, our interest is mainly in partial differential expressions which are elliptic everywhere. We assume that the compact region $\Omega$ is bounded by its surface $\partial\Omega$, possessing smoothness properties sufficient to justify the treatment that follows.

Next, we consider the differential expression formally adjoint to $T$. We will broadly constrain the functions on which $T$ acts as originating from $\mathcal{L}_2(\Omega)$

---

†See "An Abstract Formulation of the Method of Separation of Variables," by B. Friedman, *Proceedings of the Conference on Differential Equations*, University of Maryland, College Park, Md., 1956, pp. 209–226.

‡See Section 2.12.

with the inner product

$$\langle \mathbf{u}, \mathbf{v} \rangle = \int_{\Omega} u(\mathbf{x}) v(\mathbf{x})\, dV \tag{10.2.2}$$

where $dV = dx_1, dx_2, \ldots, dx_n$ is an infinitesimal volume in $\mathcal{R}_n$. The adjoint expression $T^*$ is defined such that, for any two functions $\mathbf{u}, \mathbf{v}$, the difference

$$\langle T\mathbf{u}, \mathbf{v} \rangle - \langle \mathbf{u}, T^*\mathbf{v} \rangle$$

must depend only on the values of $\mathbf{u}$ and $\mathbf{v}$ on the *boundary* $\partial\Omega$. In Chapter 9 the adjoint expression was determined by integration by parts. Here we make use of the divergence theorem† and a vector identity concerned with the divergence of products of functions. The procedure, although cumbersome, is not difficult and leads to the result that

$$T^*v \equiv -\left\{\sum_{i=1}^{n}\sum_{j=1}^{n} a_{ij}\frac{\partial^2 v}{\partial x_i \partial x_j} + \sum_{i=1}^{n}\frac{\partial v}{\partial x_i}\left(2\sum_{j=1}^{n}\frac{\partial a_{ij}}{\partial x_j} - b_i\right)\right.$$
$$\left. + \left[\sum_{i=1}^{n}\sum_{j=1}^{n}\frac{\partial^2 a_{ij}}{\partial x_i \partial x_j} - \sum_{i=1}^{n}\frac{\partial b_i}{\partial x_i} - c\right]v\right\} \tag{10.2.3}$$

It is clear from (10.2.3) that for a *self-adjoint differential* expression, we must have

$$\sum_{j=1}^{n}\frac{\partial a_{ij}}{\partial x_j} = \frac{b_i}{2} \tag{10.2.4}$$

since all the coefficients appearing in the differential expression $T$ and $T^*$ would then be identical. A self adjoint expression $T$ may then be written as

$$Tu = -\sum_{i=1}^{n}\sum_{j=1}^{n}\frac{\partial}{\partial x_i}\left(a_{ij}\frac{\partial u}{\partial x_j}\right) + cu \tag{10.2.5}$$

Further, it is easily shown that

$$\langle T\mathbf{u}, \mathbf{v} \rangle - \langle \mathbf{u}, T\mathbf{v} \rangle = -\int_{\Omega}\sum_{i=1}^{n}\sum_{j=1}^{n}\frac{\partial}{\partial x_i}\left[a_{ij}\left(v\frac{\partial u}{\partial x_j} - u\frac{\partial v}{\partial x_j}\right)\right] dv$$

which, by virtue of the divergence theorem, yields

$$\langle T\mathbf{u}, \mathbf{v} \rangle - \langle \mathbf{u}, T\mathbf{v} \rangle = -\oint_{\partial\Omega}\sum_{i=1}^{n}\sum_{j=1}^{n} a_{ij}\left(v\frac{\partial u}{\partial x_j} - u\frac{\partial v}{\partial x_j}\right) n_i\, dA \tag{10.2.6}$$

where $dA$ is an infinitesimal area on $\partial\Omega$ and $n_i$ are the components of the local normal vector on $\partial\Omega$ directed toward the surface from the interior of $\Omega$. If the formal operation $T$ in (10.2.5) is replaced by

$$T \equiv -\frac{1}{r(\mathbf{x})}\left[\sum_{i=1}^{n}\sum_{j=1}^{n}\frac{\partial}{\partial x_i}\left(a_{ij}\frac{\partial}{\partial x_j}\right)\right] + c \tag{10.2.7}$$

where $r(\mathbf{x})$ is positive on $\Omega$, then $T$ is formally self-adjoint and formula

†See, for example, *Foundations of Modern Analysis*, by J. Dieudonne, Academic Press, New York, 1960.

(10.2.6) is again applicable, provided that the inner product (10.2.2) is replaced by

$$\langle \mathbf{u}, \mathbf{v} \rangle = \int_{\Omega} r(\mathbf{x}) u(\mathbf{x}) \mathbf{v}(\mathbf{x}) \, dV \tag{10.2.8}$$

Similarly, if $T$ in (10.2.1) were considered, with $r(\mathbf{x})$ dividing the prevailing expression, then $T^*$ is the same as (10.2.3) except for a dividing factor of $r(\mathbf{x})$. Again the inner product to be used is given by (10.2.8).

It may be noted that the differential expression (10.2.7) is fundamental to the description of molecular transport of mass, momentum, or energy, in regard to which we consider some examples later.

Having identified formally self-adjoint partial differential operations, what remains is the association of suitable domains leading to self-adjoint operators. The domain is a subspace of $\mathfrak{L}_2(\Omega)$ defined by homogeneous boundary conditions and certain other conditions which guarantee that the range space of the operation $T$ is $\mathfrak{L}_2(\Omega)$. The boundary conditions must be specified at every point in $\partial\Omega$. The type of admissible boundary conditions depends on the category to which the partial differention expression belongs. These matters, which have been discussed in standard texts on partial differential equations,† are related to the "well-posedness" of boundary value problems and will not invite our digression here. However, it is important to recognize that parabolic and hyperbolic partial differential equations involve an "evolutionary" parameter (such as time), and must be provided with *initial conditions* (referred to in the mathematical literature as "Cauchy data") in addition to boundary conditions. Thus initial-boundary value problems are obtained in these situations. Descriptions of *transient* physical phenomena come under this category. On the other hand, elliptic partial differential equations describe *steady-state* situations and require only boundary conditions. By restricting considerations to elliptic partial differential operators, it is possible to deal with both steady and transient problems of interest to us. Steady-state problems are of the type

$$T\mathbf{u} = \mathbf{f} \tag{10.2.9}$$

where $T$ is an elliptic partial differential operation. [Note that (10.2.9) was encountered in Section 7.4 as (7.4.13).] Transient problems are generally represented by

$$\frac{d}{dt}\mathbf{u}(t) = -T\mathbf{u}(t) + \mathbf{f}, \qquad u(0) = u_0 \tag{10.2.10}$$

or as

$$\frac{d^2\mathbf{u}(t)}{dt^2} = -T\mathbf{u}(t) + \mathbf{f}, \qquad u(0) = u_0, \qquad \frac{d}{dt}u(0) = \mathbf{u}_1 \tag{10.2.11}$$

†See, for example, *Partial Differential Equations of Mathematical Physics*, by S. L. Sobolev, translated by E. R. Dawson, Pergamon Press, Oxford, 1964.

If $T$ is elliptic, the differential expression $T + d/dt$ in (10.2.10) is parabolic, while in (10.2.11) $T + d^2/dt^2$ is hyperbolic.† Equations (10.2.10) and (10.2.11) are initial-boundary value problems, the boundary conditions arising in the stipulation of **T**.

From the foregoing discussion, it should become clear that a treatment of elliptic operators can accommodate parabolic and hyperbolic operator equations by viewing the latter as evolution equations of the types (10.2.10) and (10.2.11). In this connection, one may note that in Chapter 9 we had already encountered parabolic and hyperbolic differential expressions in solving initial-boundary value problems with $T$ as an ordinary differential expression. We will therefore focus exclusively on elliptic partial differential operators and direct our attention next to the boundary conditions which constitute suitable domains for self-adjoint partial differential operators.

### 10.2.1 Boundary Conditions

Since we are concerned with domains for $T$, which must be subspaces of $\mathfrak{L}_2(\Omega)$, the boundary conditions must be homogeneous. The simplest boundary condition is given by

$$u(\mathbf{x}) = 0, \qquad \mathbf{x} \in \partial\Omega \tag{10.2.12}$$

and is referred to as the *Dirichlet* boundary condition. Now it is evident that if‡

$$D(T) = \{\mathbf{u}, T\mathbf{u} \in \mathfrak{L}_2(\Omega); u(\mathbf{x}) = 0, \mathbf{x} \in \partial\Omega\}$$

then $\mathbf{T} \equiv \{T, D(T)\}$ is a self-adjoint operator because the right-hand side of (10.2.6) vanishes identically for $\mathbf{u}, \mathbf{v} \in D(T)$.

Another boundary condition, which yields a self-adjoint operator, is given by

$$\sum_{i=1}^{n} n_i \sum_{j=1}^{n} a_{ij} \frac{\partial u}{\partial x_j} = 0 \tag{10.2.13}$$

Differential equations of either type (10.2.9) or (10.2.10) describe transport of some physical entity, in which case the boundary condition (10.2.13) is equivalent to the vanishing of the "flux" of that entity in the direction normal to the boundary $\partial\Omega$ at every point. The engineering student will recall that the differential expression (10.2.5) arises in the description of transport in anisotropic media in terms of Cartesian coordinates. For isotropic media,

---

†Here we assume that $T$ is given by (10.2.7) with all eigenvalues of $\{a_{ij}\}$ positive everywhere in $\Omega$.

‡Several additional conditions have been suppressed in stating this domain. In view of the use of the divergence theorem, the conditions for the validity of this theorem must be required of the functions in the domain.

the "transport" coefficients $\{a_{ij}\}$ are given by

$$a_{ij} = \begin{cases} a, & i = j \\ 0, & i \neq j \end{cases} \tag{10.2.14}$$

in which case the boundary condition (10.2.13) is equivalent to the vanishing of the *normal* derivative. This condition is referred to as the *Neumann* boundary condition (see also Exercise 10.2.3).

A third boundary condition leading to a self-adjoint operator is the *mixed* boundary condition

$$\alpha u + \sum_{i=1}^{n} n_i \sum_{j=1}^{n} a_{ij} \frac{\partial u}{\partial x_j} = 0 \tag{10.2.15}$$

which is sometimes referred to as the *Robin* boundary condition. Substitution of (10.2.15) into (10.2.6) makes the right-hand side of the latter vanish, thus yielding a self-adjoint operator. When (10.2.14) holds, the boundary condition (10.2.15) becomes

$$\alpha u + a\frac{\partial u}{\partial n} = 0 \tag{10.2.16}$$

where $\partial u/\partial n$ is the normal derivative of $u$. This boundary condition arises frequently when boundary fluxes can be expressed in terms of transfer coefficients.

A very general boundary condition due to Birkhoff† may be represented by

$$u(\mathbf{x}) + \oint_{\partial\Omega} K(\mathbf{x}, \mathbf{y}) \sum_{i=1}^{n} n_i \sum_{j=1}^{n} a_{ij}(\mathbf{y}) \frac{\partial}{\partial y_j} u(\mathbf{y})\, dA_y = 0, \qquad \mathbf{x} \in \partial\Omega \tag{10.2.17}$$

It is left as an exercise to show that $\mathbf{T} = \{T, D(T)\}$, where $T$ is given by (10.2.7) and

$$D(T) = \{\mathbf{u}, T\mathbf{u} \in \mathfrak{L}_2(\Omega); u \text{ satisfies (10.2.17) on } \partial\Omega\}$$

is self-adjoint with respect to the inner product (10.2.8) if and only if $K(\mathbf{x}, \mathbf{y})$ is symmetric, that is, $K(\mathbf{x}, \mathbf{y}) = K(\mathbf{y}, \mathbf{x})$. If, further, (10.2.14) holds, then the boundary condition (10.2.17) becomes

$$u(\mathbf{x}) + \oint_{\partial\Omega} K(\mathbf{x}, \mathbf{y}) a(\mathbf{y}) \frac{\partial u}{\partial n}(\mathbf{y})\, dA_y = 0 \tag{10.2.18}$$

Note that this boundary condition is in some sense a generalization of the boundary condition (9.1.11) for ordinary differential expressions (see Exercise 10.2.2).

In each of these cases, the self-adjoint operator $\mathbf{T}$ has a spectral representation in accordance with the development of Chapter 7. The nature of the

---

†See *Boundary Problems in Differential Equations*, edited by E. Langer, University of Wisconsin, Madison, Wis., 1960, pp. 163–178.

spectrum, however, is not a simple matter to establish. We will not endeavor to prove the existence of a compact inverse for **T** but instead refer the interested reader to other books.† Henceforth, we will regard the operator **T** as endowed with a discrete spectral representation with real eigenvalues for any of the boundary conditions considered above as long as **T** is self-adjoint and $\Omega$ is compact. Some examples are discussed next in which we encounter the operators that have been dealt with in this section.

*Example 1*

We have already pointed out that the formally self-adjoint operation (10.2.7) occurs in the description of molecular transport of mass, momentum, or energy. If $x_1$, $x_2$, and $x_3$ are *Cartesian* coordinates, then (10.2.7), excluding $c$, represents the expression for diffusion in an anisotropic medium. For an isotropic medium (10.2.14) holds, so that one gets

$$T = -\frac{1}{r(\mathbf{x})}\sum_{j=1}^{3}\frac{\partial}{\partial x_i}\left(a\frac{\partial}{\partial x_j}\right), \qquad a > 0$$

which expresses diffusion in the situation of a nonuniform diffusion coefficient. (In heterogeneous media $a$ may vary discontinuously across interfaces, but this possibility has not been envisaged in the treatment here.) If $a$ is constant and $r(\mathbf{x}) \equiv 1$, one obtains the familiar Laplace operation

$$T = -a\sum_{j=1}^{3}\frac{\partial^2}{\partial x_j^2}$$

*Example 2*

If $x_1$, $x_2$, and $x_3$ are curvilinear nonorthogonal coordinates with a metric tensor $g_{ij}$ (which is a positive-definite matrix), the diffusion (Laplace) operation is given by

$$T = -\frac{1}{\sqrt{g}}\sum_{i=1}^{3}\sum_{j=1}^{3}\frac{\partial}{\partial x_i}\left(\sqrt{g}\, g_{ij}\frac{\partial}{\partial x_j}\right) \qquad (10.2.19)$$

where $g$ is the reciprocal of the determinant of $g_{ij}$. For details the student is referred to Sokolnikoff.‡ The operation (10.2.19) is elliptic because $g_{ij}$ is positive definite. Self-adjoint operators are obtained in diffusion problems by specifying boundary concentrations [i.e., the Dirichlet boundary condition

†See *Methods of Mathematical Physics*, Vol. 1, by R. Courant and D. Hilbert, Interscience, New York, 1953.

‡See *Tensor Analysis: Theory and Applications*, by I. S. Sokolnikoff, Wiley, New York, 1951.

(10.2.12)] or normal fluxes [the Neumann boundary condition (10.2.13) or the Robin boundary condition (10.2.15)]. Exercise 10.2.4 shows how integral boundary conditions such as (10.2.18) arise in applications.

*Example 3*

For orthogonal curvilinear coordinate systems, one has†

$$g_{ij} = h_i^{-2}\delta_{ij}$$

where $\{h_i\}$ are the scale factors and (10.2.19) becomes

$$T = -\frac{1}{h_1 h_2 h_3}\left[\frac{\partial}{\partial x_1}\left(\frac{h_2 h_3}{h_1}\frac{\partial}{\partial x_1}\right) + \frac{\partial}{\partial x_2}\left(\frac{h_1 h_3}{h_2}\frac{\partial}{\partial x_2}\right) + \frac{\partial}{\partial x_3}\left(\frac{h_1 h_2}{h_3}\right)\frac{\partial}{\partial x_3}\right]$$

*Example 4*

For the familiar cylindrical polar coordinates with $x_1 \equiv r$, $x_2 \equiv \theta$, and $x_3 \equiv z$, one has $h_1 = h_3 = 1$, $h_2 = r$, so that $T$ is given by

$$T = -\frac{1}{r}\left[\frac{\partial}{\partial r}\left(r\frac{\partial}{\partial r}\right) + \frac{\partial}{\partial \theta}\left(\frac{1}{r}\frac{\partial}{\partial \theta}\right) + \frac{\partial}{\partial z}\left(r\frac{\partial}{\partial z}\right)\right]$$

*Example 5*

For spherical polar coordinates with $x_1 \equiv r$, $x_2 \equiv \phi$, $x_3 \equiv \theta$, the scale factors are given by $h_1 = 1$, $h_2 = r\sin\theta$, $h_3 = r$, and one obtains for $T$

$$T = -\frac{1}{r^2\sin\theta}\left[\frac{\partial}{\partial r}\left(r^2\sin\theta\frac{\partial}{\partial r}\right) + \frac{\partial}{\partial \phi}\left(\frac{1}{\sin\theta}\frac{\partial}{\partial \phi}\right) + \frac{\partial}{\partial \theta}\left(\sin\theta\frac{\partial}{\partial \theta}\right)\right]$$

*Example 6*

The foregoing Laplacian expressions are very important in engineering applications. For example, steady-state heat conduction in a heated sphere is thus described by the elliptic partial differential equation

$$-\left[\frac{1}{r^2}\frac{\partial}{\partial r}\left(r^2\frac{\partial \Theta}{\partial r}\right) + \frac{1}{r^2\sin^2\theta}\frac{\partial \Theta^2}{\partial \phi^2} + \frac{1}{r^2\sin\theta}\frac{\partial}{\partial \theta}\left(\sin\theta\frac{\partial \Theta}{\partial \theta}\right)\right] = f(r, \theta, \phi)$$

where $\Theta$ is the temperature and $f(r, \theta, \phi)$ arises from the heat source. For a sphere of radius $R$, we have $\Omega \equiv \{0 < r < R, 0 < \theta < \pi, 0 < \phi < 2\pi\}$ and $\partial\Omega \equiv \{r = R, 0 < \theta < \pi, 0 < \phi < 2\pi\}$. The boundary conditions must be specified on the surface of the sphere. The corresponding unsteady-state

†The metric tensor $g_{ij}$ here is actually the contravariant form, although in terms of standard tensor notation, it appears in the covariant form.

heat conduction problem will yield a parabolic partial differential equation. However, as pointed out earlier, this would be an evolution equation involving an elliptic partial differential operator, yielding an initial-boundary value problem.

*Example 7*

The vibration of a circular plate may be described by the hyperbolic differential equation

$$-\frac{\partial^2}{\partial t^2}u(r,\theta,t) = -\frac{1}{r}\left[\frac{\partial}{\partial r}\left(r\frac{\partial}{\partial r}\right) + \frac{\partial}{\partial \theta}\left(\frac{1}{r}\frac{\partial}{\partial \theta}\right)\right]u(r,\theta,t) + f(r,\theta,t),$$

$$0 < r < R, \quad 0 < \theta < 2\pi \qquad (10.2.20)$$

where $u(r, \theta, t)$ is the vibrational displacement, $R$ is the plate radius, and $f(r, \theta, t)$ arises from external forces. The operation on the right-hand side of the differential equation above is clearly an elliptic expression. Boundary conditions must be specified at $r = R$ and, in addition, initial values for $u$ and $\partial u/\partial t$ may be represented by elements from $\mathfrak{L}_2(\Omega)$, where $\Omega \equiv \{0 < r < R; 0 < \theta < 2\pi\}$.

## EXERCISES

**10.2.1** Show that the formally self-adjoint expression (10.2.7) obtains a self-adjoint operator with any of the boundary conditions (10.2.12), (10.2.13), (10.2.15), or (10.2.17), and other conditions needed to define the domain of the operation completely.

**10.2.2** Show how the boundary condition (10.2.18) may be viewed as a generalization of the boundary condition (9.1.11).

**10.2.3** For the special case of the operation (10.2.7) represented by (10.2.19), show that the boundary condition (10.2.15) is the same as (10.2.16). (Note that this exercise calls for some familiarity with the treatment of general curvilinear coordinate systems.)

**10.2.4** Consider steady-state heat conduction in two concentric spheres, the inner one of "infinite" thermal conductivity and comprising a nonuniform heat source. The outer surface of the outer sphere is maintained at a constant temperature. Set up the heat conduction problem for the outer sphere and identify the boundary conditions. In particular, show that the boundary condition on the inner surface is an integral condition, which is a special version of (10.2.18).

**10.2.5** If $T$ is an elliptic partial differential expression in $n$ variables, as in (10.2.1) with $\{\mathbf{a}_{ij}\}$ positive-definite everywhere, show that $T + d/dt$ and $T + d^2/dt^2$ are, respectively, parabolic and hyperbolic expressions in $(n + 1)$ variables with $t$ as the $(n + 1)$st variable.

### 10.2.2 Discontinuous Coefficients

We focus again on the formally self-adjoint operation (10.2.7) and entertain discontinuities in the coefficients $a_{ij}$ but consider only the case where the discontinuities occur together along a finite set of surfaces $\{\partial\Omega_k\}_{k=1}^m$ as shown in Figure 10.2.1. Furthermore, no discontinuity exists *along* any of the sur-

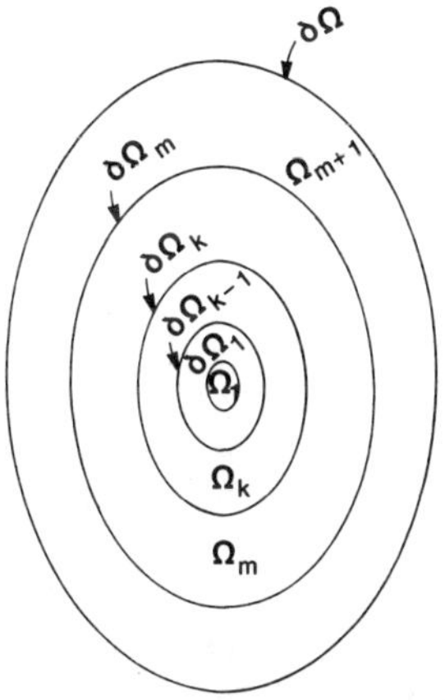


**Figure 10.2.1** Composite medium of "concentric" media.

faces $\partial\Omega_k$. This situation, which arises in composite media of different transport coefficients oriented in a "concentric" manner, is a multidimensional† generalization of the cases dealt with in Sections 9.3 and 9.4. Recall that the operators in Section 9.3 differed from those of Section 9.4 in that the boundary or discontinuity points in the latter possessed weights. We may make a similar distinction here and recognize that one or more of the surfaces $\{\partial\Omega_k\}_{k=1}^m$ may possess volumetric weights. As before, the motivation for considering such problems is physical, which will be borne out by the examples to be presented later.

Because now $a_{ij}$ may be discontinuous on the surfaces $\{\partial\Omega_k\}_{k=1}^m$, the application of the divergence theorem modifies (10.2.6) as

$$\langle T\mathbf{u}, \mathbf{v}\rangle - \langle \mathbf{u}, T\mathbf{v}\rangle = -\oint_{\partial\Omega} \sum_{i=1}^n \sum_{j=1}^n a_{ij}\left(v\frac{\partial u}{\partial x_j} - u\frac{\partial v}{\partial x_j}\right)n_i\, dA$$

$$+ \sum_{k=1}^m \oint_{\partial\Omega_k} \sum_{i=1}^n \sum_{j=1}^n \left[a_{ij}\left(v\frac{\partial u}{\partial x_j} - u\frac{\partial v}{\partial x_j}\right)\right]_{\partial\Omega_k^-}^{\partial\Omega_k^+} n_i\, dA \qquad (10.2.21)$$

†Here the word "multidimensional" refers to the space of independent variables occurring in the differential operation.

where the notation $[\cdot]_{\partial\Omega_k^-}^{\partial\Omega_k^+}$ signifies the limit (at any point on $\partial\Omega_k$ of the expression within the brackets) taken from the "interior" minus the limits taken from the exterior.

Consider two cases. In the first, assume that neither $\partial\Omega$ nor any of the surfaces $\{\partial\Omega_k\}_{k=1}^m$ have "weights" (volume measure). Here one may define domains $D(T)$ in several ways to produce self-adjoint operators. For example, one may require that the domain must consist of functions, say $u$, continuous in $\Omega$ and, in particular, across the surfaces $\{\partial\Omega_k\}_{k=1}^m$, that is,†

$$[u]_{\partial\Omega_k^-}^{\partial\Omega_k^+} = 0 \qquad \text{everywhere on } \partial\Omega_k, \quad k = 1, 2, \dots, m \tag{10.2.22}$$

Further, we may impose the condition

$$\sum_{i=1}^{n}\sum_{j=1}^{n}\left[a_{ij}\frac{\partial u}{\partial x_j}\right]_{\partial\Omega_k^-}^{\partial\Omega_k^+} n_i = 0 \qquad \text{everywhere on } \partial\Omega_k, \quad k = 1, 2, \dots, m \tag{10.2.23}$$

When the coefficients $a_{ij}$ satisfy (10.2.14), (10.2.23) becomes

$$\left[a\frac{\partial u}{\partial n}\right]_{\partial\Omega_k^-}^{\partial\Omega_k^+} = 0, \qquad k = 1, 2, \dots, m \tag{10.2.24}$$

Equations (10.2.23) and (10.2.24) are reminiscent of the continuity requirements on the energy or mass flux at material interfaces in heat or mass transfer problems. Clearly, the second expression on the right-hand side of (10.2.21) will vanish if the interface conditions (10.2.22) and (10.2.23) are imposed. If, in addition, any of the boundary conditions in Section 10.2.1 such as (10.2.12), (10.2.13), (10.2.15), or (10.2.17) are satisfied on $\partial\Omega$, the right-hand side of (10.2.21) vanishes entirely, yielding a self-adjoint operator **T**. Alternatively, (10.2.22) may be replaced by the condition

$$[\kappa_k u]^{\partial\Omega_k^-} = [\kappa_{k+1} u]^{\partial\Omega_k^+} \tag{10.2.25}$$

where $\kappa_1, \kappa_2, \dots, \kappa_{m+1}$ are positive constants. The foregoing condition occurs in mass transfer applications in which the concentrations of a diffusing solute in different phases are assumed to be in equilibrium. Conditions (10.2.23) and (10.2.25), together with boundary conditions (10.2.12), (10.2.13), (10.2.15), or (10.2.17), will yield a self-adjoint operator with **T** given by (10.2.7), provided that the inner product (10.2.8) is modified as

$$\langle \mathbf{u}, \mathbf{v} \rangle = \sum_{k=1}^{m+1} \kappa_k \int_{\Omega_k} r(\mathbf{x})u(\mathbf{x})v(\mathbf{x})\, dV \tag{10.2.26}$$

where $\Omega_1$ is the innermost region and $\Omega_k$ is the region between $\partial\Omega_{k-1}$ and $\partial\Omega_k$ for $k = 2, \dots, m$, and $\Omega_{m+1}$ is the region between $\partial\Omega_m$ and $\partial\Omega$ (see Figure 10.2.1). The inner product (10.2.26) also admits other interface conditions which yield self-adjoint operators (see Exercise 10.2.6).

---

†The reader should be familiar with this notation from page 346.

For the second case, consider the interfaces $\{\partial\Omega_k\}_{k=1}^m$ as "weighted." The formulation of the operation derives from that in Section 9.4. To simplify the discussion, let $m = 1$. We formulate first the Hilbert spaces $\mathcal{H}_V \equiv \mathcal{L}_2\{\Omega; r_V(\mathbf{x})\}$ and $\mathcal{H}_A \equiv \mathcal{L}_2\{\partial\Omega_1; r_A(\mathbf{x})\}$, where a distinction has been made between the positive weight functions $r_V(\mathbf{x})$ in the inner product for $\mathcal{H}_V$, and $r_A(\mathbf{x})$ in the inner product for $\mathcal{H}_A$. The Hilbert space $\mathcal{H}_A$ consists of functions $u_i(\mathbf{x})$ such that

$$\oint_{\partial\Omega} r_A(\mathbf{x})u_1^2(\mathbf{x})\,dA < \infty$$

The Hilbert space $\mathcal{H}$ of interest is given by $\mathcal{H} = \mathcal{H}_V \oplus \mathcal{H}_A$, whose typical element may be represented as

$$\mathbf{u} \equiv \begin{bmatrix} \{u(\mathbf{x}), \mathbf{x} \in \Omega\} \\ \{u(\mathbf{x}), \mathbf{x} \in \partial\Omega_1\} \end{bmatrix}$$

The inner product in $\mathcal{H}$ may be taken as

$$\langle \mathbf{u}, \mathbf{v} \rangle = \int_\Omega r_V(\mathbf{x})u(\mathbf{x})v(\mathbf{x})\,dV + \oint_{\partial\Omega_1} r_A(\mathbf{x})u(\mathbf{x})v(\mathbf{x})\,dA \tag{10.2.27}$$

A formally self-adjoint differential operation may be identified following a development analogous to that of Section 9.4. One obtains an expression of the type

$$L \equiv \begin{bmatrix} T & 0 \\ -\dfrac{1}{r_A(x)}\left[\sum_{i=1}^n \sum_{j=1}^n n_i a_{ij} \dfrac{\partial}{\partial x_j}\right]_{\partial\Omega_1^-}^{\partial\Omega_1^+} & 0 \end{bmatrix} \tag{10.2.28}$$

In (10.2.28) it must be understood that the limits on either side of $\partial\Omega_1$ of the expression in the lower left-hand corner are taken at the *local* point $\mathbf{x}$ (in $\partial\Omega_1$) which appears in $r_A(\mathbf{x})$. It is now readily shown that, in terms of the inner product (10.2.27),

$$\langle L\mathbf{u}, \mathbf{v} \rangle - \langle \mathbf{u}, L\mathbf{v} \rangle = -\oint_{\partial\Omega} \sum_{i=1}^n \sum_{j=1}^n a_{ij}\left(v\frac{\partial u}{\partial x_j} - u\frac{\partial v}{\partial x_j}\right)n_i\,dA \tag{10.2.29}$$

which is arrived at using (10.2.21) with $m = 1$. In order to obtain a self-adjoint operator $\mathbf{L}$, the right-hand side of (10.2.29) must vanish, the conditions for which have already appeared in Section 10.2.1, that is, any of the boundary conditions (10.2.12), (10.2.13), (10.2.15), or (10.2.17).

There are several possible variations of the situation above which yield self-adjoint operators. The nature of such variations is readily inferred from the discussion on the ordinary differential operations of Section 9.4.

Let us consider some engineering examples in which we encounter operators of the type just discussed.

*Example 1. Heat Conduction in a Composite Sphere*

Consider a composite sphere of two materials such that the inner sphere of material 1 is concentric with the outer sphere of material 2. The sphere, initially possessing some known temperature distribution, is subjected to a specified temperature field at the outer surface. The problem is to calculate the temperature field in the composite sphere at any instant $t$. Denote the physical properties of the $i$th material by $k_i$ for the thermal conductivity and $c_i$ for the volumetric heat capacity. The inner and outer spheres have radii $R_1$ and $R_2$, respectively. The heat conduction equation for the composite sphere may be written in the following dimensionless form.

$$-\frac{1}{\gamma(x)x^2 \sin\theta}\left[\frac{\partial}{\partial x}\left\{\kappa(x)x^2 \sin\theta \frac{\partial}{\partial x}\right\} + \frac{\partial}{\partial \phi}\left\{\frac{\kappa(x)}{\sin\theta}\frac{\partial}{\partial \phi}\right\} + \frac{\partial}{\partial \theta}\left\{\kappa(x)\sin\theta\frac{\partial}{\partial \theta}\right\}\right]\Theta$$

$$= -\frac{\partial}{\partial \tau}\Theta, \qquad \tau > 0, 0 < x < 1, 0 < \theta < \pi, 0 < \phi < 2\pi \qquad (10.2.30)$$

where $\Theta(x, \theta, \phi, \tau)$ is a dimensionless version of the temperature field, with $x$ as the dimensionless radial distance $r/R_2$, and $\theta$ and $\phi$ as the two azimuthal angles in spherical coordinates. Further, we have

$$\kappa(x) = \begin{Bmatrix} \dfrac{k_1}{k_2}, & 0 < x < a, & \dfrac{c_1}{c_2} \\ 1, & a < x < 1, & 1 \end{Bmatrix} = \gamma(x) \qquad (10.2.31)$$

where $a \equiv R_1/R_2$; the dimensionless time $\tau$ in (10.2.30) is expressed in terms of the actual time $t$ by $\tau = k_2 t/c_2 R_2^2$.

The differential operation

$$T \equiv -\frac{1}{\gamma(x)x^2 \sin\theta}\left[\frac{\partial}{\partial x}\left\{\kappa(x)x^2 \sin\theta \frac{\partial}{\partial x}\right\} + \frac{\partial}{\partial \phi}\left\{\frac{\kappa(x)}{\sin\theta}\frac{\partial}{\partial \phi}\right\}\right.$$

$$\left. + \frac{\partial}{\partial \theta}\left\{\kappa(x)\sin\theta\frac{\partial}{\partial \theta}\right\}\right] \qquad (10.2.32)$$

is indeed of the formally self-adjoint, elliptic type (10.2.7) with discontinuities in the coefficients occurring concurrently at $x = a$. Note that $x = a$ represents the spherical surface $\partial\Omega_1$ in the present case. Across this surface we expect the temperature and the energy flux to be continuous expressed by

$$\Theta(a-, \theta, \phi, \tau) = \Theta(a+, \theta, \phi, \tau)$$

$$-\kappa(a-)\frac{\partial}{\partial x}\Theta(a-, \theta, \phi, \tau) = -\kappa(a+)\frac{\partial}{\partial x}\Theta(a+, \theta, \phi, \tau)$$

which are identical to the interface conditions (10.2.22) and (10.2.24). The homogeneous version of the boundary condition on $\Theta$ is the Dirichlet condition (10.2.12). If we define the domain

$$D(T) = \Big\{\mathbf{u},\ T\mathbf{u} \in \mathcal{H};\ u(1, \theta, \phi) = 0, \qquad u(a-, \theta, \phi) = u(a+, \theta, \phi),$$

$$\kappa(a-)\frac{\partial u}{\partial x}(a-, \theta, \phi) = \kappa(a+)\frac{\partial u}{\partial x}(a+, \theta, \phi)\Big\}$$

where $\mathcal{H} = \mathcal{L}_2\{\Omega; \gamma(x)x^2 \sin\theta\}$, $\Omega \equiv \{0 < x < 1, 0 < \theta < \pi, 0 < \phi < 2\pi\}$. The operator $\mathbf{T} = \{T, D(T)\}$ is self-adjoint and belongs to the first of the two cases considered in this subsection.

*Example 2. Temperature Distribution in a Peripherally Cooled Composite Sphere*

This example was considered in Section 2.12† to demonstrate the occurrence of direct sums of linear spaces in applications. We introduce here some additional complications in this example by letting the sphere be a composite of the type considered in the previous example. Next, we allow a spatially nonuniform, time-dependent heat source distribution $\sigma(x, \theta, \phi, \tau)$ so that conduction must be considered in all the three dimensions. The periphery of the sphere is cooled by a finite volume of well-stirred fluid. The energy differential equation in terms of dimensionless variables defined in Example 1 becomes

$$T\Theta(x, \theta, \phi, \tau) = \sigma(x, \theta, \phi, \tau) - \frac{\partial}{\partial \tau}\Theta(x, \theta, \phi, \tau) \qquad (10.2.33)$$

where $T$ is given by (10.2.32). The dimensionless coolant temperature $\Theta_1(\tau)$ must satisfy the following dimensionless energy balance:

$$\int_0^{\pi} d\theta \sin\theta \int_0^{2\pi} d\phi\, \alpha \frac{\partial}{\partial x}\Theta(1-, \theta, \phi, \tau) = -\frac{d}{d\tau}\Theta_1(\tau) \qquad (10.2.34)$$

If there is heat transfer resistance between the fluid and the sphere, it is expressed in dimensionless form as

$$-\frac{\partial}{\partial x}\Theta(1-, \theta, \phi, \tau) = \beta[\Theta(1-, \theta, \phi, \tau) - \Theta_1(\tau)] \qquad (10.2.35)$$

where $\beta$ measures the resistance to heat transfer in the fluid relative to that in the sphere. For infinitely large values of $\beta$, (10.2.35) may be replaced by

$$\Theta(1-, \theta, \phi, \tau) = \Theta_1(\tau) \qquad (10.2.36)$$

For finite values of $\beta$, (10.2.33) and (10.2.34) may be combined to obtain

$$L\begin{bmatrix}\{\Theta(x, \theta, \phi, \tau)\} \\ \Theta_1(\tau)\end{bmatrix} = \frac{\partial}{\partial \tau}\begin{bmatrix}\{\Theta(x, \theta, \phi, \tau)\} \\ \Theta_1(\tau)\end{bmatrix} + \begin{bmatrix}\{\sigma(x, \theta, \phi, \tau)\} \\ 0\end{bmatrix}$$

---

†See page 65.

where

$$L \equiv \begin{bmatrix} T & 0 \\ -\int_0^{\pi} d\theta \sin\theta \int_0^{2\pi} d\phi\, \alpha\left[\frac{\partial}{\partial x}\right]_{\partial\Omega^-} & 0 \end{bmatrix},$$

$$\partial\Omega = \{x = 1, 0 < \theta < \pi, 0 < \phi < 2\pi\} \qquad (10.2.37)$$

The reader is alerted to the notational scheme on page 346 in interpreting the bottom left element of $L$. The boundary $\partial\Omega$ has clearly the property of being "weighted." Hence the situation here is that of the second case in the discussion of this subsection before the examples. Thus the Hilbert space of interest is given by

$$\mathcal{H} = \mathcal{L}_2\{\Omega; \gamma(x)x^2 \sin\theta\} \oplus \mathcal{L}_2\{\partial\Omega; \sin\theta\}$$

Because of the well-stirred situation in the fluid and the consequent uniformity of temperature over the surface $\partial\Omega$, we will be interested only in the subspace of $\mathcal{L}_2(\partial\Omega)$ containing the "constant" elements. (See Exercise 10.2.4 for an application that involves temperature variation along a weighted interface.) Denoting a typical element $\mathbf{u}$ of $\mathcal{H}$ by

$$\mathbf{u} \equiv \begin{bmatrix} \{u(x, \theta, \phi)\} \\ \{u_1(\theta, \phi)\} \end{bmatrix}$$

the inner product from prior discussion must be defined as

$$\langle \mathbf{u}, \mathbf{v}\rangle = \int_0^1 dx\gamma(x)x^2 \int_0^{\pi} d\theta \sin\theta \int_0^{2\pi} d\phi u(x, \theta, \phi)v(x, \theta, \phi)$$

$$+ r_1 \int_0^{\pi} d\theta \sin\theta \int_0^{2\pi} d\phi u_1(\theta, \phi)v_1(\theta, \phi), \qquad r_1 > 0 \qquad (10.2.38)$$

By recognizing suitable boundedness and continuity properties of $u$ and $v$ (especially at $\phi = 0$ or $2\pi$), and letting $u_1$ and $v_1$ be constants it is readily shown that

$$\langle L\mathbf{u}, \mathbf{v}\rangle - \langle \mathbf{u}, L\mathbf{v}\rangle = \int_0^{\pi} d\theta \sin\theta \int_0^{2\pi} d\phi \Big[\{v(1, \theta, \phi) - 4\pi\alpha r_1 v_1\}\frac{\partial u}{\partial x}(1, \theta, \phi)$$

$$- \{u(1, \theta, \phi) - 4\pi\alpha r_1 v_1\}\frac{\partial v}{\partial x}(1, \theta, \phi)\Big] \qquad (10.2.39)$$

We now take cognizance of (10.2.35) to require that

$$u(1, \theta, \phi) - u_1(\theta, \phi) = -\frac{1}{\beta}\frac{\partial u}{\partial x}(1, \theta, \phi)$$

which also must hold for $v$. The relationship above, which must enter the definition of the domain of $L$, makes the right-hand side of (10.2.39) vanish if $4\pi r_1\alpha = 1$. Thus if we define

$$D(L) = \Big\{\mathbf{u}, L\mathbf{u} \in \mathcal{H}; u(a-, \theta, \phi) = u(a+, \theta, \phi), \kappa(a-)\frac{\partial u}{\partial x}(a-, \theta, \phi)$$

$$= \kappa(a+)\frac{\partial u}{\partial x}(a+, \theta, \phi), \qquad u_1(\theta, \phi) = u_1 \equiv \text{constant},$$

$$u(1, \theta, \phi) - u_1(\theta, \phi) = -\frac{1}{\beta}\frac{\partial u}{\partial x}(1, \theta, \phi)\Big\}$$

where we have implicitly assumed several additional properties required of $u$ and $v$, then $L = \{L, D(L)\}$ is a self-adjoint operator.

The example above can be generalized further but some additional considerations are required. Wc therefore postpone this to the next section. The following exercises are strongly recommended for students who are interested in transport problems in multiphase media.

## EXERCISES

**10.2.6** Let $\Omega$ be the compact region considered in Section 10.2.1 with the family of "concentric" surface $\{\partial\Omega_k\}_{k=1}^m$ within $\Omega$ (see Figure 10.2.1). $\mathcal{H}$ is a Hilbert space of functions $\mathbf{u}$ such that

$$\|\mathbf{u}\|^2 \equiv \sum_{k=1}^m \kappa_k \int_{\Omega_k} r(\mathbf{x})u^2(\mathbf{x})\, dV$$

with the inner product defined as in (10.26). Consider the differential expression

$$T \equiv -\frac{1}{r(\mathbf{x})}\sum_{i=1}^n \sum_{j=1}^n \frac{\partial}{\partial x_i}\left(a\frac{\partial}{\partial x_j}\right) + c$$

Let $D(T)$ be a subspace of $\mathcal{H}$ with functions $\mathbf{u}$ satisfying any of the self-adjoint boundary conditions (10.2.12), (10.2.13), (10.2.15), or (10.2.17), and the interface conditions comprising (10.2.24) and

$$-a\frac{\partial u}{\partial n} = h([\kappa_k u]^{\partial\Omega_k^-} - [\kappa_{k+1}u]^{\partial\Omega_k^+}) \qquad \text{on } \partial\Omega_k$$

where $n$ is measured along the outer normal on $\partial\Omega_k$ and $h$ may vary with position on $\partial\Omega_k$. Show that $\mathbf{T} \equiv \{T, D(T)\}$ is self-adjoint.

**10.2.7** The description of molecular transport processes such as conduction, diffusion, and so on, in general three-dimensional curvilinear coordinates involves the operation

$$T \equiv -\frac{1}{\sqrt{g}}\sum_{i=1}^3 \sum_{j=1}^3 \frac{\partial}{\partial x_i}\left(D\sqrt{g}\, g_{ij}\frac{\partial}{\partial x_j}\right)$$

where $D$ is the transport coefficient, and the significance of the other symbols appears below (10.2.19). Consider transport in a heterogeneous medium contained in the region $\Omega$, as displayed in Figure 10.2.1, in which the transport coefficient undergoes discontinuous variations across the surfaces $\{\partial\Omega_j\}_{j=1}^m$. Using the same setting as in Exercise 10.2.6, identify the self-adjoint operator $\mathbf{T}$ by specifying the domain $D(T)$. (Express the normal derivative on boundaries in terms of the components of the normal, and the partial derivatives along the space coordinates.)

**10.2.8** Suppose that in Example 2 of Section 10.2.2 one were to allow for heat loss from the coolant to the surrounding ambient described by $\gamma[\Theta_1(\tau) - \Theta_f]$ in terms of dimensionless variables. Formulate a self-adjoint problem by defining $L$ and $D(L)$.

**10.2.9** In Example 1 of Section 10.2.2, assume that the inner sphere has infinite thermal conductivity. Reformulate the heat conduction problem and identify a self-adjoint operator $L$. (Pay particular attention to the condition at $x = a$.)

**10.2.10** Suppose that the composite region considered in Exercise 10.2.6 does not consist of "concentric" surfaces such as in Figure 10.2.1 but instead is as shown in Figure 10.2.2. Discuss suitable domains for self-adjoint operators based on the differential expression appearing in Exercise 10.2.6. (Note that the solution of these problems is difficult. The methods of Section 10.4 are not generally applicable even for simple situations.)

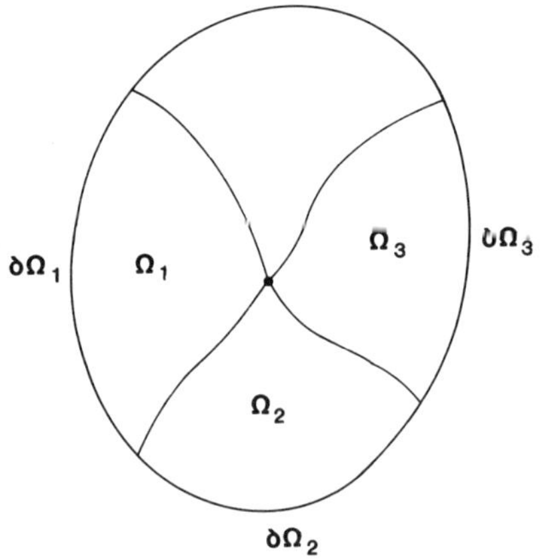


**Figure 10.2.2** Composite medium for Exercise 10.2.10.

## 10.3 Higher-Order Partial Differential Expressions

We will severely limit the scope of this discussion by accommodating only expressions obtained by taking integral powers of the operation $T$ as given by (10.2.7). The operation $T^n$ is readily shown to be formally self-adjoint in view of the formal self-adjointness of $T$. Let $b(\mathbf{u}, \mathbf{v})$ be the bilinear form whose value depends only on the boundary values of $\mathbf{u}$ and $\mathbf{v}$ defined by

$$\langle T\mathbf{u}, \mathbf{v}\rangle - \langle \mathbf{u}, T\mathbf{v}\rangle = b(\mathbf{u}, \mathbf{v})$$

By suitably restricting the domains from which $\mathbf{u}$ and $\mathbf{v}$ arise, it is readily shown that

$$\langle T^n\mathbf{u}, \mathbf{v}\rangle - \langle \mathbf{u}, T^n\mathbf{v}\rangle = \sum_{j=0}^{n-1} b(T^{n-1-j}\mathbf{u}, T_j\mathbf{v}) \tag{10.3.1}$$

which establishes the formal self-adjointness of $T^n$. Of particular interest is the operation $T^2$, which appears in problems on elasticity. Clearly,

$$\langle T^2\mathbf{u}, \mathbf{v}\rangle - \langle \mathbf{u}, T^2\mathbf{v}\rangle = b(T\mathbf{u}, \mathbf{v}) + b(\mathbf{u}, T\mathbf{v})$$

An obvious situation leading to a self-adjoint operator is that for which $\mathbf{u}$, $T\mathbf{u} \in D(T)$, where $D(T)$ is a self-adjoint domain for $T$. In this case we are dealing with the operator $\mathbf{T}^2$. However, it is easy to find other domains $D(T^2)$ such that the operator $\{T^2, D(T^2)\}$ is self-adjoint. For example, let $D(T)$ be not a self-adjoint domain for $T$ and $D^*(T)$ be the adjoint domain. Then

$$D(T^2) = \{\mathbf{u} \in D(T), T\mathbf{u} \in D^*(T)\}$$

is indeed a self-adjoint domain for $T^2$, which is left for the student to see.

## EXERCISES

**10.3.1** Consider the biharmonic partial differential expression in Cartesian coordinates

$$T \equiv \left[\frac{\partial^2}{\partial x^2} + \frac{\partial^2}{\partial y^2}\right]^2, \ (x, y) \in \Omega$$

where $\Omega = \{0 < x < 1, 0 < y < 1\}$. Investigate which of the boundary conditions lead to self-adjoint operators. Let

$$u \equiv u(x, y), \qquad \nabla^2 u \equiv \frac{\partial^2 u}{\partial x^2} + \frac{\partial^2 u}{\partial y^2}$$

(a) $$u(x, 0) = \frac{\partial u}{\partial y}(x, 0) = u(x, 1) = \frac{\partial u}{\partial y}(x, 1) = 0$$

$$\frac{\partial u}{\partial x}(0, y) = \frac{\partial}{\partial x}\nabla^2 u(0, y) = u(1, y) = \frac{\partial u}{\partial x}(1, y) = 0.$$

(b) $$u(x, 0) = \nabla^2 u(x, 0) = \frac{\partial}{\partial y}u(x, 1) = \frac{\partial}{\partial y}\nabla^2 u(x, 1) = 0$$

$$u(0, y) = \frac{\partial}{\partial x}u(0, y) = u(1, y) = \nabla^2 u(1, y) = 0.$$

(c) $$u(x, 0) = \frac{\partial}{\partial y}\nabla^2 u(x, 0) = u(x, 1) = \nabla^2 u(x, 1) = 0$$

$$u(0, y) = \frac{\partial}{\partial x}u(0, y) = u(1, y) = \frac{\partial}{\partial x}u(x, 1) = 0.$$

**10.3.2** In the vibration of a thin annular circular plate, clamped at the inside edge and free at the outer edge, one encounters the operation $L = T^2$, where

$$T \equiv \left[\frac{1}{r}\frac{\partial}{\partial r}\left(r\frac{\partial}{\partial r}\right) + \frac{1}{r^2}\frac{\partial^2}{\partial \theta^2}\right], \qquad (r, \theta) \in \Omega$$

with $\Omega \equiv \{(a, b) \times (0, 2\pi)\}$. Let $\mathcal{H} = \mathcal{L}_2\{\Omega; r\}$. For the situation on hand, we have

$$D(L) = \Big\{\mathbf{u}, L\mathbf{u} \in \mathcal{H};\ u(a+, \theta) = \frac{\partial}{\partial r} u(a+, \theta) = 0$$

$$\left[T - (1 - \nu)\left(\frac{1}{r}\frac{\partial}{\partial r} + \frac{1}{r^2}\frac{\partial^2}{\partial \theta^2}\right)\right] u(b-, \theta) = 0$$

$$\left[\frac{\partial}{\partial r}T + \frac{1 - \nu}{r^2}\frac{\partial^2}{\partial \theta^2}\left(\frac{\partial}{\partial r} - \frac{1}{r}\right)\right] u(b-, \theta) = 0\Big\}$$

Show that $\mathbf{L} = \{L, D(L)\}$ is self-adjoint.

## 10.4 Separable Partial Differential Operators

The method of separation of variables is a well-known technique for constructing explicit solutions to certain types of partial differential equations. It is applicable only subject to the availability of suitable coordinate systems under which the solution to the eigenvalue problem can be expressed as the product of functions of the separate coordinates. Even when a suitable coordinate system is available, separability depends on the boundary conditions or, in other words, the domains of the different operations appearing in the partial differential expression. We are concerned specifically with the issue of the suitable combination of domains which permit separability, granting that an appropriate coordinate system is already available. In regard to the suitability of coordinate systems, we will be content with presenting examples automatically satisfying the due requirements. As pointed out earlier, our approach is based on that of Friedman, in which the partial differential operator is expressed as the sum of tensor products of ordinary differential operators. It is then a question of expressing the spectral representation of the partial differential operator in terms of those of the ordinary differential operators.

In Section 2.12 we considered an example of how partial differential expressions may be expressed as the sum of tensor products of the type

$$T = A_1 \otimes B_1 + \ldots + A_m \otimes B_m \tag{10.4.1}$$

In the discussion below it will be assumed that the partial differential expression (in two variables) of interest to us can always be expressed in the form (10.4.1) (see Exercise 10.4.1). The ordinary differential expression $A_1, A_2, \ldots, A_m$ have domains in the same Hilbert space, say $\mathcal{H}_A$, while $B_1, B_2, \ldots, B_m$ possess domains in a Hilbert space $\mathcal{H}_B$. The domain of the partial differential operation $T$ must then originate from $\mathcal{H} \equiv \mathcal{H}_A \otimes \mathcal{H}_B$. The expressions $A_1, A_2, \ldots, A_m$ and $B_1, B_2, \ldots, B_m$ are formally self-adjoint in their respective spaces. Thus $T$ is also formally self-adjoint in $\mathcal{H}$. We further let the operators $\{A_j\}_{j=1}^m$ and $\{B_j\}_{j=1}^m$ be self-adjoint such that the domains $\{D(A_j)\}_{j=1}^m$ do not depend on the coordinate connected with $\mathcal{H}_B$ and the domains $\{D(B_j)\}_{j=1}^m$ do not depend on the coordinate connected with $\mathcal{H}_A$. Under these circumstances, we may replace (10.4.1) by

$$\mathbf{T} = \mathbf{A}_1 \otimes \mathbf{B}_1 + \ldots + \mathbf{A}_m \otimes \mathbf{B}_m. \tag{10.4.2}$$

which now features the operators instead of the expressions.

Our basic problem may be stated as follows. Given that $\mathbf{A}_1, \mathbf{A}_2, \ldots, \mathbf{A}_m$ and $\mathbf{B}_1, \mathbf{B}_2, \ldots, \mathbf{B}_m$ are self-adjoint operators† in their respective Hilbert spaces $\mathcal{H}_A$ and $\mathcal{H}_B$, and consequently endowed with spectral representations, is it possible to identify the spectral representation of $\mathbf{T}$? We state the answer to this question as a slightly modified version of Friedman's theorem.

***Theorem 10.4.1*** Let $\mathbf{T}$ be a self-adjoint operator expressed as in 10.4.2 such that $\mathbf{A}_1, \mathbf{A}_2, \ldots, \mathbf{A}_m$ have a *common* spectral representation in $\mathcal{H}_A$; that is, there exists a common resolution of the identity $\mathbf{I}^A$ in $\mathcal{H}_A$ given by an orthogonal family of self-adjoint projections $\{\mathbf{P}_n^A\}$ such that

$$\sum_{n=1}^{\infty} \mathbf{P}_n^A = \mathbf{I}^A, \qquad \mathbf{A}_j \equiv \sum_{n=1}^{\infty} \alpha_{jn} \mathbf{P}_n^A$$

where $\{\alpha_{jn}\}$ is the spectrum of $\mathbf{A}_j$.

The operator $\sum_{j=1}^m \alpha_{jn}\mathbf{B}_j$ is clearly self-adjoint for each fixed $n$. Let $\{\mathbf{P}_k^B(n)\}_{k=1}^{\infty}$ be the orthogonal family of self-adjoint projections such that

$$\sum_{j=1}^{m} \alpha_{jn}\mathbf{B}_j = \sum_{k=1}^{\infty} \lambda_k(n)\mathbf{P}_k^B(n), \qquad n = 1, 2, \ldots$$

where $\{\lambda_k(n)\}$ is the spectrum of $\sum_{j=1}^n \alpha_{jn}\mathbf{B}_j$. Then

(i) Each $\lambda_k(n)$ is an eigenvalue of $T$ with $\mathbf{z}_n^A \otimes \mathbf{z}_k^B(n)$ as the corresponding eigenvector.

(ii) The formal spectral representation

$$\mathbf{T} = \sum_{n=1}^{\infty} \sum_{k=1}^{\infty} \lambda_k(n)\mathbf{P}_n^A \otimes \mathbf{P}_k^B(n) \tag{10.4.3}$$

may be extended to functions of $\mathbf{T}$ such as $\mathbf{f}(\mathbf{T})$ provided that $f(\lambda_k(n))$ is bounded for all $k$ and $n$.

We will not bother to prove this theorem. The stipulations of the theorem are trivially satisfied in the applications of interest to us. We have in (10.4.3) the spectral representation of $\mathbf{T}$ with $\{\lambda_k(n); n, k = 1, 2, \ldots\}$ as the eigenvalues and $\{\mathbf{P}_n^A \otimes \mathbf{P}_k^B(n)\}$ as the associated self-adjoint projections. The projections are defined in terms of the eigenvectors of $\mathbf{T}$. The actual calculation

---

†Note that the operators $\mathbf{A}_1, \mathbf{A}_2, \ldots, \mathbf{A}_m$ and $\mathbf{B}_1, \mathbf{B}_2, \ldots, \mathbf{B}_m$ are of the type encountered in Chapter 9, that is, unbounded but closed.

proceeds as follows. The eigenvalue problem

$$(\sum_{j=1}^{m} \alpha_{jn}\mathbf{B}_j)\mathbf{z}^B = \lambda_k(n)\mathbf{z}^B, \qquad n = 1, 2, \ldots \tag{10.4.4}$$

must be first solved for each fixed $n$ for the eigenvalues $\{\lambda_k(n)\}_{k=1}^{\infty}$ and the corresponding eigenvectors $\{\mathbf{z}_k^B(n)\}_{k=1}^{\infty}$ of the operator $\sum_{j=1}^{m} \alpha_{jn}\mathbf{B}_j$. Note that the eigenvalues of this operator for each $n$ are also the eigenvalues of $\mathbf{T}$. On the other hand, the eigenvector of $\mathbf{T}$ corresponding to the eigenvalue $\lambda_k(n)$ is denoted as $\mathbf{z}_n^A \otimes \mathbf{z}_k^B(n)$, where $\mathbf{z}_n^A$ is the $n$th common eigenvector of the set of operators $\mathbf{A}_1, \mathbf{A}_2, \ldots, \mathbf{A}_m$.

The action of the projection operator $\mathbf{P}_n^A \otimes \mathbf{P}_k^B(n)$ may be represented as follows. If $\mathbf{w} \in \mathcal{H}$, then

$$[\mathbf{P}_n^A \otimes \mathbf{P}_k^B(n)]\mathbf{w} = \langle\langle \mathbf{w}, \mathbf{z}_n^A\rangle_A, \mathbf{z}_k^B(n)\rangle_B[\mathbf{z}_n^A \otimes \mathbf{z}_k^B(n)] \tag{10.4.5}$$

For $(\mathbf{u} \otimes \mathbf{v}) \in \mathcal{H}_A \otimes \mathcal{H}_B$,

$$[\mathbf{P}_n^A \otimes \mathbf{P}_k^B(n)](\mathbf{u} \otimes \mathbf{v}) = \langle \mathbf{u}, \mathbf{z}_n^A\rangle_A\langle \mathbf{v}, \mathbf{z}_k^B(n)\rangle_B[\mathbf{z}_n^A \otimes \mathbf{z}_k^B(n)] \tag{10.4.6}$$

where $\langle\cdot, \cdot\rangle_A$ and $\langle\cdot, \cdot\rangle_B$ are the respective inner products in $\mathcal{H}_A$ and $\mathcal{H}_B$. In attempting to understand relations (10.4.5) and (10.4.6), the student is alerted to the operations on tensor products of inner product spaces in Section 6.3.

Only operators with discrete spectra have been considered in the foregoing theorem. If the inverse of $\mathbf{T}$ exists, then we may write

$$\mathbf{T}^{-1} = \sum_{n=1}^{\infty}\sum_{k=1}^{\infty} \lambda_k^{-1}(n)[\mathbf{P}_n^A \otimes \mathbf{P}_k^B(n)] \tag{10.4.7}$$

We will also be frequently interested in the function $\mathbf{e}^{-t\mathbf{T}}$ given by

$$\mathbf{e}^{-t\mathbf{T}} = \sum_{n=1}^{\infty}\sum_{k=1}^{\infty} e^{-\lambda_k(n)t}[\mathbf{P}_n^A \otimes \mathbf{P}_k^B(n)] \tag{10.4.8}$$

where $\mathbf{T}$ is positive definite and $t$ is a positive parameter such as time.

Theorem 10.4.1 requires that the operators in one of the sets in (10.4.2), say $\mathbf{A}_1, \mathbf{A}_2, \ldots, \mathbf{A}_m$, have a common spectral representation. This condition is satisfied when the operators *commute mutually*; that is, $\mathbf{A}_i\mathbf{A}_j = \mathbf{A}_j\mathbf{A}_i$ for any $i$ and $j$ (see Exercise 7.4.11).

The spectral representations (10.4.7) or (10.4.8) are extremely useful in the solution of partial differential equations. We present below some examples of such applications.

*Example 1. The Poisson Equation on a Rectangle*

Consider the two-dimensional Poisson equation on a rectangular domain

$$-\left(\frac{\partial^2}{\partial x^2} + \frac{\partial^2}{\partial y^2}\right)v(x, y) = f(x, y), \qquad (x, y) \in \Omega \tag{10.4.9}$$

where $\Omega = \{0 < x < a, 0 < y < b\}$. On the boundary $\partial\Omega$ we impose the Dirichlet boundary condition

$$\begin{aligned} v(x, 0) &= g_1(x), & v(x, b) &= g_2(x), & 0 < x < a \\ v(0, y) &= h_1(y), & v(a, y) &= h_2(y), & 0 < y < b \end{aligned} \tag{10.4.10}$$

The operator of interest is $\mathbf{T} = \{T, D(T)\}$, where $T = -(\partial^2/\partial x^2 + \partial^2/\partial y^2)$ and

$$D(T) = \{\mathbf{u}, T\mathbf{u} \in \mathfrak{L}_2(\Omega), u = 0 \text{ on } \partial\Omega\}$$

Clearly, $\mathbf{T}$ is a self-adjoint elliptic operator. It is readily shown to be positive definite. Since $v(x, y)$ satisfies inhomogeneous boundary conditions, $\mathbf{v} \notin D(T)$. Thus (10.4.9) may be rewritten as

$$T\mathbf{v} = \mathbf{f} \tag{10.4.11}$$

We solve (10.4.11) using the spectral representation of $\mathbf{T}$. First we express $\mathbf{T}$ as a tensor product sum. Letting $\mathfrak{H}_A \equiv \mathfrak{L}_2[0, a]$, $\mathfrak{H}_B \equiv \mathfrak{L}_2[0, b]$, $\mathbf{A}_1 = \{A_1, D(A_1)\}$, $A_1 = -d^2/dx^2$, $D(A_1) = \{\mathbf{u}, A_1\mathbf{u} \in \mathfrak{H}_A; u(0) = u(a) = 0\}$, $\mathbf{A}_2 = \mathbf{I}^A$, $\mathbf{B}_1 = \mathbf{I}^B$, $\mathbf{B}_2 = \{B_2, D(B_2)\}$, $B_2 = -d^2/dy^2$, $D(B_2) = \{\mathbf{u}, B_2\mathbf{u} \in \mathfrak{H}_B, u(0) = u(b) = 0\}$, we have

$$\mathbf{T} = \mathbf{A}_1 \otimes \mathbf{B}_1 + \mathbf{A}_2 \otimes \mathbf{B}_2$$

Clearly, $\mathbf{A}_1$ and $\mathbf{A}_2$ commute and have a common spectral representation which is readily found by the methods of Section 9.2.2. Thus

$$\mathbf{A}_1 = \sum_{n=1}^{\infty} \frac{n^2\pi^2}{a^2}\mathbf{P}_n^A, \qquad \mathbf{A}_2 = \sum_{n=1}^{\infty} \mathbf{P}_n^A, \qquad \mathbf{z}_n^A = \left\{\sqrt{\frac{2}{a}} \sin \frac{n\pi}{a} x\right\}$$

so that $\alpha_{1n} = n^2\pi^2/a^2$ and $\alpha_{2n} = 1$. Next we solve the eigenvalue problem

$$\left(\frac{n^2\pi^2}{a^2} - \frac{d^2}{dy^2}\right) z^B(y) = \lambda z^B(y)$$

$$z^B(0) = z^B(b) = 0$$

to obtain

$$\lambda_k(n) = \pi^2\left(\frac{n^2}{a^2} + \frac{k^2}{b^2}\right), \qquad z_k^B(y) = \sqrt{\frac{2}{b}} \sin \frac{k\pi}{b} y, \qquad k = 1, 2, \ldots$$

and

$$\frac{n^2\pi^2}{a^2}\mathbf{I}^B + \mathbf{B}_2 = \sum_{k=1}^{\infty} \pi^2\left(\frac{n^2}{a^2} + \frac{k^2}{b^2}\right)\mathbf{P}_k^B(n)$$

Notice particularly that $\mathbf{z}_k^B$ (and hence $\mathbf{P}_k^B$) is independent of $n$. This is generally true of all the applications of interest here. We are now equipped to solve boundary value problem (10.4.11). Letting $\mathbf{P}_{nk} \equiv \mathbf{P}_n^A \otimes \mathbf{P}_k^B(n)$ and $\mathbf{z}_{nk} \equiv \mathbf{z}_n^A \otimes \mathbf{z}_k^B$, we apply $\mathbf{P}_{nk}$ on either side of (10.4.11). The result is

$$\mathbf{P}_{nk}T\mathbf{v} = \lambda\mathbf{P}_{nk}\mathbf{v} + \oint_{\partial\Omega} \left(z_{nk}\frac{\partial v}{\partial n} - v\frac{\partial}{\partial n}z_{nk}\right) dA\mathbf{z}_{nk} = \mathbf{P}_{nk}\mathbf{f}$$

which obtains on using (10.2.6). Recognizing the boundary conditions (10.4.10) and the homogeneous boundary conditions on $\mathbf{z}_{nk}$, we write

$$\lambda_k(n)\mathbf{P}_{nk}\mathbf{v} + \left[\int_0^a g_1(x)\frac{\partial}{\partial y}z_{nk}(x,0)\,dx - \int_0^b h_2(y)\frac{\partial}{\partial x}z_{nk}(a,y)\,dy \right.$$

$$\left. - \int_0^a g_2(x)\frac{\partial}{\partial y}z_{nk}(x,b)\,dx + \int_0^b h_1(y)\frac{\partial}{\partial x}z_{nk}(0,y)\,dy\right]\mathbf{z}_{n_k} = \mathbf{P}_{nk}\mathbf{f}$$

Solving for $\mathbf{P}_{nk}\mathbf{v}$ yields

$$\mathbf{P}_{nk}\mathbf{v} = -\left[\langle \mathbf{g}_1, \mathbf{z}_n^A\rangle_A \frac{\sqrt{2}\,k\pi}{b^{3/2}} - \langle \mathbf{h}_2, \mathbf{z}_k^B\rangle_B \frac{\sqrt{2}\,n\pi}{a^{3/2}}(-1)^n \right.$$

$$\left. - \langle \mathbf{g}_2, \mathbf{z}_n^A\rangle_A \frac{\sqrt{2}\,k\pi}{b^{3/2}}(-1)^k + \langle \mathbf{h}_1, \mathbf{z}_k^B\rangle_B \frac{\sqrt{2}\,n\pi}{a^{3/2}}\right]\lambda_k^{-1}(n)\mathbf{z}_{nk} + \lambda_k^{-1}(n)\mathbf{P}_{nk}\mathbf{f}$$

from which the solution is immediately identified by summing over $k$ and $n$:†

$$\mathbf{v} = \sum_{n=1}^{\infty}\sum_{k=1}^{\infty} \lambda_k^{-1}\left\{\langle\langle \mathbf{f}, \mathbf{z}_n^A\rangle_A, \mathbf{z}_k^B\rangle_B - \sqrt{2}\,\pi\left[\langle \mathbf{g}_1, \mathbf{z}_n^A\rangle_A \frac{k}{b^{3/2}} - \langle \mathbf{h}_2, \mathbf{z}_k^B\rangle_B \frac{(-1)^n n}{a^{3/2}} \right.\right.$$

$$\left.\left. - \langle \mathbf{g}_2, \mathbf{z}_n^A\rangle_A \frac{(-1)^k k}{b^{3/2}} + \langle \mathbf{h}_1, \mathbf{z}_k^B\rangle_B \frac{n}{a^{3/2}}\right]\right\}(\mathbf{z}_n^A \otimes \mathbf{z}_k^B) \qquad (10.4.12)$$

The student is apt to find the symbolism here somewhat irksome at the outset. However, once the symbolism is grasped thoroughly, the solution becomes entirely straightforward. For example, consider the solution of the transient problem

$$T\mathbf{v}(t) = -\frac{d}{dt}\mathbf{v}(t) + \mathbf{f}(t), \quad \mathbf{v}(0) = \mathbf{v}_0 \qquad (10.4.13)$$

where time dependence has been inserted in the inhomogeneous term $\mathbf{f}$. Let the boundary conditions be the same as in the previous problem except for inserting time dependence in the boundary values, $\mathbf{g}_1$, $\mathbf{g}_2$, $\mathbf{h}_1$, and $\mathbf{h}_2$. The solution of this problem is immediately written. Out of consideration for the typesetter, we denote the term in the solution (10.4.12) appearing within the braces by $G_{nk}(t)$, where the time dependence has been accommodated. The solution to the initial boundary value problem is given by

$$\mathbf{v}(t) = \sum_{n=1}^{\infty}\sum_{k=1}^{\infty}\left\{e^{-\lambda_k(n)t}\left[\int_0^t e^{\lambda_k(n)t'}G_{nk}(t')\,dt' + \langle\langle \mathbf{v}_0, \mathbf{z}_n^A\rangle_A, \mathbf{z}_k^B\rangle_B\right]\right\}(\mathbf{z}_n^A \otimes \mathbf{z}_k^B)$$

In the next example we consider a partial differential operator in three independent variables to show that the concepts just discussed can be applied here also.

---

†One may note that $\lambda_k^{-1}(n)$ is bounded for all $k$ and $n$, thus satisfying the conditions of Theorem 10.4.1.

*Example 2. Three-Dimensional Transport in a Sphere*

We return here to the partial differential expression appearing in the description of molecular transport in spherical coordinates. We will be concerned with the unit sphere and assume Dirichlet boundary conditions at $r = 1$. The differential operation $T$ is written as

$$T \equiv -\left[\frac{1}{r^2}\frac{\partial}{\partial r}\left(r^2\frac{\partial}{\partial r}\right) + \frac{1}{r^2 \sin^2\theta}\frac{\partial^2}{\partial \phi^2} + \frac{1}{r^2 \sin\theta}\frac{\partial}{\partial \theta}\left(\sin\theta\frac{\partial}{\partial\theta}\right)\right],$$
$$(r, \phi, \theta) \in \Omega \equiv \{0 < r < 1, 0 < \phi < 2\pi, 0 < \theta < \pi\} \qquad (10.4.14)$$

Letting $\mathcal{H} \equiv \mathcal{L}_2\{\Omega; r^2 \sin\theta\}$, we identify the domain as

$$D(T) = \{\mathbf{u}, T\mathbf{u}, \in \mathcal{H}; u = 0 \text{ on } \partial\Omega\}$$

Implicit in the foregoing domain is the continuity of $u$ and $\partial u/\partial\phi$ at $\phi = 0$ or $2\pi$, and the fact that $u$ cannot become unbounded. These observations are important in the calculation of the spectra. Define the following Hilbert spaces:

$$\mathcal{H}_A \equiv \mathcal{L}_2\{[0, 1]; r^2\}, \qquad \mathcal{H}_B \equiv \mathcal{L}_2[0, 2\pi], \qquad \mathcal{H}_C \equiv \mathcal{L}_2\{[0, \pi]; \sin\theta\},$$
$$\mathcal{H} \equiv \mathcal{H}_A \otimes \mathcal{H}_B \otimes \mathcal{H}_C$$

and the operators

$$A_1 = -\frac{1}{r^2}\frac{d}{dr}\left(r^2\frac{d}{dr}\right), \qquad D(A_1) = \{\mathbf{u}, A_1\mathbf{u} \in \mathcal{H}_A; u(1) = 0\},$$

$$\mathbf{A}_1 = \{A_1, D(A_1)\} \qquad A_2 = \frac{1}{r^2}\mathbf{I}^A, \qquad \mathbf{A}_3 = \frac{1}{r^2}\mathbf{I}^A = \mathbf{A}_2$$

$$\mathbf{B}_1 = \mathbf{I}^B, \qquad \mathbf{B}_2 = \{B_2, D(B_2)\},$$

$$B_2 = -\frac{d^2}{d\phi^2}, \qquad D(B_2) = \{\mathbf{u}, B_2\mathbf{u} \in \mathcal{H}_B; u(0) = u(2\pi), u'(0) = u'(2\pi)\}$$

$$\mathbf{B}_3 = \mathbf{I}^B = \mathbf{B}_1$$

$$\mathbf{C}_1 = \mathbf{I}^C, \qquad \mathbf{C}_2 = \frac{1}{\sin^2\theta}\mathbf{I}^C, \mathbf{C}_3 = \{C_3, D(C_3)\} \quad \text{where}$$

$$C_3 = -\frac{1}{\sin\theta}\frac{d}{d\theta}\left(\sin\theta\frac{d}{d\theta}\right), \qquad D(C_3) = \{\mathbf{u}, C_3\mathbf{u} \in \mathcal{H}_C; |u| < \infty\}$$

We may now write

$$\mathbf{T} = \mathbf{A}_1 \otimes \mathbf{B}_1 \otimes \mathbf{C}_1 + \mathbf{A}_2 \otimes \mathbf{B}_2 \otimes \mathbf{C}_2 + \mathbf{A}_3 \otimes \mathbf{B}_3 \otimes \mathbf{C}_3 \qquad (10.4.15)$$

The problem now is to obtain the spectral representation of $\mathbf{T}$ from those of the individual operators occurring in (10.4.15). The first step is to identify the operators (on any given Hilbert space) with a common spectral representation. Clearly, $\mathbf{B}_1$, $\mathbf{B}_2$, and $\mathbf{B}_3$ satisfy these requirements, since $\mathbf{B}_1 = \mathbf{B}_3 = \mathbf{I}^B$, and $\mathbf{B}_2$ commutes with them. Moreover, by the methods of Section 9.2.2 it is readily shown that

$$\mathbf{B}_2 = \sum_{n=0}^{\infty} n^2\mathbf{P}_n^B, \qquad \mathbf{B}_1 = \mathbf{B}_3 = \mathbf{I}^B = \sum_{n=0}^{\infty} \mathbf{P}_n^B$$

where $\mathbf{P}_n^B(\cdot) = \langle \cdot, \mathbf{z}_{n,1}^B \rangle_B \mathbf{z}_{n,1}^B + \langle \cdot, \mathbf{z}_{n,2}^B \rangle_B \mathbf{z}_{n,2}^B$ with

$$\mathbf{z}_{n,1}^B = \left\{ \begin{matrix} \left\{\frac{1}{\sqrt{2\pi}}\right\}, & n = 0 & 0 \\ \left\{\frac{1}{\sqrt{\pi}} \cos n\phi\right\}, & n = 1, 2, \dots, & \left\{\frac{1}{\sqrt{\pi}} \sin n\phi\right\} \end{matrix} \right\} = \mathbf{z}_{n,2}^B$$

Thus $T$ becomes

$$\begin{aligned} \mathbf{T} &= \mathbf{A}_1 \otimes \sum_{n=0}^{\infty} \mathbf{P}_n^B \otimes \mathbf{C}_1 + \mathbf{A}_2 \otimes \sum_{n=0}^{\infty} n^2\mathbf{P}_n^B \otimes \mathbf{C}_2 + \mathbf{A}_2 \otimes \sum_{n=0}^{\infty} \mathbf{P}_n^B \otimes \mathbf{C}_3 \\ &= \sum_{n=0}^{\infty} \{\mathbf{A}_1 \otimes \mathbf{P}_n^B \otimes \mathbf{C}_1 + \mathbf{A}_2 \otimes \mathbf{P}_n^B \otimes (\mathbf{C}_2 n^2 + \mathbf{C}_3)\} \end{aligned} \tag{10.4.16}$$

Further examination of $T$ as in (10.4.16) shows that for fixed $n$, the operators $(\mathbf{C}_2 n^2 + \mathbf{C}_3)$ and $\mathbf{C}_1 \equiv \mathbf{I}^C$ commute and hence have the same spectral representation. To find this representation, we solve the eigenvalue problem

$$(\mathbf{C}_2 n^2 + \mathbf{C}_3)\mathbf{z}^C = \mu \mathbf{z}^C$$

or

$$-\frac{1}{\sin\theta}\frac{d}{d\theta}\left(\sin\theta \frac{d}{d\theta} z^C\right) + \frac{n^2}{\sin^2\theta} z^C = \mu z^C$$

By letting $x = \cos\theta$, $p(x) \equiv z^C(\theta)$, the differential equation above transforms to

$$\frac{d}{dx}\left[(1 - x^2)\frac{dp}{dx}\right] + \left(\mu - \frac{n^2}{1 - x^2}\right)p = 0 \tag{10.4.17}$$

This equation is called *Legendre's associated equation*. When $n = 0$, (10.4.17) becomes Legendre's equation,

$$\frac{d}{dx}\left[(1 - x^2)\frac{dp}{dx}\right] + \mu p = 0 \tag{10.4.18}$$

Both differential equations (10.4.17) and (10.4.18) have regular singular points at $x = \pm 1$ (see Exercise 9.2.16) and may be readily solved by the method of Frobenius (Appendix A.3). The results are well known.† Equation (10.4.17) has eigenvalues $\{\mu_k(n) = k(k + 1); k = n, (n + 1), \dots\}$ with the eigenvector $\mathbf{z}_k^C(n)$ corresponding to $\mu_k(n)$ given by

$$\mathbf{z}_n^C(n) = \{p_k^n(\cos\theta); 0 < \theta < \pi\} \tag{10.4.19}$$

†See, for example, *Methods of Mathematical Physics*, Vol. 1, by R. Courant and D. Hilbert, Interscience, New York, 1953.

where $p_k^n(\cdot)$ is referred to as the *associated Legendre* polynomial of $n$th order. It is given by

$$p_k^n(x) = \frac{(-1)^n}{2^n n!}\sqrt{\frac{(2k+1)(k-n)!}{2(k+n)!}}(1-x^2)^n \frac{d^{n+k}}{dx^{n+k}}(1-x^2)^k \qquad (10.4.20)$$

There are several other forms of the Legendre polynomials which are available elsewhere (see footnote below). When $n = 0$ one obtains the Legendre polynomials. Letting $\mathbf{P}_k^C(n)$ be the self-adjoint projection corresponding to the eigenvector $\mathbf{z}_k^C(n)$ [given by (10.4.19)] of the operator $(\mathbf{C}_2 n^2 + \mathbf{C}_3)$, we note that for $\mathbf{u} \in \mathcal{H}_{C}$,

$$\mathbf{P}_k^C(n)\,\mathbf{u} \equiv \langle \mathbf{u}, \mathbf{z}_k^C(n)\rangle_C \mathbf{z}_k^C(n)$$

where

$$\langle \mathbf{u}, \mathbf{z}_k^C(n)\rangle_C = \int_0^\pi u(\theta) p_k^n(\cos\theta)\sin\theta\, d\theta$$

$$= \int_{-1}^1 u(\cos^{-1} x) p_k^n(x)\, dx$$

The spectral representation of $(\mathbf{C}_2 n^2 + \mathbf{C}_3)$ in terms of $\{\mathbf{P}_k^C(n)\}$ makes it possible to rewrite $\mathbf{T}$ from (10.4.19) as

$$\mathbf{T} = \sum_{n=0}^\infty \sum_{k=n}^\infty \{\mathbf{A}_1 + k(k+1)\mathbf{A}_2\} \otimes \mathbf{P}_n^B \otimes \mathbf{P}_k^C(n) \qquad (10.4.21)$$

What remains now is the determination of the spectral representation of self-adjoint operator $\mathbf{A}_1 + k(k+1)\mathbf{A}_2$, which has the formal operation

$$-\frac{1}{r^2}\frac{d}{dr}\left(r^2\frac{d}{dr}\right) + k(k+1)\frac{1}{r^2}$$

Thus we must address the eigenvalue problem

$$-\frac{1}{r^2}\frac{d}{dr}\left(r^2\frac{dz}{dr}\right) + k\frac{(k+1)}{r^2}z = \lambda z, \qquad z(1) = 0, \qquad z(0) < \infty \qquad (10.4.22)$$

Note that (10.4.22) is free from explicit dependence on the integer $n$. Let us investigate if $\lambda = 0$ is an eigenvalue of $\mathbf{A}_1 + k(k+1)\mathbf{A}_2$ and hence that of $\mathbf{T}$. Accordingly, we set $\lambda = 0$ in (10.4.22) to obtain

$$-\frac{d}{dr}\left(r^2\frac{dz_0}{dr}\right) + k(k+1)z_0 = 0 \qquad (10.4.23)$$

The two linearly independent solutions of (10.4.25) are readily found by assuming $r^p$ as a solution and determining that $p$ can take on values $k$ and $-(k+1)$. Thus

$$z_0 = C_1 r^k + C_2 r^{-(k+1)} \qquad (10.4.24)$$

In view of the condition $z(0) < \infty$, we must have $C_2 \equiv 0$. The boundary condition $z_0(1) = 0$ requires that $C_1 \equiv 0$, so that we have $z_0 \equiv 0$. Hence $\lambda = 0$ is not an eigenvalue of either $\mathbf{T}$, or $\mathbf{A}_1 + k(k+1)\mathbf{A}_2$ for any value of $k$ (including zero). The logical next step is to proceed with the differential equation (10.4.22), but we deviate from this pursuit temporarily in the interest of certain other important applications.

Suppose that we were interested in solving the Laplace equation

$$T\omega(r, \theta, \phi) = 0, \qquad a < r < b, \quad 0 < \phi < 2\pi, \quad 0 < \theta < \pi \tag{10.4.25}$$

where $T$ is the Laplacian (10.4.14) in spherical coordinates. Assume the Dirichlet boundary conditions

$$\omega(a, \theta, \phi) = f_a(\theta, \phi), \qquad \omega(b, \theta, \phi) = f_b(\theta, \phi) \tag{10.4.26}$$

subject to which we seek to solve (10.4.25). This solution is obtained readily as follows. Denote the solution by $\boldsymbol{\omega}(r)$, where the boldface symbol is used to denote it as a vector in $\mathcal{H}_B \otimes \mathcal{H}_C$ for each value of the explicit argument $r$ between $a$ and $b$; that is,

$$\boldsymbol{\omega}(r) \in \mathcal{H}_B \otimes \mathcal{H}_C \qquad \forall \quad r \in [a, b]$$

The general solution to (10.4.25) can be expressed as

$$\omega(r) = \sum_{n=0}^{\infty} \sum_{k=n}^{\infty} (c_k r^k + d_k r^{-(k+1)}) \mathbf{z}_n^B \otimes \mathbf{z}_k^c(n) \tag{10.4.27}$$

where $c_k$ and $d_k$ are constants to be determined subject to the boundary conditions (10.4.26). Since $\mathbf{f}_a, \mathbf{f}_b \in \mathcal{H}_B \otimes \mathcal{H}_C$, they are expanded in terms of the orthonormal basis $\{\mathbf{z}_n^B \otimes \mathbf{z}_k^c(n); n = 0, 1, 2, \ldots, k = n, n+1, \ldots\}$ in $\mathcal{H}_B \otimes \mathcal{H}_C$.† Subjecting the solution (10.4.27) to the boundary conditions (10.4.26) leads to

$$c_k a^k + d_k a^{-(k+1)} = \langle\langle \mathbf{f}_a, \mathbf{z}_n^B \rangle_B, \mathbf{z}_k^c(n) \rangle_C$$

$$c_k b^k + d_k a^{-(k+1)} = \langle\langle \mathbf{f}_b, \mathbf{z}_n^B \rangle_B, \mathbf{z}_k^c(n) \rangle_C$$

which are a pair of simultaneous equations in the two unknowns $c_k$ and $d_k$. The solution (10.4.27) is thus completely determined. We note in passing that the double summation in (10.4.27) is more conveniently performed by the equivalent procedure of summing over $n$ from 0 to $k$ for each integer $k$ between 0 and $\infty$.

If $a = 0$ and $0 < b < \infty$, then only the boundary condition at $r = b$ becomes relevant. In this case $d_k \equiv 0$ and the solution becomes

$$\boldsymbol{\omega}(r) = \sum_{k=0}^{\infty} \left(\frac{r}{b}\right)^k \sum_{n=0}^{\infty} \langle\langle \mathbf{f}_b, \mathbf{z}_n^B \rangle_B, \mathbf{y}_k^c(n) \rangle_C \mathbf{z}_n^B \otimes \mathbf{z}_k^c(n)$$

Thus we have the solution to the Dirichlet problem in a finite sphere. If $0 < a < \infty$ and $b = \infty$, then we only have the boundary condition at $r = a$,

†The vectors $\mathbf{z}_n^B \otimes \mathbf{z}_k^C(n)$ are known as *spherical surface harmonics*. The functions $\{r^k \mathbf{z}_n^B \otimes \mathbf{z}_k^C(n)\}$ and $\{r^{-(k+1)} \mathbf{z}_n^B \otimes \mathbf{z}_k^C(n)\}$ are *harmonic* in that they satisfy the equation $Tu(r, \theta, \phi) = 0$.

for which $c_k = 0$. The solution to this problem is given by

$$\boldsymbol{\omega}(r) = \sum_{k=0}^{\infty} \left(\frac{a}{r}\right)^{k+1} \langle\langle \mathbf{f}_a, \mathbf{z}_n^B \rangle, \mathbf{z}_k^c(n)\rangle_C \mathbf{z}_n^B \otimes \mathbf{z}_k^c(n)$$

We now return to the problem of the spectral representation of $\mathbf{T}$ given by (10.4.15). We had arrived at the intermediate result (10.4.21) and were at the stage of computing the spectral respresentation of $\mathbf{A}_1 + k(k+1)\mathbf{A}_2$, which led to the eigenvalue problem (10.4.22). Recall that $\lambda = 0$ is not an eigenvalue of $\mathbf{T}$. For nonzero $\lambda$ the two linearly independent solutions of (10.4.22) are obtained as follows. We let $x = \sqrt{\lambda}\, r$ and $u(x) = \sqrt{r}\, z(r)$ so that (10.4.22) leads to

$$x^2u'' + xu' + [x^2 - (k + \tfrac{1}{2})]u = 0$$

which is Bessel's equation. The solutions are $J_{k+1/2}(x)$ and $Y_{k+1/2}(x)$ so that

$$z(r, k) = \frac{1}{\sqrt{r}}[p_k J_{k+1/2}(\sqrt{\lambda}\, r) + q_k Y_{k+1/2}(\sqrt{\lambda}\, r)] \qquad (10.4.28)$$

where $p_k$ and $q_k$ are constants. The Bessel function of the second kind $Y_{k+1/2}(\sqrt{\lambda}\, r)$ has a singularity at $r = 0$ which is forbidden in the problem on hand so that $q_k \equiv 0$. The boundary condition at $r = 1$ leads to the characteristic equation

$$J_{k+1/2}(\sqrt{\lambda}) = 0$$

which must be solved numerically for the eigenvalues $\{\lambda_j(k)\}$. The normalized eigenvector may be denoted

$$\mathbf{z}_j^A(k) = \{r^{-1/2} p_k J_{k+1/2}(\sqrt{\lambda_j}\, r)\}$$

where

$$p_k^{-2} = \int_0^1 r^2 [r^{-1/2} J_{k+1/2}(\sqrt{\lambda_j}\, r)]^2\, dr$$

The self-adjoint projection $\mathbf{P}_j^A(k)$ is then defined by

$$\mathbf{P}_j^A(k)(\cdot) = \langle \cdot, \mathbf{z}_j^A(k)\rangle_A \mathbf{z}_j^A(k)$$

We have thus the spectral representation of $\mathbf{T}$ as

$$\mathbf{T} = \sum_{k=0}^{\infty} \sum_{n=0}^{k} \sum_{j=1}^{\infty} \lambda_j(k) \mathbf{P}_j^A(k) \otimes \mathbf{P}_n^B \otimes \mathbf{P}_k^C(n) \qquad (10.4.29)$$

As an example of the use of this spectral representation, the initial value problem

$$-\frac{d}{dt}\boldsymbol{\omega}(t) = \mathbf{T}\boldsymbol{\omega}(t), \qquad \boldsymbol{\omega}(0) = \boldsymbol{\omega}_0 \in \mathcal{H} \qquad (10.4.30)$$

has the solution

$$\boldsymbol{\omega}(t) = \sum_{k=0}^{\infty} \sum_{n=0}^{k} \sum_{j=1}^{\infty} e^{-\lambda_j(k)t} [\mathbf{P}_j^A(k) \otimes \mathbf{P}_n^B \otimes \mathbf{P}_k^C(n)]\boldsymbol{\omega}_0$$

For other examples, the student is directed to the exercises at the end of this section which cover numerous applications.

*Example 3. Temperature Distribution in a Peripherally Cooled Composite Sphere*

This example is, in essence, a generalization of Example 2 of Section 10.2. The physical setting features a composite sphere comprising two concentric spheres, the inner one of radius $a$ and the outer one of radius unity. The outer sphere loses heat to the ambient, which is uniformly at zero temperature. We assume that the outer sphere is of a thin ($a \simeq 1$) material of thermal conductivity, large relative to that of the inner sphere. Thus it is possible to neglect radial temperature gradients in the region $a < r < 1$, but the transverse gradients cannot be neglected. The unsteady-state temperature distribution is desired in the entire sphere. In terms of suitably defined dimensionless variables, we write the heat conduction equation for the inner sphere as

$$-\left[\frac{1}{x^2}\frac{\partial}{\partial x}\left(x^2\frac{\partial}{\partial x}\right)+\frac{1}{x^2\sin^2\theta}\frac{\partial^2}{\partial\phi^2}+\frac{1}{x^2\sin\theta}\frac{\partial}{\partial\theta}\left(\sin\theta\frac{\partial}{\partial\theta}\right)\right]\Theta(x,\theta,\phi;\tau)$$
$$=-\frac{\partial}{\partial\tau}\Theta(x,\theta,\phi;\tau)+\sigma(x,\theta,\phi,\tau), \qquad 0<x<1, \quad 0<\theta<\pi,$$
$$0<\phi<2\pi, \quad \tau>0 \qquad (10.4.31)$$

For the thin outer sphere we denote the dimensionless temperature distribution by $\Theta_1(\theta, \phi, \tau)$. From physical considerations the transverse heat conduction equation is given by

$$\alpha\frac{\partial}{\partial x}\Theta(1,\theta,\phi,\tau)+\left[-\frac{\gamma}{\sin\theta}\left\{\frac{1}{\sin\theta}\frac{\partial^2}{\partial\phi^2}+\frac{\partial}{\partial\theta}\left(\sin\theta\frac{\partial}{\partial\theta}\right)\right\}+\beta\right]\Theta_1(\theta,\phi,\tau)$$
$$=-\frac{\partial}{\partial\tau}\Theta_1(\theta,\phi,\tau)+\sigma_1(\theta,\phi,\tau) \qquad 0<\phi<2\pi, \quad 0<\theta<\pi, \quad \tau>0$$
$$(10.4.32)$$

where $\alpha$, $\beta$, and $\gamma$ are positive constants related to dimensionless combinations of physical quantities. Further, we must have continuity of temperature expressed by

$$\Theta(1,\theta,\phi,\tau)=\Theta_1(\theta,\phi;\tau) \qquad (10.4.33)$$

The problem requires a certain amount of reformulation before the methods of this section become applicable. First, we recognize that the surface $x = 1$ is weighted as in Example 2 of Section 10.2. We let $\Omega = \{0 < x < 1, 0 < \theta < \pi, 0 < \phi < 2\pi\}$ and $\partial\Omega = \{x = 1, 0 < \theta < \pi, 0 < \phi < 2\pi\}$ and define the required Hilbert space

$$\mathcal{H}=\mathcal{L}_2\{\Omega; x^2\sin\theta\}\oplus\mathcal{L}_2\left\{\partial\Omega;\frac{1}{\alpha}\sin\theta\right\} \qquad (10.4.34)$$

Further, let

$$T \equiv -\left[\frac{1}{x^2}\frac{\partial}{\partial x}\left(x^2\frac{\partial}{\partial x}\right) + \frac{1}{x^2\sin^2\theta}\frac{\partial^2}{\partial\phi^2} + \frac{1}{x^2\sin\theta}\frac{\partial}{\partial\theta}\left(\sin\theta\frac{\partial}{\partial\theta}\right)\right]$$
$$T_1 \equiv \left[-\frac{\gamma}{\sin\theta}\left\{\frac{1}{\sin\theta}\frac{\partial^2}{\partial\phi^2} + \frac{\partial}{\partial\theta}\left(\sin\theta\frac{\partial}{\partial\theta}\right)\right\} + \beta\right] \tag{10.4.35}$$

$$L \equiv \begin{bmatrix} T & 0 \\ -\alpha\left[\dfrac{\partial}{\partial x}\right]_{\partial\Omega^-} & T_1 \end{bmatrix} \tag{10.4.36}$$

The expression $L$ arises from the two partial differential equations (10.4.31) and (10.4.22) and must operate on $\boldsymbol{\Theta}(\tau)$ represented by

$$\boldsymbol{\Theta}(\tau) \equiv \begin{bmatrix} \{\Theta(x, \theta, \phi, \tau)\} \\ \{\Theta_1(\theta, \phi, \tau)\} \end{bmatrix}$$

Denoting a typical element $\mathbf{u} \in \mathcal{H}$ by

$$\mathbf{u} \equiv \begin{bmatrix} u(x, \theta, \phi) \\ u_1(\theta, \phi) \end{bmatrix}$$

the inner product is given by

$$\langle \mathbf{u}, \mathbf{v} \rangle = \int_0^1 dx\, x^2 \int_0^\pi d\theta \sin\theta \int_0^{2\pi} d\phi\, u(x, \theta, \phi)v(x, \theta, \phi)$$
$$+ \frac{1}{\alpha}\int_0^\pi d\theta \sin\theta \int_0^{2\pi} d\phi\, u_1(\theta, \phi)v_1(\theta, \phi) \tag{10.4.37}$$

That $L$ is formally self-adjoint is readily established. Moreover,

$$\langle L\mathbf{u}, \mathbf{v} \rangle - \langle \mathbf{u}, L\mathbf{v} \rangle = \int_0^\pi d\theta \sin\theta \int_0^{2\pi} d\phi \left[\frac{\partial}{\partial x}v(1, \theta, \phi)\{u(1, \theta, \phi) - u_1(\theta, \phi)\}\right.$$
$$\left. - \frac{\partial u}{\partial x}(1, \theta, \phi)\{v(1, \theta, \phi) - v_1(\theta, \phi)\}\right]$$

When the continuity requirement (10.4.33) is applied to $\mathbf{u}$ and $\mathbf{v}$, the right-hand side of the equation above vanishes. Thus $\mathbf{L} \equiv \{L, D(L)\}$, where

$$D(L) = \{\mathbf{u}, L\mathbf{u} \in \mathcal{H};\, u(1, \theta, \phi) = u_1(\theta, \phi)\}$$

is a self-adjoint operator.

It is now possible to represent the entire boundary-initial value problem consisting of differential equations (10.4.31) and (10.4.32) together with condition (10.4.33) as

$$\mathbf{L}\Theta(\tau) = -\frac{d}{d\tau}\boldsymbol{\Theta}(\tau) + \boldsymbol{\sigma}(\tau), \qquad \boldsymbol{\Theta}(0) = \boldsymbol{\Theta}_0 \tag{10.4.38}$$

where

$$\boldsymbol{\sigma}(\tau) \equiv \begin{bmatrix} \{\sigma(x, \theta, \phi, \tau)\} \\ \{\sigma_1(\theta, \phi, \tau)\} \end{bmatrix}$$

is the heat source distribution function in the two spheres and $\boldsymbol{\Theta}_0 \in \mathcal{H}$. The solution of (10.4.38) is readily written as

$$\boldsymbol{\Theta}(\tau) = \mathbf{e}^{-\tau\mathbf{L}}\boldsymbol{\Theta}_0 + \int_0^\tau \mathbf{e}^{-(\tau-\tau')\mathbf{L}}\boldsymbol{\sigma}(\tau')\,d\tau' \tag{10.4.39}$$

for which the spectral representation of $\mathbf{L}$ is required. We shall now apply the method of tensor products as follows.

Define the Hilbert spaces

$$\mathcal{H}_A \equiv \mathcal{L}_2\{[0, 1]; x^2\}, \qquad \mathcal{H}_B \equiv \mathcal{L}_2\{[0, 2\pi]\}, \qquad \mathcal{H}_C \equiv \mathcal{L}_2\{[0, \pi]; \sin\theta\}$$

Then we have

$$\mathcal{L}_2\{\Omega; x^2 \sin\theta\} = \mathcal{H}_A \otimes \mathcal{H}_B \otimes \mathcal{H}_C$$
$$\mathcal{L}_2\{\partial\Omega; \sin\theta\} = \mathcal{H}_B \otimes \mathcal{H}_C$$

The Hilbert space $\mathcal{H}$ defined earlier can then be represented by

$$\mathcal{H} = (\mathcal{H}_A \oplus \mathcal{R}) \otimes \mathcal{H}_B \otimes \mathcal{H}_C$$

if the inner product in the Hilbert space $\mathcal{H}_{A'} \equiv \mathcal{H}_A \oplus \mathcal{R}$ is defined by

$$\left\langle \begin{bmatrix} \{u(x)\} \\ u_1 \end{bmatrix}, \begin{bmatrix} \{v(x)\} \\ v_1 \end{bmatrix} \right\rangle_{A'} = \int_0^1 x^2 u(x) v(x)\,dx + \frac{1}{\alpha} u_1 v_1$$

At this point we may recall the operators $\mathbf{B}_1$, $\mathbf{B}_2$, $\mathbf{B}_3$, $\mathbf{C}_1$, $\mathbf{C}_2$, and $\mathbf{C}_3$ as defined in Example 2 of this section. Next we define on $\mathcal{H}_{A'}$

$$A_1' \equiv \begin{bmatrix} A_1 & 0 \\ -\alpha \left[\dfrac{d}{dx}\right]_{x=1-} & \beta \end{bmatrix}, \qquad A_2' \equiv \begin{bmatrix} \dfrac{1}{x^2} & 0 \\ 0 & \gamma \end{bmatrix} = A_3'$$

It is then a simple matter to show that $L$, as defined by (10.4.36), may be expressed as

$$L = A_1' \otimes B_1 \otimes C_1 + A_2' \otimes B_2 \otimes C_2 + A_3' \otimes B_3 \otimes C_3$$

The operations $A_1'$, $A_2'$, and $A_3'$ have a common domain in

$$D(A_1') = \left\{ \mathbf{u} \equiv \begin{bmatrix} \{u(x)\} \\ u_1 \end{bmatrix} \in \mathcal{H}_A \oplus \mathcal{R},\ A_1'\mathbf{u} \in \mathcal{H}_A \oplus \mathcal{R},\ u(1) = u_1 \right\}$$

Then we may write

$$\mathbf{L} = \mathbf{A}_1' \otimes \mathbf{B}_1 \otimes \mathbf{C}_1 + \mathbf{A}_2' \otimes \mathbf{B}_2 \otimes \mathbf{C}_2 + \mathbf{A}_2' \otimes \mathbf{B}_3 \otimes \mathbf{C}_3 \tag{10.4.40}$$

In view of all the common steps in Example 2 of this section leading to (10.4.21), we have

$$\mathbf{L} = \sum_{k=0}^{\infty} \sum_{n=0}^{k} \{\mathbf{A}_1' + k(k+1)\mathbf{A}_2'\} \otimes \mathbf{P}_n^B \otimes \mathbf{P}_k^C(n)$$

We are thus required to find the spectral representation of $\mathbf{A}_1' + k(k+1)\mathbf{A}_2'$

for which we must solve the equations

$$-\frac{1}{x^2}\frac{d}{dx}\left(x^2\frac{d}{dx}z\right) + \frac{k(k+1)}{x^2}z = \lambda z$$

$$\alpha z'(1) + \beta z(1) + k(k+1)\gamma z(1) = \lambda z(1) \tag{10.4.41}$$

As in Example 2, the first of the equations above has the linearly independent solutions $J_{k+1/2}(\sqrt{\lambda}\,x)/\sqrt{x}$ and $Y_{k+1/2}(\sqrt{\lambda}\,x)/\sqrt{x}$, the latter of which is ruled out because of the singularity at $x = 0$. Thus substituting $z(x) = J_{k+1/2}(\sqrt{\lambda}\,x)/\sqrt{x}$ into (10.4.41), we have the characteristic equation

$$\alpha\sqrt{\lambda}\,J'_{k+1/2}(\sqrt{\lambda}) + [\beta + k(k+1)\gamma - \lambda]J_{k+1/2}(\sqrt{\lambda}) = 0$$

which must be solved numerically for the eigenvalues $\{\lambda_j(k)\}$ for each $k$. The $j$th eigenvector $\mathbf{z}_j^{A'}(k)$ is given by

$$\mathbf{z}_j^{A'}(k) = c_j(k)\begin{bmatrix} \dfrac{J_{k+1/2}(\sqrt{\lambda}\,x)}{\sqrt{x}} \\ J_{k+1/2}(\sqrt{\lambda}) \end{bmatrix}$$

where $c_j(k)$ is the normalization constant obtained from

$$c_j^{-2}(k) = \int_0^1 x^2\left[\frac{J_{k+1/2}(\sqrt{\lambda_j}\,x)}{\sqrt{x}}\right]^2 dx + \frac{1}{\alpha}[J_{k+1/2}(\sqrt{\lambda_j})]^2$$

The self-adjoint projection $\mathbf{P}_j^{A'}(k)$ is defined as

$$\mathbf{P}_j^{A'}(k)(\cdot) = \langle \cdot, \mathbf{z}_j^{A'}(k)\rangle_{A'}\mathbf{z}_j^{A'}(k)$$

The spectral representation of $\mathbf{L}$ is then obtained as

$$\mathbf{L} = \sum_{k=0}^{\infty}\sum_{n=0}^{k}\sum_{j=1}^{\infty}\lambda_j(k)\mathbf{P}_j^{A'}(k)\otimes\mathbf{P}_n^B\otimes\mathbf{P}_k^C(n)$$

so that the solution (10.4.39) may be written as

$$\mathbf{\Theta}(\tau) = \sum_{k=0}^{\infty}\sum_{n=0}^{k}\sum_{j=1}^{\infty}[\mathbf{P}_j^{A'}(k)\otimes\mathbf{P}_n^B\otimes\mathbf{P}_k^C(n)]e^{-\lambda_j(k)\tau}\left[\mathbf{\Theta}_0 + \int_0^{\tau} e^{\lambda_j(k)\tau'}\boldsymbol{\sigma}(\tau')\,d\tau'\right]$$

We have considered several examples demonstrating the power of the tensor product approach. However, in all of the examples above the traditional method of the separation of variables would have worked just as well. Next we present an example due to Friedman† which shows that the tensor product approach is somewhat more general than the method of separation of variables. Thus in the next example the tensor product approach is shown to work, although the method of separation of variables does not.

---

†See "An Abstract Formulation of the Method of Separation of Variables," by B. Friedman, *Proceedings of the Conference on Differential Equations*, University of Maryland, College Park, Md., 1956, pp. 209–226.

*Example 4. The Biharmonic Equation*

Consider the rectangular strip $\Omega \equiv \{0 \leq x \leq a; 0 \leq y \leq b\}$ and the biharmonic operation

$$T \equiv \frac{\partial^4}{\partial x^4} + 2\frac{\partial^4}{\partial x^2 \partial y^2} + \frac{\partial^4}{\partial y^4}$$

which is an example of the expressions considered in Section 10.3, so that it is formally self-adjoint. Define the domain

$$D(T) = \left\{\mathbf{u}, T\mathbf{u} \in \mathfrak{L}_2(\Omega); u(x, 0) = \frac{\partial u}{\partial y}(x, 0) = u(x, b) = \frac{\partial u}{\partial y}(x, b) = 0; \right.$$
$$\left. u(0, y) = u(a, y) = \frac{\partial^2 u}{\partial x^2}(0, y) = \frac{\partial^2 u}{\partial x^2}(a, y) = 0\right\}$$

The operator $\mathbf{T} = \{T, D(T)\}$ is self-adjoint (see Exercise 10.3.1). If we define

$$\mathcal{H}_A = \mathfrak{L}_2[0, a], \qquad \mathcal{H}_B = \mathfrak{L}_2[0, b]$$

then $\mathfrak{L}_2(\Omega) = \mathcal{H}_A \otimes \mathcal{H}_B$. Further, we define the operators

$$\mathbf{A}_1 = \{A_1, D(A_1)\}, \qquad A_1 = \frac{d^4}{dx^4},$$
$$D(A_1) = \{\mathbf{f}, A_1\mathbf{f} \in \mathcal{H}_A; f(0) = f(a) = f''(0) = f''(a) = 0\}$$
$$\mathbf{A}_2 = \{A_2, D(A_2)\}, \qquad A_2 = -\frac{d^2}{dx^2},$$
$$D(A_2) = \{\mathbf{f}, A_2\mathbf{f} \in \mathcal{H}_A; f(0) = f(a) = 0\}, \qquad \mathbf{A}_3 = \mathbf{I}^A$$

Indeed, not only do the operators $\mathbf{A}_1$, $\mathbf{A}_2$, and $\mathbf{A}_3$ commute, but we have $\mathbf{A}_1 = \mathbf{A}_2^2$. The eigenvalues of $\mathbf{A}_2$ are $\{n^2\pi^2/a^2\}$ and those of $\mathbf{A}_1$ are $\{n^4\pi^4/a^4\}$. The common eigenvectors are given by

$$\mathbf{z}_n^A = \sqrt{\frac{2}{a}} \sin \frac{n\pi x}{a}$$

It is worth noting here that if a similar attempt had been made to identify the operators in $\mathcal{H}_B$, it would have been found that they do *not* commute. The eigenvalue problem corresponding to the **B** operators [see (10.4.4)] becomes

$$\left[\frac{n^4\pi^4}{a^4} - \frac{2n^2\pi^2}{a^2}\frac{d^2}{dy^2} + \frac{d^4}{dy^4}\right]z^B(y) = \lambda z^B(y)$$

with $z^B(0) = z^{B\prime}(0) = z^B(1) = z^{B\prime}(1) = 0$. We do not pursue this any further at this stage. The important thing to note here is that it is not immediately apparent how the method of separation of variables would have adapted to this situation. The student is invited to work this example out completely.

**EXERCISES**

**10.4.1** Rework Example 2 for a spherical annulus with arbitrary combinations of Dirichlet, Neumann, and Robin (also known as Fourier) boundary conditions on the two surfaces.

**10.4.2** Let $\Omega$ be the cylinder $\{0 < r < 1, 0 < \theta < 2\pi, 0 < x < l\}$,

$$T = -\left[\frac{1}{r}\frac{\partial}{\partial r}\left(r\frac{\partial}{\partial r}\right) + \frac{1}{r^2}\frac{\partial^2}{\partial \theta^2} + \frac{\partial^2}{\partial x^2}\right], \qquad (r, \theta, x) \in \Omega$$

and $\mathcal{H} \equiv \mathcal{L}_2\{\Omega; r\}$. If

$$D(T) = \left\{\mathbf{u}, T\mathbf{u} \in \mathcal{H}; \frac{\partial}{\partial r}u(0, \theta, x) = 0, \qquad u(1, \theta, x) = 0, \qquad u(r, \theta, 0) = u(r, \theta, l) = 0\right\}$$

Find the spectral representation of $\mathbf{T} \equiv \{T, D(T)\}$. Hence solve the boundary-initial value problem for a cylinder of unit radius and length $l$ whose boundary temperature and initial temperature distributions are specified.

**10.4.3** Consider heat transfer in two concentric spheres in which the inner sphere of radius unity has an infinite thermal conductivity and the outer sphere has a nonuniform volumetric heat source. The outer surface at $r = R$ is maintained at zero temperature. The unsteady-state heat conduction equations for the two spheres are given by

$$-k_0\left[\frac{1}{r^2}\frac{\partial}{\partial r}\left(r^2\frac{\partial \Theta}{\partial r}\right) + \frac{1}{r^2\sin^2\theta}\frac{\partial^2\Theta}{\partial \phi^2} + \frac{1}{r^2\sin\theta}\frac{\partial}{\partial \theta}\left(\sin\theta\frac{\partial\Theta}{\partial\theta}\right)\right] - s(r, \theta, \phi, t)$$

$$= -c_0\frac{\partial\Theta}{\partial t}, \; t > 0, \quad 1 < r < R, \quad 0 < \theta < \pi, \quad 0 < \phi < 2\pi$$

$$-\frac{4\pi}{3}c_1\frac{d\Theta_1}{dt} = -4\pi\int_0^{2\pi} d\phi \int_0^{\pi} d\theta \sin\theta \, k_0\frac{\partial\Theta}{\partial r}\bigg|_{r=1}, \qquad t > 0$$

where $\Theta$ is the temperature distribution in the outer sphere and $\Theta_1$ is the uniform temperature of the inner sphere. We must have $\Theta_1(t) = \Theta(1, \theta, \phi, t)$ at all times. The positive constants $k_0$ and $c_0$ are the thermal conductivity and volumetric heat capacity of the outer sphere, while $c_1$ is the volumetric heat capacity of the inner sphere. Complete the formulation of the initial-boundary value problem and solve it.†

## 10.5 Solution of Some Elliptic Problems by Decomposition into First-Order Systems

Our focus here is on some special elliptic systems that occur in engineering applications. The operators arising here are originally non-self-adjoint but submit to a technique of symmetrization which permits the solution of the

†See also A. K. Kulshreshta, M. S. thesis, Purdue University, West Lafayette, Ind., 1982.

boundary value problems involved via the spectral resolution of a *derived* self-adjoint operator. It is important to recognize that the original operator is essentially non-self-adjoint and cannot be symmetrized by any form of manipulation. What the symmetrization accomplishes here is to produce *another* operator that is self-adjoint and whose spectral resolution obtains a solution of the boundary value problem. The physical settings for such problems will emerge from the examples presented later.† In general, however, the operators appearing here are encountered in problems connected with convective diffusion. More specifically, we are concerned with the partial differential equation in two variables given by

$$-\frac{1}{r(y)}\left[\frac{\partial}{\partial y}\left\{p(y)\frac{\partial v}{\partial y}\right\}\right] - q(y)\frac{\partial^2 v}{\partial x^2} + s(y)\frac{\partial v}{\partial x} = f(x, y) \qquad (10.5.1)$$

where the functions $p$, $q$, and $r$ are positive functions of $y$ in some compact interval, say $[a, b]$. The independent variable $x$ is allowed to vary in the infinite interval $(-\infty, \infty)$.‡ The expression on the left-hand side of (10.5.1) is clearly elliptic, but in view of its last term, not formally self-adjoint; unless $s(y)$ is constant, it is not possible to render the expression formally self-adjoint. Thus regardless of the boundary conditions to be employed, the operator in question is inherently non-self-adjoint if $s(y)$ is not a constant. We will also be interested in boundary conditions in this section that make the operator non-self-adjoint even when the operation is rendered formally self-adjoint for the case of constant $s(y)$.

### 10.5.1 Boundary and Interface Conditions

The boundary conditions with respect to $x$ are merely boundedness requirements. With respect to $y$, we introduce the *mixed second derivative boundary condition* due to the authors,§ say at $y = a$, given by¶

$$\alpha\frac{\partial^2}{\partial x\,\partial y}v(x, a) - \beta\frac{\partial}{\partial x}v(x, a) + \frac{\partial}{\partial y}v(x, a) = g(x) \qquad (10.5.2)$$

where $\alpha$ and $\beta$ are nonnegative constants. If $\alpha = 0$, we have the *oblique derivative* boundary condition

$$-\beta\frac{\partial}{\partial x}v(x, a) + \frac{\partial}{\partial y}v(x, a) = g(x) \qquad (10.5.3)$$

---

†D. Ramkrishna, Advances in Transport Processes, Vol. III, edited by A. S. Mujumdar and R. A. Mashelkar, Wiley Eastern, New Delhi, 1983.

‡$f(x, y)$ is assumed to vanish at $x = \pm\infty$ in such a way that certain integrals to be encountered exist.

§See D. Ramkrishna and N. R. Amundson, *Chem. Eng. Sci.*, **34**, 301–308, 1979.

¶Here we assume that $\int_{-\infty}^{x} g(\xi)\, d\xi$ exists.

Note particularly that neither of the boundary conditions (10.5.2) and (10.5.3) (more precisely their homogeneous versions) can lead to self-adjoint operators with the operation in (10.5.1).† At $y = b$ a boundary condition similar to (10.5.2) or (10.5.3) could be assumed, but instead we adopt

$$v(x, b) = h(x) \tag{10.5.4}$$

or

$$p(b)\frac{\partial}{\partial y}v(x, b) = k(x) \tag{10.5.5}$$

the first of which is a Dirichlet boundary condition and the second a Neumann boundary condition. Clearly, a Robin boundary condition [e.g., (10.2.16)] may also be encountered.

In physical applications, $p, q,$ and $r$ are physical properties which are continuous in a homogeneous material. In this case the pair of boundary conditions consisting of one out of (10.5.2) and (10.5.3) and another out of (10.5.4) and (10.5.5) would make up the complete boundary value problem. In other cases where transport occurs in layered media (as, for example, in Section 9.3), discontinuities occur in $p, q,$ and $r$ along interfaces. Physically, one then encounters interface conditions at points of discontinuity of the type

$$\begin{aligned}-p(c-)\frac{\partial}{\partial y}v(x, c-) &= -p(c+)\frac{\partial}{\partial y}v(x, c+)\\ &= \gamma[x, c-) - v(x, c+)]\end{aligned} \tag{10.5.6}$$

where $c$ is the point of discontinuity. The reader will recall similar interface conditions in Section 9.3. Let us present some examples in which boundary value problems of the kind just considered are encountered.

*Example 1. Cooling of a Heated Slab*‡

Consider a slab of material of unit thickness between $y = 0$ and $y = 1$ and infinitely long along the $x$-direction extending between $x = -\infty$ and $x = \infty$. The slab is heated by a distributed source function which vanishes at $x = \pm\infty$ such that the total heat generated in the slab is a finite quantity. The face $y = 0$ is insulated, while at $y = 1$ a coolant enters at $x = -\infty$ at some inlet temperature $t_0$. Under steady-state conditions, the slab temperature distribution $v(x, y)$ must satisfy the heat conduction equation

---

†The boundary condition (10.5.2), recast in an integral form, may be shown to be a special case of Birkhoff's integral boundary condition represented by (10.2.18). However, the condition for self-adjointness is violated.

‡See also D. Ramkrishna and N. R. Amundson, *Chem. Eng. Sci.*, *34* 309–318, 1979, and D. Ramkrishna, G. Narsimhan, and N. R. Amundson, *Chem. Eng. Sci.*, *36*, 199–207, 1981.

$$-\left(\frac{\partial^2 v}{\partial x^2}+\frac{\partial^2 v}{\partial y^2}\right)=f(x,y), \qquad -\infty<x<\infty, \qquad 0<y<1 \tag{10.5.7}$$

where the thermal conductivity has been absorbed into the source function $f(x, y)$. If the coolant is assumed to be well mixed transversely, then the energy balance for the coolant temperature $t(x)$ may be written as

$$wC_p\frac{dt}{dx}=hP[v(x,1)-t] \tag{10.5.8}$$

where $w$ and $C_p$ are the flow rate and specific heat of the coolant, respectively, and $P$ is heat transfer surface per unit length. If $k$ is the thermal conductivity of the slab, then we must also have

$$-k\frac{\partial v}{\partial y}(x,1)=h[v(x,1)-t] \tag{10.5.9}$$

Eliminating $t$ between Eqs. (10.5.8) and (10.5.9), one readily finds that

$$\alpha\frac{\partial^2}{\partial x\,\partial y}v(x,1)+\beta\frac{\partial}{\partial x}v(x,1)+\frac{\partial}{\partial y}v(x,1)=0 \tag{10.5.10}$$

where $\alpha\equiv wC_p/hP$ and $\beta\equiv wC_p/kP$. Equation (10.5.8) is clearly the homogeneous version of (10.5.2) and has been obtained by tacitly requiring differentiability of $\partial v/\partial y$ with respect to $x$ at $y=1$. If the heat transfer coefficient $h$ is sufficiently large relative to $(wC_p/P)$, then $\alpha$ may be negligibly small and one obtains the oblique derivative boundary condition

$$\beta\frac{\partial}{\partial x}v(x,1)+\frac{\partial}{\partial y}v(x,1)=0$$

The boundary condition at $y=0$ is given by

$$\frac{\partial}{\partial y}v(x,0)=0 \tag{10.5.11}$$

The boundary value problem on hand now is precisely of the type laid out at the beginning of this section. Equation (10.5.7) is a special case of (10.5.1) obtained on assuming $p(y)=q(y)=r(y)\equiv 1$ and $s(y)\equiv 0$. Indeed, the operation on the right side of (10.5.7) is formally self-adjoint, but the boundary condition (10.5.10) leads to a non-self-adjoint operator.

*Example 2. The Extended Graetz Problem*†

The extension of the classical Graetz problem‡ to include the effects of axial molecular transport in heat or mass transfer in laminar, fully developed

---

†See also E. Papoutsakis, D. Ramkrishna, and H. C. Lim, *Appl. Sci. Res., 36*, 13–34, 1980, and *AIChE J., 26*, 779–788, 1980.

‡See, for example, L. Graetz, *Ann. Phys., 18*, 79, 1983.

flow through a conduit constitutes an interesting boundary value problem which submits neatly to the solution techniques of this section. In terms of suitably defined dimensionless variables the differential equation for the temperature (or concentration) distribution $v(x, y)$ is given by

$$-\frac{1}{y}\frac{\partial}{\partial y}\left(y\frac{\partial v}{\partial y}\right) - \frac{1}{\text{Pe}^2}\frac{\partial^2 v}{\partial x^2} + s(y)\frac{\partial v}{\partial x} = 0,$$

$$0 < y < 1, \qquad -\infty < x < \infty \tag{10.5.12}$$

where $x$ is the axial coordinate and $y$ is the radial coordinate; $s(y)$ is the fully developed velocity profile, which is $(1 - y^2)$ for the parabolic profile, and Pe is the Peclet number, which regulates the admission of the axial conduction term $(\partial^2 v/\partial x^2)$ into the analysis. Indeed, (10.5.12) is of the form (10.5.1) with $p(y) = r(y)$, $q(y) \equiv 1/\text{Pe}^2$ and $f(x, y) \equiv 0$. The boundary condition at $y = 0$ may be

$$\frac{\partial}{\partial y}v(x, 0) = 0$$

while that at $y = 1$ may be either the Dirichlet boundary condition

$$v(x, 1) = g(x) \tag{10.5.13}$$

or the Neumann boundary condition

$$\frac{\partial}{\partial y}v(x, 1) = h(x) \tag{10.5.14}$$

The boundary value problem above involves a formally non-self-adjoint differential expression and hence the operator is itself non-self-adjoint. Here again, as in the previous example, no interface conditions were required. Next we consider an example in which one encounters (10.5.1) with a discontinuous $p(y)$.

*Example 3. Liquid Metal Cooling of a Packed Bed*†

Consider a cylindrical bed of fissionable material (such as a packing of nuclear pellets) of, say, length $l$ sandwiched between infinite sections of inert pellets of the same physical and thermal properties as those of the nuclear pellets. The cylindrical enclosure is of radiative shield material whose thickness but not its thermal resistance is negligible. Peripheral cooling is accomplished by a liquid metal coolant in laminar, fully developed flow through the surrounding annulus. Some heat loss also occurs to the ambient.

---

†See also E. Papoutsakis and D. Ramkrishna, *Chem. Eng. Sci.*, *36*, 1381–1391, 1981.

In presenting the energy equations for the packed bed and the fluid we will use exactly the same strategy as that in Section 9.3; that is, we will present a single differential equation applicable for the bed and the fluid by allowing for the properties to vary in a discontinuous manner. Thus if $v(x, y)$ represents the temperature distribution where $x$ and $y$ are the dimensionless axial and radial coordinates, then we must have

$$-\frac{1}{r(y)}\frac{\partial}{\partial y}\left[p(y)\frac{\partial}{\partial y}v(x, y)\right] - q(y)\frac{\partial^2}{\partial x^2}v(x, y) + s(y)\frac{\partial}{\partial x}v(x, y) = f(x, y),$$

$$-\infty < x < \infty, \quad 0 < y < 1 \qquad (10.5.15)$$

where $p(y) \equiv y\kappa(y)$, $r(y) \equiv y$. The function $p(y)$ is discontinuous because $\kappa(y)$, which is the ratio of the local thermal conductivity to the thermal conductivity $k_f$ of the fluid, is discontinuous with

$$\kappa(y) = \begin{cases} 1, & c < y < 1 \\ \dfrac{k_e}{k_f}, & 0 < y < c \end{cases}$$

where $c$ is the dimensionless radius of the packed cylinder. Further,

$$q(y) = \begin{cases} \dfrac{1}{\mathrm{Pe}^2}, & c < y < 1, \\ \dfrac{k_e}{k_f \mathrm{Pe}^2}, & 0 < y < c, \end{cases} \qquad \left.\begin{matrix} 1 - y^2 + (c^2 - 1)\dfrac{\ln y}{\ln c} \\ 0 \end{matrix}\right\} = s(y) \qquad (10.5.16)$$

The heat source function $f(x, y)$ corresponds to the heat released by fissionable material in the region $0 < y < c$, $0 < x < l$, the heat released by viscous dissipation in the fluid in the region $c < y < 1$, $-\infty < x < \infty$, and vanishes in the region $0 < y < c$, $x < 0$, $x > l$.

The boundary conditions are clearly

$$v(x, 0) < \infty, \qquad -\frac{\partial v}{\partial y}(x, 1) = [v(x, 1) - v_a] \qquad (10.5.17)$$

the second representing heat loss to the ambient at dimensionless temperature $v_a$. In addition, we have the interface conditions

$$-c\kappa(c+)\frac{\partial}{\partial y}v(x, c+) = -c\kappa(c-)\frac{\partial}{\partial y}v(x, c-) = \gamma[v(x, c-) - v(x, c+)]$$

which accounts for continuity of heat flux and interface resistance to heat transfer at $c$.

The boundary value problem is now complete. It should be clear that the differential equation and the boundary and interface conditions fit the mold of the general problem of the section. We next consider the solution of such problems by a technique of decomposing the second-order partial differential equation into a pair of first-order partial differential equations.

### 10.5.2 The Method of Decomposition into First-Order Systems

We present here a method which decomposes the second-order elliptical equation into a pair of first-order equations. Interestingly, this decomposition leads to a self-adjoint problem. The normal strategy is to define a *second* variable, say $w(x, y)$ by a first-order partial differential equation involving the original variable $v(x, y)$ and to identify the second first-order partial differential equation (featuring $v$ and $w$) in such a way that the elimination of $w$ between the two first-order partial differential equations should produce the original equation (10.5.1). The identification of the variable $w(x, y)$ can be done in many different ways, but we suggest here an approach that is capable of physical interpretation in the applications of interest and leads directly to a self-adjoint problem. We define $w(x, y)$ by

$$\frac{\partial w}{\partial y} = -r(y)q(y)\frac{\partial v}{\partial x} + r(y)s(y)v \tag{10.5.18}$$

from which we may write

$$w(x, y) = \int^{y}\left[-q(y')\frac{\partial v}{\partial x} + s(y')v\right]r(y')\,dy' \tag{10.5.19}$$

where the lower limit is unspecified for the present. Through physical examples later, it will become clear that $w(x, y)$ has an interesting interpretation. Equation (10.5.1) will imply that

$$\frac{\partial w}{\partial x} = p(y)\frac{\partial v}{\partial y} + \phi(x, y) \tag{10.5.20}$$

where $\phi(x, y)$ is obtained by solving

$$\frac{\partial \phi}{\partial y} = f(x, y)r(y) \tag{10.5.21}$$

so that

$$\phi(x, y) = \int_a^y f(x, y')r(y')\,dy' \qquad \text{with} \quad \phi(x, a) = 0$$

It is now convenient to rewrite (10.5.18) and (10.5.20) as

$$\frac{\partial}{\partial x}\begin{bmatrix} v(x, y) \\ w(x, y) \end{bmatrix} = \begin{bmatrix} \dfrac{s(y)}{p(y)} & -\dfrac{1}{r(y)q(y)}\dfrac{\partial}{\partial y} \\ p(y)\dfrac{\partial}{\partial y} & 0 \end{bmatrix}\begin{bmatrix} v(x, y) \\ w(x, y) \end{bmatrix} + \begin{bmatrix} 0 \\ \phi(x, y) \end{bmatrix} \tag{10.5.22}$$

We let the matrix differential expression in (10.5.22) be given by

$$L \equiv \begin{bmatrix} \dfrac{s(y)}{q(y)} & -\dfrac{1}{r(y)q(y)}\dfrac{d}{dy} \\ p(y)\dfrac{d}{dy} & 0 \end{bmatrix} \tag{10.5.23}$$

Noting the coefficients of the derivatives in (10.5.23), a Hilbert space can be defined in which $L$ becomes formally self-adjoint as follows. Let

$$\mathcal{H} = \mathcal{H}_1 \oplus \mathcal{H}_2 \tag{10.5.24}$$

where

$$\mathcal{H}_1 = \mathcal{L}_2\{[a, b]; r(y)q(y)\}$$

which consists of functions $f_1(y)$ such that

$$\int_a^b r(y)q(y)f_1^2(y)\,dy < \infty$$

with an inner product $\langle\cdot, \cdot\rangle_1$ defined by

$$\langle \mathbf{f}_1, \mathbf{g}_1\rangle_1 = \int_a^b r(y)q(y)f_1(y)g_1(y)\,dy \tag{10.5.25}$$

and

$$\mathcal{H}_2 = \mathcal{L}_2\left\{[a, b]; \frac{1}{p(y)}\right\}$$

which consists of functions $f_2(y)$ such that

$$\int_a^b \frac{1}{p(y)} f_2^2(y)\,dy < \infty$$

with an inner product $\langle\cdot, \cdot\rangle_2$ defined by

$$\langle \mathbf{f}_2, \mathbf{g}_2\rangle_2 = \int_a^b \frac{1}{p(y)} f_2(y)g_2(y)\,dy \tag{10.5.26}$$

An element $\mathbf{f} \in \mathcal{H}$ may be denoted by $\mathbf{f} \equiv \begin{bmatrix}\mathbf{f}_1\\ \mathbf{f}_2\end{bmatrix}$, where $\mathbf{f}_1 \in \mathcal{H}_1$ and $\mathbf{f}_2 \in \mathcal{H}_2$. The inner product on $\mathcal{H}$ may be defined as

$$\langle \mathbf{f}, \mathbf{g}\rangle = \langle \mathbf{f}_1, \mathbf{g}_1\rangle_1 + \alpha\langle \mathbf{f}_2, \mathbf{g}_2\rangle_2 \tag{10.5.27}$$

where $\alpha$ is a positive number at one's disposal (see Exercise 10.5.3). Here, however, we set $\alpha = 1$. We entertain a discontinuity for $p(y)$ at, say, $y = c$.† Then it is readily perceived that $L$ is formally self-adjoint with respect to the inner product (10.5.27) and that

$$\begin{aligned}\langle L\mathbf{f}, \mathbf{g}\rangle - \langle \mathbf{f}, L\mathbf{g}\rangle &= [-f_2(y)g_1(y) + f_1(y)g_2(y)]_a^b \\ &\quad + [-f_2(y)g_1(y) + f_1(y)g_2(y)]_{c+}^{c-}\end{aligned} \tag{10.5.28}$$

What remains is to recast the boundary and interface conditions in terms of $v(x, y)$ and $w(x, y)$ and thus identify the domain of $L$. The boundary

†One may note here that formula (10.5.28) may be written if discontinuities occur at $y = c$ of $\mathbf{f}$ and $\mathbf{g}$, and has little to do with the discontinuity of $p(y)$. Reference is made to the discontinuity of $p(y)$ because physically it is that which leads to the stipulation of discontinuities in $\mathbf{f}$ and $\mathbf{g}$.

condition (10.5.4) involves $v$ and needs no further discussion. On the other hand, if (10.5.5) is the boundary condition at $y = b$, it influences (10.5.20) to yield

$$\frac{\partial}{\partial x} w(x, b) = k(x) + \phi(x, b)$$

which is readily integrated to get

$$w(x, b) = \int_{-\infty}^{x} [k(\xi) + \phi(\xi, b)]\, d\xi \equiv l(x) \tag{10.5.29}$$

where we have needfully set $w(-\infty, b) = 0$. Equation (10.5.29) is a specification of $w(x, b)$, an apt boundary condition for the system (10.5.22). Next we address the boundary condition (10.5.2) and seek to express it in terms of $v$ and $w$. At $y = a$, we may write (10.5.20) as

$$\frac{\partial}{\partial y} v(x, a) = \frac{1}{p(a)} \frac{\partial}{\partial x} w(x, a)$$

and substitute into (10.5.2) to obtain

$$\frac{\partial}{\partial x}\left[\alpha \frac{\partial v}{\partial y}(x, a) - \beta v(x, a) + \frac{1}{p(a)} w(x, a)\right] = g(x)$$

The foregoing may be integrated with respect to $x$ to produce

$$\left[\alpha \frac{\partial}{\partial y} v(x, a) - \beta v(x, a) + \frac{1}{p(a)} w(x, a)\right] = \psi(x) \tag{10.5.30}$$

where

$$\psi(x) \equiv \int_{-\infty}^{x} g(\xi)\, d\xi$$

is a known function which is stipulated to exist. When $\alpha = 0$ one obtains

$$-\beta v(x, a) + \frac{1}{p(a)} w(x, a) = \psi(x) \tag{10.5.31}$$

In view of the several alternatives that are available with respect to boundary and interface conditions we lay out several different cases to be analyzed here in Table 10.5.1 for ready reference. These cases are not obviously exhaustive but suffice to establish the methodology for others. Cases 1 to 3 involve no discontinuities and hence interfacial conditions do not arise. Case 4 assumes a discontinuity in $p(y)$ where a discontinuity in $\partial v/\partial y$ also arises such that the interfacial condition (10.5.6) holds. Consider these cases individually and establish $D(L)$ in each case and show that $\mathbf{L} = \{L, D(L)\}$ is self-adjoint.

*Case 1*

The reformulated problem consists of (10.5.22), the Dirichlet boundary condition (10.5.4), and the boundary condition (10.5.31) originating from the oblique derivative boundary condition (10.5.3). Since the domain $D(L)$

Table 10.5.1

Self-Adjoint Problems in Section 10.5 Selected for Discussion

| Problem | Original | | | | Reformulated | | | |
|---|---|---|---|---|---|---|---|---|
| Cases | 1 | 2 | 3 | 4 | 1 | 2 | 3 | 4 |
| Differential equation | (10.5.1) | (10.5.1) | (10.5.1) | (10.5.1) | (10.5.22) | (10.5.22) | (10.5.22) | (10.5.22) |
| Boundary condition 1 | (10.5.3) | (10.5.3) | (10.5.2) | (10.5.5) | (10.5.31) | (10.5.31) | (10.5.30) | (10.5.29) |
| Boundary condition 2 | (10.5.4) | (10.5.5) | (10.5.4) | (10.5.4) | (10.5.4) | (10.5.29) | (10.5.4) | (10.5.4) |
| Interface conditions | None | None | None | (10.5.6) | None | None | None | (10.5.6) |

must use homogeneous boundary conditions, we have

$$D(L) = \{\mathbf{f} \in \mathcal{H}; L\mathbf{f} \in \mathcal{H}; f_1(b) = 0, \beta p(a)f_1(a) = f_2(a)\} \qquad (10.5.32)$$

Since no discontinuities exist (10.5.28) becomes

$$\langle L\mathbf{f}, \mathbf{g}\rangle - \langle \mathbf{f}, L\mathbf{g}\rangle = [-f_2(y)g_1(y) + f_1(y)g_2(y)]_a^b \qquad (10.5.33)$$

If as usual we define $\mathbf{L} = \{L, D(L)\}$, then we obtain

$$\begin{aligned}\langle \mathbf{Lf}, \mathbf{g}\rangle - \langle \mathbf{f}, \mathbf{Lg}\rangle &= f_2(a)g_1(a) - f_1(a)g_2(a)\\ &= \beta p(a)f_1(a)g_1(a) - \beta p(a)f_1(a)g_1(a) = 0\end{aligned}$$

so that $\mathbf{L}$ is now self-adjoint. The reformulated boundary value problem may now be written as

$$\frac{d}{dx}\mathbf{V}(x) = L\mathbf{V}(x) + \mathbf{\Phi}(x) \qquad (10.5.34)$$

where we have let

$$\mathbf{V}(x) \equiv \begin{bmatrix}\{v(x, y)\}\\ \{w(x, y)\}\end{bmatrix} \quad \text{and} \quad \mathbf{\Phi}(x) \equiv \begin{bmatrix}\{0\}\\ \{\phi(x, y)\}\end{bmatrix}$$

As $\mathbf{V}(x) \notin D(L)$ it is useful to record that for $\mathbf{u} \in D(L)$,

$$\begin{aligned}\langle L\mathbf{V}(x), \mathbf{u}\rangle - \langle \mathbf{V}(x), \mathbf{Lu}\rangle &= [-w(x, y)u_1(y) + v(x, y)u_2(y)]_a^b\\ &= h(x)u_2(b) + p(a)u_1(a)\psi(x) \qquad (10.5.35)\end{aligned}$$

which obtains from the use of boundary conditions (10.5.4) and (10.5.31). We do not pursue the solution of the boundary value problem at this stage since it will be taken up later jointly for all the cases.

*Case 2*

The difference between this and the previous cases lies only in the use of boundary condition (10.5.5) reformulated as (10.5.29) in place of (10.5.4) in case 1. Thus $L$ remains as given by (10.5.23) and

$$D(L) = \{\mathbf{f} \in \mathcal{H}; L\mathbf{f} \in \mathcal{H}; f_2(b) = 0, \beta p(a)f_1(a) = f_2(a)\} \qquad (10.5.36)$$

The self-adjointness of $\mathbf{L} = \{L, D(L)\}$ follows again from (10.5.33) as in case 1. The boundary value problem is again described by (10.5.34). Again $\mathbf{V}(x) \notin D(L)$ and (10.5.35) must be replaced by

$$\langle L\mathbf{V}(x), \mathbf{u}\rangle - \langle \mathbf{V}(x), \mathbf{Lu}\rangle = -l(x)u_1(b) + p(a)u_1(a)\psi(x) \qquad (10.5.37)$$

where $\mathbf{u} \in D(L)$.

*Case 3*

This case involves the mixed derivative boundary condition (10.5.2) reformulated as (10.5.30) together with the Dirichlet boundary condition (10.5.4) of case 1. If we retain the Hilbert space (10.5.24), then (10.5.33) continues to hold. Also if $\mathbf{f} \in \mathcal{H}$ satisfies the homogeneous version of (10.5.30), then

$$\alpha f_1'(a) - \beta f_1(a) + \frac{1}{p(a)} f_2(a) = 0 \tag{10.5.38}$$

Suppose further that we set $f_1(b) = 0$, as required by the boundary condition (10.5.4); then by allowing both $\mathbf{f}$ and $\mathbf{g}$ to satisfy this and condition (10.5.38) in (10.5.33), we obtain

$$\langle L\mathbf{f}, \mathbf{g}\rangle - \langle \mathbf{f}, L\mathbf{g}\rangle = -\alpha p(a) f_1'(a) g_1(a) + \alpha p(a) g_1'(a) f_1(a)$$

yielding the nonvanishing residue on the right-hand side. However, the foregoing equation itself suggests the remedy as will become evident from the following development. We define the Hilbert space $\mathcal{H}'$ by

$$\mathcal{H}' = \mathcal{H}_1 \oplus \mathcal{H}_2 \oplus \mathcal{R} \tag{10.5.39}$$

where $\mathcal{H}_1$ and $\mathcal{H}_2$ are the Hilbert spaces appearing in (10.5.24). We let $\mathbf{f} \in \mathcal{H}'$ by

$$\mathbf{f} \equiv \begin{bmatrix} \mathbf{f}_1 \\ \mathbf{f}_2 \\ f_3 \end{bmatrix}, \qquad \mathbf{f}_1 \in \mathcal{H}_1, \quad \mathbf{f}_2 \in \mathcal{H}_2, \quad f_3 \in \mathcal{R}$$

and define the inner product $\langle \cdot, \cdot \rangle'$ on $\mathcal{H}'$ by

$$\langle \mathbf{f}, \mathbf{g}\rangle' = \langle \mathbf{f}_1, \mathbf{g}_1\rangle_1 + \langle \mathbf{f}_2, \mathbf{g}_2\rangle_2 + \delta f_3 g_3 \tag{10.5.40}$$

where $\langle \cdot, \cdot \rangle_1$ and $\langle \cdot, \cdot \rangle_2$ are the inner products on $\mathcal{H}_1$ and $\mathcal{H}_2$ defined earlier, and $\delta$ is a positive constant at our disposal. Next we define a differential expression

$$L' \equiv \begin{bmatrix} \dfrac{s(y)}{q(y)} & -\dfrac{1}{r(y)q(y)}\dfrac{d}{dy} & 0 \\ p(y)\dfrac{d}{dy} & 0 & 0 \\ \left[p(a)\dfrac{d}{dy}\right]^a & 0 & 0 \end{bmatrix} \tag{10.5.41}$$

The expression $L'$ enables one to represent (10.5.22) and (10.5.20) at $y = a$ by the single equation

$$\frac{d}{dx}\mathbf{V}(x) = L'\mathbf{V}(x) + \mathbf{\Phi}(x) \tag{10.5.42}$$

which is similar to (10.5.34) except that $\mathbf{V}(x)$ and $\mathbf{\Phi}(x)$ are now given by

$$\mathbf{V}(x) = \begin{bmatrix} \{v(x, y)\} \\ \{w(x, y)\} \\ w(x, a) \end{bmatrix}, \qquad \mathbf{\Phi}(x) = \begin{bmatrix} \{0\} \\ \{\phi(x, y)\} \\ 0 \end{bmatrix}$$

One must take particular note of (10.5.30) because it is the one which ensures that the mixed derivative boundary condition (10.5.2) in the original boundary value problem is satisfied. Remembering that the homogeneous version of (10.5.30) is given by (10.5.38), and that the third component of $\mathbf{V}(x)$ is $w(x, a)$, we collect the conditions defining $D(L')$ as

$$D(L') = \left\{\mathbf{f} \in \mathcal{H}'; L'\mathbf{f} \in \mathcal{H}'; f_1(b) = 0, \alpha f_1'(a) - \beta f_1(a) + \frac{1}{p(a)} f_2(a) = 0, \right.$$
$$\left. f_3 = f_2(a)\right\} \tag{10.5.43}$$

Letting $\mathbf{L}' = \{L', D(L')\}$, it is now readily verified that

$$\langle \mathbf{L}'\mathbf{f}, \mathbf{g}\rangle' - \langle \mathbf{f}, \mathbf{L}'\mathbf{g}\rangle' = (\delta - \alpha)p(a)[f_1'(a)g_1(a) - f_1(a)g_1'(a)]$$

which vanishes on setting $\delta = \alpha$. Thus the inner product (10.5.40) must be employed on $\mathcal{H}'$ with $\delta = \alpha$ in order that $L'$ be self-adjoint.

In view of (10.5.30), $\mathbf{V}(x) \notin D(L')$. But it is readily shown that for $\mathbf{u} \in D(L)$,

$$\langle L'\mathbf{V}(x), \mathbf{u}\rangle - \langle \mathbf{V}(x), \mathbf{L}'\mathbf{u}\rangle = p(a)u_1(a)\psi(x) \tag{10.5.44}$$

In summary let us observe that the self-adjoint formalism obtains from *appending* to the first-order system (10.5.22), Eq. (10.5.20) evaluated at $y = a$ *and* requiring the boundary condition (10.5.30).

*Case 4*

The strategy in this problem follows along lines very similar to those of case 3. Here the boundary value problem consists originally of (10.5.1), featuring a discontinuity in $p(y)$ at $y = c$, boundary condition (10.5.5), the Dirichlet boundary condition (10.5.4), and interface condition (10.5.6).

The reformulated problem consists of (10.5.22), boundary conditions (10.5.4) and (10.5.29), and the interface condition (10.5.6). Recalling (10.5.28) with $\mathbf{f}$ and $\mathbf{g}$ satisfying homogeneous versions of boundary conditions (10.5.4) and (10.5.29), that is, $f_1(a) = f_2(b) = 0$, we obtain

$$\langle L\mathbf{f}, \mathbf{g}\rangle - \langle \mathbf{f}, L\mathbf{g}\rangle = [-f_2(y)g_1(y) + f_1(y)g_2(y)]_{c+}^{c-} \tag{10.5.45}$$

where $\mathbf{f}, \mathbf{g} \in \mathcal{H}$. Since $\phi(x, y)$ is continuous at $c$, continuity of $p(y)(\partial v/\partial y)$ at $c$, implied by (10.5.6), requires $w(x, y)$ to be continuous at $c$, that is,

$$w(x, c-) = w(x, c+) \tag{10.5.46}$$

Further, (10.5.6) implies that

$$\frac{\partial}{\partial x} w(x, c) = \gamma[v(x, c+) - v(x, c-)] + \phi(x, c) \tag{10.5.47}$$

Note how (10.5.20), (10.5.45), and (10.5.46) together imply the interface conditions (10.5.6). Following the same lines of arguments as in case 3, we are again led to the Hilbert space $\mathcal{H}'$ as given by (10.5.39). In writing the first-order system, we append (10.5.47) with (10.5.22) to write

$$\frac{d}{dx}\mathbf{V}(x) = L''\mathbf{V}(x) + \mathbf{\Phi}(x) \tag{10.5.48}$$

where

$$L'' \equiv \begin{bmatrix} \dfrac{s(y)}{q(y)} & -\dfrac{1}{r(y)q(y)}\dfrac{d}{dy} & 0 \\ p(y)\dfrac{d}{dy} & 0 & 0 \\ \gamma[\cdot]_{c-}^{c+} & 0 & 0 \end{bmatrix} \tag{10.5.49}$$

$$\mathbf{V}(x) = \begin{bmatrix} v(x, y) \\ w(x, y) \\ w(x, c) \end{bmatrix}, \qquad \mathbf{\Phi}(x) = \begin{bmatrix} 0 \\ \phi(x, y) \\ \phi(x, c) \end{bmatrix}$$

The formal self-adjointness of $L''$ follows from that of $L$. The domain of $L''$ is given by

$$D(L'') = \{\mathbf{f} \in \mathcal{H}'; L''\mathbf{f} \in \mathcal{H}'; f_1(a) = f_2(b) = 0; \\ f_2(c-) = f_2(c+) = f_3\} \tag{10.5.50}$$

Using inner product (10.5.40), we obtain

$$\langle \mathbf{L}''\mathbf{f}, \mathbf{g}\rangle - \langle \mathbf{f}, \mathbf{L}''\mathbf{g}\rangle = (\gamma\delta - 1)[f_2(c)\{g_1(c-) - g_1(c+)\} \\ + g_2(c)\{f_1(c-) - f_1(c+)\}]$$

which vanishes on letting $\delta = 1/\gamma$. Thus $\mathbf{L}''$ is self-adjoint with respect to the inner product (10.5.40) with $\delta = 1/\gamma$. Finally, we recognize that $\mathbf{V}(x) \notin D(L'')$ because for $\mathbf{u} \in D(L'')$,

$$\langle L''\mathbf{V}(x), \mathbf{u}\rangle' - \langle \mathbf{V}(x), \mathbf{L}''\mathbf{u}\rangle' = -l(x)u_1(b) \tag{10.5.51}$$

Equation (10.5.51) is useful in the solution of the boundary value problem.

## EXERCISES

**10.5.1** Reformulate the boundary value problem comprising (10.5.1) with boundary condition (10.5.2) at $y = a$ and at $y = b$,

$$\alpha'\frac{\partial^2}{\partial x\,\partial y}v(x, b) + \beta'\frac{\partial}{\partial x}v(x, b) + \frac{\partial}{\partial y}v(x, b) = h(x)$$

where $\alpha'$ and $\beta'$ are different from $\alpha$ and $\beta$, respectively. Show that the operator in the reformulation is self-adjoint.

**10.5.2** Show that the problem in case 1 of Table 10.5.1 leads to a self-adjoint problem with (10.5.4) replaced by the Robin boundary condition at $y = b$,

$$\frac{\partial}{\partial y}v(x, b) + \varepsilon v(x, b) = h(x)$$

all other conditions being the same as those in case 1.

**10.5.3** Obtain a self-adjoint formulation in case 4 of Table 10.5.1 when the interface condition is changed to

$$-p(c-)\frac{\partial}{\partial y}v(x, c-) = -p(c+)\frac{\partial}{\partial y}v(x, c+) = \gamma[v(x, c-) - \kappa v(x, c+)]$$

where $\kappa > 0$.

**10.5.4** Reformulate case 4 of Table 10.5.1 with (10.5.29) replaced by (10.5.30). Identify the Hilbert space and the self-adjoint problem.

**10.5.5** Develop a self-adjoint formalism for case 4 with the interface conditions replaced by

$$-p(c-)\frac{\partial}{\partial y}v(x, c-) + p(c+)\frac{\partial}{\partial y}v(x, c+) = \beta\frac{\partial}{\partial x}v(x, c)$$

assuming that $v$ is continuous at $y = c$.

**10.5.6** Following case 4 further from Exercise 10.5.5 allow $v(x, y)$ to be discontinuous at $x = c$; the interface conditions may be replaced by the pair below.

$$p(c+)\frac{\partial}{\partial y}v(x, c+) + p(c-)\frac{\partial}{\partial y}v(x, c-) = \gamma[v(x, c+) - v(x, c-)]$$

$$p(c+)\frac{\partial^2}{\partial x\,\partial y}v(x, c+) - p(c-)\frac{\partial^2}{\partial x\,\partial y}v(x, c-)$$
$$= \gamma\left[\frac{\partial}{\partial x}v(x, c-) + \frac{\partial}{\partial x}v(x, c+)\right]$$
$$-\beta\left[p(c+)\frac{\partial}{\partial y}v(x, c+) - p(c-)\frac{\partial}{\partial y}v(x, c-)\right]$$

Identify the Hilbert space and the self-adjoint operator.

### 10.5.3 Spectral Analysis of Self-Adjoint First-Order Systems

The preceding section developed self-adjoint formalisms for the several boundary value problems of this section which were originally non-self-adjoint. Thus Table 10.5.2 summarizes for each of the different cases the differential expression and its domain, the Hilbert space of interest, and the inner product with respect to which the operator is self-adjoint.

As in Chapter 9, the discrete nature of the spectrum of each self-adjoint operator in Table 10.5.2 may be established by the existence of a compact self-adjoint inverse. We desist from this effort because in the cases of interest, this procedure is relatively straightforward and is left as an exercise. Thus the eigenvalues are real and countably infinite and become infinitely large in magnitude. Furthermore, the eigenvalues can take on both negative and positive values. That the eigenvalues become infinite in value on both the

TABLE 10.5.2

| | Reformulated Operator | | Hilbert Space | |
|---|---|---|---|---|
| Case | Operation | Domain | Space | Inner Product |
| 1 | $L$ (10.5.23) | (10.5.32) | (10.5.24) | (10.5.27) $\alpha = 1$ |
| 2 | $L$ (10.5.23) | (10.5.36) | (10.5.24) | (10.5.27) $\alpha = 1$ |
| 3 | $L'$ (10.5.41) | (10.5.43) | (10.5.39) | (10.5.40) $\delta = \alpha$ |
| 4 | $L''$ (10.5.49) | (10.5.50) | (10.5.39) | $\delta = \frac{1}{\gamma}$ |

positive and negative sides can be established by examining $\langle \mathbf{Lu}, \mathbf{u} \rangle$ and is left as an exercise. We will also see this by explicitly computing the spectra in some examples to be presented in this subsection.

The spectrum of the operator in any of the cases of Table 10.5.1 is to be computed by solving the system of differential equations

$$L\mathbf{u} = \lambda \mathbf{u}$$

and demanding that $\mathbf{u} \in D(L)$, which will lead to the characteristic equation for the eigenvalues. As in Chapter 9 (see Section 9.2 to 9.4), it is possible to derive general formulations of characteristic equations, but we will refrain from such generalities here. We denote the positive eigenvalues by $\{\lambda_j^+\}$, the corresponding eigenvectors by $\{\mathbf{u}_j^+\}$, and the negative eigenvalues by $\{\lambda_j^-\}$ with $\{\mathbf{u}_j^-\}$ as the corresponding eigenvectors. From spectral theorem 7.4.3 the set of eigenvectors $\{\mathbf{u}_j^+, \mathbf{u}_j^-\}$ form an orthonormal basis in $\mathcal{H}$ (or $\mathcal{H}'$ in cases 3 and 4). Thus for any $\mathbf{f} \in \mathcal{H}$, we have

$$\mathbf{f} = \sum_{j=1}^{\infty} [\langle \mathbf{f}, \mathbf{u}_j^+ \rangle \mathbf{u}_j^+ + \langle \mathbf{f}, \mathbf{u}_j^- \rangle \mathbf{u}_j^-] \tag{10.5.52}$$

provided that the eigenvectors have been normalized. We consider some examples below for the computation of the spectrum.

*Example 1*

We consider Example 1 of Section 10.5.1 first with the oblique derivative boundary condition obtained by putting $\alpha = 0$ in (10.5.10). The operator $\mathbf{L} = \{L, D(L)\}$ is given by

$$L = \begin{bmatrix} 0 & -\dfrac{d}{dy} \\ \dfrac{d}{dy} & 0 \end{bmatrix}$$

which is obtained from (10.5.23) by putting $s(y) \equiv 0$, $p(y) = q(y) = r(y) \equiv 1$, and

$$D(L) = \{\mathbf{f}, L\mathbf{f} \in \mathcal{H}; f_2(0) = 0, \beta f_1(1) + f_2(1) = 0\}$$

The eigenvalue problem $\mathbf{Lu} = \lambda\mathbf{u}$ leads to

$$-u_2' = \lambda u_1, \qquad u_1' = \lambda u_2$$

so that one has $u_1'' = -\lambda^2 u_1$, yielding the solution

$$u_1 = A\cos\lambda y \quad \text{and} \quad u_2 = -A\sin\lambda y$$

which satisfies $u_2(0) = 0$. The boundary condition $\beta f_1(1) + f_2(1) = 0$ obtains

$$\tan\lambda = \beta$$

as the characteristic equation. If $\theta = \arctan\beta$ in the interval $(0, \pi/2)$, then

$$\lambda_j^{\pm} = \theta \pm (j-1)\pi, \qquad j = 1, 2, \ldots$$

represents all the eigenvalues of the operator $\mathbf{L}$. The eigenvector $\mathbf{u}_j^{\pm}$ is given by

$$\mathbf{u}_j^{\pm} = A_j^{\pm}\begin{bmatrix} \cos\lambda_j^{\pm} y \\ -\sin\lambda_j^{\pm} y \end{bmatrix}$$

where $A_j^{\pm}$ is obtained by normalizing $\mathbf{u}_j^{\pm}$ as follows. Recalling inner product (10.5.27) with $\alpha = 1$, we have

$$\langle \mathbf{u}_j^{\pm}, \mathbf{u}_j^{\pm} \rangle = 1 = 1(A_j^{\pm})^2 \int_0^1 [\cos^2\lambda_j^{\pm} y + \sin^2\lambda_j^{\pm} y]\,dy = (A_j^{\pm})^2 \quad \text{or} \quad A_j^{\pm} = 1$$

Thus the eigenvalues and eigenvectors have been completely identified.

It is strongly recommended that the student work Exercise 10.5.7 to understand the nature of the expansion of vectors in $\mathcal{H}$ in terms of the set $\{\mathbf{u}_j^{\pm}\}$.

*Example 2*

We consider the situation in Example 1 again with $\alpha \neq 0$ so that the mixed second derivative boundary condition (10.5.10) applies. The operator of interest is given by $\mathbf{L}' = \{L', D(L')\}$ (see case 3), where

$$L' \equiv \begin{bmatrix} 0 & -\dfrac{d}{dy} & 0 \\ \dfrac{d}{dy} & 0 & 0 \\ \left[\dfrac{d}{dy}\right]^1 & 0 & 0 \end{bmatrix}$$

$D(L') =$

$$\{\mathbf{u} \in \mathcal{H}'; L'\mathbf{u} \in \mathcal{H}'; f_2(0) = 0, \alpha f_1'(1) + \beta f_1(1) + f_2(1) = 0, f_3 = f_2(1)\}$$

To compute the spectrum we set $\mathbf{L}'\mathbf{u} = \lambda \mathbf{u}$ and obtain $u_1 = A \cos \lambda y$, $u_2 = A \sin \lambda y$, $u_3 = A \sin \lambda$, which satisfies $u_2(0) = 0$, $u_3 = u_2(1)$. The boundary condition at $y = 1$ yields

$$-\alpha\lambda \sin \lambda + \beta \cos \lambda + \sin \lambda = 0$$

or

$$\tan \lambda = \frac{-\beta}{1 - \alpha\lambda}$$

for the characteristic equation which must be solved numerically for the positive and negative eigenvalues $\{\lambda_j^\pm\}$. The eigenvector $\mathbf{u}_j^\pm$ is given by

$$\mathbf{u}_j^\pm = A_j^\pm \begin{bmatrix} \cos \lambda_j^\pm y \\ \sin \lambda_j^\pm y \\ \sin \lambda_j^\pm \end{bmatrix}$$

where $A_j^\pm$ is to be calculated by normalizing $\mathbf{u}_j^\pm$. The inner product (10.5.40) with $\delta = \alpha$ must be used for this step as follows.

$$\langle \mathbf{u}_j^\pm, \mathbf{u}_j^\pm \rangle = 1 = (A_j^\pm)^2 \left[ \int_0^1 (\cos^2 \lambda_j^\pm y + \sin^2 \lambda_j^\pm y)\, dy + \alpha \sin^2 \lambda_j^\pm \right]$$

so that

$$A_j^\pm = (1 + \alpha \sin^2 \lambda_j^\pm)^{-1/2}$$

We refer the student to the literature for other examples of computation of such spectra.†

## EXERCISES

**10.5.7** Show that the expansion (10.5.52) implies that in case 1 or 2 of Table 10.5.1, $\forall\, \mathbf{f}_i \in \mathcal{H}_i$, $i = 1, 2$.

$$\mathbf{f}_i \delta_{ik} = \sum_{j=1}^{\infty} [\langle \mathbf{f}_i, \mathbf{u}_{i,j}^+ \rangle_i \mathbf{u}_{k,j}^+ + \langle \mathbf{f}_i, \mathbf{u}_{i,j}^- \rangle_i \mathbf{u}_{k,j}^-], \qquad i, k = 1, 2$$

where $\mathbf{u}_{i,j}^\pm \in \mathcal{H}_i$, $i = 1, 2$, are the component vectors occurring in

$$\mathbf{u}_j^\pm = \begin{bmatrix} \mathbf{u}_{1,j}^\pm \\ \mathbf{u}_{2,j}^\pm \end{bmatrix} \quad \text{and} \quad \delta_{ik} = \begin{cases} 1, & i = k \\ 0, & i \neq k \end{cases}$$

Establish similar results for cases 3 and 4.

†See E. Papoutsakis, D. Ramkrishna, and H. C. Lim, *Appl. Sci. Res.*, *36*, 13–34, 1980, and *AIChE J.*, *26*, 779–788, 1980. For problems of the type of Case 4, see E. Papoutsakis and D. Ramkrishna, *Chem. Eng. Sci.*, *36*, 1381–1391, 1981.

**10.5.8** Consider Example 1 of Section 10.5.2. Using the results of Fourier sine and cosine series given by

$$\delta(y-y') = 2\sum_{j=1}^{\infty} \sin j\pi y \sin j\pi y' = 1 + 2\sum_{j=1}^{\infty} \cos j\pi y \cos j\pi y', \qquad 0 < y, y' < 1$$

where $\delta(y-y')$ is the Dirac delta function, show that

$$\cos\theta(y-y')\delta(y-y') = \sum_{j=1}^{\infty} [\sin\lambda_j^+ y \sin\lambda_j^+ y' + \sin\lambda_j^- y \sin\lambda_j^- y']$$

$$= \sum_{j=1}^{\infty} [\cos\lambda_j^+ y \cos\lambda_j^+ y' + \cos\lambda_j^- y \cos\lambda_j^- y']$$

$$-\sin\theta(y-y')\,\delta(y-y') = \sum_{j=1}^{\infty} [\sin\lambda_j^+ y \cos\lambda_j^+ y' + \sin\lambda_j^- y \cos\lambda_j^- y']$$

where $\lambda_j^{\pm} = -\theta \pm (j-1)\pi$, and hence independently establish the result of Exercise 10.5.7 for this example.

**10.5.9** For the different cases in Table 10.5.1, show that $\langle \mathbf{Lu}, \mathbf{u}\rangle$ can take on arbitrarily large positive and negative values. Thus establish that the eigenvalues can be positive and negative becoming infinite in value on either side of the real axis.

**10.5.10** For Example 2 of Section 10.5.1 with $s(y) = 1 - y^2$, obtain the characteristic equation in terms of Kummer's functions for the Dirichlet and Neumann problems.

**10.5.11** Obtain the characteristic equations for the operators encountered in Exercises 10.5.1 to 10.5.6.

### 10.5.4 Solution of the Boundary Value Problem

The solution of the boundary value problem for all of the cases in Table 10.5.1 may be considered together as an expansion of the form (10.5.52). Denote the reformulated problem by

$$\frac{d}{dx}\mathbf{V}(x) = L\mathbf{V}(x) + \mathbf{\Phi}(x) \tag{10.5.53}$$

which represents (10.5.34) for cases 1 and 2, (10.5.42) for case 3, and (10.5.48) for case 4. Since $\mathbf{V}(x) \notin D(L)$ in all the cases considered, it is essential to recall that

$$\langle L\mathbf{V}(x), \mathbf{u}\rangle = \langle \mathbf{V}(x), \mathbf{Lu}\rangle + l(\mathbf{u}; x) \tag{10.5.54}$$

which is a restatement of (10.5.35) for case 1, (10.5.37) for case 2, (10.5.44) for case 3, and (10.5.51) for case 4 with their right-hand sides jointly represented by $l(\mathbf{u}; x)$, a linear functional mapping elements of $D(L)$ into functions of $x$. Forming the inner product of (10.5.53) with $\mathbf{u}_j^{\pm}$, one obtains in view of (10.5.54)

$$\frac{d}{dx}\langle \mathbf{V}(x), \mathbf{u}_j^{\pm}\rangle = \lambda_j^{\pm}\langle \mathbf{V}(x), \mathbf{u}_j^{\pm}\rangle + \langle \mathbf{\Phi}(x), \mathbf{u}_j^{\pm}\rangle + l(\mathbf{u}_j^{\pm}; x) \qquad (10.5.55)$$

which must be solved separately for $\langle \mathbf{V}(x), \mathbf{u}_j^{+}\rangle$ and $\langle \mathbf{V}(x), \mathbf{u}_j^{-}\rangle$. Remembering the boundedness of $\mathbf{V}(x)$ at $x = \pm\infty$, we may write

$$\langle \mathbf{V}(x), \mathbf{u}_j^{+}\rangle = -\int_x^{\infty} [\langle \mathbf{\Phi}(\xi), \mathbf{u}_j^{+}\rangle + l(\mathbf{u}_j^{+}; \xi)]e^{\lambda_j^{+}(x-\xi)}\, d\xi \qquad (10.5.56)$$

$$\langle \mathbf{V}(x), \mathbf{u}_j^{-}\rangle = \int_{-\infty}^{x} [\langle \mathbf{\Phi}(\xi), \mathbf{u}_j^{-}\rangle + l(\mathbf{u}_j^{-}; \xi)]e^{\lambda_j^{-}(x-\xi)}\, d\xi \qquad (10.5.57)$$

The solution to the boundary value problem may then be obtained as

$$\mathbf{V}(x) = \sum_{j=1}^{\infty} \left\{-\int_x^{\infty} [\langle \mathbf{\Phi}(\xi), \mathbf{u}_j^{+}\rangle + l(\mathbf{u}_j^{+}; \xi)]e^{\lambda_j^{+}(x-\xi)}\, d\xi\, \mathbf{u}_j^{+} \right.$$
$$\left. + \int_{-\infty}^{x} [\langle \mathbf{\Phi}(\xi), \mathbf{u}_j^{-} + l(\mathbf{u}_j^{-}; \xi)]e^{\lambda_j^{-}(x-\xi)}\, d\xi\, \mathbf{u}_j^{-}\right\} \qquad (10.5.58)$$

An alternative way to represent the solution above is given by

$$\mathbf{V}(x) = \int_{-\infty}^{\infty} [\langle \mathbf{K}(x, \xi), \mathbf{\Phi}(\xi)\rangle + l\{\mathbf{K}(x, \xi); \xi\}\, d\xi \qquad (10.5.59)$$

where

$$\mathbf{K}(x, \xi) = \begin{cases} -\sum_{j=1}^{\infty} (\mathbf{u}_j^{+} \otimes \mathbf{u}_j^{+})e^{\lambda_j^{+}(x-\xi)}, & x < \xi < \infty \\ \sum_{j=1}^{\infty} (\mathbf{u}_j^{-} \otimes \mathbf{u}_j^{-})e^{\lambda_j^{-}(x-\xi)}, & -\infty < \xi < x \end{cases} \qquad (10.5.60)$$

In interpreting the solution (10.5.59), the reader must note that $l$ maps $\mathcal{H} \otimes \mathcal{H}$ into $\mathcal{H}$ so that $l\{\mathbf{K}(x, \xi); \xi\}$ is a vector in $\mathcal{H}$ for each value of $x$ and $\xi$.

The solution to all of the boundary value problems is now known in terms of $\mathbf{K}(x, \xi)$ whose evaluation requires the spectral data on $\mathbf{L}$. Clearly, the solution itself is straightforward once the spectral information has been obtained on the operator.

## EXERCISE

**10.5.12** Set up and solve the boundary value problem arising in cooling the bed in Example 3 of Section 10.5.1 with a second fluid flowing through a narrow, concentric annular gap within the bed concurrently with the peripheral coolant. Neglect axial conduction in this fluid. Formulate the interface conditions at $y = b$ $(b < c)$, the radial location of the gap and compare with those in Exercise 10.5.6.

## 10.6 Concluding Remarks

We have in this chapter dealt with self-adjoint, elliptic partial differential expressions and their spectra. Although Sections 10.1 to 10.3 were concerned with general expressions, our treatment in regard to the evaluation of spectra was restricted to situations amenable to the method of separation of variables. Where variables cannot be separated, numerical approximation procedures are unavoidable.

The tensor product treatment of the method of separaton of variables, due to Friedman, is a convenient mold of the method, although the symbolism is somewhat elaborate. In particular, the consequence of Theorem 10.4.1 is to provide an instant route to the solution of initial value problems featuring parabolic or hyperbolic equations in which the partial differential operator is elliptic with a separable spectral representation.

Section 10.5 presents problems generally not encountered in the literature. The problems are physically very relevant and the methods richly draw on the resources of abstract operator formulations to enable symmetrization of operators that are otherwise non-self-adjoint. It is this, then, that has been the main motivation of this book. Equally important, however, is the need to recognize that operators may be encountered which are unsymmetrizable and hence in this sense are incurably non-self-adjoint. For example, the operators of Section 10.5 are indeed incurably non-self-adjoint. The solutions to the boundary value problems did emerge from spectral representations of self-adjoint operators but ones that were *derived* from the main non-self-adjoint operator. The spectral resolution of the derived, self-adjoint, first-order system cannot solve, for example, initial value problems featuring the original non-self-adjoint operator. Thus the unsteady-state version of (10.5.1) cannot be solved by the spectral resolution of the first-order system.

## FURTHER READING

For a good account of the classical theory of partial differential equations, see

*Methods of Mathematical Physics*, Vol. 2, by R. Courant and D. Hilbert, Interscience, New York, 1962, and

*Lectures on Partial Differential Equations*, by I. G. Petrovsky, translated from the Russian by A. Shenitzer, Interscience, New York, 1954.

The modern theory of partial differential equations requires machinery normally not covered in engineering mathematics and most books on the subject would require considerable background. Some simple introductory remarks on page 65 of Petrovsky's book above are available on the notion of generalized solutions to partial differential equations, which is preliminary to the modern theory. The more adventurous reader may consult

> *Abstract Methods in Partial Differential Equations*, by R. W. Carroll, Harper & Row, New York, 1969.

For numerical methods in partial differential equations, see

> *Numerical Methods for Partial Differential Equations*, by W. F. Ames, Nelson, London, 1969.

There are several books on mathematical physics that make useful further reading material. In particular, see

> *Equations of Mathematical Physics*, by A. N. Tikhonov and A. A. Samarskii, translated by A. R. M. Robson and P. K. Basu, edited by D. M. Brink, Pergamon Press, Oxford, 1963.

# Non-Self-Adjoint Operators

# 11

## 11.0 Foreword

This chapter is a brief excursion into the analysis of non-self-adjoint operators. Because the theory takes a leap in complexity beyond the scope of this book, our treatment is essentially superficial and confined to rather simple examples.

In Chapter 7 we alluded to a *normal* operator that has the property of commuting with its adjoint and has all the powerful properties of a self-adjoint operator (see Exercises 7.4.10 and 7.4.12). Hence the complexity to which we have just referred actually arises for *nonnormal* operators. Thus our attention will be on nonnormal operators of the differential type.

## 11.1 Introduction

Let us begin by recalling the properties of a self-adjoint operator **T** which is either compact or has a compact inverse. The eigenvalues of **T** are real and its eigenvectors are mutually orthogonal and complete in the sense that every vector in the Hilbert space can be expanded in terms of the eigenvectors. These properties have been seen to be extremely useful in the solution of equations in which self-adjoint operators occur.

In the case of non-self-adjoint (or nonnormal) operators, it is difficult to make assertions about the nature of the spectrum with respect to whether it is real or complex, discrete or continuous, and so on. Further, the eigenvectors are not mutually orthogonal and, even more seriously, the completeness of the eigenvectors is not always guaranteed. In dealing with non-self-adjoint operators, one must not only work with eigenvectors but also what we shall presently define and call as *root vectors*. Thus in considering expansions of arbitrary vectors, one must include the root vectors together with the eigenvectors. The completeness question would then pertain to the joint set of eigenvectors and root vectors.

We will only be concerned with differential, non-self-adjoint operators. The development in Section 9.1 is an essential prerequisite to the material in this chapter both in regard to definitions and notations. Section 9.1 deals with second-order differential operators, which also constitute the scope of the present chapter. However, insofar as most of the results herein also carry over to higher-order differential operators, familiarity with the contents of Section 9.5 will be useful.

We recall the differential operation (9.1.1) below as

$$T \equiv p_0(x)\frac{d^2}{dx^2} + p_1(x)\frac{d}{dx} + p_2(x), \qquad p_0 \neq 0 \tag{11.1.1}$$

whose adjoint operation is given by

$$T^* \equiv p_0\frac{d^2}{dx^2} + (2p_0' - p_1)\frac{d}{dx} + (p_0'' - p_1' + p_2) \tag{11.1.2}$$

We represent the domain $D(T)$ of operator $\mathbf{T} = \{T, D(T)\}$ by

$$D(T) = \left\{\mathbf{u} \in \mathfrak{L}_2[a, b];\, T\mathbf{u} \in \mathfrak{L}_2[a, b];\, [\boldsymbol{\alpha} \quad \boldsymbol{\beta}]\begin{bmatrix}\mathbf{w}(u(a))\\ \mathbf{w}(u(b))\end{bmatrix} = \mathbf{0}\right\} \tag{11.1.3}$$

which originates from (9.1.11) of Chapter 9. Thus $[\boldsymbol{\alpha} \quad \boldsymbol{\beta}]$ is a *real* $2 \times 4$ matrix of rank 2 used to express two independent boundary conditions. The adjoint operator $\mathbf{T}^* = \{T^*, D(T^*)\}$ has

$$D(T^*) = \left\{\mathbf{u} \in \mathfrak{L}_2[a, b];\, T^*\mathbf{u} \in \mathfrak{L}_2[a, b];\, [\boldsymbol{\alpha}^* \quad \boldsymbol{\beta}^*]\begin{bmatrix}\mathbf{w}(u(a))\\ \mathbf{w}(u(b))\end{bmatrix} = \mathbf{0}\right\} \tag{11.1.4}$$

where $[\boldsymbol{\alpha}^* \quad \boldsymbol{\beta}^*]$ is a real $2 \times 4$ matrix satisfying the condition

$$\boldsymbol{\alpha}\mathbf{P}^{-1}(a)\boldsymbol{\alpha}^{*t} = \boldsymbol{\beta}\mathbf{P}^{-1}(b)\boldsymbol{\beta}^{*t} \tag{11.1.5}$$

which is a recall of (9.1.18).† In view of our interest in physical applications, we restrict $T$ and $T^*$ to consist only of real-valued coefficients. However, because the eigenvalues and eigenvectors of $\mathbf{T}$ and $\mathbf{T}^*$ could be complex-valued, the space $\mathfrak{L}_2[a, b]$ is a complex Hilbert space in which the inner

†We alert the reader to the footnote on page 272 in regard to the asterisked notation on $\boldsymbol{\alpha}^* \quad \boldsymbol{\beta}^*$.

product is defined by

$$\langle \mathbf{u}, \mathbf{v} \rangle = \int_a^b u(x)v^*(x)\,dx \tag{11.1.6}$$

where $v^*(x)$ is the complex conjugate of $v(x)$. Accordingly, $D(T)$ and $D(T^*)$ must also entertain complex-valued functions.† The recall of (9.1.13) must take the form

$$\langle \mathbf{Tu}, \mathbf{v} \rangle - \langle \mathbf{u}, \mathbf{T^*v} \rangle = [\mathbf{w}^t(v^*(x))\mathbf{P}(x)\mathbf{w}(u(x))]_a^b \tag{11.1.7}$$

In the development that follows we will be concerned with the operators $\mathbf{T}$ and $\mathbf{T}^*$ as laid out in the foregoing.

## 11.2 Eigenvalues, Eigenvectors, and Root Vectors. Biorthogonal Expansions

We first consider the null spaces of $\mathbf{T}$ and $\mathbf{T}^*$ and establish the following theorem concerning them. Familiarity with Exercise 9.1.7 is advised before reading this theorem. Further, the theorems of this section hold also for the general $n$th-order operators of Section 9.5 in which no singularities are present.

***Theorem 11.2.1*** Let $\mathbf{T}$ be given by (11.1.1) and (11.1.3), so that $\mathbf{T}^*$ is given by (11.1.2) and (11.1.4). Then $N(\mathbf{T})$ and $N(\mathbf{T}^*)$ have the same dimension.

*Proof*: Let $\mathbf{U}(x) = \begin{bmatrix} u_1(x) \\ u_2(x) \end{bmatrix}$, where $u_1$ and $u_2$ are the two linearly independent solutions of $T\mathbf{u} = \mathbf{0}$. Then form the $2 \times 2$ matrix [see also (9.2.23)]

$$\mathbf{D}_0 \equiv \boldsymbol{\alpha}\mathbf{W}(\mathbf{U}(a)) + \boldsymbol{\beta}\mathbf{W}(\mathbf{U}(b))$$

If $r_{D_0}$ is the rank of $\mathbf{D}_0$ which is a $2 \times 2$ matrix, then clearly $N(\mathbf{T})$ has dimension $(2 - r_{D_0})$. Let $\mathbf{V}(x) = \begin{bmatrix} v_1(x) \\ v_2(x) \end{bmatrix}$, where $v_1$ and $v_2$ are the linearly independent solutions of $T^*\mathbf{v} = \mathbf{0}$. We also form the matrix

$$\mathbf{D}_0^* \equiv \boldsymbol{\alpha}^*\mathbf{W}(\mathbf{V}^*(a)) + \boldsymbol{\beta}^*\mathbf{W}(\mathbf{V}^*(b))$$

where again the asterisk on $\mathbf{D}_0^*$ (as on $\boldsymbol{\alpha}^*$ and $\boldsymbol{\beta}^*$) is not used to denote the complex conjugate but to represent its association with $T^*$. The dimension of $N(\mathbf{T}^*)$ is $(2 - r_{D_0^*})$, where $r_{D_0^*}$ is the rank of $\mathbf{D}_0^*$. From the nonsingular matrices

$$\mathbf{H} \equiv \begin{bmatrix} \boldsymbol{\alpha} & \boldsymbol{\beta} \\ \boldsymbol{\alpha}_c & \boldsymbol{\beta}_c \end{bmatrix}, \qquad \mathbf{J} \equiv \begin{bmatrix} \boldsymbol{\alpha}_c^* & \boldsymbol{\beta}_c^* \\ \boldsymbol{\alpha}^* & \boldsymbol{\beta}^* \end{bmatrix} \tag{11.2.1}$$

---

†Note that if $\mathbf{u} \in D(T)$ is complex-valued, both the real and imaginary parts must satisfy all the conditions in $D(T)$.

where $[\boldsymbol{\alpha}_c \quad \boldsymbol{\beta}_c]$ and $[\boldsymbol{\alpha}_c^* \quad \boldsymbol{\beta}_c^*]$ are the respective complements of $[\boldsymbol{\alpha} \quad \boldsymbol{\beta}]$ and $[\boldsymbol{\alpha}^* \quad \boldsymbol{\beta}^*]$ (see Exercise 9.1.7). Defining

$$\mathbf{D}_c \equiv \boldsymbol{\alpha}_c \mathbf{W}(\mathbf{U}(a)) + \boldsymbol{\beta}_c \mathbf{W}(\mathbf{U}(b)), \qquad \mathbf{D}_c^* \equiv \boldsymbol{\alpha}_c^* \mathbf{W}(\mathbf{V}^*(a)) + \boldsymbol{\beta}_c^* \mathbf{W}(\mathbf{V}^*(b))$$

we may write from the result of Exercise 9.1.7 [see (9.1.21)]

$$\mathbf{D}_c^* \mathbf{D}_0 = -\mathbf{D}_0^* \mathbf{D}_c \tag{11.2.2}$$

Let $\mathbf{u} = \{u(x)\} \in N(\mathbf{T})$ be given by $\mathbf{c}^t \mathbf{U}(x)$, where $\mathbf{c}^t \equiv [c_1 \quad c_2]$ is a vector of constants. Then $\mathbf{u} \in D(T)$ implies that

$$\mathbf{D}_0 \mathbf{c} = \mathbf{0} \tag{11.2.3}$$

which from (11.2.2) leads to the result

$$\mathbf{D}_0^* \mathbf{D}_c \mathbf{c} = \mathbf{0} \tag{11.2.4}$$

From (11.2.3) the dimension of $N(\mathbf{T})$ is the same as that of $N(\mathbf{D}_0)$. Similarly, $N(\mathbf{T}^*)$ has the same dimension as that of $N(\mathbf{D}_0^*)$. We must now show that $N(\mathbf{D}_0)$ and $N(\mathbf{D}_0^*)$ have the same dimension to prove the present theorem.

Let $\{\mathbf{c}_i\}_{i=1}^n \subset N(\mathbf{D}_0)$ be linearly independent, where $n \equiv 2 - r_{D_0}$ is the dimension of $N(\mathbf{D}_0)$. Clearly, from (11.2.4) $\mathbf{D}_c \mathbf{c}_i \in N(\mathbf{D}_0^*)$ for each $i$. We will show that the set $\{\mathbf{D}_c \mathbf{c}_i\}_{i=1}$ is linearly independent. Set

$$\mathbf{0} = \sum_{i=1}^n \alpha_i \mathbf{D}_c \mathbf{c}_i$$

so that

$$\mathbf{D}_c \left( \sum_{i=1}^n \alpha_i \mathbf{c}_i \right) = \mathbf{0} \tag{11.2.5}$$

But $\mathbf{D}_0 \mathbf{c}_i = \mathbf{0}$ for each $i$, so that

$$\mathbf{D}_0 \left( \sum_{i=1}^n \alpha_i \mathbf{c}_i \right) = \mathbf{0} \tag{11.2.6}$$

Let $\bar{\mathbf{c}} \equiv \sum_{i=1}^n \alpha_i \mathbf{c}_i$. If $\bar{u}(x) \equiv \bar{\mathbf{c}}^t \mathbf{U}(x)$, then (11.2.5) and (11.2.6) yield

$$\mathbf{H} \begin{bmatrix} \mathbf{w}(\bar{u}(a)) \\ \mathbf{w}(\bar{u}(b)) \end{bmatrix} = \mathbf{0}$$

where $\mathbf{H}$ is defined by (11.2.1). Since $\mathbf{H}$ is nonsingular we have $\mathbf{w}(\bar{u}(a)) = \mathbf{w}(\bar{u}(b)) = \mathbf{0}$. Furthermore, since $T\bar{u}(x) = -0$, its solution subject to $\bar{u}(a) = \bar{u}'(a) = \bar{u}(b) = \bar{u}'(b) = 0$ can only be zero. Thus $\bar{u}(x) \equiv 0$, so that $\bar{\mathbf{c}} = \mathbf{0}$. Hence $\alpha_i = 0$ for each $i$ (why?), so that $\{\mathbf{D}_c \mathbf{c}_i\}_{i=1}^n$ is a linearly independent set. Because $\mathbf{D}_c \mathbf{c}_i \in N(\mathbf{D}_0^*)$ for each $i$ the dimension of $N(\mathbf{D}_0^*)$, say $n^*$, is at least $n$; that is, $n^* \geq n$. By an exactly similar procedure we can show that $n \geq n^*$. Thus we conclude that $n = n^*$ and that $N(\mathbf{T}) = N(\mathbf{T}^*)$. ∎

The foregoing theorem leads to the following important corollary.

***Theorem 11.2.2*** Let **T** and **T*** be as in Theorem 11.2.1. If $\lambda$ is an eigenvalue of **T**, then $\lambda^*$ is an eigenvalue of **T***. The number of eigenvectors of **T** corresponding to eigenvalue $\lambda$ is the same as the number of eigenvectors of **T*** corresponding to eigenvalue $\lambda^*$.

*Proof*: The proof follows immediately from Theorem 11.2.1 on recognizing that $(\mathbf{T} - \lambda\mathbf{I})^* = \mathbf{T}^* - \lambda^*\mathbf{I}$. ∎

Next we establish that the eigenvectors of **T** and **T*** form a pair of biorthogonal sets in the sense of the following theorem.

***Theorem 11.2.3*** Let $\lambda_1$ and $\lambda_2$ be distinct eigenvalues of **T** with $\mathbf{u}_1$ and $\mathbf{u}_2$ as the corresponding eigenvectors. Further, let $\mathbf{v}_1$ and $\mathbf{v}_2$ be the eigenvectors of **T*** corresponding to its eigenvalues $\lambda_1^*$ and $\lambda_2^*$. Then

$$\langle \mathbf{u}_1, \mathbf{v}_2 \rangle = \langle \mathbf{u}_2, \mathbf{v}_1 \rangle = 0$$

*Proof*:

$$\begin{aligned}\lambda_1\langle \mathbf{u}_1, \mathbf{v}_2 \rangle &= \langle \mathbf{T}\mathbf{u}_1, \mathbf{v}_2 \rangle = \langle \mathbf{u}_1, \mathbf{T}^*\mathbf{v}_2 \rangle = \langle \mathbf{u}_1, \lambda_2^*\mathbf{v}_2 \rangle \\ &= \lambda_2\langle \mathbf{u}_1, \mathbf{v}_2 \rangle\end{aligned}$$

or $(\lambda_1 - \lambda_2)\langle \mathbf{u}_1, \mathbf{v}_2 \rangle = 0$. Since $\lambda_1 \neq \lambda_2, \langle \mathbf{u}_1, \mathbf{v}_2 \rangle = 0$. Similarly, we have $\langle \mathbf{u}_2, \mathbf{v}_1 \rangle = 0$. ∎

The relationship between $\mathbf{u}_1$ and $\mathbf{v}_1$ (or $\mathbf{u}_2$ and $\mathbf{v}_2$) is a more complicated issue and is contained in the next theorem, which we state without proof. The following definition, however, takes precedence.

***Definition 11.2.1*** An eigenvalue of **T** is said to be *simple* if it does not occur as a *multiple* root of the characteristic equation.

If we let $\mathbf{D}_\lambda = \boldsymbol{\alpha}\mathbf{W}(\mathbf{U}(a, \lambda)) + \boldsymbol{\beta}\mathbf{W}(\mathbf{U}(b, \lambda))$, then the characteristic equation for **T** is readily seen to be

$$D_\lambda = 0 \tag{11.2.7}$$

where $D_\lambda$ is the determinant of $\mathbf{D}_\lambda$. If $\lambda_0$ is a multiple root of (11.2.7), occurring $n$ times, then we regard the eigenvalue as *repeated* $(n - 1)$ times. This occurs when at $\lambda = \lambda_0$,

$$D_\lambda = \frac{dD_\lambda}{d\lambda} = \dots = \frac{d^{n-1}}{d\lambda^{n-1}}D_\lambda = 0, \qquad \frac{d^n}{d\lambda^n}D_\lambda \neq 0 \tag{11.2.8}$$

For a simple eigenvalue none of the derivatives of $D_\lambda$ vanish at the root. It is left as an exercise (see Exercise 11.2.6) to show that if $\lambda$ is a simple eigenvalue of **T**, $\lambda^*$ is a simple eigenvalue of **T***.

***Theorem 11.2.4*** If the spectrum of **T** is discrete, consisting entirely of simple eigenvalues $\{\lambda_j\}$ and corresponding eigenvectors $\{\mathbf{u}_j\}$, then the set $\{\mathbf{u}_j\}$ is complete in $\mathfrak{L}_2[a, b]$. Likewise, the set of eigenvectors $\{\mathbf{v}_j\}$ of **T*** is also complete in $\mathfrak{L}_2[a, b]$. Moreover, $\forall\, \mathbf{f} \in \mathfrak{L}_2[a, b]$,

$$\mathbf{f} = \sum_{j=1}^{\infty} \langle \mathbf{f}, \mathbf{v}_j \rangle \mathbf{u}_j = \sum_{j=1}^{\infty} \langle \mathbf{f}, \mathbf{u}_j \rangle \mathbf{v}_j \tag{11.2.9}$$

where $\mathbf{u}_j$ and $\mathbf{v}_j$ have been "normalized" to yield $\langle \mathbf{u}_j, \mathbf{v}_j \rangle = 1$.

The proof of Theorem 11.2.4 depends on an approach based on the Green's function representing the inverse of $(\mathbf{T} - \lambda \mathbf{I})^{-1}$ and may be found elsewhere.† Note that it is the completeness property of the sets $\{\mathbf{u}_j\}$ and $\{\mathbf{v}_j\}$ that makes $\langle \mathbf{u}_j, \mathbf{v}_j \rangle \neq 0$ (why?).

We consider below some examples, one in which Theorem 11.2.4 becomes applicable and others where the premises of the theorem are violated.

*Example 1*

Consider $T = -d^2/dx^2$, $D(T) = \{\mathbf{u}, T\mathbf{u} \in \mathfrak{L}_2[0, 1];\, u'(0) = \alpha u(0),\, u'(0) + u'(1) = 0\}$. Clearly, $\mathbf{T} \equiv \{T, D(T)\}$ is non-self-adjoint. To find the eigenvalues, we solve $Tu(x) = \lambda u(x)$ to get

$$u(x) = A \sin \sqrt{\lambda}\, x + \beta \cos \sqrt{\lambda}\, x$$

Substituting the boundary conditions, one obtains the characteristic equation

$$\begin{vmatrix} \sqrt{\lambda} & -\alpha \\ \sqrt{\lambda}(1 + \cos \sqrt{\lambda}) & -\sqrt{\lambda} \sin \sqrt{\lambda} \end{vmatrix} = 0$$

which yields

$$\cos \frac{\sqrt{\lambda}}{2} \left( -\sqrt{\lambda} \sin \frac{\sqrt{\lambda}}{2} + \alpha \cos \frac{\sqrt{\lambda}}{2} \right) = 0$$

so that $\cos(\sqrt{\lambda}/2) = 0$ or $\sqrt{\lambda} \tan(\sqrt{\lambda}/2) = \alpha$. The eigenvalues are all simple and given by the union of the sets $\{(2j-1)^2\pi^2, j = 1, 2, \ldots\}$ and $\{\lambda_k : \sqrt{\lambda_k} \tan(\sqrt{\lambda_k}/2) = \alpha, k = 1, 2, \ldots\}$. The eigenvectors of $\mathbf{T}$ are given by

$$\mathbf{u}_j = \left\{ A_j \left( \sin \sqrt{\lambda_j}\, x + \frac{\sqrt{\lambda_j}}{\alpha} \cos \sqrt{\lambda_j}\, x \right) \right\}$$

$$T^* = -\frac{d^2}{dx^2},$$

$$D(T^*) = \{\mathbf{v}, T^*\mathbf{v} \in \mathfrak{L}_2[0, 1];\, v'(1) = 0',\, v'(0) = \alpha v(0) + \alpha v(1)\}$$

The characteristic equation for $\mathbf{T}^*$ is the same as that for $\mathbf{T}$. Thus the eigenvalues of $\mathbf{T}$ and $\mathbf{T}^*$ are the same, which is expected from Theorem 11.2.2 since they are real. The eigenvector $\mathbf{v}_j$ of $\mathbf{T}^*$ is

$$\mathbf{v}_j = \{\cos \sqrt{\lambda_j}\,(1 - x)\}$$

---

†See, for example, *Theory of Ordinary Differential Equations*, by E. A. Coddington and N. Levinson, McGraw-Hill, New York, 1955, p. 311.

Since the eigenvalues are simple, Theorem 11.2.4 decrees that $\{\mathbf{u}_j\}$ and $\{\mathbf{v}_j\}$ are both complete in $\mathfrak{L}_2[0, 1]$. The constant $A_j$ may be evaluated by requiring that $\langle \mathbf{u}_j, \mathbf{v}_j \rangle = 1$.

$$1 = \langle \mathbf{u}_j, \mathbf{v}_j \rangle = A_j \int_0^1 \left( \sin \sqrt{\lambda_j}\, x + \frac{\sqrt{\lambda_j}}{\alpha} \cos \sqrt{\lambda_j}\, x \right) \cos \sqrt{\lambda_j}(1 - x)\, dx$$

the integration of which is straightforward.

$$A_j = 2\Bigg[\frac{1}{\sqrt{\lambda_j}}\left(\cos \sqrt{\lambda_j} + \frac{\sqrt{\lambda_j}}{\alpha} \sin \sqrt{\lambda_j}\right) \sin^2 \sqrt{\lambda_j}$$

$$+ \frac{1}{\sqrt{\lambda_j}}\left(\frac{\sqrt{\lambda_j}}{\alpha} \cos \sqrt{\lambda_j} - \sin \sqrt{\lambda_j}\right) \sin \sqrt{\lambda_j} \cos \sqrt{\lambda_j}$$

$$+ \frac{\sqrt{\lambda_j}}{\alpha} \cos \sqrt{\lambda_j} + \sin \sqrt{\lambda_j}\Bigg]^{-1}$$

*Example 2*

Although Example 1 conformed to Theorem 11.2.4, it is readily possible to run into situations where the premises of the theorem are violated. Let

$$T = -\frac{d^2}{dx^2}, \qquad D(T) = \{\mathbf{u}, T\mathbf{u} \in \mathfrak{L}_2[0, 1];\, u(0) = 0,\, u'(0) + u'(1) = 0\}$$

The operator $\mathbf{T} = \{T, D(T)\}$ is non-self-adjoint. The eigenvalue problem is given by

$$-u'' = \lambda u, \qquad u(0) = 0, \qquad u'(0) + u'(1) = 0$$

so that $u(x) = A \sin \sqrt{\lambda}\, x + B \cos \sqrt{\lambda}\, x$, which leads to the characteristic equation

$$\begin{vmatrix} 0 & 1 \\ \sqrt{\lambda}\,(1 + \cos \sqrt{\lambda}) & -\sqrt{\lambda} \sin \sqrt{\lambda} \end{vmatrix} = 0$$

or

$$1 + \cos \sqrt{\lambda} = 0$$

The eigenvalues are $\{(2j - 1)^2\pi^2\}$, but each eigenvalue is repeated once because

$$\frac{d}{d\lambda}(1 + \cos \sqrt{\lambda}) = \frac{1}{2\sqrt{\lambda}} \sin \sqrt{\lambda}\,\Big|_{\lambda=\lambda_j} = \frac{1}{2(2j - 1)\pi} \sin (2j - 1)\pi = 0$$

Thus the eigenvalues are not simple, so Theorem 11.2.4 cannot be applied.

*Example 3*

We consider here another situation in which the premises of Theorem 11.2.4 are violated. Define $\mathbf{T} = \{T, D(T)\}$, where

$$T = -\frac{d^2}{dx^2}, \qquad D(T) = \{\mathbf{u}, T\mathbf{u} \in \mathfrak{L}_2[0, 1];\, u(0) = u(1),\, u'(0) + u'(1) = 0\}$$

Again **T** is clearly non-self-adjoint, but the eigenvalue problem

$$-u'' = \lambda u, \qquad u(0) = u(1), \qquad u'(0) + u'(1) = 0$$

leads to

$$\begin{vmatrix} \sin\sqrt{\lambda} & \cos\sqrt{\lambda} - 1 \\ \sqrt{\lambda}\,(1 + \cos\sqrt{\lambda}) & -\sqrt{\lambda}\sin\sqrt{\lambda} \end{vmatrix} = 0$$

from which $\sqrt{\lambda} = 0$ or $\cos^2\sqrt{\lambda} + \sin^2\sqrt{\lambda} = 1$. Thus *every* complex number is an eigenvalue. Hence the spectrum is not discrete, so **T** does not come within the purview of Theorem 11.2.4.

We have seen from Examples 2 and 3 that the behavior of non-self-adjoint operators can become very complicated with rather small variations in the boundary conditions. Some further examples may be encountered in the following exercises.

### EXERCISES

**11.2.1** Show that $\mathbf{T} = \{T, D(T)\}$, where $T = -d^2/dx^2$,

$$D(T) = \{\mathbf{u}, T\mathbf{u} \in \mathcal{L}_2[0, 1]; u(0) = 0, u'(1) = 2u'(0)\}$$

has simple, complex eigenvalues

$$\lambda_j = 4j^2\pi^2 - [\ln(2 \pm \sqrt{3})]^2 + i4j\pi \ln(2 \pm \sqrt{3}) \quad \text{and} \quad \lambda_j^*$$

Identify the biorthogonal expansion of $\mathbf{f} \in \mathcal{L}_2[0, 1]$.

**11.2.2** Let $T = -d^2/dx^2$, $D(T) = \{\mathbf{u}, T\mathbf{u} \in \mathcal{L}_2[0, 1]; u'(0) = u(1), u'(1) = u(0)\}$. Show that the eigenvalues of **T** are simple and given by 1, and $j^2\pi^2, j = 1, 2, \ldots$.

**11.2.3** Consider the operator $\mathbf{T} = \{T, D(T)\}$, where $T \equiv -d^2/dx^2$ and

$$D(T) = \{\mathbf{u}, T\mathbf{u} \in \mathcal{L}_2[0, 1]; u'(0) = u'(1), u'(1) = u(1) - u(0)\}$$

Find the characteristic equation and the eigenvalues. Show that zero is an eigenvalue repeating once with an eigenspace of dimension 2; determine the two eigenvectors.

**11.2.4** Let

$$T = -\frac{d^2}{dx^2} + \beta\frac{d}{dx},$$

$$D(T) = \{\mathbf{u}, T\mathbf{u} \in \mathcal{L}_2[0, 1]; u'(1) = 0, u'(0) = \alpha u(0) - \beta u(1)\}$$

Determine the characteristic equation. Note that this operator arises in the description of a tubular reactor with recycle if $\alpha < \beta$.†

**11.2.5** Furnish the details of the proof of Theorem 11.2.2.

---

†See, for example, W. R. Schmeal and N. R. Amundson, *AIChE J.*, 1966, Vol. 12, 1202–1211.

**11.2.6** Consider the general non-self-adjoint operator **T** of Section 11.2. Suppose that $\lambda$ is a repeated eigenvalue of **T**. Show that $\lambda^*$ is a repeated eigenvalue of **T***.

**11.2.7** Consider

$$T = -\frac{d^2}{dx^2}, \quad D(T) = \{\mathbf{u}, T\mathbf{u} \in \mathcal{L}_2[0, 1]; u(0) + u(1) = 0, u'(0) = u'(1)\}$$

Determine the spectrum of **T**.

**11.2.8** Show that the operator $\mathbf{T} = \{T, D(T)\}$ with $T = -d^2/dx^2$ and $D(T) = \{\mathbf{u}, T\mathbf{u} \in \mathcal{L}_2[0, 1]; u(0) = u'(0) = 0\}$ has no eigenvalues.

Although Theorem 11.2.4 is a useful result, we have seen that one rather readily runs into situations where the theorem becomes inapplicable. One such instance is provided by Example 2, in which each eigenvalue of the operator concerned was found to repeat once. On solving for the eigenvector corresponding to a given eigenvalue, one finds that only one eigenvector exists; the set of eigenvectors must then be clearly incomplete. In such circumstances it is necessary to admit the concept of a *root vector*, which is defined next.

***Definition 11.2.2*** Let $\lambda$ be an eigenvalue of operator **T**. A vector $\mathbf{u} \in D(T)$ is called a *root vector*† of **T** if there exists a positive integer $n$ for which $(\mathbf{T} - \lambda\mathbf{I})^n\mathbf{u} = \mathbf{0}$. We refer to $n$ as the order of the root vector.

Since we are dealing with differential operators it is necessary to identify their operations and domains. The operator $(\mathbf{T} - \lambda\mathbf{I})^n$ is defined by $\{(T - \lambda)^n, D[(T - \lambda)^n]\}$, where

$$D[(T - \lambda)^n] = \{T^j\mathbf{u} \in D(T), j = 0, 1, 2, \ldots, n - 1; T^n\mathbf{u} \in \mathcal{L}_2[0, 1]\}$$

A root vector of a given order is also one of any *higher* order. Thus an eigenvector is also a root vector of any order. Indeed, the converse is not true. The quest for root vectors arises when the concerned eigenvalue is not simple and the dimension of $N(\mathbf{T} - \lambda\mathbf{I})$ is less than the multiplicity of the eigenvalue (see Exercise 11.2.3 for a situation where the multiplicity of the eigenvalue and the dimension of the eigenspace are the same).

It is not difficult to establish that if $\lambda$ is a repeated eigenvalue of **T**, then $\lambda^*$ is a repeated eigenvalue of **T*** (see Exercise 11.2.6). Furthermore, the application of Theorem 11.2.1 to the null spaces of $(\mathbf{T} - \lambda\mathbf{I})^n$ and $(\mathbf{T}^* - \lambda^*\mathbf{I})^n$ yields the result that if **T** has a root vector of order $n$ corresponding to eigenvalue $\lambda$, **T*** also has a root vector of order $n$ corresponding to eigenvalue $\lambda^*$.

---

†The term "generalized eigenvector" is also used in the literature in place of "root vector," but we prefer the latter.

It turns out that the root vectors can be obtained in a straightforward way. We consider first the root vector of order 2 from which the generalization to higher-order root vectors, required for higher-order differential operators, is self-suggestive. Let $u_1(x, \lambda)$ and $u_2(x, \lambda)$ be the linearly independent solutions of $(T - \lambda)u = 0$. It is readily seen that

$$(T - \lambda)\frac{\partial}{\partial \lambda}u_i(x, \lambda) = u_i(x, \lambda), \qquad i = 1, 2 \tag{11.2.10}$$

Operating on both sides with $(T - \lambda)$, one has

$$(T - \lambda)^2\frac{\partial}{\partial \lambda}u_i(x, \lambda) = 0, \qquad i = 1, 2 \tag{11.2.11}$$

Note that $u_1$ and $u_2$ also satisfy (11.2.10), but $\partial u_1/\partial\lambda$ and $\partial u_2/\partial\lambda$ are linearly independent of $u_1$ and $u_2$, for let

$$0 = k_1u_1 + k_2u_2 + k_3\frac{\partial u_1}{\partial \lambda} + k_4\frac{\partial u_2}{\partial \lambda} \tag{11.2.12}$$

Operating with $T$ on either side of the above and using (11.2.10), one obtains

$$0 = (k_1\lambda + k_3)u_1 + (k_2\lambda + k_4)u_2 + k_3\lambda\frac{\partial u_1}{\partial \lambda} + k_4\lambda\frac{\partial u_2}{\partial \lambda} \tag{11.2.13}$$

Multiplying (11.2.12) by $\lambda$ and subtracting from (11.2.13) yields

$$k_3u_1 + k_4u_2 = 0$$

so that $k_3 = k_4 = 0$, which converts (11.2.12) into

$$k_1u_1 + k_2u_2 = 0$$

yielding $k_1 = k_2 = 0$. Thus $u_1, u_2, \partial u_1/\partial\lambda, \partial u_2/\partial\lambda$ are linearly independent. Let $\lambda_k$ be the $k$th eigenvalue of $\mathbf{T}$ and suppose that we seek a root vector of order 2 corresponding to $\lambda_k$. Since $u_i$ and $\partial u_i/\partial\lambda$, $i = 1, 2$, satisfy (11.2.11) for all $\lambda$, we enquire if a linear combination of them can be found to lie in $D[(T - \lambda)^2]$ (see Definition 11.2.2). This is readily done as follows.

We let $u(x, \lambda) \equiv \mathbf{U}^t(x, \lambda)\mathbf{c}$, satisfying $(T - \lambda)u = 0$ and *one* of the boundary conditions in (11.1.3), say the first one. If $D_{ij}(\lambda)$ represents the $ij$th coefficient of $D_\lambda$, then we must have

$$D_{11}(\lambda)c_1(\lambda) + D_{12}(\lambda)c_2(\lambda) = 0 \tag{11.2.14}$$

for all $\lambda$. Let $D_{11}(\lambda) \neq 0$.† By setting $c_2(\lambda) \equiv 1$,

$$c_1(\lambda) = -\frac{D_{12}(\lambda)}{D_{11}(\lambda)} \tag{11.2.15}$$

---

†Note that the implied assumption here is that when $\lambda = \lambda_k$, at least one of the boundary conditions of $\mathbf{T}$ will lead to a nonzero element of $\mathbf{D}_\lambda$. This means that $\mathbf{D}_\lambda$ has rank 1 when $\lambda = \lambda_k$, so that only one eigenvector exists.

Equations (11.2.14) and (11.2.15) obviously hold at $\lambda = \lambda_k$. Further, we also have

$$D_{21}(\lambda_k)c_1(\lambda_k) + D_{22}(\lambda_k)c_2(\lambda_k) = 0$$

which represents the *second* boundary condition to be satisfied by the eigenvector $\{u(x, \lambda_k)\}$. Now we consider as a root vector

$$\frac{\partial}{\partial \lambda}u(x, \lambda) = \frac{\partial}{\partial \lambda}[\mathbf{U}^t(x, \lambda)\mathbf{c}(\lambda)] \tag{11.2.16}$$

since it clearly satisfies (11.2.11). Further, from (11.2.10) we obtain

$$\begin{aligned} T\frac{\partial}{\partial \lambda}u(x, \lambda) &= \lambda\frac{\partial}{\partial \lambda}u(x, \lambda) + u(x, \lambda) \\ &= \frac{\partial}{\partial \lambda}[\lambda \mathbf{U}^t(x, \lambda)\mathbf{c}(\lambda)] \end{aligned} \tag{11.2.17}$$

In order that $\partial \mathbf{u}/\partial \lambda$ is a root vector we must have $[T(\partial \mathbf{u}/\partial \lambda)]_{\lambda=\lambda_k} \in D(T)$, which obtains from (11.2.17)

$$\frac{d}{d\lambda}[\mathbf{D}_\lambda \mathbf{c}(\lambda)]_{\lambda=\lambda_k} = \mathbf{D}'_{\lambda_k}\mathbf{c}(\lambda_k) + \mathbf{D}_{\lambda_k}\mathbf{c}'(\lambda_k) = 0 \tag{11.2.18}$$

where the prime is used to denote differentiation with respect to the argument $\lambda$. The first component of the vector on the left-hand side of (11.2.18) is

$$D'_{11}(\lambda_k)c_1(\lambda_k) + D'_{12}(\lambda_k)c_2(\lambda_k) + D_{11}(\lambda_k)c'_1(\lambda_k) + D'_{12}(\lambda_k)c_2(\lambda_k)$$

Substituting for $\mathbf{c}$ and $\mathbf{c}'$ from (11.2.15), the expression above becomes

$$\frac{1}{D_{11}(\lambda_k)}[-D'_{11}(\lambda_k)D_{12}(\lambda_k) + D'_{12}(\lambda_k)D_{11}(\lambda_k) + D'_{11}(\lambda_k)D_{12}(\lambda_k) - D'_{12}(\lambda_k)D_{11}(\lambda_k)]$$

which is clearly zero, so that nothing new emerges. However, the second component of (11.2.18) yields

$$\frac{1}{D_{11}(\lambda_k)}[-D'_{21}(\lambda_k)D_{12}(\lambda_k) + D'_{22}(\lambda_k)D_{11}(\lambda_k) + D'_{11}(\lambda_k)D_{22}(\lambda_k) - D'_{12}(\lambda_k)D'_{21}(\lambda_k)] = 0$$

leading to

$$D'(\lambda_k) = 0 \tag{11.2.19}$$

so that we reach the important conclusion that for (11.2.16) to be a root vector the eigenvalue $\lambda_k$, as required by (11.2.19), must repeat once. We will not pause to generalize the result above for higher-order differential operators because the possibilities are too numerous for convenient discussion.

The root vectors and eigenvectors of an operator $\mathbf{T}$ are clearly a linearly independent set. The main issue then facing us is in regard to whether or not the set is complete in the Hilbert space of interest. It turns out that this issue is a very complicated one and has been the subject of extensive investigation in the mathematical literature.† Several constraints are available under which completeness of the root vectors is guaranteed. An operator is said to be *spectral* when its root vectors are complete, so that the issue just identified is one of identifying when an operator is spectral.

For differential operators to be spectral, the boundary conditions must play a crucial role. Thus certain *regularity conditions* have been obtained for the boundary conditions of the operation $i^{-n}(d^n/dx^n)$, so that the resulting operator is spectral.‡ It is useful to recount here the results for the operation $-d^2/dx^2$ in particular. These are included in the following theorem, which we state without proof.

***Theorem 11.2.5*** Let $T$ be the operation $-d^2/dx^2$, where $0 < x < 1$. The operator $\mathbf{T} = \{T, D(T)\}$ is spectral if $D(T)$ consists of functions $\mathbf{u}$ such that $\mathbf{u}, T\mathbf{u} \in \mathfrak{L}_2[0, 1]$ and *any* of the following pairs of boundary conditions:

(i) $$u'(0) + \alpha_1 u(0) + \beta_1 u(1) = 0$$
$$u'(1) + \alpha_2 u(0) + \beta_2 u(1) = 0.$$

(ii) $$\left.\begin{aligned} u'(0) + ku'(1) + \alpha_1 u(0) + \beta_1 u(1) &= 0 \\ u(0) + k'u(1) &= 0 \end{aligned}\right\} \quad \text{where } k + k' \neq 0.$$

The proof of this theorem depends on the fact that the boundary conditions above satisfy the required regularity conditions. It is interesting to examine the examples considered in this section in the light of the foregoing theorem. Example 1 is readily seen to feature a boundary condition of type (i) in Theorem 11.2.5 so that $\mathbf{T}$ is a spectral operator. Example 2 also yields a spectral operator since the boundary conditions therein are of type (ii) in Theorem 11.2.5. Clearly, Theorem 11.2.5 is able to accomplish what Theorem 11.2.4 does not since Example 2 does not satisfy the conditions required by the latter theorem. Finally, in Example 3 the boundary conditions are found to be "irregular," being of the kind forbidden by type (ii) in Theorem 11.2.5, so that $\mathbf{T}$ does not emerge as spectral from the theorem.

The extension of results obtained for the operators derived from the expression $i^{-n}(d^n/dx^n)$ has been accomplished for more general operators based on perturbation methods.§ The discussion of these results is outside the scope of this book.

---

†See, for example, *Introduction to Theory of Non-self-adjoint Operators*, by I. C. Gohberg and M. G. Krein, American Mathematical Society, Providence, R. I., 1969.

‡See *Linear Operators, Part III*, by N. Dunford and J. T. Schwartz, Wiley-Interscience, New York, 1971, pp. 2310–2350. The regularity conditions are applied after "renormalizing" the boundary conditions in a prescribed way.

§See *Linear Operators, Part III*, by N. Dunford and J. T. Schwartz, Wiley-Interscience, New York, 1971, pp. 2290–2374. See Theorem 16 in particular.

We turn our attention to the details concerning expansions of functions in terms of the root vectors and solutions of boundary value problems featuring spectral non-self-adjoint operators with discrete spectra. The next theorem addresses this issue.

***Theorem 11.2.6*** Let $\mathbf{T}$ be a non-self-adjoint, spectral operator with a discrete spectrum $\{\lambda_j\}$ such that $N(\mathbf{T} - \lambda_j \mathbf{I})$ has dimension 1 and $\lambda_j$ is repeated $(\nu_j - 1)$ times. The root vectors corresponding to eigenvalue $\lambda_j$, given by $d^r\mathbf{u}(\lambda_j)/d\lambda^r$, $r = 0, 1, 2, \ldots, \nu_j - 1$, are denoted $\mathbf{u}_j^{(r)}$ ($\mathbf{u}_j^{(0)}$ being the eigenvector). The root vectors of $\mathbf{T}^*$ corresponding to eigenvalue $\lambda_j^*$ are $\mathbf{v}_j^{(r)}$, $r = 0, 1, 2, \ldots, \nu_j - 1$. The following properties hold.

(i) For each $\lambda_j$,

$$\langle \mathbf{u}_j^{(r)}, \mathbf{v}_j^{(s)} \rangle = \langle \mathbf{u}_j^{(s)}, \mathbf{v}_j^{(r)} \rangle = 0; \qquad r = 0, 1, 2, \ldots, \nu_j - 2; \quad s = 0, 1, 2, \ldots, \nu_j - r - 2$$

(ii) For $\lambda_j \neq \lambda_k$,

$$\langle \mathbf{u}_j^{(r)}, \mathbf{v}_k^{(s)} \rangle = 0; \qquad r = 0, 1, 2, \ldots, \nu_j - 1; \quad s = 0, 1, 2, \ldots, \nu_k - 1$$

(iii) For each $\mathbf{f} \in \mathcal{H}$, the expansion in terms of root vectors of $\mathbf{T}$ is given by

$$\mathbf{f} = \sum_{j=1}^{\infty} \sum_{r=0}^{\nu_j - 1} \alpha_{j,r} \mathbf{u}_j^{(r)} \tag{11.2.20}$$

where $\alpha_{j,r}$ are obtained for each $j$ iteratively as

$$\left.\begin{aligned} \alpha_{j,\nu_j - r} &= \frac{\langle \mathbf{f}, \mathbf{v}_j^{(0)} \rangle}{\langle \mathbf{u}_j^{(\nu_j - 1)}, \mathbf{v}_j^{(0)} \rangle} \\ \alpha_{j,\nu_j - r} &= \frac{\langle \mathbf{f}, \mathbf{v}_j^{(r-1)} \rangle - \sum_{s=1}^{r-1} \alpha_{j,\nu_j - s} \langle \mathbf{u}_j^{(\nu_j - s)}, \mathbf{v}_j^{(r-1)} \rangle}{\langle \mathbf{u}_j^{(\nu_j - r)}, \mathbf{v}_j^{(r-1)} \rangle} \end{aligned}\right\} \quad r = 2, \ldots, \nu_j$$

*Proof*: To prove (i), we note that for $r, s = 0, 1, 2, \ldots, \nu_j - 1$,

$$\mathbf{T}\mathbf{u}_j^{(r)} = \lambda_j \mathbf{u}_j^{(r)} + r\mathbf{u}_j^{(r-1)}$$
$$\mathbf{T}^*\mathbf{v}_j^{(s)} = \lambda_j^* \mathbf{v}_j^{(s)} + s\mathbf{v}_j^{(s-1)}$$

which is easily established by induction. Then

$$0 = \langle \mathbf{T}\mathbf{u}_j^{(r)}, \mathbf{v}_j^{(s)} \rangle - \langle \mathbf{u}_j^{(r)}, \mathbf{T}^*\mathbf{v}_j^{(s)} \rangle = r\langle \mathbf{u}_j^{(r-1)}, \mathbf{v}_j^{(s)} \rangle - s\langle \mathbf{u}_j^{(r)}, \mathbf{v}_j^{(s-1)} \rangle \tag{11.2.21}$$

Setting $r = 0$ in the above, one obtains

$$\langle \mathbf{u}_j^{(0)}, \mathbf{v}_j^{(s-1)} \rangle = 0, \qquad s = 1, 2, \ldots, \nu_j - 1$$

By progressively setting $r - 1, 2, \ldots, \nu_j - 2$ in (11.2.21) and using the resulting orthogonality relations successively, (i) is established.

To prove (ii), it is readily shown that for $r, s = 0, 1, 2, \ldots, \nu_j - 1$,

$$0 = \langle \mathbf{T}\mathbf{u}_j^{(r)}, \mathbf{v}_k^{(s)} \rangle - \langle \mathbf{u}_j^{(r)}, \mathbf{T}^*\mathbf{v}_j^{(s)} \rangle$$
$$= (\lambda_j - \lambda_k)\langle \mathbf{u}_j^{(r)}, \mathbf{v}_k^{(s)} \rangle + r\langle \mathbf{u}_j^{(r-1)}, \mathbf{v}_k^{(s)} \rangle - s\langle \mathbf{u}_j^{(r)}, \mathbf{v}_k^{(s-1)} \rangle \qquad (11.2.22)$$

If $r = s = 0$ in (11.2.22), we have $\langle \mathbf{u}_j^{(0)}, \mathbf{v}_k^{(0)} \rangle = 0$. Next set $r = 1, s = 0$, which yields $\langle \mathbf{u}_j^{(1)}, \mathbf{v}_k^{(0)} \rangle = 0$, and so on. Clearly, such progressive substitution will yield (ii).

Using (i) and (ii) in (11.2.20), it is easy to prove (iii). ∎

It also follows from Eq. (11.2.21) by setting $r = s$ that

$$\langle \mathbf{u}_j^{(r-1)}, \mathbf{v}_j^{(r)} \rangle = \langle \mathbf{u}_j^{(r)}, \mathbf{v}_j^{(r-1)} \rangle$$

which is noteworthy in computations. The expansion formula (11.2.20) is extremely important in the solution of boundary value problems involving the operator **T**. For example, consider the solution of the initial-boundary value problem

$$-\frac{d}{dt}\mathbf{u}(t) = \mathbf{T}\mathbf{u}(t), \qquad \mathbf{u}(0) = \mathbf{u}_0 \qquad (11.2.23)$$

where **T** is a spectral, non-self-adjoint differential operator by Theorem 11.2.6. The strategy of solution is to express the solution $\mathbf{u}(t)$ in terms of formula (11.2.20) rewritten as

$$\mathbf{u}(t) = \sum_{j=1}^{\infty} \sum_{r=0}^{\nu_j - 1} \alpha_{j,r}(t)\mathbf{u}_j^{(r)}$$

where $\alpha_{j,r}(t)$ is given by the same formulas as those appearing below (11.2.20) with **f** replaced by $\mathbf{u}(t)$. Thus the solution is obtained when for each $\lambda_j$, the inner products $\langle \mathbf{u}(t), \mathbf{v}_j^{(r)} \rangle$, $r = 0, 1, 2, \ldots, \nu_j - 1$, are identified. To accomplish this we form the inner product of (11.2.23) with $\mathbf{v}_j^{(r)}$ to obtain

$$\left.\begin{aligned} -\frac{d}{dt}\langle \mathbf{u}(t), \mathbf{v}_j^{(r)} \rangle &= \langle T\mathbf{u}(t), \mathbf{v}_j^{(r)} \rangle \\ \langle \mathbf{u}(0), \mathbf{v}_j^{(r)} \rangle &= \langle \mathbf{u}_0, \mathbf{v}_j^{(r)} \rangle \end{aligned}\right\} \quad r = 0, 1, 2, \ldots, \nu_j - 1 \qquad (11.2.24)$$

the first of which becomes

$$-\frac{d}{dt}\langle \mathbf{u}(t), \mathbf{v}_j^{(r)} \rangle = \lambda_j\langle \mathbf{u}(t), \mathbf{v}_j^{(r)} \rangle + r\langle \mathbf{u}(t), \mathbf{v}_j^{(r-1)} \rangle$$

so that

$$\langle \mathbf{u}(t), \mathbf{v}_j^{(r)} \rangle = \langle \mathbf{u}_0, \mathbf{v}_j^{(r)} \rangle e^{-\lambda_j t} - r\int_0^t e^{-\lambda_j(t-t')}\langle \mathbf{u}(t'), \mathbf{v}_j^{(r-1)} \rangle \, dt' \qquad (11.2.25)$$

If Eq. (11.2.25) is successively applied using

$$\langle \mathbf{u}(t), \mathbf{v}_j^{(0)} \rangle = \langle \mathbf{u}_0, \mathbf{v}_j^{(0)} \rangle e^{-\lambda_j t}$$

one obtains

$$\langle \mathbf{u}(t), \mathbf{v}_j^{(r)} \rangle = e^{-\lambda_j t} \sum_{s=0}^{r} \binom{r}{s} \langle \mathbf{u}_0, \mathbf{v}_j^{(r-s)} \rangle (-t)^s, \qquad r = 0, 1, 2, \ldots, \nu_j - 1 \tag{11.2.26}$$

where $\binom{r}{s} \equiv r!/(r-s)!\,s!$. The foregoing inner products obtain the solution $\mathbf{u}(t)$ via the expansion formula (11.2.20). Since the eigenvalues are eventually unbounded, expressions such as (11.2.26) will become unbounded unless the eigenvalues have positive real parts. Notice how the solution to the problem (11.2.23) is a combination of nonexponential functions of time. This is characteristic of non-self-adjoint situations in which repeated eigenvalues are encountered. Self-adjoint problems, however, cannot yield such solutions even when the eigenvalues are repeated. We discuss an example below in the application of problems of the type just considered to a situation in one-dimensional heat conduction.

*Example. Controlled Heating of a Slab*†

A slab of finite thickness (along the $x$-direction) and infinite in the other two directions $y$ and $z$ is initially at a temperature which is uniform with respect to $y$ and $z$. The slab is to be heated from both ends in such a way that the heating rate at one end is to be adjusted depending on that at the other end whose temperature is specified. The dimensionless unsteady-state energy equation is then given by

$$\frac{\partial}{\partial t} u(x, t) = \frac{\partial^2}{\partial x^2} u(x, t), \qquad 0 < x < 1 \tag{11.2.27}$$

The temperature at $x = 0$ is specified as

$$u(0, t) = g(t) \tag{11.2.28}$$

where $g(t)$ is a known function while the boundary condition at the other end ($x = 1$) is given by

$$\frac{\partial}{\partial x} u(0, t) + \frac{\partial}{\partial x} u(1, t) = h(t) \tag{11.2.29}$$

The initial temperature distribution is expressed by

$$u(x, 0) = f(x) \tag{11.2.30}$$

In defining the operator, one drops the inhomogeneities in (11.2.28) and (11.2.29) and readily recognizes the operator **T** to be that discussed in Example

†See also "A Non-self-adjoint Problem in Heat Conduction," by D. Ramkirshna and N. R. Amundson, *J. Heat Transfer*, **104**, 185–190, 1982.

2 of this section. Its spectral analysis shows that $\mathbf{T}$ has eigenvalues $\lambda_j = (2j - 1)^2\pi^2$, $j = 1, 2, \ldots$, each repeated once. In the discussion following Theorem 11.2.5, the operator $\mathbf{T}$ is inferred to be spectral. We have seen that the eigenvector corresponding to $\lambda_j$ is given by

$$\mathbf{u}_j^{(0)} = \sin\sqrt{\lambda_j}\, x$$

The root vector $\mathbf{u}_j^{(1)}$ is given by

$$\mathbf{u}_j^{(1)} = \left[\frac{d}{d\lambda}\sin\sqrt{\lambda}\, x\right]_{\lambda=\lambda_j} = \frac{x}{2\sqrt{\lambda_j}}\cos\sqrt{\lambda_j}\, x$$

The adjoint operator $\mathbf{T}^*$ is readily seen to be given by

$$T^* = -\frac{d^2}{dx^2}, \qquad D(T^*) = \{\mathbf{v}, T\mathbf{v} \in \mathfrak{L}_2[0, 1], v'(1) = 0, v(0) + v(1) = 0\}$$

which has root vectors

$$\mathbf{v}_j^{(0)} = \{\cos\sqrt{\lambda_j}(1 - x)\}, \qquad \mathbf{v}_j^{(1)} = \left\{-\frac{1-x}{2\sqrt{\lambda_j}}\sin\sqrt{\lambda_j}(1 - x)\right\}$$

The biorthogonal expansion of an arbitrary vector $\mathbf{f}$ of $\mathfrak{L}_2[0, 1]$ is then given by (11.2.20) with $\nu_j = 2$ for each $j$.

The initial-boundary value problem is given by

$$-\frac{d}{dt}\mathbf{u}(t) = T\mathbf{u}(t), \qquad \mathbf{u}(0) = \mathbf{f}$$

where no boldface notation appears on $T$ because $\mathbf{u}(t) \notin D(T)$. However,

$$\langle T\mathbf{u}(t), \mathbf{v}_j^{(r)}\rangle - \langle\mathbf{u}(t), \mathbf{T}^*\mathbf{v}_j^{(r)}\rangle = v_j^{(r)}(0)h(t) - v_j^{(r)\prime}(0)g(t), \qquad r = 0, 1$$

so that

$$-\frac{d}{dt}\langle\mathbf{u}(t), \mathbf{v}_j^{(r)}\rangle = \lambda_j\langle\mathbf{u}(t), \mathbf{v}_j^{(r)} + r\langle\mathbf{u}(t), \mathbf{v}_j^{(r-1)}\rangle + v_j^{(r)}(0)h(t) - v_j^{(r)\prime}(0)g(t), \qquad r = 0, 1$$

which are solved readily to get

$$\langle\mathbf{u}(t), \mathbf{v}_j^{(0)}\rangle = \langle\mathbf{f}, \mathbf{v}_j^{(0)}\rangle e^{-\lambda_j t} - \int_0^t [v_j^{(0)}(0)h(t') - v_j^{(0)\prime}(0)g(t')]e^{-\lambda_j(t-t')}\, dt$$

$$\langle\mathbf{u}(t), \mathbf{v}_j^{(1)}\rangle = \langle\mathbf{f}, \mathbf{v}_j^{(1)}\rangle e^{-\lambda_j t} - \int_0^t e^{-\lambda_j(t-t')}\Big\{[v_j^{(1)}(0)h(t') - v_j^{(1)\prime}(0)g(t')] + \langle\mathbf{f}, \mathbf{v}_j^{(0)}\rangle e^{-\lambda_j t'} - \int_0^{t'} [v_j^{(0)}(0)h(t'') - v_j^{(0)\prime}(0)g(t'')]e^{-\lambda_j(t'-t'')}\, dt''\Big\}\, dt'$$

The inner products above, on substitution into the expansion formula (11.2.20), yield the solution $u(x, t)$ to the initial boundary value problem.

**EXERCISES**

**11.2.9** Let $\mathbf{T}$ be a spectral operator with a system of root vectors $\{\mathbf{u}_j^{(r)}, r = 0, 1, 2, \ldots, \nu_j - 1\}_{j=1}^{\infty}$. Show that the operator $\mathbf{e}^{\mathbf{T}}$ may be defined by

$$\mathbf{e}^{\mathbf{T}} = \sum_{j=1}^{\infty} \sum_{r=0}^{\nu_j - 1} \mathbf{u}_j^{(r)} l_{j,r}$$

where $l_{j,r}$ *are linear functionals* defined iteratively by

$$l_{j,\nu_j - 1} = \frac{e^{\lambda_j}\langle \cdot, \mathbf{v}_j^{(0)} \rangle}{\langle \mathbf{u}_j^{(\nu_j - 1)}, \mathbf{v}_j^{(0)} \rangle}$$

$$l_{j,\nu_j - r} = \frac{d^{r-1}}{d\lambda^{r-1}} [e^{\lambda} \langle \cdot, \mathbf{v}(\lambda) \rangle]_{\lambda = \lambda_j} - \sum_{s=1}^{r-1} l_{j,\nu_j - s} \frac{\langle \mathbf{u}_j^{(\nu_j - s)}, \mathbf{v}_j^{(r-1)} \rangle}{\langle \mathbf{u}_j^{(\nu_j - r)}, \mathbf{v}_j^{(r-1)} \rangle}, \qquad r = 2, \ldots, \nu_j$$

in which $\mathbf{v}(\lambda)$ satisfies $T^*\mathbf{v} = \lambda^*\mathbf{v}$ and all but *one* of the boundary conditions.

**11.2.10** Extend the result of Exercise 11.2.9 to define a more general function $\mathbf{f}(\mathbf{T})$.

## 11.3 Concluding Remarks

This chapter has provided a sketchy treatment of differential non-self-adjoint operators. It has relied on the availability of theorems for the completeness of the root vectors under conditions not always known to be satisfied but are nevertheless useful in relatively simple situations. Where the operator is spectral, that is, its root vectors are complete in the Hilbert space, Theorem 11.2.6 represents the crucial result for the expansion of arbitrary vectors in terms of the system of root vectors and hence in the solution of boundary value problems featuring the operator. The expansion formula (11.2.20) is not represented in the form of a resolution of the identity operator as in the treatment of self-adjoint operators in Chapter 7. This is indeed possible through the relation

$$\mathbf{I} = \frac{1}{2\pi i} \oint_C (\mathbf{T} - \lambda \mathbf{I})^{-1} \, d\lambda$$

where $C$ is a curve in the complex plane enclosing all the singularities of $(\mathbf{T} - \lambda\mathbf{I})^{-1}$; for a spectral operator with a discrete spectrum the singularities are at most multiple poles. The operator $(\mathbf{T} - \lambda\mathbf{I})^{-1}$ is represented by the Green's function, whose development was the subject of Section 9.2.1. The methods of contour integration would then lead to the result

$$\mathbf{I} = \sum_{j=1}^{\infty} \frac{1}{(\nu_j - 1)!} \frac{d^{\nu_j - 1}}{d\lambda^{\nu_j - 1}} [(\lambda - \lambda_j)^{\nu_j} (\mathbf{T} - \lambda\mathbf{I})^{-1}]_{\lambda = \lambda_j} \tag{11.3.1}$$

which obtains via the theorem of residues applied to the pole $\lambda_j$ of order $\nu_j$. While the foregoing resolution of the identity provides an interesting perspective and can be shown to yield the biorthogonal expansion (11.2.20), the latter is a considerably more usable relation in the solution of problems. Note that (11.3.1) can be extended to define a function **f**(**T**) of operator **T** by

$$\mathbf{f(T)} = \sum_{j=1}^{\infty} \frac{1}{(\nu_j - 1)!} \frac{d^{\nu_j - 1}}{d\lambda^{\nu_j - 1}} [f(\lambda)(\lambda - \lambda_j)^{\nu_j}(\mathbf{T} - \lambda\mathbf{I})^{-1}]_{\lambda=\lambda_j}$$

The implication of the above for the more conveniently used biorthogonal expansion in terms of root vectors is contained in Exercises 11.2.8 and 11.2.9.

Although the theory of non-self-adjoint operators is far from complete, considerably more is known about them than what we have been able to cover in this book. Some further reading as listed below is recommended in this regard.

## FURTHER READING

The contributions to non-self-adjoint operator theory from the Russian school are comprehensively covered in

*Introduction to the Theory of Non-self-adjoint Operators*, by L. C. Gohberg and M. G. Krein, American Mathematical Society. Providence, R.I., 1969.

Another excellent treatment of non-self-adjoint operators is contained in

*Linear Operators, Part III: Spectral Operators*, by N. Dunford and J. T. Schwartz, Wiley-Interscience, New York, 1971.

Both these books provide a healthy coverage of the literature. In general, the material is not easily understood by engineers with the usual background of mathematics.

# *Appendix*

## A.0 Foreword

The objective of this appendix is to furnish details which, although not logically essential for the development in many sections of Chapter 9, have great calculational importance. We are mainly concerned with constructing solutions to homogeneous differential equations.

## A.1 Introduction

We are concerned with generating the linearly independent solutions of ordinary linear differential equations encountered in Chapter 9, and, in particular, Sections 9.2 to 9.5. Recall the definitions of ordinary and regular singular points of a differential expression of the type (9.2.1) (p. 283). More generally, for an $n$th-order differential expression, a point $x_0$ in the interval of the independent variable $x$ is an *ordinary point* if it is possible to write the expression in the form

$$T \equiv \frac{d^n}{dx^n} + \sum_{r=0}^{n-1} F_r(x) \frac{d^r}{dx^r} \tag{A.1.1}$$

and each function $F_r(x)$ may be expanded as

$$F_r(x) = \sum_{j=0}^{\infty} f_{r,j}(x - x_0)^j, \qquad r = 0, 1, \ldots, n-1 \tag{A.1.2}$$

for which a common radius of convergence may be identified.

The point $x_0$ is a *regular singular point* of a differential expression $T$ if $T$ may be expressible as

$$T \equiv (x - x_0)^n \frac{d^n}{dx^n} + \sum_{r=0}^{n-1} (x - x_0)^r F_r(x) \frac{d^r}{dx^r} \tag{A.1.3}$$

where again (A.1.2) is assumed to hold.

In succeeding sections it will be our interest to construct the $n$ linearly independent solutions of

$$Ty = 0 \tag{A.1.4}$$

(which will obviously include the case of $Ty = -\lambda y$) near $x_0$ for the cases of both when $x_0$ is an ordinary point of $T$ and a regular singular point of $T$. Without loss of generality we assume $x_0$ to be 0 in what follows.

## A.2 MacLaurin's Series Solution

In this section we assume that $x = 0$ is an ordinary point of $T$ expressed in the form (A.1.1). The solution of (A.1.4) is expressed as

$$y(x) = \sum_{k=0}^{\infty} a_k x^k \tag{A.2.1}$$

where the coefficients $\{a_k\}$ must be determined. The strategy calls for substituting (A.2.1) into (A.1.4). We observe that

$$\frac{d^r}{dx^r} y = \sum_{k=0}^{\infty} \frac{(k+r)!}{k!} a_{k+r} x^k, \qquad r = 0, 1, 2, \ldots, n \tag{A.2.2}$$

Further observe that

$$\begin{aligned} F_r(x) \frac{d^r}{dx^r} y &= \sum_{j=0}^{\infty} f_{r,j} x^j \sum_{k=0}^{\infty} \frac{(k+r)!}{k!} x^k a_{k+r} \\ &= \sum_{k=0}^{\infty} x^k \sum_{j=0}^{k} \frac{(k+r-j)!}{k!} a_{k+r-j} f_{r,j} \end{aligned} \tag{A.2.3}$$

The use of (A.2.2) and (A.2.3) in (A.1.4) with $T$ given by (A.1.1) leads to

$$\sum_{k=0}^{\infty} x^k \left[ \frac{(k+n)!}{k!} a_{k+n} + \sum_{r=0}^{n-1} \sum_{j=0}^{k} \frac{(k+r-j)!}{k!} a_{k+r-j} f_{r,j} \right] = 0$$

for each $x$ in the interval of convergence. Thus we must have

$$\frac{(k+n)!}{k!} a_{k+n} + \sum_{r=0}^{n-1} \sum_{j=0}^{k} \frac{(k+r-j)!}{k!} a_{k+r-j} f_{r,j} = 0, \qquad k = 0, 1, 2, \ldots \tag{A.2.4}$$

For $k = 0$, (A.2.4) becomes

$$a_n = -\sum_{r=0}^{n-1} a_r f_{r,0} \tag{A.2.5}$$

The first $n$ coefficients $a_0, a_1, a_2, \ldots, a_{n-1}$ in the solution (A.2.1) are chosen arbitrarily in terms of which the subsequent coefficients $a_n, a_{n+1}, \ldots$ are computed through (A.2.4). If the $p$th independent solution $y_p(x)$ is expressed as

$$y_p(x) = \sum_{k=0}^{\infty} a_k^{(p)} x^k \tag{A.2.6}$$

the coefficients $a_0^{(p)}, a_1^{(p)}, \ldots, a_{n-1}^{(p)}$ may now be chosen as

$$a_k^{(p)} = \delta_{kp}, \qquad k, p = 0, 1, 2, \ldots, n-1 \tag{A.2.7}$$

where $\delta_{kp}$ is the Kronecker delta function. Clearly, for each $p$, (A.2.7) generates a solution for $a_k^{(p)}$, $k = n, n+1, \ldots$ through (A.2.4) and thus the $p$th solution (A.2.6) of (A.1.1), (A.1.4) is identified by successive calculations.

As an example consider the solution of

$$y'' + \lambda xy = 0$$

Here $n = 2, f_{1,j} = 0, f_{2,j} = \lambda\delta_{1j}$ for all $j$. Equation (A.2.4) becomes for this example

$$a_{n+3}(n+3)(n+2) + \lambda a_n = 0$$

with $a_0$ and $a_1$ as arbitrary parameters. By choosing $a_0 = 1$, $a_1 = 0$, we obtain by solving (A.2.9) successively

$$y_1(x) = 1 + \sum_{j=1}^{\infty} \frac{(-\lambda)^j}{(3j)!} \prod_{r=0}^{j-1} (1 + 3r)x^{3j}$$

With $a_0 = 0$, $a_1 = 1$, the other solution is found to be

$$y_2(x) = x + \sum_{j=1}^{\infty} \frac{(-\lambda)^j}{(3j+1)!} \prod_{r=0}^{j-1} (2 + 3r)x^{3j+1}$$

## A.3 The Method of Frobenius

This method is applicable when $x = 0$ is a regular singular point. We seek the $n$ linearly independent solutions of (A.1.4) with $T$ given by (A.1.3), where $x_0 = 0$. Express the solution $y$ as

$$y(x, s) = \sum_{k=0}^{\infty} a_k x^{k+s} \tag{A.3.1}$$

where $s$ is called an *indicial parameter*. Let us first write

$$x^r \frac{d^r}{dx^r} y = \sum_{k=0}^{\infty} a_k \frac{(k+s)!}{(k+s-r)!} x^{k+s}, \qquad r = 0, 1, 2, \ldots, n \tag{A.3.2}$$

so that

$$x^r F_r(x) \frac{d^r}{dx^r} y = \sum_{k=0}^{\infty} x^{k+s} \sum_{j=0}^{k} \frac{(k+s-j)!}{(k+s-j-r)!} a_{k-j} f_{r,j} \tag{A.3.3}$$

Substituting (A.3.1) to (A.3.3) into (A.1.4), (A.1.3), we obtain

$$\sum_{k=0}^{\infty} x^{k+s}\left[a_k\frac{(k+s)!}{(k+s-n)!}+\sum_{r=0}^{n-1}\sum_{j=0}^{k}\frac{(k+s-j)!}{(k+s-j-r)!}a_{k-j}f_{r,j}\right]=0$$

which implies that

$$a_k\frac{(k+s)!}{(k+s-n)!}+\sum_{r=0}^{n-1}\sum_{j=0}^{k}\frac{(k+s-j)!}{(k+s-j-r)!}a_{k-j}f_{r,j}=0 \tag{A.3.4}$$

For $k = 0$ (A.3.4) becomes

$$a_0 p(s) = 0 \tag{A.3.5}$$

where

$$p(s) \equiv \frac{s!}{(s-n)!}+\sum_{r=0}^{n-1}\frac{s!}{(s-r)!}f_{r,0} \tag{A.3.6}$$

is an $n$th-degree polynomial in the indicial parameter $s$. We let $a_0$ be an arbitrary constant so that (A.3.5) implies that

$$p(s) = 0 \tag{A.3.7}$$

Equation (A.3.7) is called the *indicial equation* and must possess $n$ roots, although not necessarily district. The crux of the Frobenius method is to demand that (A.3.4) hold for $k = 1, 2, \ldots$ at *all* values of the parameter $s$ [i.e., not necessarily for the roots of (A.3.7)]. It is convenient to rewrite (A.3.4) as

$$a_k p(s+k)+\sum_{j=1}^{k} a_{k-j}p_j(s+k) = 0 \tag{A.3.8}$$

where $\{p_j(s)\}_{j=1}^{k}$ are polynomials given by

$$p_j(s) \equiv \sum_{r=0}^{n-1}\frac{(s-j)!}{(s-j-r)!}f_{r,j} \tag{A.3.9}$$

Since (A.3.8) holds for all values of $s$, we conclude that

$$Ty(x, s) = a_0 p(s)x^s \tag{A.3.10}$$

It would seem from (A.3.10) that the $n$ linearly independent solutions of $Ty = 0$ are obtained by substituting the $n$ roots of the indicial equation (A.3.7) into (A.3.1). This is indeed the case if all the roots of $p(s)$ are *distinct* with no two roots differing by an integer. Then the $j$th linearly independent solution $y_j(x) = y(x, s_j)$. If the roots differ by integral values, (A.3.8) turns out to be a source of trouble. We will deal with this situation later and turn instead to the obvious trouble created by repeated roots. (This is because only one solution is obtained for the set of repeated roots.)

Suppose that root $s_1$ is repeated $r - 1$ times.† To identify the related $r$

†Note that if a root occurs $r$ times, we say that it is repeated $(r-1)$ times. Thus if a root occurs only once, it is not repeated.

linearly independent solutions of $Ty = 0$, we first observe that one is obtained by letting $s = s_1$ in (A.3.1). For the remaining $(r - 1)$ solutions we differentiate (A.3.10) with respect to $s$ to obtain

$$\frac{\partial^k}{\partial s^k} Ty(x, s) = T\left[\frac{\partial^k}{\partial s^k} y(x, s)\right] = a_0 \sum_{i=0}^{k} \binom{k}{i} \frac{d^{k-i}}{ds^{k-i}}[p(s)] \frac{\partial^i}{\partial s^i}[x^s]$$

$$= a_0 \sum_{i=0}^{k} \binom{k}{i} \frac{d^{k-i}}{ds^{k-i}}[p(s)][\ln x]^i x^s, \qquad k = 1, 2, \ldots, r - 1$$

On letting $s = s_1$ in the above and recognizing that

$$\frac{d^k}{ds^k}[p(s)]_{s=s_1} = 0, \qquad k = 0, 1, 2, \ldots, r - 1$$

because $s = s_1$ is repeated $r - 1$ times, we conclude that

$$\left[\frac{\partial^k}{\partial s^k} y(x, s)\right]_{s=s_1}, \qquad k = 0, 1, 2, \ldots, r - 1 \tag{A.3.11}$$

are the $r$ linearly independent solutions of $Ty = 0$ corresponding to $s = s_1$.

Now suppose that there are roots $s_1$ and $s_2$ with $s_1 > s_2$ and $(s_1 - s_2)$ a positive integer. Further, let $s_1$ be repeated $r - 1$ times. We have seen how to generate the solutions corresponding to $s = s_1$. For $s = s_2$, (A.3.8) may run into trouble since $p(s_2 + k)$ becomes zero for some $k = k_1$ such that $s_2 + k_1 = s_1$. Then we may not be able to solve for $a_{k_1}$ if

$$b_k(s) \equiv \sum_{j=1}^{k} a_{k-j} p_j(s) \tag{A.3.12}$$

is such that $b_{k_1}(s_1) \neq 0$, because from (A.3.8)

$$a_k(s) = -\frac{b_k(s + k)}{p(s + k)} \tag{A.3.13}$$

The calculation of $a_k(s)$ for $s = s_2$ runs into trouble when $k = k_1$ and $b_{k_1}(s_1) \neq 0$, for then $a_{k_1}(s_2)$ is unbounded. However, if $b_{k_1}(s_1) = 0$ *and* the limit of the right-hand side of (A.3.13) for $k = k_1$ exists as $s \longrightarrow s_2$, then $a_{k_1}(s_2)$ is obtained as this limit and the trouble vanishes.

Consider the case when $b_{k_1}(s_1) \neq 0$. It is then not possible to obtain a solution as $y(x, s_2)$ because $a_{k_1}(s_2)$ is unbounded. Let us note first that $p(s + k_1)$ has $s_2$ as a root repeated $r - 1$ times. In order to cancel this effect in (A.3.13), we must also seek to render $s_2$ as a root of $b_{k_1}(s + k_1)$ repeated $r - 1$ times so that $a_{k_1}(s_2)$ is defined. Reference to (A.3.12) shows that if each of the coefficients $a_j$, for $j = 0, 1, 2, \ldots, k_1$, were multiplied by $(s - s_2)^r$, then $b_{k_1}(s + k_1)$ has $s_2$ as a root repeated $r - 1$ times. The remedy is immediately suggested in the modification of the first term in the series (A.3.1) by multiplying it by $(s - s_2)^r$. Thus define

$$Y(x, s) = a_0(s - s_2)^r x^s + \sum_{k=1}^{\infty} a_k x^{k+s} \tag{A.3.14}$$

Then retaining (A.3.8), we obtain

$$TY(x, s) = a_0(s - s_2)^r p(s) x^s \tag{A.3.15}$$

For $k > k_1$, the successive solution for the coefficients $a_k(s_2)$ is carried out readily to obtain not necessarily trivial values. Note that $a_k(s_2) = 0$, $k = 1, 2, \ldots, k_1$. Thus $Y(x, s_2)$ may appear as the required solution corresponding to $s = s_2$. However, it is not difficult to show that the solution so obtained is merely a multiple of $y(x, s_1)$ and hence not a linearly independent one. Differentiating (A.3.15) $r$ times, we have

$$\frac{\partial^r}{\partial s^r} TY(x, s) = T\left[\frac{\partial^r}{\partial s^r} Y(x, s)\right] = a_0 r! p(s) x^s + q(s) \tag{A.3.16}$$

where $q(s)$ has $(s - s_2)$ as a factor. Letting $s = s_2$ in (A.3.16), it follows that

$$\left[\frac{\partial^r}{\partial s^r} Y(x, s)\right]_{s=s_2} \tag{A.3.17}$$

is the required solution corresponding to the indicial root $s_2$. If $p(s)$ has $s_2$ repeated $r' - 1$ times, the other $(r' - 1)$ solutions of $Ty = 0$ are given by

$$\left[\frac{\partial^{r+j}}{\partial s^{r+j}} Y(x, s)\right]_{s=s_2}, \qquad j = 1, 2, \ldots, r' - 1$$

We believe that the case of more than two indicial roots differing by integral values is now readily handled. Some standard examples follow.

### A.3.1 Bessel's Differential Equation

$$x^2 \frac{d^2 y}{dx^2} + x \frac{dy}{dx} + (x^2 - p^2) y = 0 \tag{A.3.18}$$

Since (A.3.18) is of order 2 the indicial equation is quadratic and is given by

$$p(s) = s(s - 1) + s - p^2 = s^2 - p^2 = 0$$

Thus the roots are $s_1 = p$ and $s_2 = -p$ with $(s_1 - s_2) = 2p$. If $2p$ is an integer, the first solution $y_1(x)$ is obtained as

$$y_1(x) = \sum_{k=0}^{\infty} a_k x^{k+p} \tag{A.3.19}$$

where the $a_k$'s are to be identified by solving (A.3.8) for (A.3.18). Since $F_1(x) = 1$, $F_0(x) = x^2 - p^2$, we have $f_{1,0} = 1$; $f_{1,j} = 0, j \geq 1$; $f_{0,0} = -p^2$; $f_{0,1} = 0$; $f_{0,2} = 1$; $f_{0,j} = 0, j \geq 3$, so that $p_1(s) \equiv 0$; $p_2(s) = 1$; $p_j(s) \equiv 0$,

$j \geq 3$. Thus one obtains

$$a_k = -\frac{a_{k-2}}{(p^2 + k^2) - p^2} = -\frac{a_{k-2}}{k(2p + k)} \tag{A.3.20}$$

By choosing $a_0 = 1/p!2^p$, (A.3.20) may be readily solved to obtain

$$a_{2k} = \frac{(-1)^k}{2^p(p + k)!k!2^{2k}} \tag{A.3.21}$$

If $p$ is not an integer, $p!$ and $(p + k)!$ are to be understood in the sense of the Γ-function

$$y! \equiv \Gamma(y + 1) = \int_0^\infty t^y e^{-t}\, dt$$

which is extended to negative arguments by $\Gamma(y) = \Gamma(y + 1)/y$. Substituting (A.3.20) in (A.3.19), we obtain the *Bessel function* of the *first kind* of $p$th *order*, $J_p(x)$.

$$y_1(x) = J_p(x) \equiv \sum_{k=0}^{\infty} \frac{(-1)^k}{k!(p + k)!}\left(\frac{x}{2}\right)^{2k+p} \tag{A.3.22}$$

If $p = 0$, the root $s = p$ is repeated and we obtain one of the solutions as

$$y_1(x) = J_0(x) \equiv \sum_{k=0}^{\infty} \frac{(-1)^k}{(k!)^2}\left(\frac{x}{2}\right)^{2k}$$

The second solution is obtained by differentiating with respect to the indicial parameter as in (A.3.11). We state only the final result as

$$y_2(x) = Y_0(x) \equiv \frac{2}{\pi}\left[\ln\left(\frac{1}{2}x\right) + \gamma\right]J_0(x) - \frac{2}{\pi}\sum_{k=1}^{\infty}\frac{(-1)^k}{(k!)^2}\left(\frac{x}{2}\right)^{2k}\sum_{j=1}^{k}\frac{1}{j}$$

where $\gamma$ is the Euler constant 0.57721 . . . . The function $Y_0(x)$ is known as *Bessel function of the second kind and zeroth order.*

If $p \neq 0$, the indicial roots are distinct. Moreover, if $2p$ is not an integer, the first solution is $J_p(x)$ and the second solution is $J_{-p}(x)$, obtained by inserting $-p$ in place of $p$ in (A.3.22). If $2p$ is an integer, the procedure is contained in (A.3.14) and (A.3.17). Again we give only the final result for $p$ an integer as

$$y_2(x) = Y_p(x) \equiv -\frac{1}{\pi}\sum_{k=0}^{p-1}\frac{(p - k - 1)!}{k!}\left(\frac{x}{2}\right)^{2k-p} + \frac{2}{\pi}\ln\left(\frac{1}{2}x\right)J_p(x)$$
$$-\frac{1}{\pi}\sum_{k=0}^{\infty}[\psi(k + 1) + \psi(p + k + 1)]\frac{(-1)^k}{k!(p + k)!}\left(\frac{x}{2}\right)^{2k+p}$$

where

$$\psi(1) = -\gamma, \qquad \psi(k) = -\gamma + \sum_{j=1}^{k-1}\frac{1}{j}$$

The function $Y(p)$ is called a *Bessel function* of the *second kind* and $p$th *order*.

### A.3.2 The Hypergeometric Differential Equation

We will confine our attention to the confluent hypergeometric equation, called *Kummer's equation*, and given by

$$x\frac{d^2y}{dx^2} + (b - x)\frac{dy}{dx} - ay = 0 \tag{A.3.23}$$

Since $x = 0$ is a regular singular point, we express the solution as (A.3.1). The indicial equation (A.3.7) becomes the quadratic equation

$$s(s + b - 1) = 0$$

whose roots are $s = 0$ and $s = 1 - b$. Equation (A.3.8) becomes

$$a_k(s + k)(s + b + k - 1) = a_{k-1}(s + a + k - 1)$$

The solution $y_1(x)$ of (A.3.23), corresponding to $s = 0$, is denoted ${}_1F_1[a; b; x]$ and is given by

$${}_1F_1[a; b; x] = 1 + \frac{a}{b}x + \frac{a(a+1)}{b(b+1)}\frac{x^2}{2!} + \frac{a(a+1)(a+2)}{b(b+1)(b+2)}\frac{x^3}{3!} + \cdots \tag{A.3.24}$$

The solution above makes sense provided that $b \neq 0$ or a negative integer. If $b = 1$ then $s = 0$ is repeated and the second solution is obtained by parametric differentiation as in (A.3.11). If $b \neq 2, 3, 4, \ldots$, then the difference between the indicial roots is not an integer and the second solution $y_2(x)$ is given by

$$y_2(x) = x^{1-b}\, {}_1F_1[1 + a - b; 2 - b; x]$$

If $b$ is a positive integer, the second solution is generated as in (A.3.17) jointly with (A.3.14). Note, however, that the second solution in this case is unbounded at $x = 0$.

Consider the solution of the differential equation

$$\frac{d^2z}{dx^2} + \frac{1}{x}\frac{dz}{dx} + \lambda(1 - x^2)z = 0 \tag{A.3.25}$$

which arose in Chapter 9 [see (9.2.84)]. By letting $t = \sqrt{\lambda}\, x^2$, $z = y(t)e^{-(1/2)t}$. It is readily shown that (A.3.25) yields

$$t\frac{d^2y}{dt^2} + (1 - t)\frac{dy}{dt} - \left(\frac{1}{2} - \frac{\sqrt{\lambda}}{4}\right)y = 0 \tag{A.3.26}$$

which is the same as (A.3.23) with $b = 1$, $a = \frac{1}{2} - \sqrt{\lambda}/4$, so that the solution of (A.3.25) bounded at $x = 0$ is given by ${}_1F_1[\frac{1}{2} - \sqrt{\lambda}/4; 1; \sqrt{\lambda}\, x^2)$.

## A.4 Properties of Bessel Functions

The formulas recounted here are useful in computations of spectral solutions with Bessel functions of the first and second kind. The arguments of the Bessel functions could be real or complex. We let $Z_p(x)$ represent either $J_p(x)$ or $Y_p(x)$. Primes denote differentiation with respect to the argument.

### A.4.1 Recurrence Relationships

$$Z_{p-1}(x) + Z_{p+1}(x) = \frac{2p}{x} Z_p(x)$$

$$Z_{p-1}(x) - Z_{p+1}(x) = 2Z'_p(x)$$

$$Z'_p(x) = Z_{p-1}(x) - \frac{p}{x} Z_p(x)$$

$$Z'_p(x) = -Z_{p+1}(x) + \frac{p}{x} Z_p(x)$$

### A.4.2 Formulas for Derivatives

$$\left.\begin{aligned} \left(\frac{1}{x}\frac{d}{dx}\right)^k \{x^p Z_p(x)\} &= x^{p-k} Z_{p-k}(x) \\ \left(\frac{1}{x}\frac{d}{dx}\right)^k \{x^{-p} Z_p(x)\} &= (-1)^k x^{-p-k} Z_{p+k}(x) \end{aligned}\right\} \quad k = 0, 1, 2, \ldots$$

### A.4.3 Integral Formulas

$$\int_0^x tJ_0^2(t)\, dt = \frac{x^2}{2}[J_0^2(x) + J_1^2(x)]$$

$$\int_a^b xZ_p(\lambda_m x)Z_p(\lambda_n x)\, dx$$

$$= \begin{cases} 0, & m \neq n \\ \left[\dfrac{1}{2}x^2\left\{\left(1 - \dfrac{p^2}{\lambda_n^2 x^2}\right)Z_p^2(\lambda_n x) + Z_p'^2(\lambda_n x)\right\}\right]_a^b, & m = n, \quad 0 < a < b \end{cases}$$

provided that (i) $\lambda_n$ is a real zero of

$$h_1 \lambda Z_{p+1}(\lambda b) - h_2 Z_p(\lambda b) = 0$$

and (ii) there are numbers $k_1$ and $k_2$ (both not zero) so that for all $n$

$$k_1 \lambda_n Z_{n+1}(\lambda_n a) - k_2 Z_n(\lambda_n a) = 0$$

If $a = 0$, we have

$$\int_0^1 xJ_p(\mu_m x)J_p(\mu_n x)\,dx = \begin{cases} 0, & m \neq n, \quad p > -1 \\ \frac{1}{2}[J_p'(\mu_n)]^2, & m = n, \quad b = 0, \quad p > -1 \\ \dfrac{1}{2\mu_n^2}\left[\dfrac{a^2}{b^2} + \mu_n^2 - p^2\right][J_p(\mu_n)]^2, & m = n, \quad b \neq 0, \quad p \geq -1 \end{cases}$$

where $\mu_1$ and $\mu_2$ are the positive roots of

$$aJ_p(x) + bxJ_p'(x) = 0$$

in which $a$ and $b$ are real constants.

For a more comprehensive set of formulas involving Bessel functions, the reader must consult *Handbook of Mathematical Functions, with Formulas. Graphs, and Mathematical Tables*, edited by M. Abramowitz and I. A. Stegun, National Bureau of Standards, Applied Mathematics Series 55, 1965.

# *Index of Symbols*

| ***Symbol*** | ***Meaning*** |
|---|---|
| $\in$ | is an element of |
| $\notin$ | is not an element of |
| $\subset$ | contained in |
| $\supset$ | contains |
| $\cup$ | union |
| $\cap$ | intersection |
| $\exists$ | there exists |
| $\forall$ | for every |
| $\ni$ | such that |
| $\rightarrow$ | implies |
| ▌ | end of proof |
| $A \times B$ | Cartesian product of sets $A$ and $B$ |
| $f: A \longrightarrow B$ | $f$ maps set $A$ into $B$ |
| $A + B$ | algebraic sum of sets $A$ and $B$ in a linear space |
| $A \oplus B$ | direct sum of sets in a linear space |
| $A \otimes B$ | direct product of sets in linear space |
| $\|\cdot\|$ | norm |
| $\langle \cdot, \cdot \rangle$ | inner product |
| $A \perp B$ | set $A$ is orthognal to $B$ |
| $A^{\perp}$ | orthogonal complement of $A$ |
| $*$ | complex conjugate, adjoint |
| $A^*$ | set of accumulation points of $A$ |
| $\bar{A}$ | closure of $A$ $(= AUA^*)$ |
| $[a, b]$ | $\{x: a \leq x \leq b\}$ |
| $(a, b)$ | $\{x: a < x < b\}$ |
| $B_\varepsilon(x_0)$ | $\varepsilon$-ball in a metric space |

| ***Symbol*** | ***Meaning*** |
|---|---|
| $B_\varepsilon(x_0-)$ | deleted $\varepsilon$-ball (not containing $x_0$) |
| $b(\cdot, \cdot)$ | bilinear functional |
| $D(\mathbf{T})$ | domain of operator **T** |
| $H$ | hyperplane |
| $L(\mathbf{x}_1, \mathbf{x}_2, \ldots)$ | linear manifold spanned by vectors in the argument |
| $l(\cdot)$ | linear functional |
| $l_p$ | $p$th power summable vectors of $\mathcal{R}_\infty$ |
| $N(\mathbf{T})$ | null space of operator **T** |
| $N_\lambda$ | eigenspace corresponding to eigenvalue $\lambda$ |
| $\mathbf{P}$ | projection operator |
| $R(\mathbf{T})$ | range space of operator **T** |
| $\mathbf{T}$ | linear transformation or operator (also **L**) |
| $W$ | Wronskian |
| $\mathbf{W}$ | Wronskian matrix |
| $\mathbf{w}$ | Wronskian vector |
| $\mathbf{x}, \mathbf{y}, \mathbf{z}$ | vectors |
| $X_E$ | characteristic function of set $E$ |
| $\mathcal{B}$ | Banach space |
| Blt | set of bounded linear transformations |
| $\mathcal{C}$ | complex field |
| $\mathcal{C}_n$ | $n$ dimensional complex vector space |
| $\mathcal{C}_\infty$ | infinite-dimensional complex vector space |
| $\mathcal{C}[a, b]$ | space of functions continuous in $[a, b]$ |
| $\mathcal{C}^{(n)}[a, b]$ | space of functions with $n$ continuous derivatives in $[a, b]$ |
| $\mathcal{C}^{(\infty)}[a, b]$ | space of infinitely differentiable functions in $[a, b]$ |
| $\mathcal{F}$ | abstract field |
| $\mathcal{H}$ | Hilbert space |
| $\mathcal{I}$ | inner product space |
| $\mathcal{K}$ | space of compact operators |
| $\mathcal{L}$ | linear space |
| $\mathcal{L}[a, b]$ | space of Lebesgue-integrable functions on $[a, b]$ |
| $\mathcal{L}_p[a, b]$ | space of $p$th-power Lebesgue-integrable functions on $[a, b]$ |
| $\mathcal{N}$ | normed linear space |
| $\mathcal{R}$ | real number field |
| $\mathcal{R}_n$ | $n$-dimensional real vector space |
| $\mathcal{R}_\infty$ | infinite-dimensional real vector space |
| $\mathcal{R}[a, b]$ | space of Riemann-integrable functions |
| $\mathcal{R}_p[a, b]$ | space of $p$th-power Riemann-integrable functions |
| $\lambda$ | eigenvalue |
| $\mu E$ | measure of set $E$ |
| $\mu_i E$ | interior measure of set $E$ |
| $\rho$ | pseudometric |
| $\sigma(\mathbf{T})$ | spectrum of operator **T** |
| $C_\sigma(\mathbf{T})$ | continuous spectrum of operator |
| $P_\sigma(\mathbf{T})$ | point spectrum of operator **T** |

# Author Index

D



E



F



G



H



I



J



K

# Subject Index

**E**



**F**

G



H



I



J

K



L



M